STATISTICAL PROCEDURES FOR ENGINEERING, MANAGEMENT, AND SCIENCE

McGraw-Hill Series in Industrial Engineering and Management Science

Consulting Editor

James L. Riggs, *Department of Industrial Engineering, Oregon State University*

Barish and Kaplan: *Economic Analysis: For Engineering and Managerial Decision Making*
Blank: *Statistical Procedures for Engineering, Management, and Science*
Gillett: *Introduction to Operations Research: A Computer-oriented Algorithmic Approach*
Hicks: *Introduction to Industrial Engineering and Management Science*
Love: *Inventory Control*
Riggs: *Engineering Economics*
Riggs and Inoue: *Introduction to Operations Research and Management Science: A General Systems Approach*

STATISTICAL PROCEDURES FOR ENGINEERING, MANAGEMENT, AND SCIENCE

Leland Blank, PE

Department of Industrial Engineering
Texas A&M University

McGraw-Hill Book Company

New York St. Louis San Francisco Auckland Bogotá Hamburg
Johannesburg London Madrid Mexico Montreal New Delhi
Panama Paris São Paulo Singapore Sydney Tokyo Toronto

This book was set in Times Roman by Science Typographers, Inc.
The editors were Julienne V. Brown and Stephen Wagley;
the production supervisor was Charles Hess.
The drawings were done by ANCO/Boston.
The cover was designed by John Hite.
Fairfield Graphics was printer and binder.

STATISTICAL PROCEDURES FOR ENGINEERING, MANAGEMENT, AND SCIENCE

Copyright © 1980 by McGraw-Hill, Inc. All rights reserved.
Printed in the United States of America. No part of this publication
may be reproduced, stored in a retrieval system, or transmitted, in any
form or by any means, electronic, mechanical, photocopying, recording, or
otherwise, without the prior written permission of the publisher.

234567890 FGFG 83210

Library of Congress Cataloging in Publication Data

Blank, Leland T
    Statistical procedures for engineering, management
and science.

    (McGraw-Hill series in industrial engineering and
management science)
    Bibliography: p.
    Includes index.
    1. Mathematical statistics. 2. Distribution
(Probability theory) I. Title.
QA276.B613   519.5   79-14876
ISBN 0-07-005851-2

This book is dedicated
to Sallie,
person who
shares with me,
that we may share
with others.

CONTENTS

Preface · xvii

Level One Introduction to Data Analysis

1 Basic Concepts · 3
    Criteria · 3
    Study Guide · 4
1.1 Deterministic and Probabilistic Data · 4
1.2 Equally Likely Events · 4
1.3 Definitions of Probability · 5
1.4 Phases of Probability Analysis · 7
1.5 Populations and Samples · 7
1.6 Discrete and Continuous Variables · 8
1.7 Definition and Phases of Statistics · 9

2 Data Presentation Formats · 14
    Criteria · 14
    Study Guide · 15
2.1 Ungrouped Data, Grouped Data, and the Frequency Tally Sheet · 15
2.2 Frequency Distributions (Histograms) for Ungrouped Data · 17
2.3 Determination of Cells to Group Data · 18
2.4 Frequency Distribution for Grouped Data · 21
2.5 Cumulative Frequency Distribution · 22
2.6 Relative Frequency and Cumulative Relative Frequency · 28
2.7 Fractiles · 33

3 Properties of Data: Part 1 · 46
    Criteria · 46
    Study Guide · 46
3.1 Average of Data · 46

	3.2	Average of Frequency Data	47
	3.3	Mode and Modal Cell	49
	3.4	Median and Median Cell	51
	3.5	Standard Deviation and Variance	53
	3.6	Standard Deviation and Variance of Frequency Data	57
	3.7	Range	59
	3.8	Average, Median, Mode, and Standard Deviation on a Frequency Polygon	60

4 Properties of Data: Part 2 — 65

	Criteria	65
	Study Guide	66
4.1	Introduction to Moment Terminology	66
4.2	Skewness	67
4.3	Peakedness	70
4.4	Mean Absolute Deviation	71
4.5	Weighted Average	73
4.6	Coefficient of Variation	75
4.7	Coding Techniques for the Average and Standard Deviation	76
4.8	Coding Techniques for Grouped Data	78

5 Probability Computations — 86

	Criteria	86
	Study Guide	87
5.1	Fundamentals of Sets	87
5.2	Venn Diagrams and Set Operations	87
5.3	Sample Spaces and Events	89
5.4	Axioms of Probability	90
5.5	Probability Rules	90
5.6	Conditional Probability and Independent Events	92
5.7	Bayes' Formula	95
5.8	Probability Trees	98

Level Two Distributions and Their Uses

6 Introduction to Probability Density Functions — 109

	Criteria	109
	Study Guide	110
6.1	Definitions	110
6.2	Probability Density Function (pdf) Terminology	112
6.3	Discrete pdf, Continuous pdf, and Graphs	114
6.4	Properties of a Discrete pdf	117
6.5	Computation of the Expected Value, Variance, and Standard Deviation of a Discrete pdf	119
6.6	Properties of a Continuous pdf	122
6.7	Computation of Expected Value, Variance, and Standard Deviation of a Continuous pdf	124

6.8	Cumulative Distribution Function (cdf)	127
6.9	Best Estimators for Parameters	128
*6.10	Moment Generating Functions	129

7 Counting Rules and Basics of the Hypergeometric Distribution — 140

	Criteria	140
	Study Guide	141
7.1	Basic Counting Rule	141
7.2	Factorials and Stirling's Approximation	141
7.3	Permutations	142
7.4	Combinations	144
7.5	Derivation of the Hypergeometric pdf	146
7.6	Graph of the Hypergeometric pdf and cdf	148
7.7	Parameters and Properties of the Hypergeometric	151
7.8	Solving Problems with the Hypergeometric	153
*7.9	Derivation of the Hypergeometric Mean	154

8 Basics of the Binomial Distribution — 161

	Criteria	161
	Study Guide	162
8.1	Formulation of the Binomial pdf	162
8.2	Graph of the Binomial pdf and cdf	165
8.3	Parameters and Properties of the Binomial	167
8.4	Use of a Binomial Distribution Table	169
8.5	Solving Problems with the Binomial	171
8.6	Comparing Observed Data and Binomial Predictions	172
8.7	Exact and Approximate Uses for the Binomial	175
8.8	Distribution of Proportions	177
*8.9	Derivation of the Binomial pdf and Its Mean	179

9 Basics of the Poisson Distribution — 187

	Criteria	187
	Study Guide	188
9.1	Formulation of the Poisson Distribution	188
9.2	Graph of the Poisson pdf and cdf	190
9.3	Parameter and Properties of the Poisson	191
9.4	Use of the Poisson Distribution Table	193
9.5	Solving Problems Using the Poisson	194
9.6	Comparing Observed Data and Poisson Probabilities	195
9.7	Exact and Approximate Uses of the Poisson	196
*9.8	Derivation of the Poisson pdf, Mean, and Variance	198

10 Other Discrete Distributions — 205

	Criteria	205
	Study Guide	206
10.1	Geometric Distribution pdf and Graph	206
10.2	Geometric Parameter and Properties	207

*Sections marked with a star may be omitted without loss of continuity.

	10.3	Geometric Distribution Application	207
	10.4	Pascal Distribution pdf and Graph	208
	10.5	Pascal Parameters and Properties	209
	10.6	Pascal Distribution Application	210
	10.7	Discrete Uniform Distribution pdf and Graph	211
	10.8	Discrete Uniform Parameters and Properties	212
	10.9	Discrete Uniform Distribution Application	212
	10.10	Discrete Uniform and the Use of Random Numbers	213

11 Basics of the Normal Distribution — 218

	Criteria	218
	Study Guide	219
11.1	Definition of the Normal pdf	219
11.2	Parameters and Properties of the Normal	220
11.3	Derivation of the Standard Normal Distribution (SND)	221
11.4	Graph and Areas of the SND	222
11.5	Use of the Standard Normal Table	225
11.6	Solving Problems with the Normal	227
11.7	Fitting the Normal to Observed Data	229
11.8	Exact and Approximate Uses of the Normal	231
*11.9	Moments of the Normal	234

12 The Central Limit Theorem — 242

	Criteria	242
	Study Guide	242
12.1	Purpose and Statement of the CLT	242
12.2	How the CLT Works	245
12.3	Sample Sizes and the CLT	246
12.4	Common Uses of the CLT	247

13 Basics of the χ^2 Distribution — 253

	Criteria	253
	Study Guide	254
13.1	Concept of Degrees of Freedom	254
13.2	Formulation of the χ^2 pdf	255
13.3	Graphs, Parameter, and Properties of the χ^2 pdf	257
13.4	Use of a χ^2 Distribution Table	258
13.5	Solving Problems with the χ^2 Distribution	260
*13.6	The pdf Derivation, mgf, and Additivity Property of χ^2	262

14 Basics of the t Distribution — 267

	Criteria	267
	Study Guide	268
14.1	Formulation of the t pdf	268
14.2	Graphs, Parameter, and Properties of the t pdf	269
14.3	Use of a t Distribution Table	271
14.4	Solving Problems with the t Distribution	273
*14.5	Derivation of the t Distribution	274

15 Basics of the F Distribution — 277

    Criteria — 277
    Study Guide — 277
15.1 Formulation of the F pdf — 277
15.2 Graphs, Parameters, and Properties of the F pdf — 279
15.3 Use of an F Distribution Table — 280
15.4 A Simple Application of the F pdf — 282
*15.5 Derivation of the F Distribution — 283

16 Other Continuous Distributions — 286

    Criteria — 286
    Study Guide — 287
16.1 The Uniform Distribution — 287
16.2 Formulation of the Exponential Distribution — 289
16.3 Parameters and Properties of the Exponential Distribution — 290
16.4 Solving Reliability Problems with the Exponential — 291
*16.5 Relation between the Exponential and Poisson Distributions — 293
16.6 The Gamma Distribution — 293
*16.7 Relationship of Gamma to Other Distributions — 296

17 Sample Size Determination and Interval Estimation — 305

    Criteria — 305
    Study Guide — 306
17.1 Confidence and Significance Levels — 306
17.2 Sample Size Determination — 306
17.3 Point Estimates and Interval Estimates — 310
17.4 Interval Estimates for Population Mean — 311
17.5 Interval Estimates for Population Standard Deviation — 313
17.6 Interval Estimates for Population Proportion — 317

18 Distributions of More than One Variable and the Transformation of Variables — 325

    Criteria — 325
    Study Guide — 326
18.1 Formation and Properties of a Bivariate pdf — 326
18.2 Marginal Distributions — 328
18.3 Independent Random Variables — 330
18.4 Conditional Distributions — 331
18.5 Two Specific Distributions of More than One Variable — 335
18.6 Means of the Functions of Variables — 337
18.7 Variances of Functions of Variables — 339
*18.8 Transformation of a Single Discrete Variable — 340
*18.9 Transformation of a Single Continuous Variable — 343
*18.10 Transformation of Two Variables — 344

Level Three Statistical Inference

19 Introduction to Statistical Inference — 357
    Criteria — 357
    Study Guide — 358
  19.1 Rationale of Hypothesis Testing — 358
  19.2 Types of Errors and Their Probabilities — 359
  19.3 One-Sided and Two-Sided Tests — 361
  19.4 Steps of the Hypothesis Testing Procedure — 362
 *19.5 Dependence between α, β, and Sample Size — 364
 *19.6 Operating Characteristic Curves — 366

20 Statistical Inferences for Means — 372
    Criteria — 372
    Study Guide — 373
  20.1 Types of Tests for Means — 373
  20.2 Means Test for One Sample with Standard Deviation Known — 373
  20.3 Means Test for One Sample with Standard Deviation Unknown and a Small Sample — 377
  20.4 Means Test for Two Independent Samples with Standard Deviation Known — 378
  20.5 Means Test for Two Independent Samples with Standard Deviation Unknown and Small Samples — 381
  20.6 Means Test for Paired Samples — 384

21 Statistical Inferences for Variances — 391
    Criteria — 391
    Study Guide — 391
  21.1 Types of Tests for Variances — 391
  21.2 Variance Test for One Sample — 392
  21.3 Variance Test for Two Samples — 395
 *21.4 Relation between the Tests for Means and Variances — 397

22 Statistical Inferences for Proportions — 401
    Criteria — 401
    Study Guide — 401
  22.1 Types of Tests for Proportions — 401
  22.2 Proportion Test for One Sample — 402
  22.3 Proportion Test for Two Samples — 403

23 Statistical Inferences for Goodness of Fit — 409
    Criteria — 409
    Study Guide — 410
  23.1 Types of Goodness-of-Fit Tests — 410
  23.2 χ^2 Goodness-of-Fit Test for Discrete Distributions — 411
  23.3 Minimum Frequency Correction to the χ^2 Statistic — 413
  23.4 χ^2 Goodness-of-Fit Test for Continuous Distributions — 414

23.5	χ^2 Test for Independence of Factors	415
*23.6	Derivation of the χ^2 Goodness-of-Fit Statistic	418
23.7	K-S Goodness-of-Fit Test	419

24 Graphical Procedure for Fitting a Distribution — 428

	Criteria	428
	Study Guide	428
24.1	Probability Plotting Procedure	428
24.2	Probability Plotting for the Normal Distribution	431
24.3	Probability Plotting for the Exponential Distribution	435
24.4	Probability Plotting for the Uniform Distribution	437
24.5	Probability Plotting for the Poisson Distribution	439

25 Statistical Inference Using Nonparametric Tests — 445

	Criteria	445
	Study Guide	446
25.1	Introduction to Nonparametric Statistics	446
25.2	One-Sample Runs Test of Randomness	447
25.3	One-Sample Sign Test for the Mean	449
25.4	Two-Sample Sign Test for Means	452
25.5	Two-Sample Rank Sum Test for Means (Mann-Whitney U Test)	453
25.6	Two-Sample Signed Rank Test for Means (Wilcoxon Test)	456

26 Statistical Inferences Using Bayesian Estimates — 470

	Criteria	470
	Study Guide	471
26.1	Classical and Bayesian Methods	471
26.2	Loss and Risk Functions	472
26.3	Prior and Posterior Distribution Relations	476
26.4	Bayesian Estimates for Quadratic Loss Functions	479

Level Four   Statistical Analysis Techniques

27 Curve Fitting by Least-Squares Regression — 487

	Criteria	487
	Study Guide	488
27.1	Introduction to Least-Squares Regression	488
*27.2	Derivation of the Linear Regression Estimators	490
27.3	Determination of the Regression Line	491
27.4	Hypothesis Tests of the Intercept and Slope Values	494
27.5	Confidence Intervals for Regression Lines	496
27.6	Multiple Linear Regression	500
*27.7	Stepwise Multiple Linear Regression	502
*27.8	Curvilinear Regression for Two Variables	503
*27.9	Partitioning of the Sum of Squares	507

28 Correlation Analysis — 516
    Criteria — 516
    Study Guide — 517
28.1 Introduction to Correlation Analysis — 517
28.2 The Correlation Coefficient and Its Properties — 518
28.3 Hypothesis Test for the Correlation Coefficient — 521
*28.4 Coefficient of Determination and the Generalized Correlation Coefficient — 523
*28.5 Development of the Population Correlation Coefficient Using Covariance — 526
*28.6 Multiple and Partial Correlation Coefficients — 528

29 Quality Control Analysis — 536
    Criteria — 536
    Study Guide — 537
29.1 Introduction to Quality Control — 537
29.2 Control of the Process Mean and Dispersion — 538
29.3 Process Capability and Natural Process Limits from $\bar{X}$ and R Charts — 544
29.4 Control of the Process Fraction Defective — 547
29.5 Control of the Number of Defects per Standard Unit — 549
29.6 Acceptance Sampling Plans for Attribute Data — 551
29.7 Published Sampling Plans — 555

30 Analysis of Variance — 565
    Criteria — 565
    Study Guide — 566
30.1 The Analysis of Variance Rationale — 566
30.2 Single-Factor (One-Way) ANOVA—Completely Randomized — 568
30.3 Analysis of Variance for a Linear Regression Equation — 571
30.4 Single-Factor (Two-Way) ANOVA—Randomized Blocks — 573
30.5 ANOVA for Multifactor Experiments — 577

Epilog: What Is Next in Statistics? — 588
    Further Statistics Applications — 588
    Further Statistical Theory — 589

Appendixes — 590
A Final Answers to Selected Problems — 618
B Statistical Tables — 618
B–1 Cumulative Distribution Function for the Binomial Distribution — 620
B–2 Cumulative Distribution Function for the Poisson Distribution — 623
B–3 Cumulative Distribution Function for the Standard Normal Distribution (SND) — 627
B–4 The χ^2 Distribution — 629
B–5 The t Distribution — 630
B–6 The F Distribution — 632
B–7 D Distribution for the Kolmogorov-Smirnov Goodness-of-Fit Test — 635

| B-8 | r Distribution for the Runs Test of Randomness for $\alpha = 0.05$ | 636 |
| B-9 | T Distribution of the Sum of Signed Ranks for the Wilcoxon Test of Means | 637 |

Bibliography 638

Index 641

PREFACE

The field of statistics is very extensive in both its theory and applications. Basically, the uses of statistics are applications of mathematics and rational decision making by analysts, be they engineers, managers, or scientists. This book presents the fundamentals of probability analysis and statistical tools which assist in decision making, using a language that is understandable to the person learning statistics, not in a style that simply impresses those already expert in statistics.

Uses of the Text

The modularized learning technique used in this text to present the distributions and tools of statistics allows the reader to first become very familiar with the fundamentals of statistical distributions and then learn the logic and procedures of the appropriate statistical tools. Specific questions are posed for each tool to determine when the procedure should and should not be used. Clearly written examples, which are separated from the explanatory text material, are included for every procedure to illustrate it and explain the conclusions. All examples are written in the context of real-world situations; that is, the "ball and urn" approach is not used.

This text may be used to cover statistics in one three-semester-hour course or a two-course sequence for engineers, scientists, or quantitatively oriented business majors. If the text is used in a one-semester course, some optional sections (explained later) should be omitted; and if it is used in a two-semester sequence, some supplementation with material listed in the Epilog is suggested. In the use of this material in the classroom it has been found that virtually all the material can be covered in the one-semester course. The chapters which present the relatively simple fundamentals about distributions may be assigned as reading because the presentation format requires only minimal classroom presentation by the instructor. This frees valuable hours for coverage of the more difficult concepts and applications with which the students will surely need assistance.

An operational knowledge of differential and integral calculus is necessary to master the material, so in general this text is usable for undergraduate students at the sophomore level or above. It may also be used for first-year graduate students taking a "core" course in statistics in a master's program.

Because of the building block approach used in its design, a practitioner unacquainted with statistics can easily use this text to learn, understand, and correctly apply the distributions and procedures of statistical decision making.

Chapter Composition

Each chapter starts with an objective statement and several specific learning criteria, which should be considered goals to the reader. Each criterion explains a skill which the reader should be able to demonstrate, given certain, specified information, once the section has been studied. These criteria are section-keyed: for example, the material for criteria 3 in Chapter 3 is discussed in Section 3.3. There are optional (starred) sections in most chapters. These sections, which present a summary or an expansion of the statistical theory applied in the chapter, may be omitted with no loss of understanding of how to correctly use the procedures. Further, subsequent chapters and sections do not require material from earlier optional sections unless they are themselves optional.

Within each section the most important equation or equations are emphasized with the symbol ●. The reader should pay particular attention to these relations and commit to memory as many of them as possible as they are commonly used in subsequent sections.

At the end of each section there are problems which utilize the concepts developed in the section. This provides the opportunity either to apply the material on a section-by-section basis or to wait until an entire chapter is completed. The final answers to most problems are included in Appendix A. Cross reference to the most appropriate section in the chapter is also given in this appendix.

If the section covers a statistical procedure that can be performed in a step-by-step fashion, the steps are presented and the accompanying example details these steps. Many chapters include solved problems after the last section containing examples which further illustrate the material in one or more sections of the chapter. These examples are often presented in the form of case studies which apply several techniques to one situation and set of data.

Finally, each chapter includes an Additional Material list which directs the reader to similar and advanced material in other texts on applied and mathematical statistics listed in the Bibliography.

Text Overview

The book is composed of 30 chapters collected into four learning levels. A flowchart or prerequisite chapter table at the beginning of Levels Two, Three, and Four gives the reader a good idea of what chapters are needed to understand the material presented in each chapter.

Level One (Chapters 1 to 5) covers basic computations, probability, and data presentation techniques for collected data. All chapters in this level should be covered in order. This is the only level which deals exclusively with collected data. Level Two (Chapters 6 to 18) presents the fundamentals of the most commonly used discrete and continuous probability density functions. Chapter 6 is an introduction to be referenced throughout the Level Two chapters as a source of the formulas used in working with statistical distributions. For each distribution the reader will learn the equation, graphs, parameters and their estimates, properties, use of the table, and some simple applications. Confidence interval determination and functions of more than one variable are also covered in this level.

Statistical inferences for population means, variances, proportions, and goodness of fit are discussed in Level Three (Chapters 19 to 26). Again the first chapter is introductory to the rest and may be used as reference material for all hypothesis-testing procedures. The use of operating characteristic curves is explained in this chapter and applied in the appropriate testing procedure. Chapter 24 presents the technique of graphical goodness-of-fit and parameter estimation for several common distributions. The final chapter of this level introduces the use of Bayesian statistics to estimate population parameters.

The final level, Four (Chapters 27 to 30), introduces the reader to several techniques used by analysts: regression and correlation analysis, quality control, and analysis of variance. These techniques, which use many of the distributions and procedures of the prior levels, are each an entire field of statistics in themselves. Thus only the major applications are presented here.

I greatly appreciate the efforts of the typists who prepared the manuscript and the students who have helped make this a more usable text. The suggestions of Dr. John Hunsucker have assisted in making the problems clear and understandable. In addition, I thank the administration of the Department of Industrial Engineering at Texas A&M University, especially Drs. Newton Ellis and Joseph Foster, for their support in this project.

Leland Blank

STATISTICAL PROCEDURES FOR ENGINEERING, MANAGEMENT, AND SCIENCE

LEVEL ONE

INTRODUCTION TO DATA ANALYSIS

The five chapters in this level will give you a basic understanding of the terminology and techniques needed to graphically and computationally analyze collected data. Virtually all the information contained here may be used for the presentation of statistical results in engineering, laboratory, and management reports. Chapter 4 may be considered optional, at least on the first reading.

You have more than likely seen some of these simple analytical procedures used somewhere, or maybe you have already used them yourself. Even so, be sure you understand them thoroughly, because this is the only level of this text that is completely devoted to the handling of actual data: the rest include some theory so that statistical conclusions may be formulated.

CHAPTER
ONE

BASIC CONCEPTS

This chapter introduces some of the basic terminology and concepts of probability and statistics. The meaning of the different phases of probability and statistics are discussed. A case study covering the subjects of several sections is included as the Solved Problem.

CRITERIA

To complete this introductory chapter, you must be able to do the following:

1. Define and give examples of *deterministic data* and *probabilistic data*.
2. Define the terms *event* and *equally likely event*, and state why an event is or is not equally likely, given a description of the event.
3. Define the term *probability*, and state and compute probability by three methods, given the event and the number of times it occurs.
4. Define the two phases of probability analysis and state how they are related.
5. State the meaning of the terms *population* and *sample*. Define *population parameter* and *sample statistic*, and state how they are related.
6. Define the terms *discrete variable* and *continuous variable*, and give an example of each.
7. State the definition of *statistics*, and name and define the two phases of statistics.

STUDY GUIDE

1.1 Deterministic and Probabilistic Data

Only two types of data are used to classify all numerical results:

Deterministic data. There is no variation from one fixed value. A scalar value such as $10 or a 5-kilogram (kg) weight is deterministic data.
Probabilistic data. This is also called stochastic data. There is a possibility that any one of several values can be observed. One fixed value is not sufficient to describe probabilistic data. For example, the number of vehicles needed to fill a river ferry fluctuates with vehicle size, weight, closeness to one another, etc. The number will therefore range from some lower to some upper value, such as 15 to 22 vehicles. The number of vehicles is probabilistic data. Of course, on any particular river crossing a count of vehicles will yield only one of the data values.

We will concentrate on probabilistic data in this book. We will describe it, compute with it, make inferences about it, and make decisions using it. Actually, when deterministic data is used, it is often because the probabilistic aspects are not known or understood.

Try to describe each of the following as probabilistic data: number of people in an elevator, yield strength of an aluminum casting, number of car wrecks per hour on a freeway.

Problems P1.1–P1.3

1.2 Equally Likely Events

Many computations in probability and statistics are based on the occurrence of equally likely events. First, an *event* is one possible outcome of an experiment. Several events can be combined to form another event. For example, if an engineer were looking for defective electronic components, the outcome of one defective component would be an event. Similarly, the occurrence of two defectives is an event. Computations can be performed for either of these events.

Equally likely events are events which have the same chance of occurring under stated conditions. The tossing of a fair coin is an example. Both heads and tails have a 50 percent chance of occurring. In tossing a fair die, the equally likely events of 1, 2, 3, 4, 5, and 6 on the top side have a 1/6 possibility of appearing. The events are not equally likely if the coin or die is unbalanced (loaded). In general, if there are n possible events and each is equally likely, each event has a 1 in n chance of occurring.

Example 1.1 Two engineers have decided to toss to see who goes on a trip to Hawaii to collect data. George wants to toss a coin and have Carol match it for her to win. Carol proposes that she toss a die and George match the number for him to go.
(a) Describe the events that will make Carol the winner for each method.
(b) Is each event equally likely? Why or why not?

SOLUTION (a) Carol wins if either of two events occurs for the coin method. These are heads on George's toss and heads on Carol's toss (HH) or tails for both (TT). For the die method only six events are winners for George. Listing Carol's result first, these are (1, 1), (2, 2), (3, 3), (4, 4), (5, 5), and (6, 6). The remaining 30 events which make Carol the winner are (1, 2), (1, 3), ... , (6, 4), (6, 5).
(b) For the coin method there are a total of four possible outcomes: (HH), (TT), (HT), and (TH). Carol's winning events each have a 1 in 4 chance of occurring, thus making them equally likely. For the die method there are 36 possible outcomes:

(1, 1)	(1, 2)	(1, 3)	(1, 4)	(1, 5)	(1, 6)
(2, 1)	(2, 2)	(2, 3)	(2, 4)	(2, 5)	(2, 6)
(3, 1)	(3, 2)	(3, 3)	(3, 4)	(3, 5)	(3, 6)
(4, 1)	(4, 2)	(4, 3)	(4, 4)	(4, 5)	(4, 6)
(5, 1)	(5, 2)	(5, 3)	(5, 4)	(5, 5)	(5, 6)
(6, 1)	(6, 2)	(6, 3)	(6, 4)	(6, 5)	(6, 6)

Each event which makes Carol the winner has a 1 in 36 chance of occurring. These events are also equally likely. If either the coin or die is loaded, not all events will be equally likely.

COMMENT The chance of equally likely events does not reflect the chance of a desirable outcome until they are somehow combined. It is not until we notice that Carol has a 2 in 4 chance of winning by coin and a 30 in 36 chance by die that we can conclude anything about the desirability of a method.

Problems P1.4–P1.7

1.3 Definitions of Probability

Probability is a number between 0 and 1 which expresses the chance that a specific event occurs under stated conditions. A probability statement may be: There is a 0.25 probability that a dimension exceeds the specification. Any probability value times 100 percent makes the number range from 0 to 100 and is called a percent chance. The word *probability* should be used only in statements with values from 0 to 1.

There are several different ways to obtain a probability value, each serving a different purpose. In this book we will commonly use one of the first two methods below.

1. *Classical, or a priori, method.* If it is known that event A can occur in m ways and a total of n equally likely ways is possible, the true probability $P(A)$ is

$$P(A) = \frac{\text{number of ways for } A}{\text{total number of ways}} = \frac{m}{n} \quad (1.1)$$

[margin note: most accurate]

Note that m and n are known a priori, that is, without experimentation. The toss of a coin is an example. We all know that if A is the event heads, then $m = 1$ and $n = 2$.

2. *Frequency, or a posteriori, method.* If an equally likely event A occurs m times in a total of n trials, the observed probability $P'(A)$ is

$$P'(A) = \frac{\text{times } A \text{ occurred}}{\text{total trials}} = \frac{m}{n} \quad (1.2)$$

[margin note: Experimentation Method]

In the limit as n becomes large, $P'(A)$ approaches the true probability $P(A)$:

$$\lim_{n \to \infty} P'(A) = P(A)$$

Here the true m and n values are unknown, but the observed times (frequency) of m estimate $P(A)$ using a finite n value. This is an a posteriori (after experimentation) method.

3. *Subjective method.* This is the best educated estimate or guess of the probability of event A. This method is necessary and legitimate, especially when insufficient numerical data is available. Actually, many business decisions are based on subjectively evaluated probabilities made in the absence of historical data.

[margin note: least accurate]

Example 1.2 State whether each of the following are examples of the classical, frequency, or subjective probability methods.
(a) Based on past marketing data there is a 0.10 chance of selling more than 1000 new passenger aircraft.
(b) An engineer guesses that there is a 50 percent chance that electricity usage will be mandatorily reduced by 30 percent in 5 years.
(c) It is incorrect to state that 1 out of 4 times you will draw a card of the heart suit from a standard deck if the cards are not replaced after each draw.
(d) Even though 15 out of 250 items were defective, the probability of a defective item will be assumed to be 0.20.

SOLUTION (a) The frequency method was used because historical data helped determine the 0.10 probability.

(b) The 50 percent chance is a subjective evaluation based on unknown, future events.

(c) The classical definition of probability is used here. If the cards are replaced, the event of a heart draw is equally likely, but without replacement it is not, so the probability cannot remain at the a priori probability of 0.25.

(d) By the frequency method Eq. (1.2) gives the probability as $15/250 = 0.06$. However, for some reason the value of 0.20 is assumed. Therefore, the subjective method is used.

Problems P1.8–P1.12

1.4 Phases of Probability Analysis

There are two closely related phases of probability analysis. *Descriptive probability* involves the explanation of how probability is distributed over different values. For example, the toss of a die can result in any number from 1 to 6, each with a 1/6 probability. The value of 1/6 for each number describes how probability is distributed. As seems natural, the sum of the probability values must add to exactly 1; otherwise, some events would not be accounted for correctly.

The manipulation of probability values is *computational probability*, which is used to determine how often one or more events should occur. Computation results are important, but they require the descriptive phase for determination.

Engineers and all types of technical analysts use probability in many ways. The lay person uses it in everyday life; however, these uses are usually more subjective and nonquantitative than in professional work.

1.5 Populations and Samples

The terms *population* and *sample* must be clearly understood. A *population* is the entire selection or universe of the characteristic which is studied. There are two sizes of populations—*finite* and *infinite*. Finite population size is measurable but can be so large that it is considered infinite for computational, and theoretical, purposes. For example, to analyze some property of a gallon of regular gas, the population is every gallon of gas produced. This value is finite, but its size is immense. The true value of any population property is called a *population parameter*, which is symbolized by some Greek letter such as μ, σ, π, etc. One parameter of the population of regular gas is the average lead content per gallon.

A *random sample*, or simply a *sample*, is a portion of the population used to estimate a population parameter value. Sample size is commonly finite and quite small in order to reduce data collection time and cost. The larger the sample, the more accurate the estimate of the parameter. Random samples are taken from

8 BASIC CONCEPTS

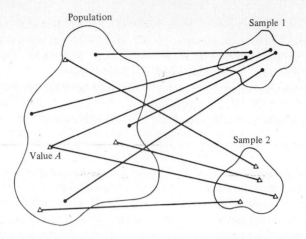

Figure 1.1 Samples taken from a population are random selections from equally likely population values.

the population in a random fashion. Figure 1.1 illustrates the fact that sample values are selected so that each and every population value is an equally likely event at the time of sample selection. It is possible for one value (value A) to be in two separate samples. A *sample statistic*, or *statistic*, is a single number calculated from sample data and used to estimate a population parameter whose exact value is usually unknown. A statistic is usually symbolized by a Roman letter such as s, t, p, X, etc. For example, the statistic X (sample average) is an estimate of the parameter μ (population mean).

Problems P1.13–P1.15

1.6 Discrete and Continuous Variables

A *variable* is the population characteristic studied. It is symbolized by a capital letter, it can take on any one of several values, and it can be classified as discrete or continuous. A *discrete variable* can assume only certain isolated values. For example, the number of persons working for a company may be called X and can equal any integer 0, 1, 2, and so on. We abbreviate the possible values as $X = 0, 1, 2, \ldots$ It is not necessary that a discrete variable always have an integer value.

A *continuous variable* can assume any value between two limits. The circumference of a person's waist, for example, is a continuous variable. The fact that a continuous variable may seem to be discrete occurs because we do not measure the variable as accurately as possible. If you measure four waists and obtain 35, 36, 39, and 42 inches (in), the variable is not discrete—it is continuous. If the 35-in circumference is actually 35.32 in, you did not measure accurately enough. You *could* measure 35.32104 in. In other words, for continuous variables the accuracy is due to the measuring scale, while for discrete a certain level of accuracy is dictated by the variable values.

If a continuous variable Y can take on any value between 10 and 20, we write it as $10 \leq Y \leq 20$. If Y must exceed 0, we write $Y > 0$. The following example explains how to recognize and write variable values.

Example 1.3 Describe each of the following variables as discrete or continuous and write the values that each variable can assume.
 (a) Number of spelling errors per page of a telephone book
 (b) Kilometers (km) between any two randomly selected cities
 (c) Time for you to complete an in-class test if a maximum of 50 minutes (min) is allowed for the test.
 (d) Number of thumb tacks in a handful
 (e) Weight of a horse
 (f) Time that it takes to repair a broken machine
 (g) Number of defective items in a sample size of 10

SOLUTION

Part	Type of variable	Possible values
a	Discrete	0, 1, 2, ..., total words
b	Continuous	>0
c	Continuous	0 < time ≤ 50
d	Discrete	0, 1, 2, ...
e	Continuous	>0
f	Continuous	>0
g	Discrete	0, 1, 2, ..., 10

Problems P1.16–P1.17

1.7 Definition and Phases of Statistics

Statistics is a branch of mathematics which uses formulas to collect, describe, analyze, and interpret quantitative data. Do not become confused between the terms *sample statistic* and *statistics*! In most cases at least one sample statistic is used to make conclusions in statistics.

Statistics is an applications field to an engineer, a scientist, a business person, or some other analyst that is used to make conclusions about the real world. Some statistical tools are regression analysis, analysis of variance, and quality control. These tools are applied to areas such as engineering, chemistry, biology, geology, botany, marketing, and numerous other fields in every discipline.

There are two phases to statistics—descriptive and inferential. Once sample data is collected, *descriptive statistics* describes and analyzes the data by computing at least one sample statistic, by constructing graphs and tables, and by comparing the results with other data. Once this phase is complete, *inferential statistics* interprets the results by using statistical tools, experience, common sense, and a general understanding of the process studied. This latter phase is more important, because without it no further action can be taken to correct or maintain the process.

The two phases of statistics are necessary to decide what to do. For example, assume a sample of 15 pieces of aluminum is taken from each of two possible vendors and the electrical conductivity measured. The computation of average conductivity per square centimeter for each sample is part of the descriptive statistics phase, while the use of a hypothesis-testing procedure to show that the two conductivity values are statistically equal is the inferential statistics phase. The conclusion of these two phases may be that either manufacturer can be used as a vendor.

Some people believe statistics is difficult to learn. The mathematics of statistics is not difficult. Most students of statistics have their biggest problems in knowing how and when to correctly apply the relatively simple computations in order to get the desired results. This simply means you must know and appreciate the concepts and assumptions of descriptive and inferential statistics. If they are applied incorrectly, you will get wrong answers; if they are applied correctly, you will have a statistically accurate answer which, of course, has a probability of being incorrect but is much better than a guess.

Example 1.4

Problems P1.18–P1.20

SOLVED PROBLEM (Case Study)

Example 1.4 A mechanical engineer (ME) is interested in the average repair time required for sensitive electronic testing sets. From the 55 units in use, a sample of 100 breakdowns and adjustments of any kind was used to construct Fig. 1.2, showing repair time in 0.25-hour (h) increments versus number of repairs. For example, 17 repairs took from 0.25 to 0.50 h, 24 took from 0.50 to 0.75 h, and so on. The average repair time is 0.84 h, which was compared by a statistical test to the average time of 0.75 h for similar equipment at a different plant. The two times were shown to be statistically unequal. Answer the following questions.

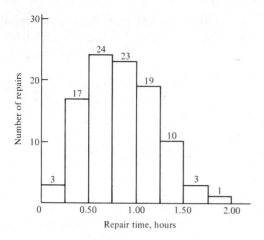

Figure 1.2 Graph of 100 repair times observed in a study of electronic equipment, Example 1.4.

(a) What is the population of data?
(b) What aspect of probability analysis does Fig. 1.2 represent?
(c) How should the value 0.84 h be used in future repair time studies?
(d) How are the two phases of statistics used?
(e) Is the data discrete or continuous?

SOLUTION (a) Since breakdowns occur in time, the population is the collection of all repair times on all 55 testing sets throughout the useful life of each unit. The population is finite in size, but it is large enough (probably 5000, 10,000, or more) to use infinite population analysis.
(b) The descriptive aspect of probability analysis includes the construction of graphs which explain how data or probability is distributed.
(c) The sample average of 0.84 h is a sample statistic usable as an estimate of a population parameter, in this case the average repair time. As long as the same test sets are used in similar ways and the repair crews do not change their repair procedures and inherent personnel skills, this estimate is correctly used as an estimate of average repair time.
(d) The descriptive phase of statistics involves the computation of the statistic average repair time for the 100 breakdowns. The statistical test and the conclusion that the two sample averages are unequal are the phase of inferential statistics, because actual inferences about populations are made based on sample data.
The data collected and graphed is continuous because any repair time greater than 0 is recordable. There is more on discrete and continuous data in the next chapter.

ADDITIONAL MATERIAL

(The section numbers in parentheses are the sections in this chapter covering the subject area and the numbers in brackets are references in the bibliography. This list is not meant to be exhaustive.)

Definitions of probability (Secs. 1.2 and 1.3): Duncan [7], pp. 15–26; Kirkpatrick [15], pp. 55–64; Lipson and Sheth [17], pp. 1–6.
Statistics (Secs. 1.5 and 1.7): Kirkpatrick [15], pp. 1–6; Lipson and Sheth [17], pp. 10, 510–514 (glossary).

PROBLEMS

Final answers to selected problems are given in App. A. You should refer to the answer only after you have completed all parts of a problem.

P1.1 State whether the following are deterministic or probabilistic:
(a) Number of days in a randomly named month.
(b) The factor of safety used in the design of an I beam of specific dimensions is 2.
(c) Number of defective bolts in a lot of 1000 bolts.
(d) The product of the dots on two fair dice.
(e) Today's cost quote of a specific machine from the XY Company.
(f) The cost of the machine in (e) from any manufacturer.

P1.2 Why is it sometimes impossible to consider the probabilistic aspects of data?

P1.3 Give two examples each of deterministic and probabilistic data from other courses you are presently attending.

P1.4 (a) Define the term *event*.
(b) Define and give three examples of equally likely events.

P1.5 The demand for an inventoried item varies from 25 to 35 per month. If each demand value is an equally likely event, what is the chance that it occurs?

P1.6 If you toss a coin 100 times and you get 95 heads, would you think that heads and tails for this coin are equally likely events? Why or why not?

P1.7 An engineer is going to test 10 resistors from 100 without replacement after each test. The event is defined as a defective resistor on any one of the 10 draws. Is the event equally likely over the 10 tests? Why or why not?

P1.8 What is the difference between the classical and the frequency definition of probability?

P1.9 A bacterium must grow at a predetermined rate, or else it cannot be used in a certain drug. Over a 3-month period the total number of specimens which exceeded the required rate was updated and recorded weekly.

(a) Use the frequency definition and the data shown in the table to compute the weekly probability that a specimen is usable.

(b) Plot the computed probability for each week and estimate the expected, long-run probability value.

Week	Total specimens	Total usable specimens	Week	Total specimens	Total usable specimens
1	10	8	7	83	61
2	25	18	8	90	69
3	34	27	9	110	80
4	46	32	10	125	93
5	57	44	11	140	106
6	71	55	12	150	113

P1.10 Compute the probability for the following and state whether you used the classical or frequency definition on each.
   (a) The grain size for an alloy was too large 3 out of 13 times.
   (b) The toss of a die will result in a 6.
   (c) In 100 die tosses the number 3 occurred 21 times.

P1.11 Two equally likely events can result from an experiment. In 1000 trials one event occurred 627 times. Would you suspect that the events are no longer equally likely? Why or why not?

P1.12 Why is it not possible to use the a priori definition of probability if events are not equally likely and the bias is not definitely known?

P1.13 Consider the telephone directory for your city.
   (a) Do all the phone numbers on one particular page represent a population or a sample?
   (b) Are the numbers in (a) random if they are used to select people from the entire city by their last name?
   (c) Suggest a statistic you can compute for this page and give the corresponding parameter.
   (d) Define an equally likely event for the phone numbers on one page.
   (e) Define an unequally likely event for the names on one page.

P1.14 A precision instruments manufacturer made a total of 25 test sets on contract for an airline company. The average impact strength for the plastic cover was computed from tests on all 25 covers.
   (a) What is the population and what is the sample?
   (b) Is the average a statistic or a parameter value?

P1.15 When is a sample not random?

P1.16 State whether the following are true or false. If you answer false, correct the statement.
   (a) If X is product demand in number of units, X is discrete.
   (b) The range $990 < Y \leq 10{,}000$ h represents a discrete variable.
   (c) If W is the weight of an item and with a bathroom scale I observe weights of 85, 101, and 72 lb, the variable W is discrete.
   (d) The time it takes for a bearing to wear out is a continuous variable.
   (e) It is always possible and meaningful to turn a discrete variable into a continuous variable.

P1.17 Give three examples of discrete variables and three examples of continuous variables from particular engineering, science, or business courses.

P1.18 What is the difference between statistics and a statistic?

P1.19 A technician took a sample of 250 processing times on the shop floor and did the following with the data. In each case determine what phase of probability or statistics was involved in the computations.
   (a) Constructed a graph similar to Fig. 1.2
   (b) Determined that there is a 78 percent chance that processing time exceeds 15 min
   (c) Computed the average processing time and the fraction of the times greater than the average
   (d) Compared the average with a similar result on another processing line and found the difference to be 1.75 min
   (e) From (d) concluded that the two lines are similar in processing time

P1.20 Here is a little harder one for you. The discrete variable X represents the number of defects per 10-meter (m) section of cold-rolled steel sheets. Therefore, $X = 0, 1, 2, \ldots$ From a sample of 25 sections the average number of defects was computed as 5.68. Does this average imply that X should actually be a continuous variable specified as $X \geq 0$? Why or why not?

CHAPTER
TWO

DATA PRESENTATION FORMATS

This chapter teaches you how to present collected data in a form that is understandable to you and to others. Different types of tabular and graphical formats are illustrated. A clear understanding of these formats is essential to the beginner in statistical analysis because data is manipulated throughout this book by these methods.

Probably the most difficult material in this chapter is the terminology. When you first encounter a new term, carefully analyze the connection between the name and the purpose. In this way, you should soon be able to work with the graphical formats using the correct names.

CRITERIA

To complete the material in this chapter, you must be able to:

1. Define the terms *ungrouped* and *grouped* data and *frequency*. Construct a *frequency tally sheet*, given the observed values.
2. Define the term *frequency distribution* (*histogram*); tabulate and graph a frequency distribution for *ungrouped* data, given the observed values of the variable.
3. Calculate the *cell width*, *cell boundaries*, and *cell midpoints* to group data, given the observed values and the number of cells.
4. Tabulate and graph a frequency distribution (histogram) for *grouped* data; and define and graph a *frequency polygon*, given the cell data and frequencies.

5. Define, tabulate, and graph a *cumulative frequency distribution*, given the values or cell definitions and frequencies.
6. State the equation for, compute, and graph *relative frequency* and *cumulative relative frequency*, given the values or cells and frequencies.
7. Determine any *fractile* values, given the data values or cell definitions and frequencies.

STUDY GUIDE

2.1 Ungrouped Data, Grouped Data, and the Frequency Tally Sheet

It is important to remember that if data has been collected in some specific order, the formats in this chaper will probably not be of much help. For example, sales data collected over a 5-year period for the purpose of forecasting future production needs has an important time order which should be retained. The graphical formats of this chapter will destroy this order.

All data, discrete or continuous, will be stated in one of two forms—ungrouped or grouped. *Ungrouped data* is merely a listing of the individual data values. *Grouped data* means that the data is lumped together into collections of several values. For example, if a variable lies between 5 and 19, the analyst may study the variable as values between 5 and 9, between 10 and 14, and between 15 and 19. A summary of discrete, continuous, ungrouped, and grouped data is given in Fig. 2.1. You should give your own examples of each type of data to be sure you understand all the differences.

Data may be listed by individual value or by frequency. *Frequency* is the number of times that a value is observed. Either ungrouped or grouped data may be presented in frequency form.

If the values are listed as observed, they are called *raw data*. Once new data is arranged in increasing or decreasing order, it is called *ordered data*. If a frequency is given for each value, it is called *frequency data*. The actual frequency values are determined by preparing a *frequency tally sheet*, which is illustrated here.

Example 2.1 An energy use analysis is being made for the city of Guzzler by a consulting firm. The number of persons in car pools was collected for 50 cars with two or more occupants. The raw data is presented in Table 2.1. Prepare a frequency tally sheet for the consulting firm.

SOLUTION The ungrouped values of the variable are 2, 3, 4, 5, and 6. Table 2.2 is a frequency tally sheet for the car pool data of Table 2.1. Note that all 50 values are accounted for.

16 DATA PRESENTATION FORMATS

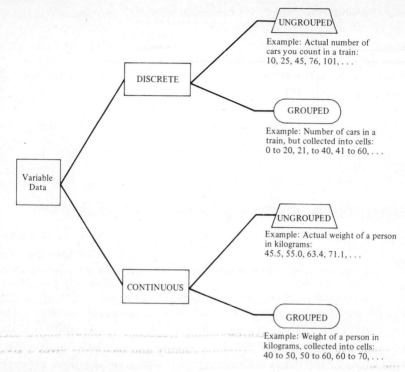

Figure 2.1 Summary of different types of data.

COMMENT You can get a good idea of the general shape of the frequency data by carefully observing the tally sheet. The general shape is more correctly presented if you are careful in making the tally marks by using approximately equal spacing between marks.

Example 2.12

Problem P2.1

Table 2.1 Number of persons per car pool for 50 cars

2	4	3	4	3	6	5	3	4	2
3	5	2	3	5	2	2	4	4	5
2	2	2	2	2	2	4	3	2	4
4	2	4	2	2	3	2	2	6	2
5	2	2	5	3	4	3	3	2	3

Table 2.2 Frequency tally sheet of car pool data, Example 2.1

Number in car pool	Frequency tally	Frequency number
2	⊥⊥⊥⊥ ⊥⊥⊥⊥ ⊥⊥⊥⊥ ⊥⊥⊥⊥ ⊥	21
3	⊥⊥⊥⊥ ⊥⊥⊥⊥ ⊥	11
4	⊥⊥⊥⊥ ⊥⊥⊥⊥	10
5	⊥⊥⊥⊥ ⊥	6
6	⊥⊥	2
		50

2.2 Frequency Distributions (Histograms) for Ungrouped Data

A *frequency distribution* is an elementary graphical format used to present the distribution of the frequencies over the observed values of the variable. The frequency distribution, also called a *histogram*, may be used for ungrouped or grouped data. Here we study the histogram for ungrouped data, which is usually used for discrete variables. The variable values are placed on the abscissa (x axis) and the frequency scale on the ordinate (y axis). For ungrouped data a vertical line the height of the corresponding frequency is drawn above each observed value. The frequency distribution rapidly and pictorially gives a good idea of the shape of the data's distribution. The histogram is always drawn to scale.

Example 2.2 Construct the frequency distribution for the ungrouped car pool data of Example 2.1.

SOLUTION Figure 2.2 is the frequency distribution, or histogram. It uses the frequency results which are tabulated in Table 2.2. Be sure to note how the axes are labeled—the abscissa with the variable name (number in car pool) and the ordinate with the frequency scale. You must always label both axes of any graph.

COMMENT Since the only values of this discrete variable are 2, 3, 4, 5, and 6, absolutely nothing should be drawn between these points. Therefore, it is incorrect to connect the top of the vertical lines to form a polygon. This practice is reserved for continuous data only and will be covered in the appropriate section.

Problems P2.2, P2.3

18 DATA PRESENTATION FORMATS

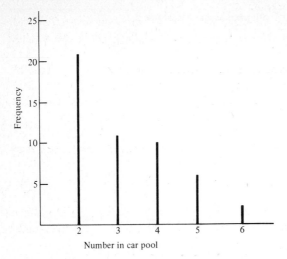

Figure 2.2 Histogram for discrete, ungrouped car pool data, Example 2.2.

2.3 Determination of Cells to Group Data

Often the ungrouped frequency distribution does not give a good idea of how the data is distributed because most observed values are different, so the frequency is very low. To improve the efficiency of studying variables, the data may be grouped into cells (also called intervals or classes). A *cell* is a grouping of successive data values using a specific upper and lower limit. You have a lot of flexibility in developing the cells; however, you must determine each of the following:

Number of cells k. You should use 10 to 20 cells for about 50 or more data points and 6 to 10 for less than 50 values. This guideline is quite general, because too many or too few cells can very easily change the general shape of the histogram.

Cell width c. This is the numerical distance between the two limits of any cell. All cells should be of equal width. Cell width is computed using the relation

$$c = \frac{\text{maximum value} - \text{minimum value}}{\text{number of cells}}$$

$$= \frac{X_{max} - X_{min}}{k} \qquad (2.1)$$

Often some extreme data values are omitted from the computation of c because they are not representative of the rest of the data.

Cell boundaries. The cell boundaries are values one-half measurement unit more accurate than the observed data. Therefore, if the observed accuracy is one decimal, the boundaries are written in two decimal values. This is necessary so that no value can be observed exactly on a boundary. The boundaries

2.3 DETERMINATION OF CELLS TO GROUP DATA

should make the cells cover all observed values, but no two cells should numerically overlap. A typical set of integer cell boundaries is 51–60, 61–70, 71–80.

Cell midpoint X_i. The cell midpoints X_i ($i = 1, 2, \ldots, k$ cells) are one-half of the distance between cell boundaries. Each X_i is assumed to be the concentration point of all values in the cell. This makes computations somewhat approximate; however, the method is efficient and the error is small.

You can use the following step-by-step procedure to determine cell boundaries and the midpoints.

1. Select the desired number of cells k.
2. Eliminate any unwanted extreme data values and compute the cell width c using Eq. (2.1). The value of c should be rounded to make it a conveniently usable number.
3. Determine the cell boundaries by starting below the lowest observed value with one-half greater measurement accuracy than the observed data.
4. Determine the cell midpoint values by starting $c/2$ above the lowest cell boundary and adding c until the last cell midpoint is determined.

If at any time you are dissatisfied with the cells (number, boundaries, or midpoints), return to step 1 with a new k value. Example 2.3 is a thorough study of cell determination for a continuous variable.

Example 2.3 A sample of 37 residential water pressure readings P, in pounds per square inch (lb/in^2), was taken by a utility company technician. The ordered readings were as follows:

$$34, 37, 37, 39, \ldots, 48, 48, 49$$

All 37 values were between 34 and 49 lb/in^2. Determine the cell boundaries and midpoint values for the technician using $k = 8$ cells.

SOLUTION We will follow the steps outlined above for cell determination.

1. The value of k is 8 cells.
2. We compute the cell width from Eq. (2.1), using $P_{max} = 49$ and $P_{min} = 34$:

$$c = \frac{49 - 34}{8} = 1.875 \; \frac{\text{lb}}{\text{in}^2}$$

For computational ease use $c = 2.0$ lb/in^2.

3. The accuracy of the boundaries will be one decimal value (because all values are integers) beginning at 33.5 lb/in^2. The cell boundaries are listed in Table 2.3.
4. Cell midpoints P_i start at 34.5 and increase by 2.0 lb/in^2 (Table 2.3).

Table 2.3 Cell boundaries and midpoints, Example 2.3

Cell i	Cell boundaries	Midpoints P_i
1	33.5–35.5	34.5
2	35.5–37.5	36.5
3	37.5–39.5	38.5
4	39.5–41.5	40.5
5	41.5–43.5	42.5
6	43.5–45.5	44.5
7	45.5–47.5	46.5
8	47.5–49.5	48.5

The upper boundary on each cell is actually not included in the cell. That is, if the cell is 33.5 to 35.5, the values in the cell are 33.5 and up to, but not including, 35.5 lb/in². To simplify cell listing, the cells and midpoints are often written in terms of the measured unit accuracy themselves, and it is assumed that the *upper cell boundary is not included in the cell*.

Example 2.4 Write the cell boundaries and midpoints for Example 2.3 in terms of the measured unit accuracy and state where a value of 36 lb/in² would be recorded.

SOLUTION The cells are listed in Table 2.4 using a width $c = 2$. Note that the midpoints are now integer values. A value of 36 lb/in² is recorded in the 36 to 38 lb/in² cell, because the value 36 is not actually included in the 34 to 36 cell.

Table 2.4 Cell boundaries in terms of measured units, Example 2.4

Cell i	Cell boundaries	Cell midpoints P_i
1	34–36	35
2	36–38	37
3	38–40	39
4	40–42	41
5	42–44	43
6	44–46	45
7	46–48	47
8	48–50	49

Problems P2.4–P2.8

2.4 Frequency Distribution for Grouped Data

A frequency distribution, or histogram, for grouped data may be constructed for continuous or discrete variables. The cell midpoint values are scaled on the abscissa, and a *rectangle* the height of the observed frequency is centered on each midpoint. Cell boundaries are also indicated on the graph. We will use the symbol f_i to represent the frequency of cell i ($i = 1, 2, \ldots, k$).

A simplified presentation format is the *frequency polygon*, which is a graph of midpoint values versus frequency. Straight lines are drawn at the observed frequency level between the midpoint values to form a $k + 2$-sided polygon (k = number of cells). This polygon may be drawn for a continuous variable, but it is incorrect for a discrete variable.

Example 2.5 The following grouped weight data for 165 garbage cans has been collected by a sanitation department technician in an attempt to study the human-strength needs of a sanitation worker.

Weight (lb)	Cell midpoint	Frequency
6–10	8	3
10–14	12	5
14–18	16	22
18–22	20	17
22–26	24	35
26–30	28	40
30–34	32	34
34–38	36	9

Construct the histogram and frequency polygon for this data.

SOLUTION Figure 2.3*a* presents the histogram with cell boundaries marked. Figure 2.3*b* is a graph of the frequency polygon with midpoints indicated. Study the two figures carefully.

COMMENT The frequency level lines drop to 0 outside the range of the observed data. Often the frequency polygon is plotted over the histogram for convenience.

The weight data used here is continuous. If the data had been discrete with values 6, 7, 8, ..., the histogram would be drawn the same as Fig. 2.3*a*. However, it would be incorrect to construct a frequency polygon, because only integer values are observable. However, you will find that

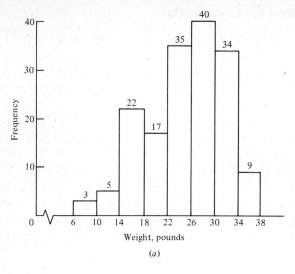

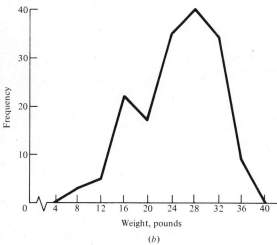

Figure 2.3 Graphs of the (a) histogram and (b) frequency polygon for continuous grouped data, Example 2.5.

people draw the frequency polygon for both discrete and continuous variables simply because it is easier to construct than the histogram.

Problems P2.9–P2.13

2.5 Cumulative Frequency Distribution

There are many occasions when it is necessary to know the total number of data values above or below a specific value. The *cumulative frequency distribution* is used for this purpose. The cumulative frequency distribution is a graph of the accumulated frequency that is *less than or equal to a stated value*. This graph for a discrete variable is much different from that of a continuous variable (Fig. 2.4). The discrete cumulative frequency distribution is a *step function* because frequency is accumulated only at the isolated, observed values. This curve has points of discontinuity at the steps and always increases or remains horizontal (never decreases). The continuous cumulative frequency distribution is not discontinuous because frequency is accumulated at all variable values. Figure 2.4*b* shows the two ways to draw this continuous cumulative curve. Curve 1 is the actual distribution because it connects each cumulative frequency point. Curve 2 is a *smooth*-curve approximation to curve 1. The approximation is commonly used because it is easier to work with.

The cumulative frequency is plotted on the ordinate at individual data values for ungrouped data or at the upper cell boundary for grouped data. The curve, which is also called an ogive (oh'-jive) curve, extends from the value of 0 to the sum of all frequency values. Subtracting the curve value from the total of all frequency values gives the accumulated frequency greater than the variable value.

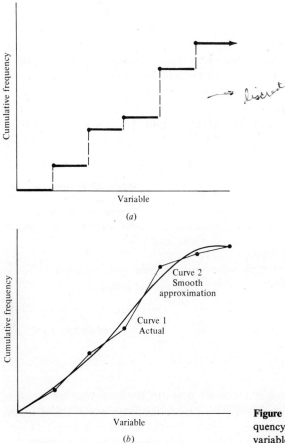

Figure 2.4 General shape of cumulative frequency distribution curves for (*a*) discrete variables and (*b*) continuous variables.

24 DATA PRESENTATION FORMATS

The following steps may be used to construct a cumulative frequency distribution curve for ungrouped or grouped data:

1. Tabulate the data values or cells.
2. Tabulate the frequency for each value or cell.
3. Compute the cumulative frequency by adding all frequency values prior to and including the present value or cell.
4. Construct a graph with the variable or cells on the abscissa and cumulative frequency on the ordinate.
5. Plot the cumulative frequency against the value for ungrouped data or the upper cell boundary for grouped data. Construct the curve.

You might get confused about how to plot the curve for grouped data. Just remember this: *Always plot* **cumulative** *information at the* **upper cell boundary**. The step increase will always take place at lower cell limits.

The following examples show how to construct a cumulative frequency distribution for discrete ungrouped and continuous grouped variables. An example of discrete grouped is given in Solved Problems.

Example 2.6 Use the car pool frequency data in Table 2.2 to (*a*) plot a cumulative frequency distribution and (*b*) estimate the percentage of car pools with less than four persons.

SOLUTION (*a*) This data is discrete and ungrouped. Use the steps above for individual data values.

1 and 2. Table 2.2.
3. The cumulative frequency values are shown here:

Number in car pool	Frequency	Cumulative frequency
2	21	21
3	11	32
4	10	42
5	6	48
6	2	50
	50	

The cumulative frequency must stop at the same value as the sum of all frequency values (50 in this example), or else an addition mistake has been made.

2.5 CUMULATIVE FREQUENCY DISTRIBUTION

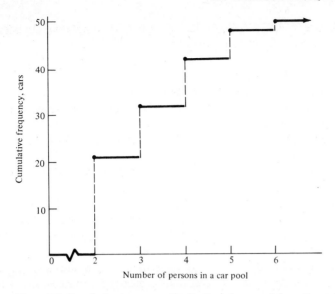

Figure 2.5 Cumulative frequency distribution of car pool data, Example 2.6.

4 and 5. Figure 2.5 is a plot of cumulative frequency. Because the data is discrete, increases occur only at the integer values 2, 3, ... , 6. Cumulative frequency below two persons is 0, and for six persons and above it is 50 cars.

(b) Less than four persons is the same as three or less. From Fig. 2.5 the cumulative frequency for three persons is 32 cars. On the basis of this 50-car sample, $32/50 = 0.64$, or 64 percent of the car pools have less than four people.

COMMENT Cumulative frequency is for less than *or equal to* a stated value. It is easier to read a cumulative frequency distribution from the right. For example, in (b) start at six persons in Fig. 2.5 and go down to three persons; then read across to 32 cars. This ensures that you include three persons. If you read from zero up to three persons, you risk stopping at 21 cars, which does not include three persons. Therefore, read a cumulative frequency distribution right to left.

Example 2.7 An environmental engineer is developing a water pollution model. The following grouped frequency data shows the number of counties having a given percentage of their freshwater supply too contaminated to use.

26 DATA PRESENTATION FORMATS

Percentage of water supply	Number of counties	Percentage of water supply	Number of counties
0– 2	44	10–12	5
2– 4	23	12–14	8
4– 6	18	14–16	3
6– 8	9	16–18	4
8–10	7	18–20	1

The engineer has asked you to do the following:
(a) Tabulate and graph the cumulative frequency distribution.
(b) Use the graph to determine the number of counties with less than 8 percent and less than 15 percent contamination.
(c) Determine how many counties have a contamination of at least 11 percent.
(d) Determine the maximum percentage of contamination for 50 counties.
(e) Draw a smooth curve to approximate the distribution and then answer (b), (c), and (d).

SOLUTION (a) This is grouped continuous data with cell limits in measured unit accuracy of integer percents. The steps for cells are used.

1 and 2. The cells and frequencies are given in the problem statement.
3.   Table 2.5 shows the cumulative frequency values.
4.   Figure 2.6 shows the scales.
5.   The cumulative frequency values are plotted at the upper cell boundaries and connected (Fig. 2.6).

All values of pollution above 20 percent have a cumulative frequency of 122. The cumulative frequency values are actually "*less than*" values because the upper cell boundary is not a recorded value, it is only a limit for the cell. As an example, from Table 2.5 we can state that 67 counties have less than 4 percent pollution. However, since percentage is a continuous variable, the cumulative frequency points are connected as in Fig. 2.6.

(b) From Fig. 2.6 we have 94 counties with less than 8 percent and 116 counties with less than 15 percent pollution.
(c) There are 103 counties with less than 11 percent. From this data we conclude there are $122 - 103 = 19$ counties with at least 11 percent.
(d) To determine a percentage value for the cumulative frequency of 50, locate 50 on the ordinate, go to the curve, and read a value of 2.5 percent.

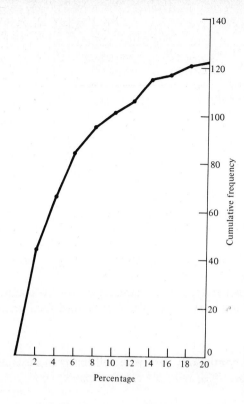

Figure 2.6 Graph of the cumulative frequency distribution for pollution data, Example 2.7.

Table 2.5 Cumulative frequency tabulation, Example 2.7

Percentage of water supply	Frequency	Cumulative frequency
0– 2	44	44
2– 4	23	67
4– 6	18	85
6– 8	9	94
8–10	7	101
10–12	5	106
12–14	8	114
14–16	3	117
16–18	4	121
18–20	1	122
	122	

28 DATA PRESENTATION FORMATS

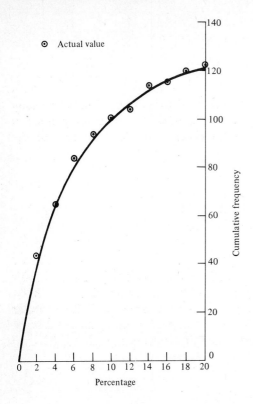

Figure 2.7 Smoothed cumulative frequency distribution curve, Example 2.7.

(e) Figure 2.7 shows a smooth, "eye-balled" curve for the data of Table 2.5. The answers to the above questions are:

(b) 92 counties have less than 8 percent, and 113 counties have less than 15 percent.
(c) $122 - 104 = 18$ counties have at least 11 percent.
(d) The value of 2.7 percent has a cumulative value of 50 counties.

Example 2.13

Problems P2.14–P2.18

2.6 Relative Frequency and Cumulative Relative Frequency

The *relative frequency* $P(X_i)$ is a value between 0 and 1 which represents the fraction of the time that a certain variable value X_i is observed. It is computed by using the ratio of frequency to sample size:

$$P(X_i) = \frac{f_i}{n} \tag{2.2}$$

2.6 RELATIVE FREQUENCY AND CUMULATIVE RELATIVE FREQUENCY

where $i = 1, 2, \ldots, k$ cells
f_i = frequency of the ith value
n = sample size = $\sum_{i=1}^{k} f_i$

The relative frequency is empirical because it uses observed data. Actually $P(X_i)$ is computed in the same way as probability, by using the frequency approach (Sec. 1.3). Also, the relative frequencies add to 1 because of the identity

$$\sum_{i=1}^{k} P(X_i) = \sum_{i=1}^{k} \frac{f_i}{n} = \frac{\sum_{i=1}^{k} f_i}{\sum_{i=1}^{k} f_i} = 1$$

You should study the next name carefully so that you do not confuse it with similar terms. *Cumulative relative frequency* $F(X_i)$ is the accumulation of $P(X_i)$ values for all values of X less than or equal to a particular value. $F(X_i)$ is similar to the cumulative frequency discussed in Sec. 2.5, except the relative frequency is added. Computationally, for grouped data $F(X_i)$ is the sum of all $P(X_i)$ for all cells up to and including X_i:

- $$F(X_i) = \sum_{X=1}^{X_i} P(X) \qquad (2.3)$$

The fraction of data exceeding X_i is found by using the complement relation:

- $$1 - F(X_i) = \sum_{X=X_{i+1}}^{X_k} P(X) \qquad (2.4)$$

Example 2.8 tabulates and interprets $P(X_i)$ and $F(X_i)$.

Example 2.8 The percent Y of inert ingredients for 22 different manufacturers of a certain chemical has been grouped and placed into percentage cells (Table 2.6).
(a) Tabulate the relative frequency.
(b) Find the cumulative relative frequency.
(c) Interpret the value $F(14.3)$ and $1 - F(14.3)$.

SOLUTION (a) Equation (2.2) is used to compute $P(Y_i)$ values, as shown in column 5 of Table 2.6. An example computation at the midpoint of cell $i = 2$ is $P(Y_2 = 13.1) = 3/22 = 0.136$. Note that the sum of all $P(Y_i)$ values is 1.0.

Table 2.6 Computation of $P(Y_i)$ and $F(Y_i)$, Example 2.8

i (1)	Cell boundaries (2)	Midpoint Y_i (%) (3)	f_i (4)	Relative frequency $P(Y_i)$ (5)	Cumulative relative frequency $F(Y_i)$ (6)
1	12.2–12.8	12.5	8	0.364	0.364
2	12.8–13.4	13.1	3	0.136	0.500
3	13.4–14.0	13.7	2	0.091	0.591
4	14.0–14.6	14.3	5	0.227	0.818
5	14.6–15.2	14.9	3	0.136	0.954
6	15.2–15.8	15.5	0	0.000	0.954
7	15.8–16.4	16.1	1	0.046	1.000
			22	1.000	

(b) Equation (2.3) is used to determine the values of $F(Y_i)$ in Table 2.6. The $F(Y_6)$ value is the same as $F(Y_5)$ because the relative frequency at Y_6 is 0. Note that the last $F(Y_i)$ value is 1.0.

(c) $F(14.3) = 0.818$ is the cumulative relative frequency value at the midpoint of cell $i = 4$. We can state that the fraction of the data less than or equal to 14.3 is 0.818 (or 81.8 percent). The value of 0.818 is a correct fraction for the entire cell $i = 4$, that is, $F(14.0$ through $14.6) = 0.818$. The value $1 - F(14.3) = 1 - 0.818 = 0.182$ is the fraction of data greater than cell $i = 4$, identified by the midpoint $Y_4 = 14.3$. The value 0.182 can also be obtained by adding $P(Y_5) + P(Y_6) + P(Y_7) = 0.182$.

It is usually easier to work with and visualize the results of a relative frequency or cumulative relative frequency tabulation if they are graphed. The graphs are similar to a histogram or cumulative frequency distribution, respectively. There is a distinct difference in the graphs for a discrete and continuous variable, as discussed in Sec. 2.5. The abscissa is scaled for the variable X, and the $P(X_i)$ or $F(X_i)$ values are placed on the ordinate. Actually, it is more revealing if the information from the frequency and relative frequency graphs is combined into one graph, using these scales:

What is graphed	Abscissa	Left ordinate	Right ordinate
Frequency data	Variable	Frequency	Relative frequency
Cumulative data	Variable	Cumulative frequency	Cumulative relative frequency

2.6 RELATIVE FREQUENCY AND CUMULATIVE RELATIVE FREQUENCY

Example 2.9 Use the data of Example 2.8 (Table 2.6) to plot frequency and relative frequency on one graph, and plot cumulative frequency and cumulative relative frequency on another graph.

SOLUTION The data in Table 2.6 is continuous and presented in cell format. We can present the graphs in histogram or polygon form. Both are presented here. Figure 2.8 shows the histogram for frequency and relative frequency with the midpoint marked. Figure 2.9 presents the same results in polygon form. Plots in Fig. 2.9 are at midpoints for $P(Y_i)$ and at the upper cell boundaries for $F(Y_i)$.

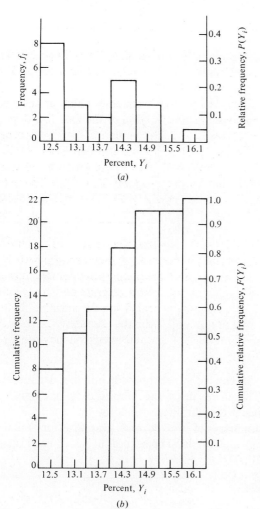

Figure 2.8 Graphs of (a) $P(Y_i)$ and (b) $F(Y_i)$ in histogram form, Example 2.9.

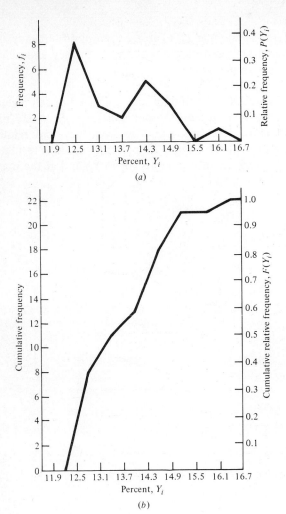

Figure 2.9 Graphs of (a) $P(Y_i)$ and (b) $F(Y_i)$ in polygon form, Example 2.9.

COMMENT Figures 2.8 and 2.9 can be superimposed, but this usually adds confusion to the picture. Therefore, draw the graphs only one way: the polygon form is usually faster if the data is continuous.

Figure 2.10 gives a summary of the correct graphical presentation formats for ungrouped and grouped data. The section in which each format is discussed is also shown.

Problems P2.19–P2.24

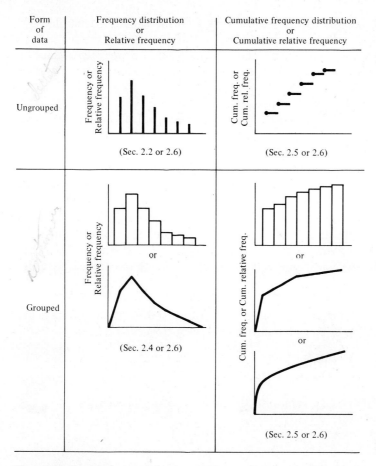

Figure 2.10 Graphical presentation formats for ungrouped and grouped data.

2.7 Fractiles

The analyst often wants to know what fraction of observed data is below or above a specific value. For this reason, *fractiles* are used to divide data into percentiles (100 parts) or some other convenient division. For example, an engineer may find that 25 percent of all data fell below 35°C in a certain experiment. Then 35°C is the 25th percentile, or first quartile, of the data. Common fractile names and symbols are:

Fractile name	Number of divisions k	Symbol	Range of i
Percentile	100	P_i	1 to 100
Decile	10	D_i	1 to 10
Quartile	4	Q_i	1 to 4

These properties are useful because they help the engineer determine how much of the data is included between certain values. However, the fractiles are obtained differently for ungrouped and grouped data. For *ungrouped* data the steps to find any fractile are:

1. Order the data from smallest to largest.
2. Determine the fractile desired and the associated k value (maximum number of divisions).
3. Compute the value

$$F = \frac{i}{k}(n + 1) \qquad (2.5)$$

where n is the sample size and F is the fractile indicator P_i, D_i, or Q_i.
4. Count through the ungrouped, ordered data until the value in step 3 is reached. This observed value is the desired fractile. If the result of Eq. (2.5) is not an integer, use linear interpolation to approximate the needed fraction.

Example 2.10 An electrical engineer has collected the following voltage gain values for 15 single-stage amplifiers in a given frequency range:

15.1	16.5	17.0	18.2	19.1
21.4	21.4	24.0	24.3	25.0
26.9	30.4	31.1	33.0	35.3

Determine the voltage values for the fractiles in (*a*) through (*e*).
(*a*) 50th percentile
(*b*) 35th percentile
(*c*) 7th decile
(*d*) 3d quartile
(*e*) Between the 1st and 3d quartile
(*f*) Determine the fractile for the gain value of 24.0 volts (V).

SOLUTION This data is ungrouped. We use the steps above to determine the desired fractiles. Since the gain values are already ordered, we can start with step 2. The steps are shown in (*a*) only.
(*a*) Step 2. For a percentile basis, $k = 100$.
   Step 3. From Eq. (2.5) for the 50th percentile

$$F = \frac{50}{100}(15 + 1) = 8$$

Step 4. The eighth value is $P_{50} = 24.0$ V.
(*b*) We have $F = \frac{35}{100}(16) = 5.6$. By linear interpolation

$$P_{35} = 19.1 + 0.6(21.4 - 19.1) = 20.5 \text{ V}$$

(c) From Eq. (2.5)

$$F = \tfrac{7}{10}(15 + 1) = 11.2$$
$$D_7 = 26.9 + 0.2(30.4 - 26.9) = 27.6 \text{ V}$$

Note that the D_7 value is the same as the P_{70} value.

(d) We have $F = \tfrac{3}{4}(15 + 1) = 12.0$. Then $Q_3 = 30.4$ V.

(e) The values between the first and third quartile (Q_1 and Q_3) represent the middle 50 percent of the data. For Q_1, we have $F = \tfrac{1}{4}(16) = 4$, so $Q_1 = 18.2$ V. From (d), $Q_3 = 30.4$ V. Therefore, 50 percent of the data occurred between 18.2 and 30.4 V.

(f) To determine a specific fractile represented by an observed value, we can always solve Eq. (2.5) for i using a percentile basis ($k = 100$). Since 24.0 V is the 8th value,

$$F = 8 = \frac{i}{100} 16$$
$$i = 50$$

The value 24.0 V is P_{50}, as found in part (a).

If the data is grouped into cells, a mathematical linear interpolation method or graphical method can be used to *approximate* the desired fractile. The graphical method illustrated here is faster and almost as accurate as the mathematical method. The steps in the graphical method are:

1. Construct a polygon graph of the cumulative relative frequency.
2. At the desired fractile, draw a horizontal line to the graph.
3. Draw a vertical line down to the abscissa. This number is the value corresponding to the desired fractile.

If you want to find the data values that were observed between two fractile values, use the steps above for both fractile values. This technique is presented in Example 2.11.

Example 2.11 Figure 2.11 is the polygon for the cumulative relative frequency tabulated in Example 2.8. Find the (a) third decile value, (b) third quartile value, and (c) values between the 5th and 95th percentiles.

SOLUTION We will express all fractiles as percentiles so we can use the scales of Fig. 2.11.

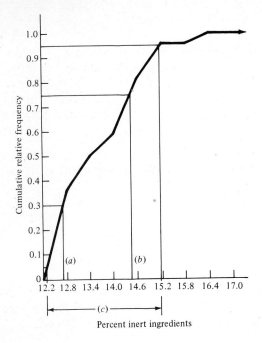

Figure 2.11 Polygon graph of cumulative relative frequency used in Example 2.11.

(a) The value for D_3 is the same as P_{30}. Use the three steps above:

1. Figure 2.11.
2. At the ordinate value of 0.3, a horizontal line is drawn to the graph.
3. The vertical line labeled a indicates that $D_3 = 12.7$ percent inert ingredients.

(b) The same procedure as above for $Q_3 = P_{75}$ results in a value of 14.4 percent inert ingredients as the third quartile value.

(c) We apply the steps twice to obtain the values $P_5 = 12.3$ percent and $P_{95} = 15.1$ percent. These values are indicated by (c) in Fig. 2.11. Now we can state that, based on this sample, 90 percent of the data is between 12.3 and 15.1 percent. We can further conclude that 5 percent of the data is less than 12.3 and 5 percent is greater than 15.1 percent inert ingredients.

COMMENT Fractiles allow you to use observed data to "predict" what to expect in future observations. The confidence you have in results increases as the sample size becomes larger. Here we used only 22 data points. However, regardless of the sample size, fractiles can help you understand the observed data.

Problems P2.25–P2.28

SOLVED PROBLEMS

Example 2.12 Mr. Thomas, a civil engineer with the Hi-Strength Company, collected 30 specimen values of concrete compressive strength on 15-centimeter (cm) cubes. He gave Mr. Jones the ungrouped data by increasing values (Table 2.7) and asked him to do some analysis. Mr. Jones wants to group the data into nine groups of length 20 kg/cm^2 each. Do this grouping for Mr. Jones.

Table 2.7 Ungrouped values of compressive strength, kg/cm^2

435	443	450	475	510	512
517	518	519	519	521	522
522	524	531	534	537	540
540	541	551	552	553	553
559	560	571	573	580	598

SOLUTION The grouped data and frequencies are presented in Table 2.8. The first group begins at 430 as a matter of convenience. Since all data values must be included, any value of 435 or less could have been used.

Section 2.1

Table 2.8 Compressive strength grouped, Example 2.12

	Group	Frequency tally	Observed frequency
Number	Boundaries		
1	430–449	||	2
2	450–469	|	1
3	470–489	|	1
4	490–509		0
5	510–529	╫╫ ╫╫	10
6	530–549	╫╫ |	6
7	550–569	╫╫ |	6
8	570–589	|||	3
9	590–609	|	1

Example 2.13 A group of professional engineering organizations has formed a taskforce to study the possibility of starting a political lobby in Washington, D.C. One piece of information considered important is the age of the U.S. senators. The ages of 100 senators are presented in Table 2.9.
(*a*) Construct a histogram using cell widths of 5 and 10 years. Compare these.

38 DATA PRESENTATION FORMATS

Table 2.9 Age of 100 U.S. senators, Example 2.13

Age	Frequency	Age	Frequency	Age	Frequency
33	1	50	10	68	3
35	1	51	4	69	1
36	1	52	3	71	1
37	2	54	4	72	2
39	3	55	5	73	2
40	2	58	7	74	1
42	1	59	9	75	1
43	1	61	8	76	1
45	1	64	2	77	1
47	4	65	2	78	1
48	5	66	2	79	1
49	3	67	3	80	1

(b) Construct the cumulative frequency distribution and cumulative relative frequency graphs for a 10-year cell width.
(c) What fraction of the senators are under 45? At least 75?

SOLUTION (a) The ages are grouped into cells of 5 years in Table 2.10 and 10 years in Table 2.11 using measured unit accuracy. The histograms are graphed in Fig. 2.12. Comparison indicates that the 10-year cell width generates a more symmetric-appearing histogram. This is in contradiction to the general guide given in Sec. 2.3 for the number of cells to use. The guide suggests 10 to 20 cells for over 50 data values. You see how easy it is to make data "look better" by manipulation!
(b) Table 2.11 includes the computations for the cumulative frequency and cumulative relative frequency, which are shown in Fig. 2.13. The values are plotted at the upper cell boundaries; and since the data is discrete, the graph is drawn as a step function.
(c) From Fig. 2.13, 25 percent of the senators are under 45 and 1 percent are above 75 years old. Note that these answers will be the same for any age within the cells 40 to 50 and 70 to 80 because the data is grouped.

Table 2.10 Grouped frequency using a cell width of 5 years, Example 2.13

Cell	Frequency	Cell	Frequency
30–35	1	60–65	10
35–40	7	65–70	11
40–45	4	70–75	6
45–50	13	75–80	5
50–55	21	80–85	1
55–60	21		

Table 2.11 Grouped frequency using a cell width of 10 years, Example 2.13

Cell	Frequency	Cumulative frequency	Cumulative relative frequency
30–40	8	8	0.08
40–50	17	25	0.25
50–60	42	67	0.67
60–70	21	88	0.88
70–80	11	99	0.99
80–90	1	100	1.00

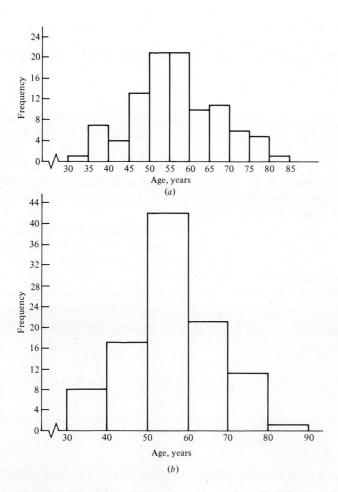

Figure 2.12 Histograms of U.S. senators' ages using cell widths of (*a*) 5 years, and (*b*) 10 years, Example 2.13.

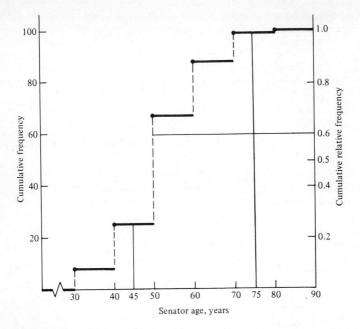

Figure 2.13 Cumulative frequency and cumulative relative frequency distributions for Example 2.13.

COMMENT The answers will be different for each cell width. You should answer parts (*b*) and (*c*) for a cell width of 5 years.

Section 2.5

ADDITIONAL MATERIAL

Benjamin and Cornell [3], Chap. 1; Duncan [7], pp. 41–49; Kirkpatrick [15], pp. 4–17; Miller and Freund [18], pp. 136–149; Neville and Kennedy [19], Chap. 2; Newton [20], Chap. 4.

PROBLEMS

The following data sets are used by the indicated problems in this chapter.

A: The number of items processed in the heat-treating department of a job shop was recorded for 22 days.

Days	Number of items heat-treated				
1 to 5	30	79	65	51	52
6 to 10	19	42	70	31	28
11 to 15	52	48	24	63	75
16 to 20	81	27	34	49	62
21 to 22	57	40			

Problems P2.1, P2.20, P2.25

B: The baking time in minutes of 150 different types of surface coatings has been developed for use with a new oven. A frequency breakdown by type is shown here:

Time (min)	Number of types	Time (min)	Number of types
1–4	2	16–19	22
4–7	8	19–22	35
7–10	15	22–25	12
10–13	23	25–28	3
13–16	30	≥ 28	0

Problems P2.9, P2.15, P2.21

C: Two MEs are redesigning a process to reduce the scrap of a precious metal that requires expensive reworking prior to reuse in the process. A sample of 600 scrap pieces was weighed prior to redesign. The weights have been multiplied by 100 and are grouped here:

Weight ($\times 100$ g)	Cell midpoint ($\times 100$ g)	Number of scrap pieces
.0–0.5	0.25	30
0.5–1.0	0.75	150
1.0–1.5	1.25	193
1.5–2.0	1.75	143
2.0–2.5	2.25	50
2.5–3.0	2.75	20
3.0–3.5	3.25	12
3.5–4.0	3.75	2
		600

Problems P2.10, P2.16, P2.22, P2.27

D: The machinability rating for 66 American Iron and Steel Institute (AISI) steels is grouped into cells of 10 percentage points:

Rating (%)	Number of steels	Rating (%)	Number of steels
30–40	2	90–100	1
40–50	8	100–110	1
50–60	23	110–120	0
60–70	19	120–130	0
70–80	6	130–140	1
80–90	5	140–150	0

Problems P2.11, P2.17, P2.23

P2.1 Do the following for data set A.
   (a) Order the values in increasing order.
   (b) Lump the data into equal groups of 10 items, starting with the smallest observed value.

P2.2 Construct (a) the frequency tally sheet and (b) the frequency distribution for the following bacteria count data taken on 30 industrial wastewater specimens. All values have been divided by 100.

Bacteria per liter ($\times 0.01$)

25	32	34	31	24
32	27	32	24	28
31	29	27	25	30
32	30	28	25	30
24	31	25	30	29
34	30	27	25	31

P2.3 Tabulate and graph the ungrouped frequency distribution for the number of defective switches found in samples of size 10.

0	1	1	2	1	1	3
1	0	0	4	2	1	0
3	0	2	5	0	0	2
1	5	0	2	1	1	3

P2.4 An engineer has collected 175 frequency values ranging from 128 to 913 hertz (Hz). Determine the correct cell boundaries and midpoints using (a) 10 cells, (b) 15 cells.

P2.5 Repeat Prob. P2.4a using measured unit accuracy.

P2.6 Thirty different degree values have been computed for angular rotation R after given torques are applied. When ordered, the numbers are 1.41, 1.43, 1.52, ..., 2.01, 2.12 degrees.
   (a) Determine the correct cell boundaries and midpoints for $k = 6$ cells.
   (b) How will the use of measured unit accuracy alter these limits?

P2.7 A student who was having difficulty with measured-unit-accuracy cell boundaries showed five lists to the instructor and asked which one was correct. Observed values are in integers.

Cell	1	2	3	4	5
1	10–20	9.5–19.5	10–20	$\geq 10 - < 20$	10.5–20.5
2	20–30	19.5–29.5	20–35	$\geq 20 - < 30$	20.5–30.5
3	30–40	29.5–39.5	35–40	$\geq 30 - < 40$	30.5–40.5
4	40–50	39.5–49.5	40–50	$\geq 40 - < 50$	40.5–50.5
5	50–60	49.5–59.5	50–60	$\geq 50 - < 60$	50.5–60.5

In answer the instructor posed the following questions. Please answer them for the student.
  (a) Which lists use the units of the observed values?
  (b) Which lists have the same cell width for all five cells in a list?
  (c) For which lists would you place the value 20 in the same cell? The value 32?
  (d) Which lists have the same midpoint values for all cells in a list?
  (e) You may use any list included in the answers to all these questions.
Which lists are these?

P2.8 What is the difference between the cell boundary lists 1 and 2 in Prob. P2.7?

P2.9 For data set B do the following.
  (a) Graph the frequency distribution.
  (b) Graph the frequency polygon.
  (c) What percentage of the times is between 7 and 12 min?

P2.10 (a) Plot the frequency distribution and frequency polygon for data set C on one graph.
  (b) Use the frequency definition of probability to determine the chance that the scrap is between 0.025 and 0.040 grams (g).

P2.11 For data set D, (a) construct the histogram and (b) combine cells such that the width is 20 percent and construct the histogram. What effect did combining the cells have on the shape of the histogram?

P2.12 A new gasoline mix was tested in 80 mid-size cars. The average kilometers per liter (km/L) is recorded in Table 2.12.
  (a) Tabulate the frequency distribution using a cell width of 0.5 km/L and a lower boundary of 5.5 km/L for the first cell. Use measured unit accuracy.
  (b) Plot the frequency distribution.
  (c) Plot the frequency polygon at cell midpoints.

Table 2.12 Average kilometers per liter for 80 test cars, Prob. P2.12

8.5	10.1	6.8	7.5	8.4	10.0	6.3	8.3
9.6	6.9	8.6	7.4	10.3	9.4	8.7	7.9
8.7	9.4	9.1	8.8	7.4	8.4	9.0	9.6
7.1	5.7	8.5	9.7	6.8	9.7	10.4	9.8
7.9	9.2	9.0	10.4	6.9	8.1	8.4	7.8
6.9	8.0	9.8	9.1	7.6	6.4	8.3	8.5
5.6	6.4	10.4	9.3	8.4	8.3	9.2	9.7
8.6	7.1	6.4	6.9	6.1	6.4	9.3	8.8
7.4	9.6	8.6	8.1	7.9	9.2	7.4	6.9
8.9	7.8	9.2	7.9	8.0	6.4	8.5	10.0

P2.13 (a) Work Prob. P2.12a and c using a cell width of 0.3 km/L.
(b) Did the frequency polygon shape change relative to the polygon of Prob. P2.12c?

P2.14 The maintenance schedule for small motors at a rock-crushing facility states that lubrication must be performed weekly (every 5 days) or more often. An ME checked the maintenance records for 72 motors and found that the days between lubrication varied from 3 to 20 days.
(a) Is the data discrete or continuous? Ungrouped or grouped?
(b) Plot the cumulative distribution function.
(c) What percentage of the motors have been correctly lubricated?

Days between lubrication	Number of motors
3	5
4	4
5	10
7	15
10	9
13	12
18	7
20	10

P2.15 Plot the cumulative frequency distribution for data set B and use it to answer the following.
(a) How many types have a baking time in excess of 10 min?
(b) How many types have a baking time less than 13 min?
(c) There are 50 types with a baking time less than what value?
(d) How many types have a baking time between 19 and 24 min?

P2.16 Plot the cumulative frequency distribution for data set C, sketch a smooth approximation curve, and use it to determine how many scrap pieces had a weight (a) less than or equal to 0.0225 g, (b) exceeding 0.030 g, and (c) between 0.018 and 0.024 gs.
(d) Use the frequency definition of probability to compute the chance that scrap weight exceeds 0.025 g.
(e) By what weight is 50 percent of the total accumulated frequency obtained?

P2.17 (a) Plot the cumulative frequency distribution for data set D and sketch a smooth approximation curve.
(b) If you want to develop a steel that is machined as easily as 90 percent of the steels included, what is the minimum rating you can accept?

P2.18 (a) Plot the cumulative frequency distribution for Table 2.12 as tabulated in Prob. P2.12a.
(b) What mileage value represents the 10 top-performing cars?

P2.19 Use the data in Prob. P2.14.
(a) Tabulate relative frequency and cumulative relative frequency.
(b) Prepare one graph for frequency and relative frequency.
(c) Prepare one graph for cumulative frequency and cumulative relative frequency.

P2.20 Group data set A into cells of width 10 items and tabulate the cumulative relative frequency.

P2.21 (a) Plot the cumulative relative frequency of data set B in histogram form.
(b) What fraction of the types require a baking time of more than 16 min?

P2.22 Compute the cumulative relative frequency for data set C and place this scale on the graph from Prob. P2.16. Use the smooth approximation curve to answer the questions in Prob. P2.16 for relative frequency rather than actual frequency.

P2.23 Plot the (a) polygon form and (b) smooth approximation of the cumulative relative frequency for data set D.

(c) Use both curves to determine the machinability rating exceeded by 25 percent of the AISI steels.

P2.24 Gloria has collected frequency data on the number of times T that a computerized, preprogrammed statics package for steel construction was used in a 1-year period by the 18 people in the engineering group. Use the results shown in Table 2.13 to plot relative frequency $P(T_i)$ and cumulative relative frequency $F(T_i)$.

Table 2.13 Data on use of preprogrammed statics package, Prob. P2.24

Number of times t_i	Frequency	$P(T_i)$	$F(T_i)$
1	10	0.56	0.56
2	5	0.28	0.84
4	2	0.11	0.95
5	1	0.05	1.00

P2.25 For data set A determine the number of items which represent the following fractiles: (a) 10th percentile, (b) 3d quartile, (c) 5th decile, and (d) middle 80 percentile.

P2.26 For the data of Prob. P2.2 determine the (a) 25th percentile, (b) middle 50 percentile, and (c) percentile for a bacteria value of 2900 per liter.

P2.27 Use the graphical method to determine the following fractiles for data set C: (a) 1st quartile, (b) 95th percentile, (c) between the 1st and 5th decile.

(d) What fractile is defined by the cell midpoint 0.0225 g?

P2.28 Here is a short case study. Table 2.14 gives the results of 175 chemical concentration tests in a synthetic fertilizer plant. Do what you think is necessary to answer the following questions.

(a) How many tests show an excess concentration if 0.08 percent is the limit?
(b) What concentration is exceeded by 25 percent of the tests?
(c) What concentration occurred in 90 percent of the tests such that 5 percent is below and 5 percent above these values?
(d) What fraction of the tests indicate less than 0.04 percent concentration?

Table 2.14 Chemical concentration percentages in synthetic fertilizer, Prob. P2.28

Concentration (%)	Number of tests	Concentration (%)	Number of tests
0.00–0.01	2	0.06–0.07	18
0.01–0.02	3	0.07–0.08	21
0.02–0.03	2	0.08–0.09	30
0.03–0.04	9	0.09–0.10	18
0.04–0.05	22	0.10–0.11	15
0.05–0.06	30	0.11–0.12	5

CHAPTER
THREE

PROPERTIES OF DATA: PART 1

The properties of collected data are studied in two chapters. In this chapter you learn how to compute and interpret commonly used measures for central tendency (mean, mode, and median) and dispersion (standard deviation, variance, and range). All the properties discussed here are frequently used to describe the outcome of experiments or tests.

CRITERIA

To complete this chapter, you must be able to:

1. Define and compute the *average*, given the variable values.
2. Compute the *average* of frequency data, given either the data and the frequencies or the cell boundaries and cell frequencies.
3. Define and determine the *mode* or the *modal cell* of grouped data, given the observed values or frequency data.
4. Define and determine the *median* or the *median cell* of frequency data, given the observed values.
5. Define and compute the *standard deviation* and *variance* of ungrouped data, given the observed values.
6. Compute the *standard deviation* and *variance* of frequency data, given either the data and frequencies or the cell boundaries and cell frequencies.
7. Define and compute or state the *range*, given the observed values or observed cell boundaries.
8. Plot the average, median, mode, and standard deviation on a histogram or frequency polygon, given the histogram and values of the measures.

STUDY GUIDE

3.1 Average of Data

The *average, mean*, or *expected value* are all terms used to describe a very common measure of the *central tendency* of data. Central tendency is a definite indication that the data seems to *cluster* around some central value. For example, an engineer who has studied a product's processing might comment to another engineer, "We try to keep the average temperature around 500°F [260°C]." This tells the listener nothing about the low and high temperature values, but now the central tending value is known.

The average of a variable X and a sample of size n is computed by summing the values and dividing by n:

$$\bar{X} = \frac{\Sigma X_i}{n} \tag{3.1}$$

The sum is taken over all values of X_i ($i = 1, 2, \ldots, n$). The dimension of $\bar{X}$ is the same as that of the data. (A short comment on notation is appropriate at this point. We will not confuse equations with the index limits on Σ. Once we define i as having the limits $i = 1, 2, \ldots, n$, you should automatically know that Σ is taken over all n values in the sample. Any exceptions to this basic format will be explained by using specific index values in the equation.)

Example 3.1 Compute $\bar{X}$ for the strength of the concrete specimens in Example 2.12.

SOLUTION We use Eq. (3.1) and simply add the 30 values and divide by 30:

$$\bar{X} = \frac{15{,}860}{30} = 528.667 \ \frac{\text{kg}}{\text{cm}^2}$$

COMMENT There is no reason to expect $\bar{X}$ to be an observed or observable value of X. All compressive strengths were recorded in integer values of kilograms per centimeter squared. Therefore, it would be impossible to record a value of 528.667. This value is simply a central tendency.

Problem P3.1

3.2 Average of Frequency Data

The only difference in the computation of $\bar{X}$ for individual values and frequency data is the following: $\bar{X}$ *for frequency data must account for the number of times that a data value is observed*. If the data point 505 kg was observed 5 times in a sample of 100, then the number 505 must be included in the sum of Eq. (3.1) five different times. Therefore, for frequency data we include the frequency in the formula by inserting the symbol f_i ($i = 1, 2, \ldots, k$) into the numerator of Eq.

(3.1):

$$\bar{X} = \frac{\sum f_i X_i}{n} \qquad (3.2)$$

Here $n = \sum f_i$ is the number of data values. Two comments are appropriate at this time. First, the upper limit on i is now k, where $k \le n$ because there are only k different values of X. Second, the sample size n is determined by adding all the frequency values f_i, not by counting the number of different values of X.

There are two ways in which data can be used in Eq. (3.2). The X_i values may be individual points, say, 150, 175, and so on, with a frequency f_i for each value X_i. Or the data may be presented in cells, say, 140 to 160, 160 to 180, with a frequency f_i for each cell. If cells are used, X_i is the midpoint, and it is assumed that all data in the cells is concentrated at this value. Example 3.2 illustrates these two forms. Part (b) is especially important since it pertains to cell grouped data and much future work will use this form.

Example 3.2 Compute the average of the data in Example 2.12 using (a) the data in frequency form for exact values and (b) the grouped format as presented by Mr. Thomas.

SOLUTION (a) If you look back to the example, you can count 26 different strength values. The individual values are grouped in Table 3.1. We now have

$$n = \sum_{i=1}^{26} f_i = 30$$

From Eq. (3.2)

$$\bar{X} = \frac{\sum_{i=1}^{26} f_i X_i}{30}$$

$$= \frac{1(435) + 1(443) + \cdots + 2(519) + \cdots + 1(580) + 1(598)}{30}$$

$$= 528.667 \ \frac{\text{kg}}{\text{cm}^2}$$

This is, as it should be, the same answer obtained in Example 3.1 using Eq. (3.1).

(b) When cell grouped data is used, the $\bar{X}$ value will not have the same value as in (a), because we are now only approximating the actual values by using the midpoint value. Table 3.2 presents a convenient format for the computation of $\bar{X}$ using the cell grouped data. Computation of n and use of Eq. (3.2) give us

$$n = \sum_{i=1}^{9} f_i = 30 \qquad \bar{X} = \frac{15{,}960}{30} = 532.0 \ \frac{\text{kg}}{\text{cm}^2}$$

Table 3.1 Compressive strength data of Example 2.12 in frequency form

Value i	Strength X_i	Frequency f_i	Value i	Strength X_i	Frequency f_i
1	435	1	14	534	1
2	443	1	15	437	1
3	450	1	16	540	2
4	475	1	17	541	1
5	510	1	18	551	1
6	512	1	19	552	1
7	517	1	20	553	2
8	518	1	21	559	1
9	519	2	22	560	1
10	521	1	23	571	1
11	522	2	24	573	1
12	524	1	25	580	1
13	531	1	26	598	1

Table 3.2 Computation of the mean of grouped data, Example 3.2

Cell i	Cell midpoint X_i	Cell frequency f_i	$f_i X_i$
1	440	2	880
2	460	1	460
3	480	1	480
4	500	0	0
5	520	10	5,200
6	540	6	3,240
7	560	6	3,360
8	580	3	1,740
9	600	1	600
		30	15,960

COMMENT Use of the midpoints has caused an error of 0.63 percent above the actual $\overline{X}$ value of 528.667 kg/cm². This error is very small, as is the usual case. Because of the ease and time efficiency of using data grouped in cell format, the method of part (b) is favored.

Problems P3.2–P3.5

3.3 Mode and Modal Cell

The *mode m* is another measure of central tendency which occurs at the most frequently observed value of the variable. The mode for grouped data is the midpoint of the *modal cell*, which is the most frequently observed cell. The mode is stated as a value of the variable, not the frequency of the variable value. For example, suppose an engineer recorded the following number of defects in newly manufactured mattresses:

Mattress number	1	2	3	4	5	6	7
Number of defects	3	4	1	2	4	0	4

The variable observed is defects, and four defects were observed three different times, which is more than any other defect number. Therefore $m = 4$. Note that m is not 3, the frequency of four defects.

It is possible for there to be more than one mode. If there are two values observed with the same frequency, the data is referred to as *bimodal*. Data with one mode is *unimodal*. If all values are observed with the same frequency, there is no mode for the data.

Example 3.3 A mechanical engineer has recorded the flow rate of a liquid chemical in liters per second (L/s). Since this flow is regulated by a computer monitoring a continuous process, the flow varies. Readings were recorded in cell form as shown. The mean has been 8.35 L/s in the past, but the ME is now interested in the most commonly occurring flow value. What is this value?

Flow rate (L/s)	Frequency
7.50–7.80	1
7.80–8.10	5
8.10–8.40	35
8.40–8.70	17
8.70–9.00	12
9.00–9.30	10

SOLUTION The most commonly observed flow value is the modal cell, which is 8.10 to 8.40 (Fig. 3.1). Therefore, the mode is the midpoint of this cell, or $m = 8.25$ L/s.

COMMENT Note that m is less than the previous $\overline{X} = 8.35$ and m is also less than the $\overline{X}$ for the data recorded here, which is $\overline{X} = 8.49$ L/s. When we study the shape of frequency distributions, we will look at the relative locations of $\overline{X}$ and m.

If the frequency of the cell 8.10 to 8.40 had been 17, there would have been two modal cells. This would not be bimodal data, since the two modal cells 8.10 to 8.40 and 8.40 to 8.70 are adjacent. However, if the frequency of

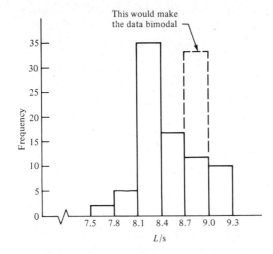

Figure 3.1 Plot of flow data, Example 3.3.

the cell 8.10 to 8.40 were 35, as recorded, and the frequency of the cell 8.70 to 9.00 were 33 (Fig. 3.1), the data could be referred to as bimodal, even though there is only one true modal cell (8.10 to 8.40).

Problems P3.6–P3.9

3.4 Median and Median Cell

The *median M* is a measure of central tendency which divides the data into two equal halves. The value of the variable which is the median is found by ordering the data in increasing size and counting until the middle observation is reached. If there are an odd number of ungrouped data points, M is the center observed value; however, if there are an even number of observations, M is the average of the two center values.

If the data is grouped into cells, the *median cell* is determined by computing the cumulative frequency values until the cell containing M is included. A single value of M may be approximated by taking a proportion of the median cell, rather than simply using the cell boundary values.

The median is a useful central tendency measure because it is not biased by extreme values as much as the average. Therefore, if there are a few extremely small or large values, M gives a better idea of the central value.

Example 3.4 Determine the median or median cell for each of the following.
(a) The maximum shearing stress was computed for seven $\frac{1}{2}$-in-diameter, high-strength shafts. The computed values in pounds per square inch were:

| 8560 | 8430 | 8640 | 8545 |
| 8210 | 9120 | 8735 | |

Table 3.3 Salary data for Example 3.4c

Salary range ($/month) (1)	Number of engineers (2)	Cumulative frequency (3)
900–1000	5	5
1000–1100	10	15
1100–1200	15	30
1200–1300	35	65
1300–1400	85	150
1400–1500	45	195
1500–1600	30	225
1600–1700	5	230

(b) The data in (a) plus an eighth reading of 8130 lb/in².
(c) Monthly starting salary data for 230 new graduates with a B.S. in engineering was calculated and grouped as presented in Table 3.3.

SOLUTION (a) We order the seven shearing stress values:

8210   8430   8545   8560   8640   8735   9120

The median value is the fourth data point, $M = 8560$ lb/in².
(b) The eighth value (8130 lb/in²) is the first in order. This causes M to occur half the distance between 8545 and 8560 lb/in². Thus

$$M = \frac{8545 + 8560}{2} = 8552.5 \; \frac{\text{lb}}{\text{in}^2}$$

(c) Table 3.3 presents the cumulative frequency of the salary for new graduates. The median occurs in the cell with 115 values on each side. This is the cell containing the salary between the cumulative frequency of 115 and 116. This value is included in the median cell $M = \$1300$ to $\$1400$.

COMMENT Note the effect of the eighth stress value added in (b). The relatively low stress value of 8130 lb/in² causes these changes:

	Number of values	
	7	8
Median M	8560.0	8552.5
Average $\bar{X}$	8605.7	8546.3

As you see, $\bar{X}$ is reduced by 59.4 lb/in², but M decreases by only 7.5 lb/in².

Therefore, an extreme value has a much greater effect on $\overline{X}$ than it does on M. The median should be used to avoid the biasing effect of extreme data values that is present in $\overline{X}$.

Problems P3.10, P3.11

3.5 Standard Deviation and Variance

The *standard deviation* is a commonly used measure of dispersion. It gives a numerical value, in the units of the variable itself, to the clustering tendency of the data. A large standard deviation is an indication that the histogram will be flat and wide, with some relatively extreme values possibly observed. If the standard deviation is small, the data points will be more closely clustered around the center of the data. Figure 3.2 graphs the values and histograms for two sets

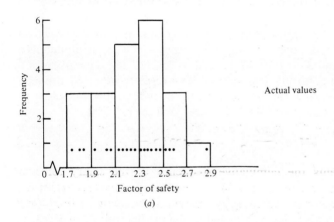

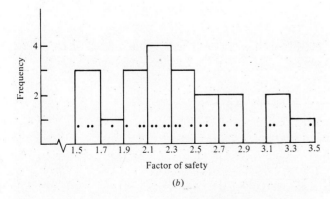

Figure 3.2 Examples of (*a*) small and (*b*) large standard deviations.

54 PROPERTIES OF DATA: PART 1

of safety factor data, Fig. 3.2a with a small standard deviation and Fig. 3.2b with a larger value. The histogram in Fig. 3.2a is narrow and tall relative to that of Fig. 3.2b, and in Fig. 3.2a the data is more closely massed around some central value. The standard deviation uses the central tendency measure of the average $\bar{X}$ as a guide about which to measure the clustering. The formula used to compute standard deviation is

$$s = \left[\frac{\Sigma(X_i - \bar{X})^2}{n - 1} \right]^{1/2} \tag{3.3}$$

The deviation from $\bar{X}$ can have a plus or minus sign. To accurately measure this two-direction deviation, the difference $X_i - \bar{X}$ is squared. However, to return to the dimension of the variable itself, the square root is extracted. Actually Eq. (3.3) is the average of the *mean squared deviation* $(X_i - \bar{X})^2$. The value $n - 1$ is used in the computation of s to obtain an unbiased estimate. If n, the sample size, is used, the result is a slightly decreased s. As n increases, this bias effect diminishes. A further study of biased and unbiased computations is made later.

Figure 3.3 is a graphical illustration of how s is computed for the X_i values and located in relation to the mean $\bar{X} = 17$. The value $s = 8.211$ is plotted as a *deviation from* $\bar{X}$ so that $\bar{X} + 1s = 17 + 8.211 = 25.211$. The $e_i = (X_i - \bar{X})^2$ values increase as X_i gets farther from $\bar{X}$. Thus, the more dispersed the data points are from $\bar{X}$, the larger s will be.

i	X_i	$X_i - \bar{X}$	$e_i = (X_i - \bar{X})^2$	
1	4	−13	169	
2	8	−9	81	
3	13	−4	16	
4	18	+1	1	$\bar{X} = 136/8 = 17$
5	19	+2	4	$s = (472/7)^{1/2} = 8.211$
6	21	+4	16	
7	25	+8	64	
8	28	+11	121	
	136		472	

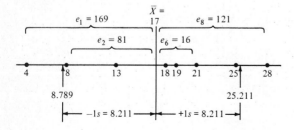

Figure 3.3 Illustration of the computation and graphical interpretation of the standard deviation.

3.5 STANDARD DEVIATION AND VARIANCE

The standard deviation has a companion computation in engineering mechanics. The moment of inertia I is computed as

$$I = \text{constant} \int_A d^2 \, dA$$

where d is the distance from an axis representing the center of gravity and dA is an area element. If the average is the axis and X_i is the distance from the axis to a weight, the mean squared deviation is d^2, that is,

$$d^2 = (X_i - \bar{X})^2$$

You should note that s is the positive square root in Eq. (3.3). Therefore we know that $s \geq 0$. If $s = 0$, we have data which has absolutely no deviation from $\bar{X}$; that is, all values are the same. In this case we would be dealing not with probabilistic data but with deterministic values. Therefore, we expect that $s > 0$ in the real world.

All computations in Eq. (3.3) performed prior to taking the square root are called the *variance s^2;* that is,

$$s^2 = \frac{\Sigma(X_i - \bar{X})^2}{n - 1} \qquad (3.4)$$

The dimension of s^2 is the square of the variable itself. We must compute s^2 to obtain s, but s is commonly used in practice. Physically the variance has no direct interpretation.

Computationally, it is usually easier to algebraically simplify Eqs. (3.3) and (3.4). We show the algebraic progression for Eq. (3.4):

$$s^2 = \frac{1}{n-1} \Sigma (X_i - \bar{X})^2$$

$$= \frac{1}{n-1} \Sigma (X_i^2 - 2X_i\bar{X} + \bar{X}^2)$$

$$= \frac{1}{n-1} \Sigma X_i^2 - \frac{2\bar{X}}{n-1} \Sigma X_i + \frac{1}{n-1} \bar{X}^2 \Sigma 1$$

$$= \frac{1}{n-1} \Sigma X_i^2 - \frac{2n\bar{X}}{n-1} \frac{\Sigma X_i}{n} + \frac{n\bar{X}^2}{n-1}$$

$$= \frac{1}{n-1} \Sigma X_i^2 - \frac{2n\bar{X}^2}{n-1} + \frac{n\bar{X}^2}{n-1}$$

$$= \frac{1}{n-1} \Sigma X_i^2 - \frac{n}{n-1} \bar{X}^2$$

Thus, we have the following computational form to obtain an unbiased estimate of sample variance and standard deviation:

- Unbiased s^2 and s

$$s^2 = \frac{\Sigma X_i^2}{n-1} - \frac{n}{n-1} \bar{X}^2 \qquad (3.5)$$

$$s = \left[\frac{\Sigma X_i^2}{n-1} - \frac{n}{n-1} \bar{X}^2\right]^{1/2} \qquad (3.6)$$

If we had defined s^2 in Eq. (3.4) as

$$s^2 = \frac{\Sigma(X_i - \bar{X})^2}{n}$$

we would have obtained slightly biased computational results:

Biased s^2 and s

$$s^2 = \frac{\Sigma X_i^2}{n} - \bar{X}^2 \qquad (3.7)$$

$$s = \left[\frac{\Sigma X_i^2}{n} - \bar{X}^2\right]^{1/2} \qquad (3.8)$$

Equations (3.7) and (3.8) are often used for computational simplicity and because the bias becomes very small as n increases. Unless so mentioned, we will use the unbiased form for s^2 and s.

Example 3.5 An engineer has collected a sample of 10 readings of strength in a spot weld (Table 3.4) over a period of one day. Specifications state that no more than 25 percent of the readings may deviate from the average by more than one standard deviation.
(a) Does this sample meet the specification?
(b) Restate the specification statement in terms of variance.

SOLUTION Table 3.4 gives the squares of the readings (X_i^2). Using Eq. (3.6),

$$s = \left[\frac{460{,}978}{9} - \frac{10}{9}\left(\frac{2146}{10}\right)^2\right]^{1/2}$$

$$= (51{,}219.78 - 51{,}170.18)^{1/2}$$

$$= 7.04 \text{ kg}$$

(a) To answer the question posed, we compute $\bar{X} = 214.6$ and the values

$$\bar{X} + 1s = 214.6 + 7.04 = 221.64 \text{ kg}$$

$$\bar{X} - 1s = 214.6 - 7.04 = 207.56 \text{ kg}$$

3.6 STANDARD DEVIATION AND VARIANCE OF FREQUENCY DATA

Table 3.4 Computation of standard deviation, Example 3.5

i	Strength X_i (kg)	X_i^2
1	217	47,089
2	213	45,369
3	201	40,401
4	220	48,400
5	210	44,100
6	225	50,625
7	214	45,796
8	223	49,729
9	210	44,100
10	213	45,369
	2146	460,978

Table 3.4 has three values, or 30 percent of the sample, outside this range. Since 25 percent is allowed outside $\overline{X} \pm 1s$, the sample does not meet the specification.

(b) The variance is

$$s^2 = (7.04 \text{ kg})^2 = 49.6 \text{ kg}^2$$

The only way to correctly state the specification in terms of s^2 is as follows: No more than 25 percent of the readings may be outside the range $(\overline{X} + 1s)^2$ to $(\overline{X} - 1s)^2$. Hence, we have

$$(\overline{X} \pm 1s)^2 = (214.6 \pm 7.04)^2 = 49{,}124.29 \quad \text{and} \quad 43{,}081.15 \text{ kg}^2$$

A check of X_i^2 in Table 3.4 shows the same X_i values outside this range. The specification is not met, as concluded before.

COMMENT In (b) it is completely incorrect to state that no more than 25 percent of the readings should be outside the range $\overline{X} \pm 1s^2$ since you can compute $\overline{X} \pm 1s^2 = 264.16$ and 165.04 and no X_i values exceed this incorrect specification.

Remember, s is a measure of spread about the average, and s^2 is the square of s. The standard deviation and mean tell the engineer a great deal about the observed data. In fact, $\overline{X}$ and s give more information than any other quoted data.

Problems P3.12–P3.15

3.6 Standard Deviation and Variance of Frequency Data

If the data is in frequency form by individual values or cells, the formulas for s and s^2 are slightly adjusted for frequency. Since the frequency f_i indicates the

58 PROPERTIES OF DATA: PART 1

number of times X_i is observed, we use the following forms:

• Unbiased s^2 and s

$$s^2 = \frac{\sum f_i X_i^2}{n-1} - \frac{n}{n-1}\bar{X}^2 \qquad (3.9)$$

$$s = \left[\frac{\sum f_i X_i^2}{n-1} - \frac{n}{n-1}\bar{X}^2\right]^{1/2} \qquad (3.10)$$

Pay particular attention to the fact that in Eqs. (3.9) and (3.10) $f_i X_i^2$ is used, not $(f_i X_i)^2$. The value of X_i is squared and then multiplied by f_i; that is, f_i is not multiplied by X_i and then squared.

The biased forms of s and s^2 are similar to Eqs. (3.7) and (3.8) with f_i multiplying X_i^2.

Example 3.6 Ms. Ferrar, a traffic technician with the state highway department, has been asked to determine the most desirable time settings for a series of three traffic signals of a 1-mile (mi) strip of access road to an in-town throughway system. After devising a timing system, the technician observed the number of vehicles going through all three signals without stopping or rushing a yellow signal. Table 3.5 presents the results in cell format. Compute the (a) mean, (b) standard deviation, and (c) variance of this data.

SOLUTION Table 3.5 gives all values necessary to answer the questions. The midpoint is assumed to be the point of concentration for each cell.

(a) The sample size is $n = \sum f_i = 84$. Column 4 presents $\sum f_i X_i$, where X_i is the midpoint of each cell. Then, using Eq. (3.2), we get

$$\bar{X} = \frac{2076}{84} = 24.71 \text{ cars}$$

Table 3.5 Traffic data used to compute $\bar{X}$, s, and s^2, Example 3.6

i	Boundaries (1)	Midpoint X_i (2)	Frequency f_i (3)	$f_i X_i$ (4) = (3)(2)	X_i^2 (5)	$f_i X_i^2$ (6) = (3)(5)
1	15.5–18.5	17	12	204	289	3,468
2	18.5–21.5	20	18	360	400	7,200
3	21.5–24.5	23	10	230	529	5,290
4	24.5–27.5	26	13	338	676	8,788
5	27.5–30.5	29	20	580	841	16,820
6	30.5–33.5	32	8	256	1,024	8,192
7	33.5–36.5	35	2	70	1,225	2,450
8	36.5–39.5	38	1	38	1,444	1,444
			84	2,076	6,428	53,652

(b) We use Eq. (3.10) to compute the standard deviation:

$$s = \left[\frac{53{,}652}{83} - \frac{84}{83}(24.71)^2\right]^{1/2}$$

$$= (646.41 - 617.94)^{1/2}$$

$$= 5.34 \text{ cars}$$

(c) The variance can be computed from Eq. (3.9), or by squaring s:

$$s^2 = (5.34 \text{ cars})^2 = 28.47 \text{ cars}^2$$

COMMENT In Table 3.5 the values in column 6 are obtained by computing $f_i X_i^2$, not $(f_i X_i)^2$. The results below show what happens if $(f_i X_i)^2$ is incorrectly used:

Midpoint X_i	$(f_i X_i)^2$	Midpoint X_i	$(f_i X_i)^2$
17	41,616	29	336,400
20	129,600	32	65,536
23	52,900	35	4,900
26	114,244	38	1,444

$$\Sigma (f_i X_i)^2 = 746{,}640$$

Incorrectly, $$s = \left[\frac{746{,}640}{83} - \frac{84}{83}(24.71)^2\right]^{1/2}$$

$$= (8995.66 - 61.94)^{1/2}$$

$$= 91.53 \text{ cars}$$

This would mean that the number of cars passing through the three signals has an average of 24.71 and a standard deviation of 91.53 cars. But the entire set of data (Table 3.5) is included in the range 16 to 39 cars. Therefore, the s computation above is not only wrong, but completely meaningless!

Problems P3.16–P3.18

3.7 Range

The *range* R is a measure of dispersion defined as the difference between the largest and smallest data values:

- $$R = X_{max} - X_{min} \qquad (3.11)$$

The range is commonly quoted because it is easy to calculate. It is an inaccurate

60 PROPERTIES OF DATA: PART 1

measure of dispersion because (1) it uses only two of the data points and neglects all others and (2) it gives no idea of how the data is clustered. Whether the sample size is 5 or 5000, only the largest and smallest values are used, so R is very much influenced by extreme data values. If you desire to use all data points in the dispersion measure, use the standard deviation.

Equation (3.11) is usable for all types of data; however, in the case of grouped data R is approximate because X_{max} is the upper limit of the largest cell and X_{min} the lower limit of the smallest cell.

Example 3.7 (*a*) Compute the range for the traffic data of Example 3.6.
(*b*) Compare the result with the value of s for the same data.

SOLUTION (*a*) From Table 3.5 and Eq. (3.11),

$$R = 39.5 - 15.5 = 24 \text{ cars}$$

The value $R = 24$ cars is a *maximum* for the actual range, since all values must be observed within the cell boundaries of 15.5 and 39.5.

(*b*) The actual traffic population has only one true dispersion value. We have two estimates: $R = 24$ and $s = 5.34$ cars. R is so large because it represents only the extreme cell boundaries; s is a much better estimate.

COMMENT Actually R and s do not estimate dispersion in the same way. In fact, s can be estimated from R by the simple formula $s = \bar{R}/d_2$, where $\bar{R}$ is the average of several R readings and d_2 is a factor which varies with sample size. More on this approximation later.

It is common to hear an engineer make a statement such as, "The range varies from 10 to 15 lb/in². " This means that the difference $X_{max} - X_{min}$ will change over time and from sample to sample.

Problems P3.19–P3.22

3.8 Average, Median, Mode, and Standard Deviation on a Frequency Polygon

It is often convenient to plot $\bar{X}$, M, m, or s on a histogram or frequency polygon. The points $\bar{X} \pm ts$ ($t = 1, 2, 3$) are plotted to give an idea of how much of the data is included between $\bar{X} - ts$ and $\bar{X} + ts$.

Example 3.8 Mr. Jones, an industrial engineer, has been studying the weight of the fluid in a bottle with specifications 32 ± 0.10 ounces (oz). Mr. Jones has computed the following measures:

Average: $\bar{X} = 31.99$ oz
Median: $M = 31.98$ oz
Mode: $m = 31.96$ oz
Standard deviation: $s = 0.07$ oz

3.8 AVERAGE, MEDIAN, MODE, AND STANDARD DEVIATION ON A FREQUENCY POLYGON 61

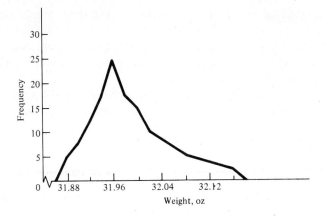

Figure 3.4 Frequency polygon of bottle weight, Example 3.8.

Use the frequency polygon of Fig. 3.4 to do the following.
(a) Plot $\bar{X}$, M, m, and $\bar{X} \pm 1s$.
(b) Shade the area to indicate weights between $\bar{X} + 1s$ and $\bar{X} + 2s$.

SOLUTION (a) Figure 3.5 shows the points $\bar{X}$, M, and m. The two values

$$\bar{X} \pm 1s = 31.99 \pm 0.07 = 31.92 \quad \text{and} \quad 32.06 \text{ oz}$$

are also indicated.
(b) The area between $\bar{X} + 1s = 32.06$ and $\bar{X} + 2s = 32.13$ is hatched on the polygon.

Problems P3.23, P3.24

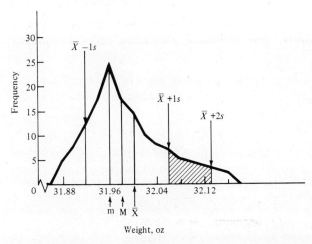

Figure 3.5 Graph of average, median, mode, and standard deviation, Example 3.8.

ADDITIONAL MATERIAL

Central tendency measures (Secs. 3.1 to 3.4): Duncan [7], pp. 51–57; Kirkpatrick [15], pp. 17–28; Neville and Kennedy [19], Chap. 3; Newton [20], pp. 122–143; Volk [24], pp. 63–68.

Dispersion measures (Secs. 3.5 to 3.7): Duncan [7], pp. 57–64; Kirkpatrick [15], pp. 28–34; Neville and Kennedy [19], Chap. 4; Newton [20], pp. 144–156; Volk [24], pp. 68–76.

PROBLEMS

The following data sets are used by the indicated problems of this chapter.

A: The emitter voltage for a certain type of transmitter measured during 24 in-service tests is given in volts:

5.8	5.8	6.1	5.9	6.3	5.8	6.0	6.1
6.1	6.0	5.9	6.4	5.7	6.0	6.1	5.8
5.7	6.2	6.4	6.1	6.0	5.6	5.9	5.8

Problems P3.2, P3.3, P3.7, P3.10, P3.13, P3.19

B: A city has a minor league baseball team. A record of paid attendance for four seasons is summarized here:

Number of spectators	Number of games
500– 750	6
750–1000	6
1000–1250	3
1250–1500	5
1500–1750	10
1750–2000	2
2000–2250	2
	34

Problems P3.4, P3.8, P3.10, P3.16, P3.20, P3.24

P3.1 Compute the average of the following waiting times in hours recorded at a company shipping dock: 12, 19, 4, 35, 6, 12, 21, 10, 17.

P3.2 Compute the average of data set A by two methods: (*a*) individual X values using Eq. (3.1), and (*b*) frequency for each X value using Eq. (3.2).

(*c*) Should your two answers have the same value? Why or why not?

P3.3 (*a*) Group the values of data set A into cells of width 0.3 volt (V) and compute the average.
(*b*) Why is the answer different from the averages in Prob. P3.2?

P3.4 Two city politicians want to use data set B as reason for keeping the ball club. What should they use as average attendance per game?

P3.5 An engineer has computed the averages of four samples of size 10 each from a prestressing operation. She wants the average of the 40 individual values. Show how the averages from the four samples may be used to compute the average for the 40 values. Remember that the sample sizes are equal.

P3.6 Consider the number of days in each month of the year. What is the mode of these day values?

P3.7 (*a*) For data set A develop the frequency for each voltage value. Determine the mode.
(*b*) Group the data into cells of 0.3 V and determine the modal cell.
(*c*) Do the answers above indicate different shapes for the frequency distribution? If so, what is the difference?

P3.8 What is the mode for data set B?

P3.9 If a histogram is distinctly bimodal, where would you expect the average to fall? Sketch a histogram that will verify your answer.

P3.10 Make the calculations necessary to show that these statements below are true or false:
(*a*) The median of data set A is 6.0 V.
(*b*) The median of these numbers is 4 because this is the frequency of the value 3 which occurs most often:

6   3   4   3   3   4   3   5   1

(*c*) The median cell of data set B is 1500–1750 spectators.
(*d*) A single value estimate of the median of data set B is 1350 spectators.

P3.11 Why do extremely large or small data values affect the median less than the average?

P3.12 The temperature in degrees Celsius for 10 industrial coolers is recorded as follows:

3.0	4.2	−1.1	0.3	−2.0
0.1	−0.6	2.1	−0.2	0.0

Compute the standard deviation using (*a*) Eq. (3.3) and (*b*) Eq. (3.6). Determine which form you prefer.
(*c*) Compute the standard deviation using Eq. (3.8) and determine the percent error present in this biased computational form.

P3.13 (*a*) Compute the unbiased s for data set A.
(*b*) Use this s value and the $\bar{X}$ value (Prob. P3.2) to determine what percentage of the data is between $\bar{X} \pm 3s$.
(*c*) Compute the variance for data set A.

P3.14 If the biased variance Eq. (3.7) is used, will the variance be overestimated or underestimated?

P3.15 The data for two samples of five each is given below. Compute the average, variance, and standard deviation for each sample and determine which sample is more widely dispersed about the average.

Sample 1:	5	12	13	19	21
Sample 2:	7	7	14	15	23

P3.16 Compute the variance and standard deviation of data set B.

P3.17 (*a*) Compute the unbiased s value for the data in Table 3.2.
(*b*) A specification requires that no more than 30 percent of the values deviate from the average by more than one standard deviation. Is this specification met?

P3.18 Exactly why is it incorrect to use $(f_i X_i)^2$ in computing s by Eq. (3.9)?

P3.19 Compute the range of data set A.

P3.20 Compute the range of data set B.

P3.21 A student measured the noise level in a plant every 5 min for only one day. The results were grouped into the standard government decimal intervals.
    (a) Compute $\bar{X}$, s, and R for this data.
    (b) Is R a better measure of dispersion than s in this case?
    (c) Look at the frequency data. How should the student explain its shape using graphical and numerical techniques?

Noise level interval (dBa)	Number of readings in interval
82– 85	10
85– 87	14
87– 90	7
90– 92	6
92– 95	8
95– 97	6
97–100	6
100–102	14
102–105	15
105–107	10

P3.22 Assume you have only two data points X_1 and X_2.
    (a) Show that the biased $s = R/2$.
    (b) Show that the biased s equals the absolute value of either X_1 or X_2 minus $\bar{X}$.

P3.23 For Prob. P3.21 plot the frequency polygon and indicate the average, median, mode, and one standard deviation from the average on the graph.

P3.24 (a) For data set B, plot $\bar{X}$, M, and m on the histogram.
    (b) Construct the frequency polygon and shade the area two standard deviations on each side of the average.
    (c) Shade the area outside the limits $\bar{X} \pm 1.5s$.

CHAPTER
FOUR

PROPERTIES OF DATA: PART 2

The objective of this chapter is to teach you several additional properties of collected data. These properties will help you further investigate the central tendency and variability of data. Also presented is a method by which you can code observed data to make many computations less cumbersome.

CRITERIA

To complete all the material in this chapter, you must be able to:

1. State the formula for any *moment about the origin* or *central moment* and state the relationship between the average, the variance, and these moments.
2. Define *skewness*; use a frequency distribution to explain skewness; and compute and interpret the coefficient of skewness, given the observed data.
3. Do the same as for 2, except for *peakedness*.
4. Define and compute *mean absolute deviation* and use it to approximate the standard deviation, given the data and the approximating relation.
5. State the formula for and compute the *weighted average*, given the variable values and the weighting factors.
6. Define and compute the *coefficient of variation*, given the sample average and standard deviation.
7. Compute *coded* data values and the average and standard deviation for ungrouped data, given the observed data.
8. Do the same as 7 for ungrouped data.

STUDY GUIDE

4.1 Introduction to Moment Terminology

An important property in the study of strength of materials is the moment, which is a force tending to bend a beam or bar. An analogy exists in the study of statistics. *Moment about the origin* is a term used to describe the computation

- $$\mu_k = \frac{\Sigma X_i^k f_i}{n} \tag{4.1}$$

where k = the power 1, 2, ...
X_i = observed values
f_i = frequency of X_i for grouped data
n = sample size = Σf_i

The word *about* as used in moment terminology actually means around some base value, which is 0 for the moments about the origin. Therefore, we could correctly express μ_k as

$$\mu_k = \frac{\Sigma (X_i - 0)^k f_i}{n}$$

If $k = 1$ in Eq. (4.1), we have the first moment about the origin, which is the average or mean. Higher moments about the origin are used for computations you will learn later.

If the moment is taken about the *average (mean)*, it is called the *moment about the mean*, or *central moment*. The formula for the kth central moment m_k is

- $$m_k = \frac{\Sigma (X_i - \bar{X})^k f_i}{n} \tag{4.2}$$

where $k = 1, 2, \ldots$ For the first central moment $k = 1$ and $m_1 = 0$. You should be able to show that $m_1 = 0$ from Eq. (4.2). If $k = 2$, we have the biased variance

$$\sigma^2 = m_2 = \frac{\Sigma (X_i - \bar{X})^2 f_i}{n}$$

In Sec. 3.5 we drew an analog between the standard deviation and the *moment of inertia*; so $\sigma = \sqrt{m_2}$ has a physical interpretation. Central moments larger than 2 cannot easily be physically interpreted, but they are important to the computation of certain frequency distribution properties.

There is a simple relation between the two types of moments discussed. If the $\bar{X}$ in the kth central moment is replaced by 0, the result is the moment about the origin with the same k value.

Problems P4.1–P4.4

4.2 Skewness

When a frequency distribution (histogram or polygon) is equally distributed around the mean, it is called bell-shaped, symmetric, or nonskewed. Figure 4.1a shows a perfectly symmetric polygon. If the polygon has a longer tail on the right than on the left, it is *skewed right*. A longer left tail makes the data *skewed left*. See Fig. 4.1b and c, respectively. Many types of data occur naturally as skewed. For example, data on wealth distribution and age at marriage is skewed right, while age of death due to heart attack is skewed left. The distribution of errors on a manufacturing process and IQ values are very close to symmetric.

The skewness of data is measured in relation to some vertical axis. We can relate skewness to the placement of the average, median, and mode.

Symmetric data: Average, median, and mode are coincident.
Skewed-right data: Median is between mode and average with the mode on the left as in Fig. 4.2a. The greater the distance between these, the more pronounced the skewness.
Skewed-left data: Median is between the mode and the average with the average on the left as in Fig. 4.2b.

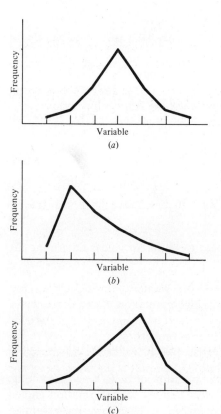

Figure 4.1 General shape of polygons which are (a) symmetric, (b) skewed right, and (c) skewed left.

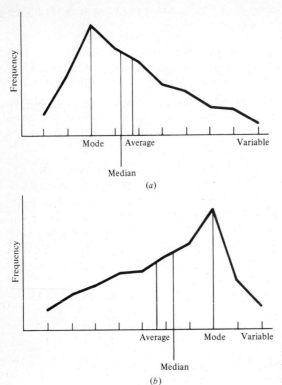

Figure 4.2 Relative location of the median, mode, and average for data which is (a) skewed right and (b) skewed left.

A dimensionless coefficient of skewness a_3 may be computed by using the relation

$$a_3 = \frac{\text{third central moment}}{(\text{standard deviation})^3} = \frac{m_3}{s^3}$$

$$= \frac{\sum f_i(X_i - \overline{X})^3 / n}{s^3} \tag{4.3}$$

The coefficient a_3 is a comparative value because different values determine the relative amount of skewness.

$a_3 = 0$ symmetric data
$a_3 > 0$ skewed right
$a_3 < 0$ skewed left

If two positive a_3 values are compared, the frequency distribution with the larger a_3 value is more skewed right.

You can see from Eq. (4.3) that the computation of a_3 is time-consuming, even with a sophisticated, hand-held calculator. Besides, the information gained from the quantification is often considered marginal. Therefore, the extent of skewness is usually observed visually. It is also possible to numerically compare different degrees of skewness by computing

$$\frac{\text{average} - \text{mode}}{\text{standard deviation}} = \frac{\bar{X} - m}{s} \qquad (4.4)$$

for each data set and comparing the values. A value of 0 means that the data is symmetric, since the mode would equal the average.

Example 4.1 Delta Associates is a corporate holding company for 27 manufacturing concerns. The executives of Delta have been concerned about the quality control spending in these companies. Data has been collected on the percentage of total quality dollars spent by the company to prevent quality defects. Table 4.1 (columns 1 and 2) presents these percentages for the 27 companies. The frequency distribution for this data is shown in Fig. 4.3. Analyze the data by answering the following questions.
(a) In which direction is the data skewed?
(b) Compute the average, median, and mode and plot these on the histogram. Comment on their placement.
(c) Compute the coefficient of skewness. What does it tell you?

SOLUTION (a) The data is skewed right, as can be seen in Fig. 4.3.
(b) You can verify that the following are correct.

$$\bar{X} = \text{average} = 4.963 \text{ percent}$$
$$M = \text{median} = 4.85 \text{ percent}$$
$$m = \text{mode} = 4.0 \text{ percent}$$

These are indicated in Fig. 4.3. As expected for a skewed-right histogram, the mode is to the left of the median and the average.

Table 4.1 Percentage of quality dollars spent on prevention of defects, Example 4.1

Percentage value X_i (1)	Frequency f_i (2)	$f_i X_i$ (3)	$f_i X_i^2$ (4)	$f_i(X_i - \bar{X})^3$ (5)
3	5	15	45	− 37.820
4	10	40	160	− 8.930
5	6	30	150	0.000
7	3	21	147	25.356
8	2	16	128	56.022
12	1	12	144	348.468
	27	134	774	383.096

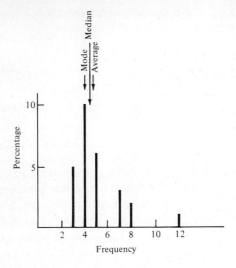

Figure 4.3 Histogram of percentage of quality dollars spent on defect prevention, Example 4.1.

(c) From the computation in Table 4.1 and Eqs. (3.10) and (4.3),

$$a_3 = \frac{383.096/27}{\left[\frac{774}{28} - \frac{28}{27}(4.963)^2\right]^{3/2}} = \frac{14.189}{(2.047)^3}$$

$$= +1.654$$

This verifies that the data is skewed right.

Problems P4.5–P4.8

4.3 Peakedness

Another measure for unimodal frequency distributions is the property of *peakedness*, or *kurtosis*. Peakedness is a comparative measure of the height of the peak in two different frequency distributions. Figure 4.4 shows two polygons, one which is considerably more peaked than the other. Since the measure is com-

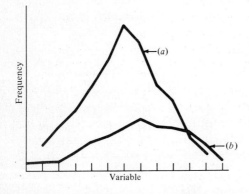

Figure 4.4 Frequency distributions showing (a) large peakedness and (b) small peakedness.

parative, a coefficient and norm value are defined. The coefficient of peakedness a_4 is dimensionless and computed as

$$a_4 = \frac{\text{fourth central moment}}{(\text{standard deviation})^4} = \frac{m_4}{s^4}$$

$$= \frac{\Sigma f_i(X_i - \bar{X})^4/n}{s^4} \qquad (4.5)$$

For comparison purposes, the norm value $a_4 = 3$ is used. If $a_4 < 3$, the frequency data is less peaked than a symmetric, bell-shaped frequency distribution. An $a_4 > 3$ indicates that the data is more peaked than this distribution. You can see that a_4 is difficult to compute and interpret. Observation of frequency data to be compared is often sufficient for the initial data analysis.

The concepts of skewness and peakedness are very useful, because they help the analyst make general conclusions about the shape of frequency distributions.

Example 4.7

Problems P4.9, P4.10

4.4 Mean Absolute Deviation

The *mean absolute deviation* d_m is the average deviation from the mean over all observations of the data. The formula is

$$d_m = \frac{\Sigma f_i |X_i - \bar{X}|}{n} \qquad (4.6)$$

The absolute value of each difference is taken to overcome the canceling effect of X_i values less than $\bar{X}$ when compared to X_i values greater than $\bar{X}$. This problem is overcome in standard deviation computations by squaring the difference. In a practical sense, the engineer finds that d_m can be used to approximate the more time-consuming computations for s by using the transformation

$$s = 1.25 d_m \qquad (4.7)$$

which is correct for a symmetric, bell-shaped frequency distribution. Actually, the factor 1.25 can be altered slightly in individual cases so that Eq. (4.7) can be used. Example 4.2 illustrates the application of d_m to the area of inventory control.

Example 4.2 JR, a mechanical engineer at the Reynolds Number Company, is responsible for the inventory control of all high-pressure vessels used in the plant. JR has collected 12 months of demand data for the two most used vessels (Table 4.2). To obtain control of the inventory, JR must compute the standard deviation of the monthly demand. To do this, the engineer wants to use the formula $s = 1.25 d_m$ for both demands. How accurate will this equation be for this data?

Table 4.2 Monthly demand data for two pressure vessels, Example 4.2

Month	Demand X_i Type A	Demand X_i Type B	$\|X_i - \bar{X}\|$ Type A	$\|X_i - \bar{X}\|$ Type B
Jan	22	15	2	2.75
Feb	27	17	3	0.75
Mar	21	16	3	1.75
Apr	39	24	15	6.25
May	14	17	10	0.75
Jun	23	14	1	3.75
Jul	26	15	2	2.75
Aug	25	28	1	10.25
Sep	25	21	1	3.25
Oct	24	15	0	2.75
Nov	19	17	5	0.75
Dec	23	14	1	3.75
Totals	288	213	44	39.50

SOLUTION Table 4.3 is a summary of the actual $\bar{X}$, s, d_m, and approximated s values for both types of pressure vessels. The d_m values are computed using the absolute deviation values from Table 4.2 and Eq. (4.6) where $f_i = 1$ for each month and $n = 12$. If JR uses the d_m values, he will have these approximated s values:

Type A: $s_A = 1.25(3.667) = 4.584$ vessels

Type B: $s_B = 1.25(3.292) = 4.115$ vessels

Accuracy can be measured by computing the percent error introduced by the approximating relation. Standard deviations are underestimated by 22 percent for A and 6 percent for type B.

COMMENT JR could compute the factor values needed to get the correct s from d_m, but future use would incorrectly assume that the demand data had the same values. Look at the monthly demand data plotted in Fig. 4.5. The data is obviously not symmetrical; thus the factor value of 1.25 will introduce error into the approximation. You should realize in the future that all approximating relations have error inherent to them. The assumptions

Table 4.3 Summary of computations for Example 4.2

Vessel type	$\bar{X}$	s	d_m	Approximated s
A	24.00	5.878	3.667	4.584
B	17.75	4.372	3.292	4.115

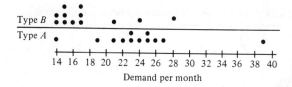

Figure 4.5 Plot of demand data, Example 4.2.

used must be known and investigated in each instance to ensure that you understand the implications for specific data.

Problems P4.11–P4.13

4.5 Weighted Average

For several values of a variable X the average is computed by the formula $\bar{X} = \Sigma X_i/n$. If each value of X is observed several times with a frequency f_i, the average is computed by $\bar{X} = \Sigma f_i X_i/n$. The multiplication of X_i by f_i is called weighting with the frequency. The general *weighted average* $\bar{X}_w$ uses the formula

$$\bar{X}_w = \frac{\Sigma w_i X_i}{n} \qquad (4.8)$$

where w_i ($i = 1, 2, \ldots, k$) are the weighting factors. $\bar{X}_w$ is useful when the variable has been observed from several different-sized samples. The average must take this variable sample size into account. Expanding Eq. (4.8), we have

$$\bar{X}_w = \frac{w_1}{n} X_1 + \frac{w_2}{n} X_2 + \cdots + \frac{w_k}{n} X_k$$

where $n = \Sigma w_i$. Then each w_i/n value is the weighting for the observed X_i value and $\Sigma w_i/n = 1$. For the frequency form of data the weights become $w_i/n = f_i/n$ and $n = \Sigma f_i$.

Example 4.3 Carlos Johnson is a metallurgical engineer with a firm which owns and leases barges throughout the world. Management believes that because of increased pollution in bay and port area waters, barges have to be repaired much more quickly at some sites because of corrosion. Johnson was asked to investigate this possibility. He collected some data for 1 year to determine the number of barges operating and the number repaired because of corrosion prior to the mean time to repair. Table 4.4 presents this data for six port locations.
(a) Compute the average number of barges prematurely repaired for each port.
(b) Compute the average number of barges owned at each port.

Table 4.4 Barges repaired prematurely because of corrosion, Example 4.3

Port number	Port location	Barges owned	Premature repairs
1	Houston	75	10
2	New Orleans	103	12
3	Tampa	52	8
4	San Francisco	38	3
5	Seattle	19	4
6	New York	98	28

SOLUTION (a) Since the sample sizes (barges owned) vary at each location, the weighted average must be computed. We use the following terms in Eq. (4.8):

$$w_i = \text{barges owned}$$
$$X_i = \text{premature repairs}$$
$$n = \text{fleet size} = \Sigma w_i$$

The average for premature repairs is

$$\bar{X}_w = \frac{75(10) + 103(12) + \cdots + 98(28)}{385}$$

$$= \frac{5336}{385}$$

$$= 13.860 \text{ barges per location}$$

(b) The average number of barges owned at each port site is a simple, unweighted average. So, the average $\bar{X}$ of the "Barges owned" column is taken for the $n = 6$ sites:

$$\bar{X} = \frac{75 + 103 + \cdots + 98}{6} = \frac{385}{6}$$

$$= 64.167 \text{ barges}$$

COMMENT In part (a) a simple average of the "Premature repairs" column can be computed as $(10 + 12 + \cdots + 28)/6 = 10.833$ barges. This average is wrong because it does not recognize the fact that a different fraction of the entire fleet is stationed at each port. If it were possible to station one-sixth of the fleet, or 64.167 barges, at each site, the average premature repairs would be

$$\bar{X}_w = \frac{64.167(10) + 64.167(12) + \cdots + 64.167(28)}{385} = 10.833 \text{ barges}$$

Now both $\bar{X}$ and $\bar{X}_w$ are the same because of equal weighting for each X_i value.

Example 4.8

Problems P4.14–P4.16

4.6 Coefficient of Variation

The *coefficient of variation* v is used to express the standard deviation as a percentage of the average. The formula is

$$v = \frac{s}{\bar{X}} 100\% \qquad (4.9)$$

Since s and $\bar{X}$ have the same units, v is a measure independent of units. The coefficient is of most use when a norm value is needed or when the coefficient is compared for two or more sets of data. For example, if an engineer collects some data on a chemical process today and again in one month, comparison of the two v values indicates the stability of s with respect to $\bar{X}$. Figure 4.6 shows what happens if v increases from polygon 1 to polygon 2. Note that even though both s and $\bar{X}$ have increased, s has increased more than $\bar{X}$.

Example 4.4 Jim took a sample of size 15 on a certain chemical content at the Gasup refinery in Houston and found that $\bar{X} = 51.2$ percent and $s = 8.9$ percent. Jane took a similar sample of 100 in Oklahoma City and found $\bar{X} = 53.8$ percent and $s = 5.2$ percent. Compare and comment on the values of the coefficient of variation.

SOLUTION Let v_H be the coefficient at Houston and v_O be the coefficient at Oklahoma City. Using Eq. (4.9), we get

$$v_H = \frac{8.9}{51.2} 100\% = 17.4\%$$

$$v_O = \frac{5.2}{53.8} 100\% = 9.7\%$$

Since $v_H > v_O$, the variation in terms of the average is larger at the Houston

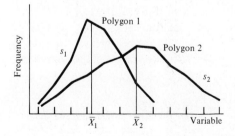

Figure 4.6 General shape for two polygons with different coefficients of variation, $v_1 < v_2$.

refinery than at Oklahoma City. However, this comparison does not consider the fact that the sample size at Houston was only 15 percent of that at Oklahoma City. Often a smaller sample will have a larger deviation because not enough data is observed to accurately account for the central tendency of the data.

Problems P4.17, P4.18

4.7 Coding Techniques for the Average and Standard Deviation

Coding is a commonly used technique to make data easier to manipulate. Coding means that you take each number and make a linear transformation by doing two things with it. You can (1) multiply or divide by some number and (2) add or subtract some number. For example, if we have two X values, $X_1 = 170.83$ and $X_2 = 181.39$, we might subtract 175 from each and multiply the result by 100. If we call the coded values d_1 and d_2, then

$$d_1 = (X_1 - 175)100 = (170.83 - 175)100 = -417$$
$$d_2 = (X_2 - 175)100 = (181.39 - 175)100 = 639$$

Coding is merely a movement of the x axis as shown by the scales in Fig. 4.7 using the relation $d = (X - 10)/5$. The same transformation must be applied to all data values. We will write any coded number d_i in the general form

$$d_i = \frac{X_i - \text{arbitrary origin}}{\text{multiplier}}$$

$$= \frac{X_i - X_0}{c} \tag{4.10}$$

The selection of X_0 is completely arbitrary. The original value is retrievable by solving Eq. (4.10) for X_i.

Coding is useful for many purposes. In this section, we see how coding can be used in the computation of $\bar{X}$ and s. The basic equations we use are

$$\bar{X} = \frac{\Sigma X_i}{n} \tag{4.11}$$

$$s = \left[\frac{\Sigma X_i^2}{n-1} - \frac{n}{n-1} \bar{X}^2 \right]^{1/2} \tag{4.12}$$

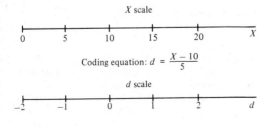

Figure 4.7 Relationship between an uncoded scale X and a coded scale d, where $d = (X - 10)/5$.

4.7 CODING TECHNIQUES FOR THE AVERAGE AND STANDARD DEVIATION

For $\overline{X}$ and s the following properties are true:

1. If you multiply (divide) each data value by a number, $\overline{X}$ and s are both affected.
2. If you add (subtract) to code each data value, $\overline{X}$ is affected but s is not.

The equations for coded $\overline{X}$ and s use these properties and the definition of d_i in Eq. (4.10):

$$\overline{X} = \text{multiplier(coded average)} + \text{arbitrary origin}$$
$$= c\overline{d} + X_0 \qquad (4.13)$$

$$s = \text{multiplier(coded standard deviation)}$$
$$= cs_d \qquad (4.14)$$

The procedure to compute $\overline{X}$ and s by using coded data is:

1. Select X_0 and c.
2. Calculate the coded data values d_i.
3. Compute the average $\overline{d}$ for the d_i values.
4. Compute $\overline{X} = c\overline{d} + X_0$.
5. Compute the standard deviation s_d for the d_i values. Use $\overline{d}$ in computing s_d.
6. Compute $s = cs_d$.

If you have a hand-held calculator with $\overline{X}$ and s preprogrammed for you, the coded forms are not time-saving. However, knowledge of coding data is important to an engineer in many ways other than $\overline{X}$ and s computations.

Example 4.5 Caroline is an industrial engineer in an aluminum fabrication plant. She has a problem. Aluminum sheets are produced with a thickness specification of 0.5 ± 0.0050 in. The following values are average thickness measurements in inches for 8 days of production:

| 0.4994 | 0.5007 | 0.4996 | 0.5001 |
| 0.4996 | 0.5003 | 0.4998 | 0.5000 |

Help Caroline determine (a) the average, using the coded data procedure; (b) the standard deviation; and (c) whether at least 99 percent of the data is within 3 standard deviations of the average.

SOLUTION (a) Use the steps outlined above to compute $\overline{X}$.

1. Select $X_0 = 0.5000$ and $c = 1/10{,}000$.
2. The d_i values are computed using Eq. (4.10):

| −6 | +7 | −4 | +1 |
| −4 | +3 | −2 | +0 |

3. The average of the coded values is
$$\bar{d} = \frac{-5}{8} = -0.625$$

4. The actual value of the average is
$$\bar{X} = c\bar{d} + X_0 = \frac{1}{10{,}000}(-0.625) + 0.5000$$
$$= 0.4999 \text{ in}$$

(b) We continue with the steps to compute s.

5. For s_d we need $\Sigma d_i^2 = 131$. Using Eq. (3.8), we get
$$s_d = \left[\frac{131}{7} - \frac{8}{7}(-0.625)^2\right]^{1/2}$$
$$= (18.268)^{1/2} = 4.274$$

6. The actual s value is
$$s = cs_d = \frac{1}{10{,}000}4.274$$
$$= 0.0004 \text{ in}$$

(c) Three standard deviations on each side of the average is
$$\bar{X} \pm 3s = 0.4999 \pm 3(0.0004)$$
$$= 0.4987 \text{ and } 0.5011 \text{ in}$$

Since all eight data values are between these limits, the 99 percent specification is met.

Problems P4.19–P4.22

4.8 Coding Techniques for Grouped Data

If the data is grouped into cells, the coding equation uses the cell interval c as the multiplier:

$$d_i = \frac{X_i - \text{arbitrary origin}}{\text{cell interval}}$$

$$= \frac{X_i - X_0}{c} \qquad (4.15)$$

The values of X_i and X_0 are cell midpoints. If the data is in *increasing* order, the d_i values will always be $d_i = \ldots -2, -1, 0, 1, 2, \ldots$ with $d_i = 0$ at X_0. This assumes that all cells are equal in width. By using the coded data, actual $\bar{X}$ and s

computations are

$$\bar{X} = \text{(cell interval)(coded average)} + \text{arbitrary origin}$$
$$= c\bar{d} + X_0 \qquad (4.16)$$
$$s = \text{(cell interval)(coded standard deviation)}$$
$$= cs_d \qquad (4.17)$$

Steps to compute $\bar{X}$ and s for grouped data are:

1. Select a cell midpoint X_0 and compute c.
2. Calculate the coded cell midpoints d_i.
3. Compute $\bar{d}$ using the frequency value f_i for each cell:

$$\bar{d} = \frac{\Sigma f_i d_i}{n}$$

4. Compute $\bar{X} = c\bar{d} + X_0$.
5. Compute

$$s_d = \left[\frac{\Sigma f_i d_i^2}{n-1} - \frac{n}{n-1} \bar{d}^2 \right]^{1/2}$$

Note that $\bar{d}$, not $\bar{X}$, is used in the computation of s_d.

6. Compute $s = cs_d$.

The coded computations for frequency data are useful because frequency data cannot be directly entered into a hand-held calculator—each value must be entered individually.

Example 4.6 An engineer has collected frequency data (Table 4.5) for the thickness of 80 steel plates made to a specification of 12.7000 ± 0.0125 millimeters (mm). Compute the average and standard deviation of this data.

Table 4.5 Computation of $\bar{X}$ and s for grouped data using the coded method, Example 4.6

Thickness midpoint X_i	f_i	d_i	$f_i d_i$	$f_i d_i^2$
12.6750	5	−3	−15	45
12.6850	15	−2	−30	60
12.6950	23	−1	−23	23
12.7050	19	0	0	0
12.7150	14	1	14	14
12.7250	4	2	8	16
	80		−46	158

SOLUTION We use the computation steps outlined earlier. See Table 4.5 for the results.

1. Let $X_0 = 12.7050$ and compute $c = 0.01$ mm.
2. The values of d_i are $-3, -2, \ldots, 2$ as computed by
$$d_i = \frac{X_i - 12.7050}{0.01}$$
3. The coded average is
$$\bar{d} = \frac{-46}{80} = -0.575$$
4. The actual average is
$$\bar{X} = c\bar{d} + X_0 = 0.01(-0.575) + 12.7050 = 12.699 \text{ mm}$$
5. The standard deviation for the d_i values is
$$s_d = \left[\frac{158}{79} - \frac{80}{79}(-0.575)^2\right]^{1/2}$$
$$= (1.665)^{1/2} = 1.290$$
6. The actual s for the data is
$$s = cs_d = 0.01(1.290) = 0.0129 \text{ mm}$$

The thickness has an average of 12.699 mm with a standard deviation of 0.0129 mm.

COMMENT There are two facts which you should always keep in mind when using coded computations for grouped data.

1. The values d_i will always be $d_i = \ldots, -2, -1, 0, 1, 2, \ldots$ for the cells. The value $d_i = 0$ will occur in the cell *arbitrarily* selected as X_0. Values of $d_i < 0$ will occur for cells with $X_i < X_0$. So, once X_0 is chosen, you can immediately write down the d_i values.
2. When s_d is computed, $\bar{d}$ is always used. If $\bar{X}$ is mistakenly used, you are mixing coded and actual data and will get the wrong answer for s_d and s.

By the way, the coded form can be used for frequency data on individual X values. Use $c = 1$ and compute the d_i values. For instance,

	X_i	f_i	d_i
	30	7	−4
	31	8	−3
$X_0 \to$	34	9	0
	39	2	5
	43	4	9

$$d_i = \frac{X_i - 34}{1}$$

Problems P4.23–P4.25

SOLVED PROBLEMS

Example 4.7 Compute the measure of peakedness for the data of Example 4.1.

SOLUTION Using the values $\overline{X} = 4.963$ percent and $s = 2.047$ percent computed in Example 4.1 and the sums in Table 4.6, compute a_4 from Eq. (4.5):

$$a_4 = \frac{2756.803/27}{(2.047)^4} = 5.815$$

The distribution for this percentage data is more peaked than a symmetric distribution with the norm value $a_4 = 3$.

COMMENT The value of a_4 is difficult to interpret when the data is skewed, because the norm of 3 is for a symmetric frequency distribution.

Section 4.3

Example 4.8 Joe Smithe, an industrial engineer (IE) with an insurance company, was doing some capital investment analysis. He collected the investment and return data shown in the table and stated to the manager of the finance department that the average rate of return was 6.5 percent. The manager told Joe the average was 4.13 percent. Who was wrong and why?

Project	Investment	Rate of return (%)
A	$100,000	6
B	50,000	9
C	20,000	10
D	175,000	1
	$345,000	26

SOLUTION Since the projects each have different amounts of investment, the average rate of return $\bar{r}$ must be computed by a weighted average with the investment being the weighting factor. If we divide all investment values by

Table 4.6 Computations for a_4, Example 4.7

X_i	f_i	$f_i(X_i - \overline{X})^4$
3	5	74.242
4	10	8.600
5	6	0.000
7	3	51.652
8	2	170.141
12	1	2452.168
	27	2756.803

$1000 and use Eq. (4.8),

$$\bar{r} = \frac{100(6) + 50(9) + 20(10) + 175(1)}{345} = 4.13\%$$

The manager is correct because the IE did not use a weighted average. The equally weighted average of the four rates is $26/4 = 6.5$ percent, as incorrectly stated by the IE.

Section 4.5

ADDITIONAL MATERIAL

Moments, skewness, and peakedness (Secs. 4.1 to 4.3): Duncan [7], pp. 584–589; Hoel [13], pp. 102–105; Wilks [26], pp. 74–77.
Data coding (Secs. 4.7 and 4.8): Kirkpatrick [15], pp. 34–38; Neville and Kennedy [19], pp. 16–18, 30–32; Newton [20], pp. 135–137, 148–152.

PROBLEMS

The following data set is used by the indicated problems.

A: A new design allows agricultural sprinkler heads to irrigate 140 m². The average area covered per head is recorded here for 42 tests under different crop and wind conditions:

Coverage (m²)	Number of tests
105–115	3
115–125	9
125–135	17
135–145	12
145–155	1

Problems P4.7–P4.9, P4.11, P4.23

P4.1 John has taken a sample of size 3 from a newly arrived batch of chlorine solution. He got three readings of percentage of inert ingredients to be X_1, X_2, and X_3. Use these to write out in expanded form (a) the relation to compute μ_2 and (b) the relation to compute m_3.
(c) What are the dimensions for these moments?

P4.2 For what values of k and $\bar{X}$ do the moment about the origin and the central moment have the same value? Write this equivalence in equation form.

P4.3 Use this data to compute (a) μ_2 and (b) m_3:

i	X_i	f_i
1	10	2
2	15	3
3	20	4
4	25	1

P4.4 Compute (a) μ_3 and (b) m_2 for the data in Prob. P4.3.

P4.5 Use dimensional analysis to show that the coefficient of skewness is dimensionless.

P4.6 Samples were taken before and after the introduction of a new billing system designed to speed up invoice preparation and mailing. Use the results in the table to compare the skewness by (a) the coefficient of skewness and (b) plotting the frequency polygons.

	Number of invoices	
Days to prepare	Old billing system	New billing system
1	4	5
2	7	9
3	16	15
4	21	11
5	12	4
6	3	0

P4.7 (a) Compute a_3 for data set A and determine how it is skewed.
 (b) Determine $\bar{X}$, M, and m and plot them on the frequency polygon to verify the conclusion above.
 (c) How is the data skewed in comparison to the two sets in Prob. P4.6?

P4.8 (a) Determine the value of Eq. (4.4) for data set A.
 (b) How does it compare with the norm value?

P4.9 (a) Compute the coefficient of peakedness for data set A.
 (b) How does it compare with the norm value?

P4.10 Compute a_4 for the two samples in Prob. P4.6 and determine which frequency distribution is flatter.

P4.11 For data set A, (a) compute the mean absolute deviation and (b) estimate the standard deviation. Compare it with the actual s value computed in Prob. P4.8.

P4.12 If the absolute value were not used in Eq. (4.6), what value would result for any set of data? Show that your answer is correct.

P4.13 Two lathes are used to machine the same product. The error for 50 readings on one measurement is given in the table for each machine. Use the mean absolute deviation to determine which lathe is more capable of machining within the tolerance $\bar{X} \pm 1s$.

Errors (mm)	Number of readings	
	Lathe 1	Lathe 2
0.5–1.0	10	8
1.0–1.5	15	10
1.5–2.0	12	8
2.0–2.5	8	10
2.5–3.0	5	5
3.0–3.5	0	6
3.5–4.0	0	3
	50	50

P4.14 The cost of a service contract for typewriter and teletype equipment is $250 per year. For the last 6 months the repairs have been paid individually.

(a) Use the costs in the table to determine which method is cheaper.
(b) What is the weighting factor for each month used to compute the average cost?

Month	Number of machines repaired	Repair cost per unit
Oct	4	$10.00
Nov	10	12.00
Dec	6	12.50
Jan	12	13.00
Feb	9	14.00
Mar	3	14.75

P4.15 A vendor of the Twi-Light Glass Company is using a new process to produce some electronic components. Alice is in quality assurance and has the results of samples on five lots received from the vendor. She wants to compute the average percent defective for these lots. Do this for her.

Lot number	Lot size	Sample size	Number of defectives	Percent defective
1	10,000	250	25	10
2	10,000	400	24	6
3	15,000	3500	175	5
4	22,000	700	63	9
5	10,000	250	10	4

P4.16 What is the difference between the average for frequency data and weighted average for different sample sizes?

P4.17 Over the past year four different methods have been used to evaluate the values of the coefficient of kinetic friction for bearings. Use the coefficient of variation to determine if the methods are equally reliable.

	Method			
Statistic	Dry with pressure	Dry in atmosphere	Lubricated with pressure	Lubricated in atmosphere
s	0.035	0.033	0.030	0.022
$\bar{X}$	0.20	0.25	0.12	0.14

P4.18 (a) Compute and compare the coefficient of variation for the two samples in Prob. P4.13 using the correct s values.
(b) Plot the frequency polygons and describe how these coefficients explain the differences in shape and location.

P4.19 Code or decode the following as requested.
(a) If $\bar{X} = 0.008596$, find an origin and a multiplier to make the coded value between 1 and 10.
(b) If $d = 3.9$, use Eq. (4.10) to find the original X value if $X_0 = 500$ and $c = 0.05$.

P4.20 The number of grams of liquid in a 400-g can of vegetables is given for 12 cans:

| 136.0 | 136.2 | 135.8 | 136.4 | 135.7 | 136.3 |
| 135.9 | 136.1 | 136.8 | 135.9 | 136.1 | 136.4 |

(a) Code the values into small integers.
(b) Compute the coded average and standard deviation.
(c) Uncode the average and standard deviation.
(d) Plot the two scales in a manner similar to Fig. 4.7. Locate the coded and uncoded averages.

P4.21 The partial composition of four solutions is listed in the table. Code the data and compute the average percentage of each chemical present.

Solution	H_2O (%)	NaCl (%)	$CaCl_2$ (%)
1	7.638	2.192	0.03
2	6.492	2.436	3.71
3	7.892	0.804	12.91
4	8.901	0.536	5.25

P4.22 Compute the average for the data below using (a) $X_0 = 0.500$ and (b) $X_0 = 0.600$. (c) Will coding with these two X_0 values affect s? Why or why not?

0.583	0.436	0.559	0.496
0.482	0.501	0.490	0.564

P4.23 Code the values in data set A and compute the actual average and standard deviation.

P4.24 Random numbers were generated for a frequency distribution of acceleration values. Compute the average using an origin of (a) 2.875 km/s^2 and (b) 5.875 km/s^2.

Acceleration (km/s^2)	Frequency
1.75–2.50	5
2.50–3.25	23
3.25–4.00	19
4.00–4.75	21
4.75–5.50	9
5.50–6.25	4
6.25–7.00	2

P4.25 The time to main-engine-cut-off (MECO) for 150 launches of a missile varies as a result of propellant mixture, combustion method, and other factors. Use the frequency data shown to determine the average burn time and its standard deviation. Do these computations by (a) the uncoded method and (b) the coded method to determine which is easier for you.

Time to MECO (s)	Number of launches
90.5–88.5	3
88.5–86.5	13
86.5–84.5	49
84.5–82.5	59
82.5–80.5	24
80.5–78.5	2

CHAPTER
FIVE

PROBABILITY COMPUTATIONS

This chapter shows you how to combine probabilities by using certain formulas. A summary of sets is included since they are basic to the correct understanding of probability computations. A case study is included in this chapter.

CRITERIA

To complete this chapter, you must be able to do the following:

1. Define and give an example of a *universe*, a *set*, and the *empty set*.
2. Define the set operations *union*, *intersection*, and *complement*. Use a Venn diagram to show these operations, given a description of the sets.
3. Define and enumerate the elements in a *sample space* and *event*, given a written description of the experiment and event.
4. State the three axioms of probability.
5. Compute the probability of an event using the probability rules, given a description of the situation.
6. Write the general probability formula for conditional and independent events, and compute probability with these formulas, given a description of the events.
7. State *Bayes' formula* and use it to compute the probability that a particular event caused an observed event, given a description and probability of each event.
8. Construct a *probability tree* and use it to compute conditional probability, given the events and their probabilities.

STUDY GUIDE

5.1 Fundamentals of Sets

A *set* is a collection of objects. Usually the elements of a set have something in common. A *subset* is a collection of objects all belonging to a set. Some examples are:

Set	Subset
All machines in a plant	Machines in department 101
Real numbers from 0 to 50	Integers from 10 to 40
Sum of any two numbers	Sum of two odd numbers

A *universe*, or universal set, is the collection of all objects of interest. The sets used are then subsets of this universe. The set of all real numbers is a common universe. We generally concentrate on some subset of this universe. A set with no elements is the *empty set*, or *null set*; it is symbolized by the Greek letter $\emptyset$ (phi).

Universe	Set	Empty set
All cars in a state	All cars with automatic transmission in a state	All operational cars with two wheels in a state
Numbers when a die is tossed—$U = \{1, 2, 3, 4, 5, 6\}$	Even numbers on toss of a die—$A = \{2, 4, 6\}$	Numbers over 8 on toss of a die—$B = \emptyset$

The elements of a set are written inside braces and identified by a capital letter. U is commonly reserved for the universe.

Problems P5.1–P5.3

5.2 Venn Diagrams and Set Operations

Figure 5.1 is a Venn diagram which shows the universe U as a rectangle and two sets which are wholly contained in U. U, A, and B can be interpreted in many ways. For example, U may be the numbers on a die as described in the previous section. A Venn diagram gives a pictorial representation of sets and how they are related.

There are three set operations useful in probability. The definitions and a Venn diagram representation are given here:

Union (Fig. 5.2a). The union of two sets is a set containing all elements that are

88 PROBABILITY COMPUTATIONS

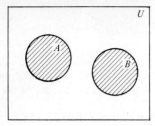

Figure 5.1 A Venn diagram of two sets.

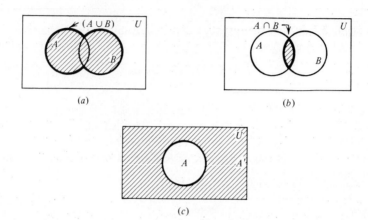

Figure 5.2 Venn diagrams showing (a) union $A \cup B$, (b) intersection $A \cap B$, and (c) complement A'.

in either set. The symbol $\cup$ is used to indicate a union. For example, if

$$A = \{1, 2, 3, 4\} \quad \text{and} \quad B = \{2, 4, 6, 8\}$$

then

$$A \cup B = \{1, 2, 3, 4, 6, 8\}$$

Intersection (Fig. 5.2b). The intersection of two sets is a set containing the elements in both sets. The symbol $\cap$ is used for intersection. For A and B above,

$$A \cap B = \{2, 4\}$$

Intersection may also be indicated as AB, that is, by omitting the $\cap$ symbol. If the intersection contains no elements, it is the empty set $\emptyset$, and the sets are called *mutually exclusive*. If $C = \{5, 6, 7\}$ and $A = \{1, 2, 3, 4\}$, then $A \cap C = \emptyset$. The Venn diagram for two mutually exclusive events will look like Fig. 5.1.

Complement (Fig. 5.2c). The complement of a set contains all elements not in the set but still in the universe. A' indicates the complement of A. If $U = \{1, 2, \ldots, 10\}$, for A and B above

$$A' = \{5, 6, 7, 8, 9, 10\}$$
$$(A \cup B)' = \{5, 7, 9, 10\}$$

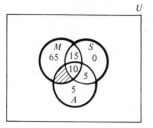

Figure 5.3 Venn diagram for switch uses, Example 5.1.

Example 5.1 Newly manufactured switching gear may be used in any of or all three modes: manual (M), semiautomatic (S), and automatic (A). In tests on 100 switches the following uses were found:

Number	Use
10	M, S, A
25	M, S
5	A only
15	S, A
Remainder	M only

(a) Define M, S, and A as sets of switches used as described. Draw the Venn diagram and show the number of switches in each use.
(b) How many switches are in M'? S? $(M \cup S)'$?

SOLUTION (a) Figure 5.3 shows the number of switches in each set. For example, $M \cap S$ has 25, but 10 are in the intersection $M \cap S \cap A$. Since the universe is $U = M \cup S \cup A = 100$, there are 65 in M only. The sets S only and $(M \cap A) \cap S'$ (shaded) are empty sets because there are no switches being used in these modes.
(b) M' has $100 - (65 + 15 + 10) = 10$ switches. Only 10 switches are never used in the manual mode. S has 30 switches in semiautomatic. $(M \cup S)$ has 95, so $(M \cup S)'$ has 5 switches.

Problems P5.4–P5.7

5.3 Sample Spaces and Events

A listing of all possible outcomes of an experiment is a universal set called the *sample space*, which we will call S. The entries in S are called *elements*. Any subset of S is called an *event*. An event is usually our main interest, but it is always necessary to know the sample space to work with different events from this space. Some examples follow.

Experiment	Sample space	Events
Toss one die	$S = \{1, 2, 3, 4, 5, 6\}$	A = even values $= \{2, 4, 6\}$ B = value ≥ 5 $= \{5, 6\}$
Inspect 10 items and count number of defectives	$S = \{0, 1, 2, \ldots, 10\}$	C = no defects $= \{0\}$ D = defectives $\leq 20\%$ of sample $= \{0, 1, 2\}$
Record the hours t until a light bulb burns out	$S = \{t \geq 0\}$	E = life exceeds 10 h $= \{t > 10\}$ F = life between 0 and 250 h $= \{0 \leq t \leq 250\}$

Problems P5.8, P5.9

5.4 Axioms of Probability

An axiom is a self-evident, widely accepted truth. The three axioms basic to all probability computations are summarized here for a sample space S and two events A and B. $P(\ )$ is used to indicate probability.

1. *Positiveness.* Every event has a nonnegative probability.
$$P(A) \geq 0$$

2. *Certainty.* The probability of the sample space is 1.
$$P(S) = 1$$

3. *Union.* The probability of the union of mutually exclusive events is the sum of each event's probability.
$$P(A \cup B) = P(A) + P(B) \quad \text{if } A \cap B = \emptyset$$

MUST KNOW

Axiom 3, which may be extended to any number of events, is called the addition law of probability.

These axioms are useful in proving the probability rules applied in the remainder of this text.

5.5 Probability Rules

There are many probability rules which may be stated and proved. The essential ones are given here for the events A, B, and $\emptyset$.

1. $0 \leq P(A) \leq 1$
2. $P(\emptyset) = 0$

KNOW

3. $P(A) \leq P(B)$ if A is a subset of B
4. $P(A') = 1 - P(A)$
5. $P(A \cup B) = P(A) + P(B) - P(A \cap B)$

You should memorize these rules, especially the last one.

Example 5.2 (*a*) Use a Venn diagram to show that $P(A \cup B) = P(A) + P(B) - P(A \cap B)$ for any events A and B.
(*b*) What happens if $A \cap B = \varnothing$?

SOLUTION (*a*) Figure 5.2*b* shows $A \cap B$. If we were to write $P(A \cup B) = P(A) + P(B)$, the area common to both events would be counted twice. So the probability $P(A \cap B)$ is subtracted once.
(*b*) If $A \cap B = \varnothing$, the events are mutually exclusive, as is Fig. 5.1 and $P(A \cap B) = 0$. In this case the rule is reduced to the third axiom, $P(A \cup B) = P(A) + P(B)$.

Example 5.3 Past experience indicates a 4 in 10 chance of finding recoverable precious metals and a 3 in 10 chance of finding them in combination with copper ore. If there is an 8 in 10 chance of copper ore deposits in a particular area, what is the probability of finding (*a*) copper or precious metals, (*b*) neither copper nor precious metals, (*c*) copper only?

SOLUTION Define the events and probabilities.

Description	Symbol	Probability
Metals	M	0.4
Copper	C	0.8
Copper or metals	$C \cap M$	0.3

You should draw the Venn diagram for these events now.
(*a*) Copper or metals is found by using rule 5:
$$P(C \cup M) = P(C) + P(M) - P(C \cap M) \qquad (5.1)$$
$$= 0.8 + 0.4 - 0.3 = 0.9$$
(*b*) The complement of $C \cup M$ is neither. By rule 4
$$P[(C \cup M)'] = 1 - P(C \cup M) = 1 - 0.9 = 0.1$$
(*c*) The event C_1 copper only means that no metals are found. By subtraction we have
$$P(C_1) = P(C) - P(C \cap M)$$
$$= 0.8 - 0.3 = 0.5$$
Since C_1 is a subset of C, rule 3 states that $P(C_1) < P(C)$.

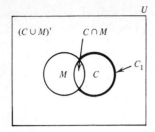

Figure 5.4 Venn diagram for discovery of copper and precious metals, Example 5.3.

COMMENT Since C and M are not mutually exclusive (they can occur together), the omission of the term $P(C \cap M)$ in Eq. (5.1) would make $P(C \cup M)$ exceed 1.0. It is best to always write down the complete form of rule 5 and substitute 0 for the intersection event probability if the events are mutually exclusive. The complete Venn diagram for this problem is shown in Fig. 5.4.

Problems P5.10–P5.15

5.6 Conditional Probability and Independent Events

The occurrence of one event is often made dependent on the occurrence of a second event. These are called *conditional events*. If A is conditional on B, the terminology and symbol are A given $B = A/B$. The sample space is reduced to the possible values of the event B. Assume the total sample space and events are

$$S = \{1, 2, 3, 4, 5, 6, 7\}$$
$$A = \{2, 4, 6\}$$
$$B = \{3, 4, 5, 6, 7\}$$

If B is known to have occurred, the five possible elements in the reduced sample space are $B = \{3, 4, 5, 6, 7\}$. An element in the conditional event $A/B = \{4, 6\}$ has a 2 in 5 chance of occurring.

Probability statements can be made for the events A, B, A/B, and $A \cap B = \{4, 6\}$:

$$P(A) = \tfrac{3}{7} \qquad P(B) = \tfrac{5}{7}$$
$$P(A/B) = \tfrac{2}{5} \qquad P(A \cap B) = \tfrac{2}{7}$$

Only the conditional event uses the reduced space of five elements. The probability for A/B assumes that the given event B has occurred:

$$P(A/B) = \frac{P(A \text{ and } B \text{ occur})}{P(B \text{ occurred})}$$

$$= \frac{P(A \cap B)}{P(B)} \qquad (5.2)$$

5.6 CONDITIONAL PROBABILITY AND INDEPENDENT EVENTS

The requirement that $P(B) > 0$ is necessary. $P(B/A)$ is written with $P(A)$ as the denominator. Solution of Eq. (5.2) for $P(A \cap B)$ gives the general conditional probability formulas

$$P(A \cap B) = P(A/B) \cdot P(B) \quad (5.3)$$

$$P(A \cap B) = P(B/A) \cdot P(A) \quad (5.4)$$

We can use Eq. (5.2) for the events above to show that

$$P(A/B) = \frac{\frac{2}{7}}{\frac{5}{7}} = \frac{2}{5}$$

Conditional probability is illustrated in Example 5.4.

Example 5.4 A total of eight cultures were checked for contaminants a and b. Four had a and six had b, but the cultures got mixed up and must be rechecked. Compute the probability that a culture (*a*) has contaminant a given it has b on recheck and (*b*) has b given it has a. (*c*) How do these compare with the possibility of finding both a and b not knowing either is present?

SOLUTION Events and probabilities are as follows.

Event	Description	Probability
A	Contaminant a	$\frac{4}{8}$
B	Contaminant b	$\frac{6}{8}$
$A \cap B$	Both a and b	$\frac{2}{8}$
A/B	a given b	Part (*a*)
B/A	b given a	Part (*b*)

(*a*) To compute $P(A/B)$, use Eq. (5.2):

$$P(A/B) = \frac{\frac{2}{8}}{\frac{6}{8}} = \frac{1}{3} = 0.33$$

(*b*) Write the probability $P(B/A)$ in the form of Eq. (5.2):

$$P(B/A) = \frac{P(A \cap B)}{P(A)} = \frac{\frac{2}{8}}{\frac{4}{8}} = \frac{1}{2} = 0.50$$

(*c*) Both conditional probability values are larger than the intersection event $P(A \cap B) = 0.25$. This is expected because a guarantee of the conditioning event gives us information which reduces the sample space.

COMMENT The intersection event probability can be computed from Eqs. (5.3) and (5.4):

$$P(A \cap B) = P(A/B) \cdot P(B) = \tfrac{1}{3} \cdot \tfrac{6}{8} = \tfrac{2}{8}$$
$$= P(B/A) \cdot P(A) = \tfrac{1}{2} \cdot \tfrac{4}{8} = \tfrac{2}{8}$$

If the chance that an event occurs is not changed by the occurrence of some other event, the two are *independent events*. Mathematically this may be stated as

$$P(B/A) = P(B) \quad \text{or} \quad P(A/B) = P(A)$$

For independent events the intersection event probability from Eqs. (5.3) and (5.4) becomes the product of the probabilities

$$P(A \cap B) = P(A) \cdot P(B) \tag{5.5}$$

This equation is called the multiplication law of probability and is used whenever you want the probability that two events occur at the same time and the two are independent.

Example 5.5 Stan plans to draw two cards from a regular card deck. Compute the probability of two kings if the first card is (*a*) not replaced and (*b*) replaced prior to the second drawing.

SOLUTION On any probability problem from now on one of your questions should be, Are the events independent? For this problem define the events

$$A = \text{king on first card}$$
$$B = \text{king on second card}$$

(*a*) If no replacement takes place, the chances change for the second draw, so A and B are definitely dependent. Use the conditional formula to find the probability of two kings:

$$P(A \cap B) = P(A) \cdot P(B/A)$$
$$= \tfrac{4}{52} \cdot \tfrac{3}{51} = \tfrac{1}{221}$$

The $P(B/A) = 3/51$ is correct because without replacement there are 51 cards left and only 3 are kings due to the conditioning event A.

(*b*) With replacement the chances of a king are the same for both draws. Are the events independent? Yes, so $P(B/A) = P(B) = \tfrac{4}{52}$. By Eq. (5.5)

$$P(A \cap B) = \tfrac{4}{52} \cdot \tfrac{4}{52} = \tfrac{1}{169}$$

To help you with probability computations, Table 5.1 summarizes the most important formulas for adding and multiplying probabilities. To find which formula to use, answer three questions for a problem statement.

1. Is the union (A or B) or the intersection (A and B) required?
2. Are the events *mutually exclusive*, that is, $A \cap B = \emptyset$?
3. Are the events *independent*, that is, $A/B = A$?

Problems P5.16–P5.21

● **Table 5.1 Summary of probability formulas**

	$P(A \cup B)$		$P(A \cap B)$
	Mutually exclusive	Not mutually exclusive	Either mutually exclusive or not mutually exclusive
Independent	$P(A) + P(B)$	$P(A) + P(B) - P(A) \cdot P(B)$	$P(A) \cdot P(B)$
Not independent	$P(A) + P(B)$	$P(A) + P(B) - P(A/B) \cdot P(B)$	$P(A/B) \cdot P(B)$

5.7 Bayes' Formula

It is possible to extend conditional probability computation in an interesting way. Consider a universe U composed of k mutually exclusive events $A_1, A_2, \ldots, A_k$ (Fig. 5.5). Assume there is an additional event B that is a subset of U, but involves part of one or more of the A_i ($i = 1, 2, \ldots, k$) events. B has been observed, and we want to compute the conditional probability that B came from A_1 or A_2 or $\cdots$ or A_k. In other words, what is the chance that B was caused by one of the A_i events, given B occurred? Using A_1 as the probable cause, the conditional probability is

$$P(A_1/B) = \frac{P(A_1 \cap B)}{P(B)} \qquad (5.6)$$

The event B may be written as the union of k mutually exclusive events (Fig. 5.5):

$$B = (A_1 \cap B) \cup (A_2 \cap B) \cup \cdots \cup (A_k \cap B)$$

The probability of each intersection can be written as a conditional probability

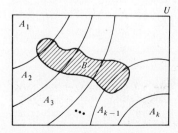

Figure 5.5 Universe composed of k mutually exclusive events with an additional event B defined on it.

by using Eq. (5.4). Also by axiom 3 these can be added.

$$P(B) = P(A_1 \cap B) + P(A_2 \cap B) + \cdots + P(A_k \cap B)$$
$$= P(B/A_1) \cdot P(A_1) + P(B/A_2) \cdot P(A_2) + \cdots + P(B/A_k) \cdot P(A_k)$$
$$= \sum_{i=1}^{k} P(B/A_i) \cdot P(A_i)$$

Substitution into Eq. (5.6) gives the probability of A_1 causing B. For any event A_i this probability can be stated:

$$P(A_i/B) = \frac{P(B/A_i) \cdot P(A_i)}{\sum_{i=1}^{k} P(B/A_i) \cdot P(A_i)} \tag{5.7}$$

Equation (5.7) is Bayes' formula. $P(A_i/B)$ are called the posterior, and $P(A_i)$ are the prior or source event probabilities.

Bayes' formula will be used in our study of Bayesian statistics and Bayesian parameter estimation later in this text.

Example 5.6 Two employees in advertising handle all the promotional campaigns for a large department store. A major appliance was advertised incorrectly, and the marketing manager wants to know who made the mistake. Employee 1 handles 75 percent of all advertising, and employee 2 does the remainder. From experience the manager knows that employee 1 makes a mistake 0.5 percent of the time, and employee 2 is incorrect 2 percent of the time. Determine the probability that the advertisement containing the mistake was made by employee 1 and by employee 2.

SOLUTION The following events may be defined.

Event	Description	Probability
A_1	Employee 1 handled advertisement	0.75
A_2	Employee 2 handled advertisement	0.25
B	Mistake made in an advertisement	Observed event
B/A_1	Mistake made given employee 1 handles an advertisement	0.005
B/A_2	Mistake made given employee 2 handles an advertisement	0.02

To determine the chance that employee 1 made the mistake, use Eq. (5.7) to compute $P(A_1/B)$, which is the conditional probability that employee 1 handled the advertisement, given that there was a mistake in the advertisement.

$$P(A_1/B) = \frac{P(B/A_1) \cdot P(A_1)}{P(B/A_1) \cdot P(A_1) + P(B/A_2) \cdot P(A_2)}$$

$$= \frac{0.005(0.75)}{0.005(0.75) + 0.02(0.25)}$$

$$= \frac{0.00375}{0.00875} = 0.43$$

There is a 43 percent chance that employee 1 made the mistake. Since all advertisements are handled by the two employees, there should be a 57 percent chance that employee 2 made the mistake. This is verified by using Eq. (5.7) to compute $P(A_2/B)$.

$$P(A_2/B) = \frac{P(B/A_2) \cdot P(A_2)}{P(B/A_1) \cdot P(A_1) + P(B/A_2) \cdot P(A_2)}$$

$$= \frac{0.02(0.25)}{0.00875} = \frac{0.005}{0.00875} = 0.57$$

COMMENT Note that even though employee 1 handles 75 percent of all the advertisements, the low percentage of mistakes reduces the chance from 75 to 43 percent that the advertisement with the error was handled by employee 1.

Example 5.7 A sensitive piece of electronic gear has a switch that fails on the average once every 100 missions. Engineers have devised a test which indicates defective 90 percent of the time if the switch is defective and not defective 99 percent of the time if it is okay. The gear has failed, and it is found that the switch was tested and found defective. What is the probability that the switch was indeed defective?

SOLUTION Define the events and probabilities.

Event	Description	Probability
A_1	Switch is defective	0.01
A_2	Switch is not defective	0.99
B	Test shows defective	Computed
B/A_1	Test shows defective given defective	0.90
B'/A_2	Test shows okay given not defective	0.99

We want the probability of A_1 given B has occurred. The events are shown

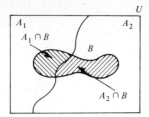

Figure 5.6 Diagram showing the intersection events used in Bayes' formula for Example 5.7.

in Fig. 5.6. Equation 5.7 is written for $P(A_1/B)$ as

$$P(A_1/B) = \frac{P(B/A_1) \cdot P(A_1)}{P(B/A_1) \cdot P(A_1) + P(B/A_2) \cdot P(A_2)} \quad (5.8)$$

where the numerator uses the general conditional probability rule. We have all values except $P(B/A_2)$, which is the complement of B'/A_2. So, $P(B/A_2)$ = 1 − 0.99 = 0.01.

$$P(A_1/B) = \frac{0.90(0.01)}{0.90(0.01) + 0.01(0.99)}$$

$$= \frac{0.009}{0.009 + 0.0099} = 0.476$$

There is a 47.6 percent chance that the switch is actually defective.

COMMENT This result is interesting. The test is 99 percent reliable; yet, there is less than a 50 percent chance that the switch is defective, even though the gear has failed and shown the test result to be correct.

You should compute $P(A_2/B) = 0.524$ at this time. Since there are only k possible $P(A_i/B)$ values, their sum is 1:

$$\sum_{i=1}^{k} P(A_i/B) = 1$$

In this example the switch must be defective (A_1) or nondefective (A_2), so $k = 2$.

Problems P5.22–P5.25

5.8 Probability Trees

When there are several options and stages in a probability computation, the probability tree is useful. It is a pictorial representation of conditional probabilities which shows all possible branches for each event. Figure 5.7 is a general tree for the events A_1, A_2, B_1, and B_2. If the probabilities along one branch are multiplied, the intersection event probability is computed by using the general conditional probability rule. For example,

$$P(A_1 \cap B_1) = P(B_1/A_1) \cdot P(A_1)$$

5.8 PROBABILITY TREES

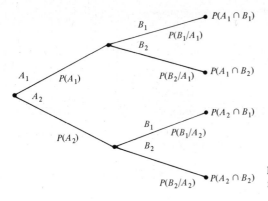

Figure 5.7 A general probability tree for four events.

All values after the first branch are conditional probabilities because they assume that branches to their left have taken place. If all possible events are evaluated on the tree, the final right-hand values should add to 1.

Example 5.8 For the data of Example 5.7 use a probability tree to (a) determine the chance that the switch is not defective and the test showed it to be defective, and (b) determine the probability that the switch was defective (A_1) given it was tested and found defective (B).

SOLUTION (a) The complete tree is given in Fig. 5.8. The chance of a nondefective switch (A_2) and a test showing defective (B) is taken directly from the tree:

$$P(A_2 \cap B) = 0.0099$$

(b) This conditional probability question is the same one answered earlier using Bayes' formula. The answer cannot be taken directly from the

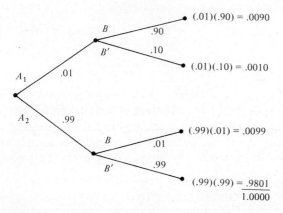

Figure 5.8 Probability tree for Example 5.8.

tree, but all probability values are already computed on it.

$$P(A_1/B) = \frac{P(A_1 \cap B)}{P(B)}$$

$$= \frac{0.009}{0.009 + 0.0099} = 0.476$$

(5.9)

$P(B)$ is the sum of all final right-hand values in which B occurs, that is, $(A_1 \cap B)$ and $(A_2 \cap B)$. Effectively, the computations in Eq. (5.9) from the tree are the same as those in Bayes' formula, Eq. (5.7).

Example 5.9

Problems P5.26–P5.30

SOLVED PROBLEM (Case Study)

Example 5.9 Two electronic switches are in parallel and one is in series (Fig. 5.9). The probability of failure is given for each individual switch. In addition, because switch 2 is old, it is estimated that if switch 1 fails, there is a 0.5 chance that switch 2 will also fail. The parallel switches 1 and 2 are redundant, so they must both fail before current cannot flow through to switch 3. Compute the probability that:
(a) The mechanism does not fail.
(b) The mechanism fails completely.
(c) Either switch 1 or 2 fails and switch 3 definitely fails.
(d) Switch 2 failed given that failure is observed.

SOLUTION Define the events and probabilities.

Event	Description	Probability
A	Switch 1 fails	0.20
B	Switch 2 fails	0.05
C	Switch 3 fails	0.10
B/A	Switch 2 fails because 1 fails	0.50

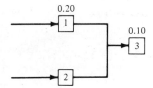

Figure 5.9 Three-switch mechanism with failure probabilities, Example 5.9.

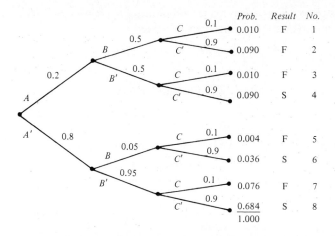

Figure 5.10 Probability tree for a three-switch mechanism, Example 5.9.

A probability tree is shown in Fig. 5.10. The events A', B', and C' indicate switch success. The final result is also shown—F for mechanism failure and S for success. The sum of all probability values is 1.0, as required.

(a) The probability of success S is the sum of results 4, 6, and 8, because C' is always present and at least one of A' or B' is true.

$$P(S) = 0.090 + 0.036 + 0.684 = 0.810$$

(b) Probability of failure F is found by subtraction:

$$P(F) = 1 - P(S) = 1 - 0.810 = 0.190$$

(c) If the event $(A \cup B) \cap C$ occurs, switch 1 or 2 and switch 3 fail. On the probability tree results 1, 3, and 5 apply.

$$P[(A \cup B) \cap C] = 0.010 + 0.010 + 0.004 = 0.024$$

If the rules summarized in Table 5.1 are used, we have the following procedure. Events A and B are not mutually exclusive or independent. Then

$$P(A \cup B) = P(A) + P(B) - P(B/A) \cdot P(A)$$
$$= 0.20 + 0.14 - 0.5(0.2)$$
$$= 0.24$$

To obtain $P(B) = 0.14$, notice in Fig. 5.10 that B occurs in results 1 and 5. Neglecting the third branch for C, we have

$$P(B) = P(A \cap B) + P(A' \cap B)$$
$$= 0.2(0.5) + 0.8(0.05) = 0.14$$

The events $A \cup B$ and C are independent.

$$P[(A \cup B) \cap C] = P(A \cup B) \cdot P(C) = 0.24(0.10)$$
$$= 0.024$$

(d) If failure results, $P(B/F)$ is a Bayesian probability that B was, in part, a cause of the failure. Bayes' formula may be stated as

$$P(B/F) = \frac{P(B \cap F)}{P(F)}$$

Failure and B occur in results 1, 2, and 5. $P(F)$ was computed in part (b).

$$P(B/F) = \frac{0.010 + 0.090 + 0.004}{0.190} = \frac{0.104}{0.190} = 0.547$$

Even though there is only a 19 percent chance that the mechanism will fail, there is a better than 50 percent chance that B has failed once failure is observed.

It is possible to compute a Bayesian probability for all events, given failure. You should verify these values.

$P(A/F) = 0.579$ $P(B/F) = 0.547$ $P(C/F) = 0.526$
$P(A'/F) = 0.421$ $P(B'/F) = 0.453$ $P(C'/F) = 0.474$

Sections 5.5–5.8

ADDITIONAL MATERIAL

Sets and probability (Secs. 5.3 to 5.6): Benjamin and Cornell [3], pp. 32–64; Hines and Montgomery [12], pp. 3–24; Hoel [13], pp. 4–20; Kirkpatrick [15], Chap. 2; Miller and Freund [18], Chap. 2; Newton [20], Chap. 2; Walpole and Myers [25], Chap. 1.

Bayes' formula (Sec. 5.7): Benjamin and Cornell [3], pp. 64–69; Hahn and Shapiro [10], pp. 18–23; Miller and Freund [18], pp. 36–40; Newton [20], pp. 47–51; Walpole and Myers [25], pp. 22–25.

PROBLEMS

P5.1 If the universal set U is all numbers from 1 to 25, specify the elements in the following sets:
(a) A is all prime numbers in U.
(b) B is all numbers evenly divisible by 4.
(c) C is all numbers containing the digit 2.

P5.2 The production capacity for an assembly is 500 per week (100 per day). If the plant is operating at 100 percent capacity, list the elements in the sets described:
(a) A is the number of defective assemblies in a sample of 10 taken from one day's production.
(b) B is the fraction defective for the sample taken in (a).
(c) C is the total number of defective assemblies in the samples for one week if a 5 percent sample is taken after each day's production is complete.

P5.3 (a) List the universal set elements for the product of dots on two dice. What are the elements in the subsets (b) product is an odd number, (c) product exceeds 40, and (d) product is an even multiple of the number 6?

P5.4 (a) For the sets A, B, and C in Prob. P5.1 construct the correct Venn diagram.
(b) Determine the elements of and indicate on the diagram the following sets: (1) $B \cap C$; (2) $(B \cup C)'$; (3) $(A \cup B \cup C)'$; (4) $A \cap B$; (5) $(B \cup C) \cap C$; (6) $A \cap B \cap C$.

P5.5 A business manager has three ways to advertise a new product. They are event N for newspaper advertisements, T for television and radio commercials, and C for a home-delivered circular. Construct the Venn diagram and indicate the following events on the diagram:
  (a) Both newspaper and circulars are used.
  (b) No advertisement will be used.
  (c) Both television and radio and newspaper advertisements are used.
  (d) Circulars only are used.
  (e) All advertisement methods are used.

P5.6 Use a Venn diagram with two intersecting events A and B to show that the DeMorgan laws are correct:
  (a) $A' \cup B' = (A \cap B)'$
  (b) $(A \cup B)' = A' \cap B'$

P5.7 A worker can fulfill the requirements for a good service award in any of three ways: approved suggestions (S), high productivity (P), and low absence (A). The number of employees that qualify are:

Qualification	Event	Number
Suggestions	S	22
Productivity	P	18
Absence	A	17
Suggestions and productivity	$S \cap P$	5
Productivity and absences	$P \cap A$	13
Suggestions and absences	$S \cap A$	2
All three	$S \cap P \cap A$	2

Construct the Venn diagram and determine how many employees qualify (a) by suggestions only, (b) by high productivity and low absences only, (c) by high productivity or low absences, and (d) by suggestions and low absences.
  (e) What is the total number of employees who qualify under any method?

P5.8 The shear strength of a spot weld must exceed 40 kg to be acceptable. List the possible outcomes for (a) the entire sample space S, (b) the event F that a weld is unacceptable, and (c) the event B that shear strength is between 50 and 75 kg.

P5.9 A shipment of 200 fire extinguishers is received. The defect rate is 2 in 100 and a sample of S extinguishers is inspected.
  (a) Write the sample space for the possible number of defectives in the sample.
  (b) Write the sample space if the defect rate increases to 4 in 100.

P5.10 The probability of events A and B are $P(A) = 0.6$ and $P(B) = 0.5$. A or B occurs with a probability of $P(A \cup B) = 0.87$.
  (a) Are events A and B mutually exclusive?
  (b) What is the probability that A and B will occur at the same time?

P5.11 A solution possibly contains two toxic chemicals. In 100 randomly mixed samples, 45 contain chemical 1, 40 contain chemical 2, and 35 contain both. Construct the Venn diagram for the following events: A, contains chemical 1; B, contains chemical 2; and $A \cap B$, contains both. If a sample is selected from the 100, what is the probability of finding (a) that the solution is toxic, (b) chemical 1 only, (c) neither chemical?

P5.12 Rework Prob. P5.11 if no samples contain both chemicals.

P5.13 Mr. and Mrs. Beluv own a cargo trailer rental agency. On an average weekend 12 percent of all persons entering the agency rent nothing and 88 percent rent trailers, trucks, or pads. A total of 75 percent rent a trailer and 15 percent rent a truck, while 2 percent rent both. In addition, 10 percent of the patrons rent pads to be used in the trailer, and 3 percent rent pads for the truck. Thus far, no one has rented all three together or pads only. Construct the Venn diagram and determine the probability that a customer will rent (a) a trailer or a truck, (b) a trailer or a truck, but no pads, (c) only a truck, (d) anything, (e) a trailer or a pad, and (f) pads.

P5.14 A couple has had a lot of trouble with an overloaded electric system. In the last 20 power failures the number of times each of three possible problems has been observed is given below. No other combinations of these sources have occurred, but one failure was caused by an unknown source. Compute the probability that the next failure is caused by (a) a fuse, (b) a circuit breaker, (c) a fuse and circuit breaker, (d) a wiring problem only, (e) all three problems, and (f) none of these three problems.

Failure source	Symbol	Number of times in 20 failures
Fuse	F	10
Circuit breaker	C	6
Wiring	W	5
Fuse and wiring	$F \cap W$	2

P5.15 If A is the event that switch a fails and B is the event that switch b fails, use the values $P(A) = 0.2$, $P(B) = 0.1$, and $P(A \cap B) = 0.05$ to compute and interpret in words the following probabilities:
(a) $P(A')$
(b) $P(A \cup B)$
(c) $P[(A \cap B)']$
(d) $1 - P(A \cup B)$
(e) $P(B \cap A')$

P5.16 The probability that a quality inspector correctly locates a defective product is 0.75 if he has not attended the error detection course and 0.95 if he has attended the course. At any one time an estimated 60 percent of the inspectors on duty have taken the course. What is the probability that (a) an inspector finds an error and has attended the course and (b) an inspector finds an error and has not attended the course?

P5.17 A new building is being constructed. Two important events are A, cost overrun, and B, passing inspection. The contractor estimates a 50 percent chance of cost overrun, a 65 percent chance of passing inspection, and a 20 percent chance of both. Compute the probability that (a) the building will pass inspection given that a cost overrun occurs and (b) either event occurs.

P5.18 A sample of size eight contains three defective and five acceptable items. If two items are drawn, compute the probability of both being acceptable if (a) the first item is replaced after observation and (b) the first item is not replaced.

P5.19 Two switches are in parallel. The probability that either one fails is 0.10. However, once one has failed, the second switch has a 0.25 chance of failing. What is the probability that both switches fail?

P5.20 Jane teaches continuing education courses in mathematics. In the past the basic course has failed to exceed the minimum enrollment criteria 20 percent of the time, and the advanced course has not made it 40 percent of the time. Since a person may enroll in both classes at once, a study was conducted of past statistics. Of the times that the basic course failed to meet the criteria, 70 percent of the time the advanced course also failed. Compute the probability that (a) both courses will fail and (b) either course will fail to meet the criteria.

P5.21 Two events are independent with probabilities $P(A) = 0.3$ and $P(B) = 0.7$.
  (a) Determine $P(A \cup B)$.
  (b) If the events are dependent and $P(A/B) = 0.5$, compute $P(A \cup B)$ and compare with the answer in (a).

P5.22 Three different processes that may be used to make a product have scrap rates of 5, 8, and 3 percent for processes 1, 2, and 3, respectively. If process 1 makes 40 percent of all this product and processes 2 and 3 make 30 percent each, and if an item must be scrapped, compute the probability it was manufactured by (a) process 1 and (b) process 2.

P5.23 An experimental test is being run on small animals with a certain disease which occurs in 25 percent of all small animals. In the last 100 trial tests, 92 animals with the disease reacted positively and 7 animals without the disease reacted positively. Compute the probability that an animal which reacts positively actually has the disease.

P5.24 The total chemical pollution in a river is composed of 80 percent chemical A and 20 percent chemical B. At the present level the probability that A will cause a fish kill is 0.15 and that B causes it is 0.25. If a fish kill has occurred, what is the chance that it was caused by chemical B?

P5.25 The probability that an airline passenger is served by attendant A or B is 0.75 or 0.25, respectively. From past experience it is known that attendant A gets the correct beverage order 98 in 100 times, while attendant B is correct 80 in 100 times. A passenger, who has just received an incorrect beverage order and does not remember which attendant took the order, has blamed attendant A for the mistake because this person took most of the orders. Which attendant has the larger probability of making the mistake?

P5.26 Solve Prob. P5.19 using the probability tree approach.

P5.27 Solve Prob. P5.20 using a probability tree.

P5.28 A job shop has a project which requires two operations in the order 1, then 2. There are three machines (A, B, and C) that can finish operation 1 in time with probabilities 0.2, 0.3, and 0.5, respectively. Operation 2 can be done on machine D or E with probabilities 0.8 or 0.2, respectively, regardless of which machine performed operation 1. Construct a probability tree and use it to determine the machine combinations which give (a) the largest and (b) the smallest chance of success.

P5.29 Use a probability tree to solve Prob. P5.24.

P5.30 Use a probability tree to solve Prob. P5.25.

LEVEL TWO

DISTRIBUTIONS AND THEIR USES

The chapters in this level cover all the common discrete and continuous probability distributions. For each distribution you learn its formula, graph, parameters, and properties. If the distribution is tabulated, you learn how to read and interpret the table.

Each chapter includes sections on elementary uses of the distribution, which is greatly expanded in the next two levels; however, when you finish this level, you should have a basic understanding of the following distributions:

Discrete	Continuous
Hypergeometric	Normal
Binomial	χ^2
Poisson	t
Geometric	F
Pascal	Uniform
Discrete uniform	Exponential
	Gamma

The topics of interval estimation of population parameters and functions of two variables are also treated. A procedure for the transformation of variables for both discrete and continuous variables is covered.

Chapter 6 should be considered as a guide for learning about any distribution, for it includes the formulas and procedures most commonly used with all distributions. The flowchart shows the relations between chapters in this level.

Level Two Flowchart

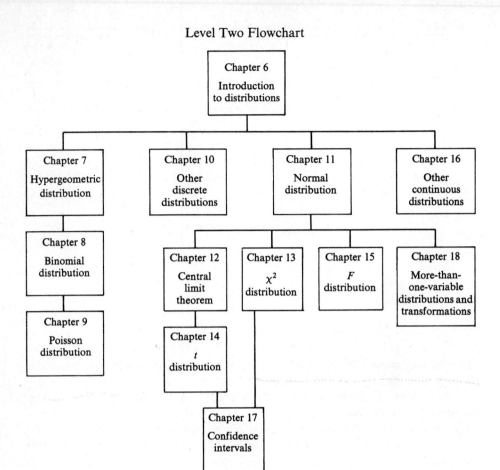

CHAPTER SIX

INTRODUCTION TO PROBABILITY DENSITY FUNCTIONS

The purpose of this chapter is to give you a general understanding of probability density functions (pdf's) and their properties.

In previous chapters we performed computations on observed data, that is, samples from a population. Here we begin to develop knowledge about populations which help us make statistical inferences about the real world.

The material in this chapter is applicable to all pdf's. Try to learn the procedure and the concept behind each procedure. Do not be concerned with the pdf's themselves, because they are only a teaching medium. The following chapters discuss particular, important pdf's and their properties. You should treat this chapter as *reference material* when you cannot remember a particular formula.

CRITERIA

To complete this chapter, you should be able to:

1. State a definition of the terms *random variable* and *probability*, and calculate probabilities using the *addition law of probability*.
2. Define the terms *probability density function, variable range, parameter*, and *parameter range*. Write these, given a complete description of the distribution of probability.
3. State and graphically demonstrate the difference between a discrete pdf and a continuous pdf, given the complete pdf form.

110 INTRODUCTION TO PROBABILITY DENSITY FUNCTIONS

4. Write and mathematically demonstrate the three properties of a discrete pdf, given the general form of the pdf.
5. State the computational form for the expected value, variance, and standard deviation of any *discrete* pdf and compute these, given the form of the pdf.
6. Write and mathematically demonstrate the three properties of a continuous pdf, given the general form of the pdf.
7. State the computational form for the expected value, variance, and standard deviation of any *continuous* pdf and compute these, given the form of the pdf.
8. Determine the complete form and graph of the cumulative distribution function for a discrete or continuous variable, given the variable and probability values (discrete) or the form of the pdf (continuous).
9. State the three properties of a *best estimator*.
*10. Define the *moment generating function*, and use it to determine the relation for the expected value and variance, given the form of the pdf.

STUDY GUIDE

6.1 Definitions

A *random variable* is any property which is investigated and has the ability to take on different numerical values. If the mineral content in the water of several towns is measured in grams per 1000 gallons (gal), the content is a random variable. Usually, the random variable is given a symbol, like X. The variable may be given a numerical value, for example, $X = 115$ g. It is also possible to give a variable an attribute designation, such as green or red. Usually our variables will be numerical.

It can be stated that the *probability* of X, $P(x)$, is a mathematical outcome between 0 and 1 which gives the chance that X will take on some particular value x:

$$P(x) = \frac{\text{number of favorable outcomes}}{\text{total number of possible outcomes}}$$

The probability is calculated for a particular value x; therefore, the value of interest must be observable for X. *The particular value will always be indicated by a small letter and the random variable by a capital letter.* Random variables may be designated by any letter desired.

Example 6.1 A young engineer has defined random variables for printing mistakes in telephone directories:

X = number of major errors (reprint page)
Y = number of minor errors (improve printing in future)
Z = number of annoyance errors (distractive to user)

In 1000 newly printed pages for the city A directory, the engineer observed $x = 5, y = 17$, and $z = 35$. If an assistant reviewed 2000 pages of the city B directory printed under identical conditions, what values of x, y, and z have the same probabilities of being observed as in the city A directory?

SOLUTION For city A we can compute probabilities of the number of major errors, minor errors, and annoyances:

$$P(x) = 5/1000 = 0.005$$
$$P(y) = 17/1000 = 0.017$$
$$P(z) = 35/1000 = 0.035$$

For city B use the probability values above and solve for the numerators, but use 2000 in the denominators:

$$P(x) = 0.005 = \frac{x}{2000} \qquad x = 10 \text{ major errors}$$

$$P(y) = 0.017 = \frac{y}{2000} \qquad y = 34 \text{ minor errors}$$

$$P(z) = 0.035 = \frac{z}{2000} \qquad z = 70 \text{ annoyances}$$

A random variable can assume different numerical values. The number of watermelons in a 100-kg container may take on values of 0, 1, ..., to infinity depending on the weight of each melon. If X is the number of melons, these possible values are indicated by $x = 0, 1, 2, \ldots$ If the probabilities that $x = 15$ and $x = 16$ are known, the probability that $x = 15$ or $x = 16$ is found by *adding* the two probabilities. The *addition law of probability* may be stated as

- $P(\text{either of two outcomes}) = \Sigma P(\text{each outcome})$

$$P(A \text{ or } B) = P(A) + P(B) \qquad (6.1)$$

where A and B are the outcomes. This law is true only when it is impossible for the two outcomes to happen at the same time, which is the usual case for random variables. Equation (6.1) can be extended to any number of outcomes, provided only one value of the variable can be observed at a time.

Example 6.2 The total package weight of a chemical is taken for 150 packages. The specification of weight W is 20 ± 0.5 oz. Ten of the packages weighed less than 19.5 oz, and five weighed more than 20.5 oz. What is the probability that a package taken at random will not meet the specification?

SOLUTION Let S be the outcome $W < 19.5$ and T be the outcome $W > 20.5$. We want $P(W < 19.5 \text{ or } W > 20.5) = P(S \text{ or } T)$. Since the weight of one particular package cannot exceed 20.5 *and* be less than 19.5 at the same

time, S and T cannot happen at the same time. Using Eq. (6.1), we get
$$P(S \text{ or } T) = P(S) + P(T) = 10/150 + 5/150 = 0.1$$

Problem P6.1

6.2 Probability Density Function (pdf) Terminology

If a variable can take on several different values, the probability for each point is less than 1 since $0 \leq P(X = x) \leq 1$, where $X = x$ means that X takes on some specific value x. It is usually possible to write the relation between X and $P(x)$ in some convenient form called a *probability density function*, or *pdf* for short. A pdf is either a listing of probability values or a mathematical relation which describes how probability is distributed over the different values of the variable. Note the use of the word *density*. This word is appropriate since it gives the weighting of probability for each value. The pdf is often called a *distribution*.

Example 6.3 Carol, a civil engineer with the Up-Strike Company, has made 100 loading measurements on a certain beam. She has categorized the loadings into low, medium, high, and excessive classes with the following results:

Loading class	Number in class
Low	12
Medium	57
High	23
Excessive	8

(a) Write the pdf of the loadings.
(b) Determine the probability that the loading will be either low or excessive.

SOLUTION (a) If the random variable Y indicates the loading class, it can assume four values: low, medium, high, and excessive. The probability $P(y)$ that Y will assume any of the four values above is computed by

$$P(y) = \frac{\text{number of measurements in a specific class}}{\text{total number of measurements}} \quad (6.2)$$

The values of $P(y)$ are

y	$P(y)$
Low	0.12
Medium	0.57
High	0.23
Excessive	0.08

The pdf may be written in a mathematical form with $y = $ low $= 1$, $y = $ medium $= 2$, etc.:

$$P(Y = y) = \begin{cases} 0.12 & y = 1 \\ 0.57 & y = 2 \\ 0.23 & y = 3 \\ 0.08 & y = 4 \end{cases}$$

(b) The classes of low and excessive cannot occur at the same time, so Eq. (6.1) is used to compute

$$P(y = 1 \text{ or } y = 4) = 0.12 + 0.08 = 0.20$$

To completely describe the pdf for a random variable, you must write the

1. probability values or mathematical relation to describe probability
2. variable and its range
3. parameters and their ranges

It is important that you understand the meaning of all components in a pdf. Probability is given for each variable value (as in Example 6.3) or by a mathematical relation. The *variable range* details the values which can be observed for the variable. For example, if the rivets in an airplane wing are defective when they are not flush with the wing surface, out of 250 rivets the number of defectives X can be any integer value in the range $x = 0, 1, 2, 3, \ldots, 250$.

The *parameter* is used in most pdf statements because it gives form and shape to the pdf. Each parameter can usually take on all real values; that is, it is continuous. An illustration is given in Example 6.4. You should remember that the parameter is not a variable, but that the value of the parameter is needed to determine probability from the pdf. If a new parameter value if determined, there is a new pdf. A pdf may have none, one, or more parameters, but only one variable is used. You should be able to recognize and write a pdf in the format presented in Example 6.4.

Example 6.4 An electrical engineer is interested in the number of cycles before a particular type of solenoid will fail. The engineer determines that the number of cycles before failure follows a formula of the form $p(1 - p)^{x-1}$, where X is the random variable representing the cycle in which failure occurs and p is the probability of failure on each cycle. Write the complete pdf for the variable X.

SOLUTION We need to determine the probability relation, the variable and its range, and the parameter and its range. The probability expression is

$$P(x) = p(1 - p)^{x-1}$$

This equation describes how probability is distributed over the values of first cycle, second cycle, etc.; that is, $x = 1, 2, \ldots$ The probability of failure p on *each* cycle is constant for a particular solenoid; therefore, p is a parameter and can assume all values between 0 and 1. Written in inequality form, $0 \leq p \leq 1$. We now have all needed information and can write the complete form of the pdf:

$$P(x) = \begin{cases} p(1 - p)^{x-1} & x = 1, 2, \ldots ; 0 \leq p \leq 1 \\ 0 & \text{elsewhere} \end{cases}$$

COMMENT $P(x)$ gives us all we need to know about the probability of X. The pdf has one parameter p; its range is 0 to 1; the variable X can take on the values $x = 1, 2, \ldots$; and if we put a value of p and x into $P(x)$, we get the probability that $X = x$. The pdf should include the value of $P(x)$ for all excluded values of the variable, thus the statement $P(x) = 0$ *elsewhere*.

Problems P6.2, P6.3

6.3 Discrete pdf, Continuous pdf, and Graphs

A pdf is commonly denoted by the symbol $f(x)$; that is, the pdf is a *function f* of the random variable X. There are two types of random variables, as you learned in Sec. 1.6:

1. *Discrete*: A variable which can assume only specific point values, for example, $x = 0, 1, 2, 3$. (Fig. 6.1). No other values of x can be observed. If $f(x)$ is calculated, the result is $P(X = x)$, the probability density at $X = x$.

Figure 6.1 Variable range of the discrete variable X, where $x = 0, 1, 2, 3$.

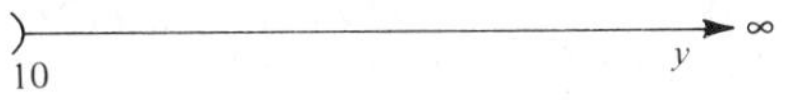

Figure 6.2 Variable range of the continuous variable Y, where $y > 10$.

2. *Continuous*: A variable which can assume any value within a specific, defined range, for example, $y > 10$ (Fig. 6.2). It is possible to observe a value of $y = 11.294$ as well as $y = 11.000$.

The form of a continuous variable pdf is similar to that of a discrete pdf, except that the *variable range is continuous*. Therefore, the continuous variable range is written in an inequality form, such as $y > 10$. We will use the symbol $f(\ )$ for a discrete or continuous pdf.

Example 6.5 An engineer has developed the pdf for (*a*) the pressure P in a vessel at any given time and (*b*) the number N of pressure vessels not working at a given time. The following is known about the pdf of P in pounds per square inch:

$$f(p) = \begin{cases} \frac{1}{50} & 250 \leq p \leq 300 \\ 0 & \text{elsewhere} \end{cases}$$

The following is known about the pdf of N in number of vessels:

$$f(n) = \begin{cases} \dfrac{e^{-\lambda}\lambda^n}{n!} & n = 0, 1, 2, \ldots; \lambda > 0 \\ 0 & \text{elsewhere} \end{cases}$$

In $f(n)$, n is the observed value of N, λ is a parameter, e is the constant 2.71828, and $n! = n(n-1)(n-2) \ldots (1)$ is n factorial (Sec. 7.2). Which is the discrete pdf and which is the continuous pdf?

SOLUTION $f(n)$ is *discrete* because N is the variable and it can assume only discrete, integer values. The pdf $f(p)$ is *continuous* since P is continuous over the range $250 \leq p \leq 300$.

COMMENT Do not get confused about $f(n)$. The letter λ is a parameter and does not make the variable N continuous. How can you tell the variable from the parameter? Use $f(n)$ as your guide. Since the letter n is in the parentheses, N is the random variable. Anything else in the mathematical form of $f(n)$ will be a parameter or a constant.

The best way to remember the difference between a discrete and a continuous variable is to picture the difference graphically. The graph of the pdf of a discrete variable will have a series of *vertical lines* only at the defined values of the variable. You plotted several discrete pdfs in Sec. 2.2. The graph of a continuous pdf will have a *continuous, nonvertical line or curve* between the extreme limits of the variable.

Example 6.6 The variables X and Y have the following pdf forms:

$$f(x) = \begin{cases} \left(\frac{1}{2}\right)^x & x = 1, 2, \ldots \\ 0 & \text{elsewhere} \end{cases}$$

$$f(y) = \begin{cases} \frac{1}{3} & 1 \le y \le 4 \\ 0 & \text{elsewhere} \end{cases}$$

State which of these variables is discrete and which is continuous and graph the pdf for each.

SOLUTION $f(x)$ is discrete because only isolated values can be assumed. The graph is shown in Fig. 6.3. Both axes are labeled; x for the variable on the abscissa and $f(x)$ for the pdf value.

The range of y in $f(y)$ is continuous; therefore, the graph (Fig. 6.4) is a continuous horizontal line at $f(y) = \frac{1}{3}$ from $y = 1$ to $y = 4$.

COMMENT It is correct to draw a line for $f(y)$ at a value $f(y) = 0$ for $1 > y > 4$, but this is usually unnecessary. Since a pdf is always 0 outside

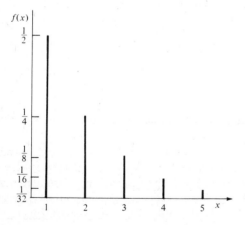

Figure 6.3 Graph of a discrete-variable pdf $f(x)$, Example 6.6.

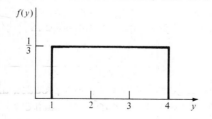

Figure 6.4 Graph of a continuous-variable pdf $f(y)$, Example 6.6.

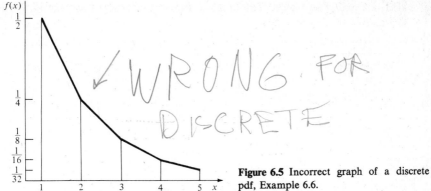

Figure 6.5 Incorrect graph of a discrete pdf, Example 6.6.

the range of the variable, we can omit the entry $f(\ ) = 0$ elsewhere from the pdf.

It is incorrect to plot the discrete pdf $f(x)$ above as in Fig. 6.5 because this indicates that $f(x)$ is defined at all values of x. But it is clear from the pdf statement that X is defined only at $x = 1, 2, \ldots$

Example 6.13

Problems P6.4–P6.7

6.4 Properties of a Discrete pdf

A discrete pdf must possess three properties:

- 1. $f(x) > 0$ for all values of x at which $f(x)$ is defined
- 2. $P(a \le x \le b) = \sum_{x=a}^{b} f(x)$
- 3. $\sum f(x) = 1$, with the sum taken over all values of x where $f(x)$ is defined

Properties 1 and 2 are easily shown. For property 1 to be true, the pdf formula must produce a positive, real number when a value of x is substituted into it. Property 2 is an application of the addition law of probability. The $f(x)$ must have only one value for each defined x value; and to find the probability that the variable is between two values, you simply add the probabilities. Both properties 1 and 2 are virtually always true and present no real problem, but property 3 must actually be shown by summing $f(x)$ over *all* values of x to see if the sum is 1. If the sum is not 1, a *constant multiplier* must be included in $f(x)$ to make property 3 true.

Example 6.7 An engineer is doing an analysis of a newly tested product, the Hang-It-Up Stapler. A total of 191 people tested the stapler by each driving

1000 staples. The number of times X that there was a problem in the 1000 drives is:

Number of problems per 1000 drives, x	Number of people observing x problems
2	125
3	37
4	16
5	8
6	5
	191

The engineer found that the number of people observing x problems per 1000 drives closely follows the equation

$$\text{Number of people} = \frac{1000}{x^3}$$

For example, $1000/6^3 = 4.63 \doteq 5$. However, this engineer wants to compute probabilities, so he uses the pdf equation

$$f(x) = \frac{10}{x^3} \qquad x = 2, 3, 4, 5, 6$$

(a) Is this a true pdf? No
(b) If not, correct it.

SOLUTION (a) The three properties of a discrete pdf must be present:

1. $f(x)$ does have a positive value at all values $x = 2, 3, 4, 5, 6$.
2. $f(x)$ is single-valued at all x values; therefore probabilities can be summed.
3. The sum of $f(x)$ over all x must be 1.

$$\sum_{x=2}^{6} f(x) = \sum_{x=2}^{6} \frac{10}{x^3}$$
$$= 10\left(\tfrac{1}{8} + \tfrac{1}{27} + \tfrac{1}{64} + \tfrac{1}{125} + \tfrac{1}{216}\right)$$
$$= 1.903 \neq 1$$

Property 3 is not true; therefore $f(x)$ is not a legitimate pdf. The engineer is using an incorrect formula to mathematically explain probability.

(b) To correct the pdf, we need only assume that the sum in property 3 is 1.

6.5 EXPECTED VALUE, VARIANCE, AND STANDARD DEVIATION OF A DISCRETE PDF

Table 6.1 Comparison of $f(x)$ and $P(x)$, Example 6.7

x	$P(x)$	$f(x)$
2	125/191 = 0.654	5.255/8 = 0.657
3	37/191 = 0.194	5.255/27 = 0.195
4	16/191 = 0.084	5.255/64 = 0.082
5	8/191 = 0.042	5.255/125 = 0.042
6	5/191 = 0.026	5.255/216 = 0.024
	1.000	1.000

This is done by placing a constant multiplier c in the sum, setting the sum equal to 1, and solving for c:

$$\sum_{x=2}^{6} cf(x) = \sum_{x=2}^{6} c\frac{10}{x^3} = 1$$

By using the result of part (a), this may be written

$$c = \frac{1}{\sum_{x=2}^{6} \frac{10}{x^3}} = \frac{1}{1.903} = 0.5255$$

After the substitution of c the pdf is

$$f(x) = \frac{5.255}{x^3} \quad x = 2, 3, 4, 5, 6$$

COMMENT To check if the results of $f(x)$ give reasonable values, we compare the outcomes of $f(x)$ with actual probability computation, using the equation

$$P(x) = \frac{\text{number of persons observing } x \text{ problems}}{\text{total number of persons}}$$

This comparison (Table 6.1) shows the results of $f(x)$ to be very close to $P(x)$. Note that the sums of the $P(x)$ and $f(x)$ columns are both 1. Often the sum of the values may not be exactly 1 because of roundoff error.

Problems P6.8–P6.12

6.5 Computation of the Expected Value, Variance, and Standard Deviation of a Discrete pdf

You learned how to compute the average $\overline{X}$, the variance s^2, and standard deviation s of actual data in Chap. 3. Since a pdf explains the distribution of probability, it is possible to obtain a general expression for these properties. All three expressions can be derived for the general pdf and then the numerical value of the parameters substituted into the general form to get the exact value of the property. See the Solved Problems for an example.

Now that we are interested in distributions rather than actual data, we will use these symbols:

Actual data symbol	Distribution symbols
$\overline{X}$	$E[X]$ or μ
s^2	$V(X)$ or σ^2
s	$\sqrt{V(X)}$ or σ

The transformation from $\overline{X}$ to $E[X]$ is easy. For observed frequency data we use Eq. (3.2):

$$\overline{X} = \frac{\sum f_i x_i}{n}$$

But f_i/n is the fraction of the time that $X = x_i$, so it is an observed probability which we can call $P(x_i)$. Then

$$\overline{X} = \sum x_i P(x_i)$$

Since $f(x_i)$ is a mathematical way of expressing $P(x_i)$, by substitution

$$\overline{X} = \sum x_i f(x_i)$$

If we drop the subscripts and rename the average as the *mean* μ, or *expected value of X*, $E[X]$, we have

$$\mu = E[X] = \sum x f(x)$$

The general expressions for all properties are easily computed by using relations for expected values, $E[u(X)]$, which take the general form

$$E[u(X)] = \sum u(x) f(x) \tag{6.3}$$

where $u(X) =$ some function of the random variable X
$f(x) =$ discrete pdf
$E[u(X)] =$ expected value of the function $u(X)$

We will use the following notation and formulas for properties of a pdf:

1. $E[X] = \mu$ is the mean of the variable X. $E[X]$ is called the *expected value*. Here the function $u(X) = X$ itself:

$$\mu = E[X] = \sum x f(x) \tag{6.4}$$

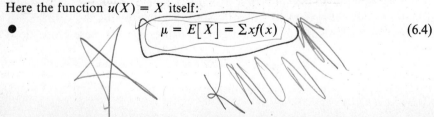

6.5 EXPECTED VALUE, VARIANCE, AND STANDARD DEVIATION OF A DISCRETE PDF

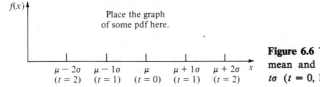

Figure 6.6 The important points of mean and standard deviation, $\mu \pm t\sigma$ ($t = 0, 1, 2$).

2. $V[X] = \sigma^2$ is the *variance* of the variable X. If $u(X) = (X - \mu)^2$, then

$$V[X] = E[(X - \mu)^2] \tag{6.5}$$

A little algebra simplifies this to the form

• $$V[X] = E[X^2] - \mu^2 \tag{6.6}$$

Expansion using Eq. (6.3) gives us the computational form of the variance:

$$V[X] = \Sigma x^2 f(x) - \mu^2 \tag{6.7}$$

3. $\sqrt{V[X]} = \sigma$ is the standard deviation of the variable X. This is commonly used to determine the probability in the range $\mu \pm t\sigma$ ($t > 0$) as pictured in Fig. 6.6. Computationally

• $$\sigma = [\Sigma x^2 f(x) - \mu^2]^{1/2} \tag{6.8}$$

or

$$\sigma = \sqrt{\sigma^2} \tag{6.9}$$

Example 6.8 The manager of a food store has collected inventory data on cling peaches for the last 30 weeks. The pdf below has been determined for X, the number of cases per week.

$$f(x) = \begin{cases} \frac{1}{3} & x = 3 \text{ cases per week} \\ \frac{1}{4} & x = 7 \text{ cases per week} \\ \frac{1}{3} & x = 10 \text{ cases per week} \\ \frac{1}{12} & x = 12 \text{ cases per week} \end{cases}$$

(a) Compute the mean number of cases per week in stock.
(b) Compute the variance and standard deviation of the number of cases per week.
(c) Plot the pdf and expected value and indicate the standard deviation on the pdf.

SOLUTION First check to see that $f(x)$ is a pdf by using the three properties of Sec. 6.4. You should demonstrate to yourself that all these properties are correct for $f(x)$.

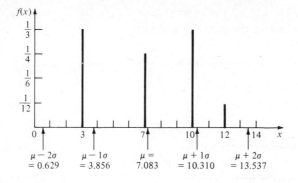

Figure 6.7 Graph of discrete $f(x)$ with $\mu = 7.083$ and $\sigma = 3.227$, Example 6.8.

(a) The expected value is computed by using Eq. (6.4):

$$\mu = \Sigma xf(x) \qquad x = 3, 7, 10, 12$$
$$= 3\left(\tfrac{1}{3}\right) + 7\left(\tfrac{1}{4}\right) + 10\left(\tfrac{1}{3}\right) + 12\left(\tfrac{1}{12}\right)$$
$$= 7.083 \text{ cases per week}$$

(b) The variance σ^2 is computed by using Eq. (6.7):

$$\sigma^2 = \Sigma x^2 f(x) - \mu^2 \qquad x = 3, 7, 10, 12$$
$$= 3^2\left(\tfrac{1}{3}\right) + 7^2\left(\tfrac{1}{4}\right) + 10^2\left(\tfrac{1}{3}\right) + 12^2\left(\tfrac{1}{12}\right) - (7.083)^2$$
$$= 60.583 - 50.169$$
$$= 10.414 \text{ (cases per week)}^2$$

Using Eq. (6.9), the standard deviation is

$$\sigma = 3.227 \text{ cases per week}$$

(c) The pdf, μ, and the points $\mu \pm t\sigma$ ($t = 1, 2$) are plotted in Fig. 6.7.

Example 6.14

Problem P6.13–P6.18

6.6 Properties of a Continuous pdf

It is simple to mentally envision the progression from a discrete to a continuous pdf. If the x axis of the variable is broken into smaller and smaller cells, there will be an increasing number of cells in the histogram. In the limit, we have an infinitesimally small interval dx, and we define a function $f(x)$ which will help compute the probability that $x \leq X \leq x + dx$. That is,

$$P(x \leq X \leq x + dx) = f(x)\, dx$$

Therefore, $f(x)\, dx$ is an expression for the pdf of the continuous random variable X. The actual value of $f(x)$ is not a probability; it is only a numerical value of the *density* at $X = x$. Using a mathematical formula, a value for $f(x)$ can be determined anywhere in the range $a \leq x \leq b$.

The three properties of a continuous pdf are similar to those of a discrete pdf, except the integral is used to accumulate area under the continuous pdf:

- 1. $f(x) \geq 0$ for $-\infty \leq x \leq \infty$
- 2. $P(a \leq x \leq b) = \int_a^b f(x)\, dx$
- 3. $\int_{-\infty}^{\infty} f(x)\, dx = 1$

The variable range does not necessarily include the entire real number line, $-\infty \leq x \leq \infty$. Therefore $f(x)$ will be 0 outside the variable range, and the integral in properties 2 and 3 will accumulate no area. Since the integral at a point is 0, the equals signs in property 2 can be excluded or included as desired. The fact that the point integral is 0 also explains why $f(x)$ at a point does *not* give actual probability for a continuous pdf.

As with the discrete pdf properties, property 2 implies a single-valued $f(x)$ and property 3 is used to be sure that the integral under the pdf is 1. If property 3 is not fulfilled, a constant multiplier c can be used to normalize the area to 1.

Example 6.9 A reliability consultant has studied the failure time T in operating hours of a certain electrical component. This consultant believes that the mathematical expression

$$\exp(-0.01t) \qquad t > 0 \qquad (6.10)$$

where exp( ) is used to indicate a power of the constant $e = 2.71828$, will give a close approximation to the pdf for failure at time $T = t$. Determine whether Eq. (6.10) can be used as a pdf, and if not, determine the form of the pdf.

SOLUTION We can look at Fig. 6.8 and see that this is not a pdf, because at $t = 0$ the expression $\exp(-0.01t) = 1.0$, thus indicating that the integral of Eq. (6.10) from 0 to infinity will exceed 1.0. This will violate property 3 for a continuous pdf.

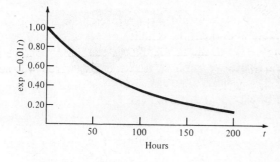

Figure 6.8 Graph of a possible probability curve for component failure time, Example 6.9.

124 INTRODUCTION TO PROBABILITY DENSITY FUNCTIONS

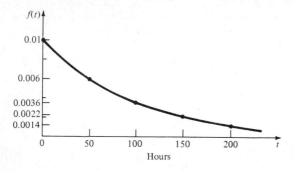

Figure 6.9 Graph of the failure pdf for an electrical component, Example 6.9.

To determine the correct pdf, introduce the multiplier c and integrate, assuming that property 3 is true.

$$\int_0^\infty ce^{-0.01t}\,dt = -\frac{c}{0.01}\left[e^{-0.01t}\right]_0^\infty = 1$$

$$-\frac{c}{0.01}(0-1) = 1$$

$$c = 0.01$$

With substitution of c, the complete pdf is

$$f(t) = \begin{cases} 0.01\exp(-0.01t) & t > 0 \\ 0 & \text{elsewhere} \end{cases} \qquad (6.11)$$

The graph is shown in Fig. 6.9. Since $f(t)$ is greater than or equal to 0 and single-valued for $t > 0$, it is a continuous pdf.

COMMENT In Eq. (6.11) there is no contribution to the integral for $-\infty \leq t \leq 0$, since the component cannot fail prior to activation. The complete $f(t)$ may be interpreted as the instantaneous chance that the component fails, that is, the chance it fails in a time t to $t + dt$.

Problems P6.19–P6.23

6.7 Computation of Expected Value, Variance, and Standard Deviation of a Continuous pdf

The formulas used to compute the measures of μ, σ^2, and σ for a continuous pdf are the same as those for a discrete pdf except that the integral is taken. In all the equations below the integrating range is shown as $-\infty$ to $+\infty$. However, positive contribution will occur only over the variable range in $f(x)$. The general

6.7 EXPECTED VALUE, VARIANCE, AND STANDARD DEVIATION OF A CONTINUOUS PDF

expected value form is

$$E[u(X)] = \int_{-\infty}^{\infty} u(x)f(x)\,dx$$

where $u(X)$ = function of the variable X
$f(x)$ = continuous pdf

The measures for a continuous pdf are defined as follows:

1. $E[X] = \mu$ is the *expected value* of X:

$$\mu = E[X] = \int_{-\infty}^{\infty} xf(x)\,dx \qquad (6.12)$$

2. $V[X] = \sigma^2$ is the *variance* of X. If we define $u(X) = (X - \mu)^2$, then

$$V[X] = E[(X - \mu)^2] = E[X^2] - \mu^2$$

Computationally,

$$\sigma^2 = \int_{-\infty}^{\infty} x^2 f(x)\,dx - \mu^2 \qquad (6.13)$$

where μ^2 is the square of the value μ in Eq. (6.12).

3. $\sqrt{V[X]} = \sigma$ is the *standard deviation* of X. Computationally,

$$\sigma = \left[\int_{-\infty}^{\infty} x^2 f(x)\,dx - \mu^2\right]^{1/2} \qquad (6.14)$$

or

$$\sigma = \sqrt{\sigma^2} \qquad (6.15)$$

It is easy to forget the x or x^2 under the integral sign when computing $E[X]$ or $E[X^2]$, respectively. This mistake will result in $\int f(x)\,dx = 1$, which is merely a verification of property 3 in Sec. 6.6.

Example 6.10 The Lit-Up Electric Company has one central computer facility and 45 terminals. An employee may use the terminal for any job that takes less than 60 s of processing time. The systems engineer wants to know the mean and standard deviation of the processing time for jobs submitted at the terminals. The engineer has determined the pdf of processing time T to take the form

$$f(t) = \frac{10 + t}{2250} \qquad 10 \leq t \leq 60 \text{ s}$$

(a) Compute the expected processing time.
(b) Compute the standard deviation.
(c) Plot $\mu \pm 2\sigma$ on the pdf.
(d) Compute the probability that processing time is between $\mu - 2\sigma$ and $\mu + 2\sigma$ s.

126 INTRODUCTION TO PROBABILITY DENSITY FUNCTIONS

SOLUTION (a) Use Eq. (6.12) to compute the expected value:

$$\mu = \int_{10}^{60} t \frac{10 + t}{2250} \, dt$$

$$= \frac{1}{2250} \left[\frac{10t^2}{2} + \frac{t^3}{3} \right]_{10}^{60}$$

$$= 39.630 \text{ s}$$

(b) Use Eq. (6.14) for the standard deviation:

$$\sigma = \left[\int_{10}^{60} t^2 \frac{10 + t}{2250} \, dt - (39.630)^2 \right]^{1/2}$$

$$= \left\{ \frac{1}{2250} \left[\frac{10t^3}{3} + \frac{t^4}{4} \right]_{10}^{60} - 1{,}570.537 \right\}^{1/2}$$

$$= (1757.407 - 1570.537)^{1/2}$$

$$= 13.670 \text{ s}$$

(c) Figure 6.10 is a plot of the pdf and the points $\mu \pm 2\sigma = 66.970$ and 12.290.

(d) Using property 2 of a continuous pdf, the probability that the processing time is in the range $\mu - 2\sigma \le t \le \mu + 2\sigma$ (Fig. 6.10, hatched area) is

$$P(12.290 \le t \le 66.970) = \int_{12.290}^{66.970} f(t) \, dt$$

The upper limit of 66.970 is out of the range of t, so the integral above $t = 60$ is 0.

$$P(12.290 \le t \le 66.970) = \int_{12.290}^{60.000} \frac{10 + t}{2250} \, dt$$

$$= \frac{1}{2250} \left[10t + \frac{t^2}{2} \right]_{12.290}^{60.000}$$

$$= 0.978$$

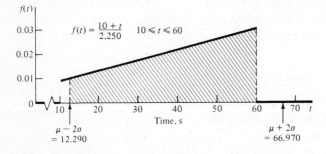

Figure 6.10 Graph of the pdf for computer terminal processing time and area $\mu - 2\sigma \le t \le \mu + 2\sigma$, Example 6.10.

There is a 97.8 percent chance that the job will deviate from the mean by no more than $2\sigma = 27.34$ s.

COMMENT It is a good idea to always check that an unfamiliar relation is a pdf. Properties 1 and 2 are true for $f(t)$ by observation. As for property 3, you should now compute the integral to see if it equals 1.

Problems P6.24–P6.27

6.8 Cumulative Distribution Function (cdf)

An equivalent way to describe the distribution of probability is to state the complete form of the *cumulative distribution function*, or *cdf*. The cdf is identified by $F(x)$. As the name implies, $F(x)$ accumulates the probability over the range of the variable. The formula is

$$F(x) = P(X \leq x)$$

where x is in the defined range of X. If the range of X is such that $a \leq x \leq b$ ($-\infty < a \leq b < \infty$), it is always true that

$$F(x < a) = 0 \quad \text{and} \quad 0 < F(a \leq x \leq b) < 1 \quad \text{and} \quad F(x \geq b) = 1$$

This is true because the probability below a or above b is 0.

The cdf is used for pdfs much as the cumulative relative frequency is used for observed data. The cdf is defined differently for discrete and continuous variables.

- Discrete:
$$F(x) = \sum_{a}^{x} P(x) \quad x = a, a+1, a+2, \ldots, b \tag{6.16}$$

- Continuous:
$$F(x) = \int_{a}^{x} f(x)\, dx \quad a \leq x \leq b \tag{6.17}$$

It is possible to make the following two statements concerning the cdf of a continuous variable X:

- $$P(x_1 \leq x \leq x_2) = F(x_2) - F(x_1)$$
- $$f(x) = \frac{d}{dx} F(x) \tag{6.18}$$

The probability that a discrete variable X is between the two values x_1 and x_2 may be computed as

$$P(x_1 \leq x \leq x_2) = F(x_2) - F(x_1 - 1) \tag{6.19}$$

$$= \sum_{x=x_1}^{x_2} f(x)$$

You should realize that in Eq. (6.19) the range $x_1 \leq x \leq x_2$ includes only the discrete points of definition of X; therefore, it is necessary to subtract the cdf value at $x_1 - 1$ to include x_1 in the probability computation. Example 6.15, which is a case study for a discrete variable, illustrates this fact.

Example 6.11 The pdf determined in Example 6.9 is
$$f(t) = 0.01 \exp(-0.01t) \quad t > 0$$
Use it to compute (a) $P(t \leq 10 \text{ h})$ and (b) $P(t \geq 250 \text{ h})$.

SOLUTION (a) Equation (6.17) is used to compute $F(t = 10)$:
$$F(t = 10) = \int_0^{10} 0.01 \exp(-0.01t) \, dt$$
$$= [-\exp(-0.01t)]_0^{10}$$
$$= 1 - 0.905 = 0.095$$

(b) We know that for a continuous pdf
$$P(t \geq 250) = 1 - P(t \leq 250) = 1 - F(t = 250)$$
We compute the cdf at $t = 250$ h using the cdf expression above:
$$F(t = 250) = [-\exp(-0.01t)]_0^{250}$$
$$= 1 - 0.082 = 0.918$$
Therefore, $\quad P(t \geq 250) = 1 - 0.918 = 0.082$

In conclusion, there is a 9.5 percent chance of failure prior to 10 h and an 8.2 percent chance of failure after 250 h. The remaining 82.3 percent chance is between 10 and 250 h.

A word of caution at this point. Most of the problems you will encounter in work with distribution properties are in algebra or simple calculus. A review of the basic relations in these areas will be very helpful.

Example 6.15 (Case Study)

Problems P6.28–P6.34

6.9 Best Estimators for Parameters

In your use of statistical distributions, you will often use observed data to estimate a parameter. The formula used should be the best possible to estimate the desired value. For example, the formula below for $\hat{\sigma}^2$ (read "sigma hat squared") is used to estimate the variance σ^2, which is a parameter in some distributions.

$$\hat{\sigma}^2 = \frac{\Sigma(X_i - \bar{X})^2}{n - 1} \tag{6.20}$$

Is this the best estimator of σ^2? If so, what makes this formula the best one? A *best estimator* must have three properties, which are summarized below using the notation

$$\theta = \text{parameter being estimated}$$
$$\hat{\theta} = \text{best estimator of } \theta$$

1. $\hat{\theta}$ must be a *consistent estimator* of θ. This means that it is possible to take a sample size n large enough so that the absolute difference $|\theta - \hat{\theta}|$ is smaller than some small value ϵ with a probability of at least β, that is,

$$P\big[|\theta - \hat{\theta}| < \epsilon\big] > \beta \qquad \text{if } n > n_0$$

Here n_0 is the minimum sample to obtain the probability β. Most estimators are consistent.

2. $\hat{\theta}$ must be an *unbiased estimator* of θ. To be unbiased, the expected value of $\hat{\theta}$ must equal θ, that is

$$E[\hat{\theta}] = \theta$$

If we take $E[\hat{\sigma}^2]$ in Eq. (6.20), we get σ^2. The use of n in the denominator (as discussed in Sec. 3.5) would result in an expected value of

$$\frac{n-1}{n}\sigma^2$$

which is a biased estimator of σ^2.

3. $\hat{\theta}$ must be an *efficient estimator* of θ. The variance of an efficient estimator $\hat{\theta}$ must be smaller than the variance of any other estimator $\hat{\theta}'$, that is,

$$V[\hat{\theta}] \leq V[\hat{\theta}']$$

An efficient estimator is both consistent and unbiased.

Therefore, if we state a formula for the best estimate, we are stating the consistent, unbiased, efficient estimator. The most popular method used to derive best estimators is the maximum likelihood method, which is discussed in many texts on mathematical statistics. We will always use the best estimator available for all computations.

Problems P6.35, P6.36

*6.10 Moment Generating Functions

In Sec. 4.1 we learned that the first moment about the origin μ_1 is the mean of observed data. Since $\mu = E[X]$ is the expected value of a pdf, we can state that

Average = expected value of X = first moment about the mean

$$\overline{X} = E[X] = \mu_1$$

Besides μ_1 we have studied the second moment, $\mu_2 = E[X^2]$, which is used in the computational form of the variance:

$$V[X] = E[X^2] - \mu^2$$
$$= \mu_2 - \mu_1^2$$

Now we will study the *moment generating function (mgf)*, which is well named because it is used to *generate moments* about the origin for a pdf. The mgf $M(t)$ is defined for all real values of t:

$$M(t) = \text{expected value of the variable } e^{tX}$$
$$= E[e^{tX}] \qquad (6.21)$$

The mgf is unique for each pdf. It is computed as follows:

Discrete:
$$M(t) = \sum_x e^{tx} f(x) \qquad (6.22)$$

Continuous:
$$M(t) = \int_x e^{tx} f(x)\, dx \qquad (6.23)$$

The kth moments about the origin are generated from $M(t)$ by taking the kth derivative with respect to t and evaluating at $t = 0$. Then, μ_1, μ_2, and $V[X]$ are

First moment:
$$\mu_1 = E[X] = M'(t = 0) \qquad (6.24)$$

Second moment:
$$\mu_2 = E[X^2] = M''(t = 0) \qquad (6.25)$$

Variance:
$$V[X] = \mu_2 - \mu_1^2$$
$$= M''(t = 0) - [M'(t = 0)]^2 \qquad (6.26)$$

Example 6.12 shows the use of the mgf for a continuous pdf.

Example 6.12 A consultant has placed 1000 switches on test and found that the pdf for failure after X h of continuous operation is

$$f(x) = 0.002 \exp[-0.002x] \qquad x > 0$$

Compute the expected value and standard deviation using the mgf, and plot them on a graph of the pdf.

SOLUTION Using Eq. (6.23), we have

$$M(t) = \int_0^\infty e^{tx} 0.002 e^{-0.002x}\, dx$$
$$= 0.002 \int_0^\infty e^{-x(0.002 - t)}\, dx$$
$$= \frac{0.002}{0.002 - t} \qquad t < 0.002$$

To determine the expected value μ, take the first derivative and evaluate at

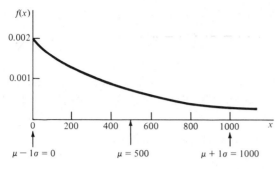

Figure 6.11 Graph of the pdf for failure of a switch, Example 6.12.

$t = 0$, as in Eq. (6.24):

$$M'(t) = 0.002(0.002 - t)^{-2}$$

$$\mu = M'(t = 0) = \frac{1}{0.002} = 500 \text{ h}$$

For the standard deviation use Eqs. (6.25) and (6.26) in succession.

$$M''(t) = 2(0.002)(0.002 - t)^{-3}$$

$$M''(0) = 0.004(0.002)^{-3} = 500{,}000$$

$$V[X] = M''(t = 0) - [M'(t = 0)]^2$$

$$= 500{,}000 - (500)^2$$

$$= 250{,}000 \text{ h}^2$$

Then

$$\sigma = \sqrt{V[X]} = 500 \text{ h}$$

The pdf, μ, and $\mu \pm 1\sigma$ are graphed in Fig. 6.11. The interpretation is as follows: The average number of operating hours to failure is $\mu = 500$ h, and the standard deviation of the operating time is $\sigma = 500$ h.

Problems P6.37–P6.39

SOLVED PROBLEMS

Example 6.13 A traffic technician has counted the number of buses that pass at two widely separated locations on an interstate highway. He has found that the number of buses B that pass per quarter hour at both locations follows the pdf

$$f(b) = e^{-\lambda}\frac{\lambda^b}{b!} \qquad b = 0, 1, 2, \ldots; \lambda > 0 \qquad (6.27)$$

At location A the parameter $\lambda = 0.3$, and at location D, $\lambda = 0.5$. Determine (a) the variable involved, (b) whether it is discrete or continuous, and (c) the graph for the pdf at both locations.

SOLUTION (*a*) The variable is B, the number of buses passing per quarter hour. The pdf will compute the probability of a certain number of buses passing in 15 min.

(*b*) Since $b = 0, 1, 2, \ldots$, the variable B is discrete. It is impossible to observe a value such as $b = 0.789$ bus.

(*c*) The different λ values do not change the variable in any way; it is still discrete. The graph of $f(b)$ will change when λ goes from 0.3 to 0.5. We first compute the value of $f(b)$ for several values of b and the two different values of λ (Table 6.2). See the comment if you are rusty on numerical substitution into $f(b)$. The graphs are shown in Fig. 6.12. Note that the probability has been shifted toward the right for $\lambda = 0.5$, but the pdf has the same general shape.

COMMENT The pdf $f(b)$ is defined for all positive integer values $0, 1, 2, 3, \ldots$, but the probability values get very small as b increases (Table 6.2). The values of $f(b)$ are found by substitution of λ and b into Eq. (6.27). For example,

$$\lambda = 0.3, b = 0: \quad f(b) = e^{-0.3} \frac{0.3^0}{0!} = 0.741$$

$$\lambda = 0.5, b = 2: \quad f(b) = e^{-0.5} \frac{0.5^2}{2!}$$

$$= 0.607 \frac{0.25}{2} = 0.076$$

Section 6.3

Example 6.14 An industrial engineer (IE) has studied the probability distribution for the occurrence time X of a machine breakdown in whole days. The pdf is shown in Fig. 6.13 where X has the range $a = 22, 23, \ldots, 44$, $b = 45$. The industrial engineer knows that this is a commonly occurring pdf, and it has an average and standard deviation which may be expressed

Table 6.2 Computation of $f(b)$ for $\lambda = 0.3$ and $\lambda = 0.5$, Example 6.13

	$f(b)$	
b	Location A $\lambda = 0.3$	Location D $\lambda = 0.5$
0	0.741	0.607
1	0.222	0.303
2	0.033	0.076
3	0.003	0.013

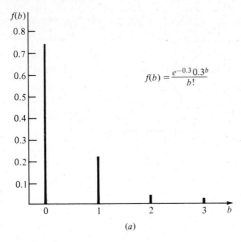

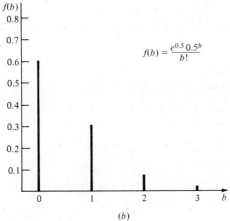

Figure 6.12 Graph of pdf $f(b)$ for (a) $\lambda = 0.3$ and (b) $\lambda = 0.5$, Example 6.13.

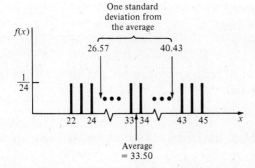

Figure 6.13 Graph of a discrete pdf $f(x)$ with the average and 1 standard deviation plotted.

by the general equations

$$\mu = \tfrac{1}{2}(a + b)$$

$$\sigma = \left[\frac{(b - a + 1)^2 - 1}{12}\right]^{1/2}$$

What values of μ and σ should the IE use?

SOLUTION The IE should substitute a and b into the equations:

$$\mu = \tfrac{1}{2}(22 + 45) = 33.50 \text{ days}$$

$$\sigma = \left[\frac{(45 - 22 + 1)^2 - 1}{12}\right]^{1/2} = 6.93 \text{ days}$$

The values of μ and $\mu \pm 1\sigma$ are plotted on the pdf (Fig. 6.13).

Section 6.5

Example 6.15 (Case Study) Samples of size 50 were taken to test the strength of a new glueing operation. The number of breaks X per sample for 24 samples is:

Number of breaks X	Samples with X breaks
0	12
1	8
2	3
3	1

Determine the following information:
(a) pdf statement in complete form
(b) Expected value and variance
(c) Graph of the cdf
(d) Probability that the number of breaks is one or two per sample. (Use the cdf to determine this probability.)

SOLUTION (a) The pdf $f(x)$ is for a discrete variable X. The probability values are determined by dividing the frequency values by 24. There are no parameters in $f(x)$.

$$f(x) = \begin{cases} 0.500 & x = 0 \\ 0.333 & x = 1 \\ 0.125 & x = 2 \\ 0.042 & x = 3 \end{cases}$$

This is a pdf because the three properties of a discrete distribution are present.
(b) The expected value is computed by Eq. (6.4):

$$\mu = E[X] = \sum_{x=0}^{3} xf(x) = 0.709$$

Use Eq. (6.7) to compute the variance:

$$V[X] = \sum_{x=0}^{3} x^2 f(x) - \mu^2$$

$$= 0(0.5) + 1(0.333) + 4(0.125) + 9(0.042) - (0.709)^2$$
$$= 1.211 - 0.503$$
$$= 0.708$$

(c) The cdf $F(x)$ is computed by using Eq. (6.16) and graphed in Fig. 6.14.

$$F(x) = \begin{cases} 0.500 & 0 \le x < 1 \\ 0.833 & 1 \le x < 2 \\ 0.958 & 2 \le x < 3 \\ 1.000 & 3 \le x \end{cases}$$

(d) From Fig. 6.14 and Eq. (6.19)

$$P(1 \le x \le 2) = F(x = 2) - F(x = 0)$$
$$= 0.958 - 0.500$$
$$= 0.458$$

This value can be verified by adding $f(x = 1)$ and $f(x = 2)$ from the pdf.

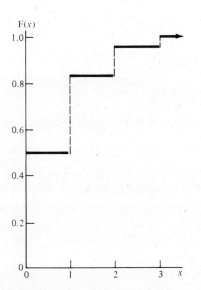

Figure 6.14 Graph of the cdf for Example 6.15.

ADDITIONAL MATERIAL

Discrete and continuous distribution fundamentals (Secs. 6.3 to 6.8): Bowker and Lieberman [4], Chap. 2; Gibra [8], Chap. 2; Hines and Montgomery [12], Chaps. 2 and 3; Walpole and Myers [25], Chap. 2; Wilks [26], Chaps. 2 and 3.

Estimator properties (Sec. 6.9): Beck and Arnold [1], Chap. 3; Bowker and Lieberman [4], Chap. 8; Gibra [8], pp. 309–321; Hines and Montgomery [12], pp. 225–234; Hoel [13], Chap. 8; Wilks [26], Chap. 12.

Generating functions (Sec. 6.10): Gibra [8], pp. 118–123; Hogg and Craig [14], pp. 40–46; Wilks [26], Chap. 5.

PROBLEMS

The following distributions are used in various problems in this chapter.

A: $\quad f(x) = \dfrac{x}{10} \quad x = 0, 1, 2, 3, 4$

Problems P6.1, P6.7, P6.11, P6.13, P6.30

B: $\quad f(x) = \begin{cases} \frac{3}{8} & x = 3 \\ \frac{2}{8} & x = 4 \\ \frac{2}{8} & x = 6 \\ \frac{1}{8} & x = 7 \end{cases}$

Problems P6.1, P6.7, P6.14, P6.31

C: $\quad f(z) = \lambda e^{-\lambda z} \quad z > 0; \lambda > 0$

Problems P6.7, P6.24, P6.33, *P6.37

D: $\quad f(y) = \frac{1}{5} \quad 5 \leq y \leq 10$

Problems P6.7, P6.26, *P6.38

E: $\quad f(y) = 2y \quad 0 \leq y \leq 1$

Problems P6.7, P6.23, P6.27

P6.1 Determine the probability that $x = 3$ or 4 using (*a*) distribution A and (*b*) distribution B.

P6.2 State the definition of a parameter as it is used in a statistical distribution.

P6.3 The distance X of actual bomb hits from the aimed-at point often follows a (Rayleigh) distribution of the form

$$\frac{x}{a} \exp\left(-\frac{x^2}{2a}\right)$$

where a must exceed 0 and $x \geq 0$. Write the complete pdf including the variable range and parameter range.

P6.4 Plot the pdf in Prob. P6.3 for (a) $a = 3$ and (b) $a = 5$ on the same graph.

P6.5 A quality manager plans to use the pdf

$$f(x) = 0.1(0.9)^x \qquad x = 0, 1, 2, \ldots$$

where x is the number of items inspected prior to finding the first defective item.
(a) What is the variable and its range? Is the variable discrete or continuous? Why?
(b) Graph $f(x)$.

P6.6 A statistical analyst in the data processing center has found that the time T in minutes to print the output for graduate student research programs follows the pdf

$$f(t) = e^{-t} \qquad t > 0$$

(a) State the variable range. (b) Graph the pdf.

P6.7 For distributions A through E:
(a) State which are discrete and which are continuous.
(b) Plot the pdf. In distribution C let $\lambda = 1.0$.

P6.8 Two engineers have developed distributions to explain the number X of incorrectly dialed telephone calls occurring each minute in a particular city. Both pdfs have the form

$$f(x) = e^{-\lambda}\frac{\lambda^x}{x!} \qquad x = 0, 1, 2, \ldots ; \lambda > 0$$

One engineer used the parameter value $\lambda = \lambda_1 = 2.7$ and the other used $\lambda = \lambda_2 = 2.0$.
(a) Compare the pdf graphs for the two λ values.
(b) Compute $P(x \leq 2)$ for each parameter value.

P6.9 An engineer wants to develop a distribution of the form cy for the range $y = 1, 2, 3, 4$. Determine the constant c for the engineer.

P6.10 Show that the $f(x)$ in Prob. P6.5 has the three properties of a pdf.

P6.11 (a) Show that distribution A is a true pdf.
(b) Compute the probability that $x \leq 3$ using it.

P6.12 Marcie has studied the number of traffic accidents per weekend on a dangerous strip of highway. The number X follows the pdf

$$f(x) = e^{-0.9}\frac{(0.9)^x}{x!} \qquad x = 0, 1, 2, \ldots$$

(a) Plot $f(x)$.
(b) Show that it is a true pdf.

P6.13 Compute the mean and standard deviation for distribution A.

P6.14 If distribution B is used to describe the usage in days per week for a piece of heavy road equipment, compute its mean and standard deviation.

P6.15 The following number X of defects per unit of incoming material was observed in 20 samples.
(a) Determine the pdf for X.
(b) Compute its expected value.

2	4	3	5	0	1	6	3	4	2
3	1	0	5	2	4	3	2	1	3

P6.16 Assume you have a 1 in 10 chance of winning $50 and a 4 in 10 chance of winning $10 in a turtle race. If your turtle loses, you forfeit the entrance fee of $5.
(a) Develop the pdf for your winnings X.
(b) Compute your expected winnings.

P6.17 A company has $25,000 invested in one project and $75,000 in a second. The first has a 50 percent chance of returning a total of $45,000 and the second a 25 percent chance of returning $125,000. If a project is not successful, half the investment is lost. Which project has the larger expected total dollar return?

P6.18 (a) Develop the pdf for the sum of the outcomes on two dice.
(b) Compute μ and σ for this pdf.

P6.19 John plans to use a pdf of the form $f(x) = c(3 - x)$ over the range $1 \leq x \leq 3$. Determine the constant c to make $f(x)$ a true pdf.

P6.20 The relation used to describe the variable P is
$$f(p) = 3(1 - p)^2 \quad 0 \leq p \leq 1$$
(a) Show that $f(p)$ is a true pdf.
(b) Plot $f(p)$.
(c) Compute $P(\frac{1}{2} \leq p \leq 1)$.

P6.21 Compute the probability that a unit will fail prior to $t = 100$ h if the instantaneous failure probability is given by Eq. (6.11).

P6.22 The state tourism department for Euphoria feels that the fraction P of its citizens taking a vacation inside the state in any year has a pdf $cp(1 - p)$, where c is a constant and $0 \leq p \leq 1$.
(a) Determine the correct value for c.
(b) Graph the resulting pdf.

P6.23 Compute the probability that $0 \leq y \leq 0.3$ for distribution E.

P6.24 Using distribution C, compute (a) μ and σ and (b) the chance that z is between $\mu \pm 1\sigma$.

P6.25 Determine $E[P]$ and $V[P]$ for $f(p)$ in Prob. P6.20.

P6.26 Plot distribution D and show that its expected value is centered on the variable range.

P6.27 (a) Compute the expected value and standard deviation for distribution E.
(b) Locate the expected value on the pdf graph.

P6.28 Write the complete cdf and construct its graph for $f(p)$ in Prob. P6.20.

P6.29 Use the distribution below to determine (a) the expected value of X, (b) the percentage of the area under the $f(x)$ curve to the left of the expected value, and (c) the equation and graph of the cdf of X.
$$f(x) = 3e^{-3x} \quad x > 0$$

P6.30 (a) Write the expression for the cdf of distribution A.
(b) Plot the pdf and cdf.

P6.31 (a) State the complete form of the cdf for distribution B.
(b) Plot this cdf.

P6.32 (a) State and graph the cdf for the distribution in Prob. P6.6.
(b) Use the cdf formula and graph to determine $P(t > 1.75 \text{ min})$.

P6.33 For distribution C with $\lambda = 0.1$:
(a) Determine $F(z)$.
(b) Plot $f(z)$ and $F(z)$.
(c) Compute $P(0 \leq z \leq 3)$ using both $f(z)$ and $F(z)$.

P6.34 The cumulative distribution function for the failure of a subassembly in T h is
$$F(t) = 1 - e^{-0.008t} \quad t > 0$$
(a) Plot $F(t)$ and estimate the probability of failure prior to 100 h.
(b) Use Eq. (6.18) to determine $f(t)$ and use it to answer the probability question posed in part (a).

P6.35 Write the mathematical relationships which describe the three properties of a best estimator and then explain the meaning of each in words.

P6.36 Consider the expression $E[(X - b)^2]$. If b is the expected value $E[X]$, this expression is the efficient estimator of the variance of X; however, if b is the median, it is not the efficient estimator. Explain why this statement is correct by using the appropriate property of a best estimator.

***P6.37** (a) Show that the moment generating function (mgf) for distribution C is $M(t) = (1 - t/\lambda)^{-1}$.

(b) Use $M(t)$ to verify that $E[Z] = 1/\lambda$.

*__P6.38__ Determine the mgf for distribution D.

*__P6.39__ Determine the mgf for the pdf

$$f(x) = 0.1 \quad 10 \leq x \leq 20$$

CHAPTER
SEVEN

COUNTING RULES AND BASICS OF THE HYPERGEOMETRIC DISTRIBUTION

This chapter has two objectives. First, you learn how to make basic computations to determine the number of ways to arrange items, using permutations and combinations. Second, you are introduced to the hypergeometric distribution and some of its applications.

CRITERIA

To complete this chapter, you must be able to do the following for permutations and combinations:

1. State and use the basic counting rule for arrangements.
2. Calculate the value of a factorial and approximate a factorial value, using Stirling's formula.
3. Define the term *permutation*, state the formula, and compute the number of permutations, given a description of the situation.
4. Define the term *combination*, state the formula, and compute the number of combinations, given a description of the situation.

You must be able to do the following for the hypergeometric distribution:

5. Derive the hypergeometric pdf, using combinations and the definition of equally probable events.

6. Determine the pdf and cdf values and graph, given the sample size, population size, number of successes, and the random variable.
7. State the parameters, and state and compute the mean and standard deviation for any hypergeometric distribution, given a statement of the situation.
8. Use the problem-solving steps to compute one or more probability values and the hypergeometric properties, given a description of the situation.
*9. Derive the expression for the hypergeometric mean, using the expected-value notation.

STUDY GUIDE

7.1 Basic Counting Rule

There is a basic principle commonly applied in determining the total number of ways to arrange items:

> If a first arrangement can be made in n ways and, after this arrangement is complete, a second can be made in r ways, the two arrangements can be made in n times r ways.

Therefore, simple multiplication of the ways to *successively* arrange items gives the total ways to arrange them all. This rule may be repeated to compute the total ways for three or more successive arrangements.

Example 7.1 Assume that machines 1 and 2 can be arranged in two ways (12 or 21), after which machines A, B, and C can be arranged in four ways (ABC, CAB, ACB, or BCA). Compute and itemize the total arrangements.

SOLUTION Since it is not possible to mix machines 1 and 2 with A, B, and C, use the principle above:

$$\text{Total ways} = 2(4) = 8 \text{ ways}$$

The possible arrangements are

12ABC	12CAB	12ACB	12BCA
21ABC	21CAB	21ACB	21BCA

7.2 Factorials and Stirling's Approximation

The term *n factorial* is written $n!$ and defined as the product of all integer values from n down to 1. Computationally,

$$n! = n(n-1)(n-2) \cdots (2)(1) \qquad (7.1)$$

By definition $0! = 1$, and if $n < 0$, the factorial does not exist.

142 COUNTING RULES AND BASICS OF THE HYPERGEOMETRIC DISTRIBUTION

The values of factorials grow rapidly as n increases. For example,

$$2! = 2 \quad 10! = 3{,}628{,}800 \quad 20! = 2.432902 \times 10^{18}$$

If factorials are divided in computations, the size of the numbers can be markedly reduced. For $r < n$, we can write

$$\frac{n!}{r!} = n(n-1)(n-2)\cdots(r+1) \tag{7.2}$$

For example, we can cancel as follows:

$$\frac{12!}{7!} = \frac{12 \cdot 11 \cdot 10 \cdot 9 \cdot 8 \cdot 7 \cdot 6 \cdot 5 \cdot 4 \cdot 3 \cdot 2 \cdot 1}{7 \cdot 6 \cdot 5 \cdot 4 \cdot 3 \cdot 2 \cdot 1} = 12 \cdot 11 \cdot 10 \cdot 9 \cdot 8$$

If you have a hand-held calculator with an $n!$ key, computation for large n values is no problem. However, actual multiplication of terms is time-consuming, so Stirling's approximation to $n!$ is used.

$$n! \cong n^n e^{-n} \sqrt{2\pi n} \tag{7.3}$$

As n increases, the error approaches 0.

Example 7.9

Problems P7.1–P7.3

7.3 Permutations

Assume that we have 10 items (e.g., balls, people, switches, etc.) labeled 1, 2, ..., 10, which are to be arranged into four different, side-by-side positions as shown:

$$\underline{\quad\quad} \quad \underline{\quad\quad} \quad \underline{\quad\quad} \quad \underline{\quad\quad}$$
$$\;\;1 \quad\quad\;\; 2 \quad\quad\;\; 3 \quad\quad\;\; 4$$

For position 1, we have a choice of 10 items. For position 2 there are nine items since the one in position 1 was not replaced. The basic counting principle of Sec. 7.1 states that there are 10(9) = 90 different arrangements possible for these two positions. We can summarize the solution for the four positions as follows:

Position number	Items remaining	Number of arrangements
1	10	10 = 10
2	9	10 · 9 = 90
3	8	10 · 9 · 8 = 720
4	7	10 · 9 · 8 · 7 = 5040

Some example arrangements of the 5040 are:

$$10\;9\;8\;7 \quad\quad 9\;10\;8\;7$$
$$3\;7\;5\;6 \quad\quad 7\;5\;3\;6$$
$$\text{etc.} \quad\quad\;\; \text{etc.}$$

7.3 PERMUTATIONS

These show that order is important, that is, 10 9 8 7 is a different arrangement than 9 10 8 7.

The ordered arrangements of $r = 4$ out of $n = 10$ items are called permutations and are symbolized by $_nP_r$. The formula used to compute the permutations of n items taken r at a time is

$$_nP_r = n(n-1)(n-2)\cdots(n-r+1)$$

This may be written

$$_nP_r = \frac{n!}{(n-r)!} \quad (7.4)$$

By definition:

A *permutation* is an *ordering* of n items without replacement into r positions with $r \leq n$.

Note that since $0! = 1$,

$$_nP_n = \frac{n!}{0!} = n!$$

Example 7.2 A chemical engineer has one each of six different styles of temperature gages to install on three new mixing tanks.
(a) How many permutations are there?
(b) How many arrangements are there if the engineer has access to a very large supply of each type of gage?

SOLUTION (a) Here $n = 6$ and $r = 3$ in Eq. (7.4):

$$_6P_3 = \frac{6!}{3!} = 6 \cdot 5 \cdot 4 = 120 \text{ arrangements}$$

(b) If a large supply of each type is available, the items are *replaced*. For each tank there are six gage types to select from. For $n = 6$ and $r = 3$,

$$\text{Total arrangements} = 6 \cdot 6 \cdot 6 = 6^3 = 216 \text{ ways}$$

COMMENT The term *permutation* is reserved to mean without replacement. The situation in (b) will be called permutations with replacement, and the number of ways is always n^r.

Example 7.3 How many license plate numbers are possible if each plate must have three nonrepeated letters followed by a three-digit number? There are 500 letter combinations which are not allowed for one reason or another.

SOLUTION For the letters we have $n = 26$ and $r = 3$ with 500 ways excluded. Total permutations for letters are

Letters: $$_{26}P_3 - 500 = \frac{26!}{23!} - 500 = 15{,}100$$

To obtain a three-digit number, we use the digits 0, 1, ..., 9, but the first position cannot be filled with a 0. The last two positions can have any of the 10 available digits.

Numbers: $9 \cdot 10 \cdot 10 = 900$

Use the counting rule for total possibilities:

$$\text{Total arrangements} = (\text{letters})(\text{numbers})$$
$$= 15{,}100(900) = 13{,}590{,}000 \text{ plates}$$

You can see from these examples that problems can easily include more than simply $_nP_r$ in their solution. The problem must be clearly understood before solution is attempted. In fact, misinterpretation of the situation is probably a more common mistake than incorrect application of formulas.

One other useful permutation formula involves the solution of all n items which are divided into k classes $(n_1, n_2, \ldots, n_k)$ in which the items are not distinguishable from one another. It is also necessary that $n_1 + n_2 + \cdots + n_k = n$. The total number of permutations is symbolized by $_nP_{n_1, n_2, \ldots, n_k}$. The formula is

$$_nP_{n_1, n_2, \ldots, n_k} = \frac{n!}{n_1! n_2! \cdots n_k!} \qquad (7.5)$$

An application is given in the Solved Problems.

Example 7.10

Problems P7.4–P7.14

7.4 Combinations

Consider the permutations of the letters a, b, c:

abc   bac   cab
acb   bca   cba

These $3! = 6$ permutations are possible because order is important, but they all include the same three letters. Therefore, neglecting order, the one arrangement is composed of a, b, and c. This is called a *combination*. A combination is a permutation with the importance of order removed. The combination of n things taken r at a time $(r \leq n)$ is

$$_nC_r = \frac{\text{number of permutations}}{\text{number of arrangements of } r \text{ items}}$$

$$= \frac{_nP_r}{r!}$$

$$_nC_r = \frac{n!}{r!(n-r)!} \qquad (7.6)$$

7.4 COMBINATIONS

By definition:

A combination is an unordered arrangement of n items into r different locations with $r \leq n$.

Before you attempt to work any counting problem, ask yourself the following question: Is order important? If you answer yes, *use a permutation*; if no, *use a combination*.

If there are to be *r or more* items in the combination, the terms can be added. For example, the number of ways to obtain either r or $r + 1$ items from n is

$$_nC_r + {}_nC_{r+1} \tag{7.7}$$

Similarly, the multiplication of combination terms

$$_{n_1}C_{r_1} \times {}_{n_2}C_{r_2} \tag{7.8}$$

will give the total ways to select r_1 items from n_1 *and* r_2 from n_2 items. Note the use of the word *or* for the plus and *and* for times. These operations are illustrated in Example 7.4.

Example 7.4 What are the number of ways to select:
(a) Three overexposed rolls of film from 10 rolls
(b) At most two defective from five rolls
(c) Seven defective from 10 rolls of film
(d) Two defective rolls from a batch of size 10 and three defective rolls from a different batch of eight
(e) Three defective rolls from a batch of 10 and at most one defective from a batch of 20 (The batch of 10 rolls is accounted for, so the three defectives are distinguishable after they are found.)

SOLUTION (a) None of the rolls is numbered, so order is not important. The combination of 3 rolls from 10 is found using Eq. (7.6):

$$_{10}C_3 = \frac{10!}{3!7!} = \frac{10 \cdot 9 \cdot 8}{3 \cdot 2 \cdot 1} = 120 \text{ ways}$$

(b) To determine the number of ways to have 0, 1, *or* 2, use the addition rule of Eq. (7.7):

$$\sum_{r=0}^{2} {}_5C_r = 1 + 5 + 10 = 16 \text{ ways}$$

(c) For $r = 7$ and $n = 10$ we have

$$_{10}C_7 = \frac{10!}{7!3!} = 120 \text{ ways}$$

This combination is the same as in (a) because it is always true that $_nC_r = {}_nC_{n-r}$ because of the exchanging of r and $n - r$.
(d) To find the ways to select $r_1 = 2$ rolls from $n_1 = 10$ *and* $r_2 = 3$ from

$n_2 = 8$, use the multiplication rule of Eq. (7.8):

$$_{10}C_2 \times {}_8C_3 = 45 \times 56 = 2520 \text{ ways}$$

(e) To arrange $r_1 = 3$ distinguishable rolls from $n_1 = 10$, use a permutation because order is important:

$$_{10}P_3 = \frac{10!}{7!} = 720 \text{ ways}$$

To select $r_2 = 0$ or 1 roll from $n_2 = 20$, add the combinations because order is not important:

$$_{20}C_0 + {}_{20}C_1 = 21 \text{ ways}$$

The total ways to observe the permutation and the combination are the product:

$$_{10}P_3 \times ({}_{20}C_0 + {}_{20}C_1) = 720 \times 21 = 15{,}120 \text{ ways}$$

Example 7.11

Problems P7.15–P7.23

7.5 Derivation of the Hypergeometric pdf

The hypergeometric pdf is derived by using combinations to determine the probability of a certain outcome in n trials from the population N ($n \leq N$). A trial is an experiment, for example, selecting one item and checking its tolerance fit, or weighing a specimen and comparing it to design weight. If we have 10 trials from a finite population, we have taken a sample of size $n = 10$. The assumptions for the hypergeometric distribution are:

1. There are n trials in a sample from a finite population of size N ($n \leq N$).
2. Only the outcomes "success" or "failure" are possible on each trial.
3. The actual number of successes S and the number of failures $F = N - S$ in the population are known.

Figure 7.1 shows that in the sample of size n there can be f failures ($f = 0, 1, \ldots, F$) and s successes ($f + s = n$), if there are F failures and S successes in N. Use combinations to determine the number of unordered ways to get f and s. For s successes use ${}_SC_s$; for failures use ${}_FC_f$; and to get both, multiply:

$$\text{Total ways to get } s \text{ and } f = {}_SC_s \cdot {}_FC_f \tag{7.9}$$

Next determine the probability that any combination of size n can be drawn from the N items. Since each sample has an equal chance,

$$P(\text{each sample}) = \frac{1}{\text{total number of samples}} = \frac{1}{{}_NC_n}$$

7.5 DERIVATION OF THE HYPERGEOMETRIC PDF

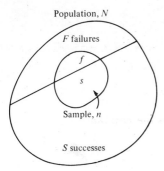

Figure 7.1 A sample of size n taken from a finite population of N items with S successes and F failures.

Because Eq. (7.9) has computed the number of ways to get s successes and f failures, we can state

$$P(s \text{ successes}, f \text{ failures}) = \frac{\text{number of samples with } s \text{ and } f}{\text{total number of samples}}$$

$$= \frac{{}_SC_s \cdot {}_FC_f}{{}_NC_n}$$

Define the variable X as the number of successes in n. The probability that $X = x$ successes with $n - x$ failures is

$$P(X = x) = \frac{{}_SC_x \cdot {}_FC_{n-x}}{{}_NC_n}$$

The variable X is *discrete* and has a *hypergeometric* distribution. To describe $P(X = x)$, we use the descriptor $h(x; n, S, N)$ to define X with a hypergeometric distribution, sample size n, and S successes in the population of N items:

$$\bullet \quad h(x; n, S, N) = \begin{cases} \dfrac{{}_SC_x \cdot {}_FC_{n-x}}{{}_NC_n} & x = 0, 1, \ldots, n \\ & F = N - S \\ 0 & \text{elsewhere} \end{cases} \quad (7.10)$$

If there are fewer successes in the lot than items in the sample ($S < n$), the variable range is $x = 0, 1, \ldots, S$.

Example 7.5 Do the following:
(a) Write the complete hypergeometric pdf relation for the number Q of ripe melons if a sample of 10 is taken from a crate of 35 which has 8 ripe melons in it.
(b) Write the descriptor for the variable W:

$$P(W = w) = \begin{cases} \dfrac{7!12!10!9!}{2!5!8!4!19!} & w = 0, 1, \ldots, 10 \\ 0 & \text{elsewhere} \end{cases}$$

SOLUTION (a) The values are $n = 10$, $N = 35$, $S = 8$, and $F = 27$. The upper limit on q is 8, because there are only 8 ripe melons in the crate.

$$h(q; 10, 8, 35) = \begin{cases} \dfrac{{}_8C_q \cdot {}_{27}C_{10-q}}{{}_{35}C_{10}} & q = 0, 1, \ldots, 8 \\ 0 & \text{elsewhere} \end{cases}$$

(b) The combinations are

$$\frac{7!}{2!5!} = {}_7C_2 \qquad \frac{12!}{8!4!} = {}_{12}C_8 \qquad \frac{10!9!}{19!} = \frac{1}{{}_{19}C_{10}}$$

Now form the hypergeometric pdf and descriptor for $w = 2$:

$$\frac{{}_7C_2 \cdot {}_{12}C_8}{{}_{19}C_{10}} = h(w = 2; 10, 7, 19)$$

You should learn the following definition of the hypergeometric distribution.

> The *hypergeometric* is a *discrete* distribution which estimates the probability that a certain outcome (called a success) will occur exactly x times in a *finite* sample of size n that is taken from a *finite* population of size N which has a known number of successes S in it.

Problems P7.24–P7.26

7.6 Graph of the Hypergeometric pdf and cdf

The graph of the hypergeometric pdf, like that for all pdf's, is obtained by substituting numerical values into the descriptor and the pdf relation. To graph the pdf for $h(x; n, S, N)$, you need (1) the hypergeometric pdf, Eq. (7.10), and (2) numerical values for n, S, and N. Substitute the values $x = 0, 1, \ldots, n$ and plot the probability value as a vertical line above the discrete x value.

In Sec. 6.8 you learned that the pdf values can be used to compute values for the cumulative distribution function (cdf) using the summation relation

$$F(x') = \sum_{x=0}^{x'} P(x)$$

The hypergeometric cdf, which may be identified as $H(x'; n, S, N)$, is determined by

$$F(x') = \sum_{x=0}^{x'} h(x; n, S, N) \tag{7.11}$$

The graph of the cdf is a step function that is obtained by using (1) values of the hypergeometric pdf for $x = 0, 1, \ldots, x'$, and (2) the hypergeometric cdf, Eq. (7.11). There is a different pdf and cdf graph for each n, S, and N value. Illustrations are given in Example 7.6 of how the pdf and cdf shape changes for different n, S, and N values.

7.6 GRAPH OF THE HYPERGEOMETRIC PDF AND CDF

Example 7.6 Plot the pdf and cdf for the following hypergeometric distributions: (a) $h(r; 4, 5, 8)$; (b) $h(r; 4, 2, 6)$; (c) $h(r; 3, 12, 40)$. (d) Comment on the shape of the pdfs.

SOLUTION (a) Values to be substituted into Eqs. (7.10) and (7.11) are $n = 4$, $S = 5$, $N = 8$, $F = 3$, and $r = 0, 1, 2, 3, 4$. The equations are

$$h(r; 4, 5, 8) = \frac{{}_5C_r \cdot {}_3C_{4-r}}{{}_8C_4} \qquad r = 0, 1, 2, 3, 4$$

$$F(r') = \sum_{r=0}^{r'} h(r; 4, 5, 8) \qquad r' = 0, 1, 2, 3, 4$$

Table 7.1 gives the computations for the pdf and cdf. The pdf is symbolized by $h(r)$ and the cdf by $H(r)$. Both are graphed in Fig. 7.2.
(b) Plots for $h(r; 4, 2, 6)$ are shown in Fig. 7.3 for $r = 0, 1, 2$.
(c) Plots for $h(r; 3, 12, 40)$ are shown in Fig. 7.4.
(d) The pdf histograms vary from symmetric to left- and right-skewed. This is characteristic of the hypergeometric as n, S, and N change.

COMMENT There is a quick-check method in the hypergeometric pdf relation to help with the combinations. If you check Eq. (7.10) and the examples above, you can see that $s + f = n$ and $S + F = N$. This check will only help you place the values in the correct combination.

The hypergeometric is not used as often as other distributions you will learn, because as N increases, some approximating pdf may be used. When the hypergeometric is to be applied, tables of the pdf are helpful. You can consult *Tables of the Hypergeometric Probability Distribution* by Lieberman and Owen (1).

Problems P7.27–P7.29

Table 7.1 Computation of pdf and cdf values for $h(r; 4, 5, 8)$

r	${}_5C_r$ (1)	${}_3C_{4-r}$ (2)	${}_8C_4$ (3)	pdf $h(r)$ (4)*	cdf $H(r)$ (5)
0	1	0	70	0.0000	0.0000
1	5	1	70	0.0714	0.0714
2	10	3	70	0.4286	0.5000
3	10	3	70	0.4286	0.9286
4	5	1	70	0.0714	1.0000

* (4) = (1)(2)/(3)

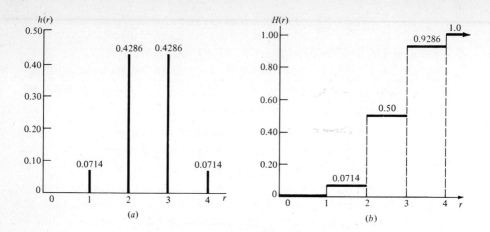

Figure 7.2 Graphs of (a) pdf and (b) cdf for $h(r; 4, 5, 8)$, Example 7.6a.

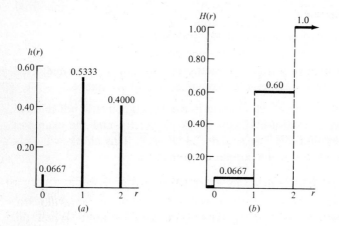

Figure 7.3 Graphs of (a) pdf and (b) cdf of $h(r; 4, 2, 6)$, Example 7.6b.

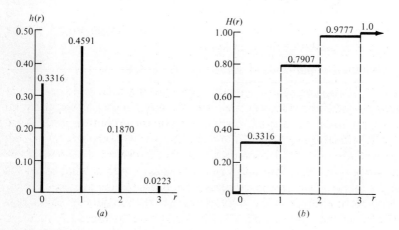

Figure 7.4 Graphs of (a) pdf and (b) cdf for $h(r; 3, 12, 40)$, Example 7.6c.

7.7 Parameters and Properties of the Hypergeometric

If you want to compute probabilities and properties of data from a *finite population*, there is no substitute for the hypergeometric distribution. The following parameters are correct for the distribution.

> Three parameters: N, S, and n
> Parameter ranges: N = positive integers
> $S = 0, 1, 2, \ldots, N$
> $n = 0, 1, 2, \ldots, N$

Rather than S, the fraction of N that is classed as successes is often stated. The fraction of success is p and $S = Np$. The distribution is the same, but the descriptor is $h(x; n, p, N)$ and the parameters are

> Three parameters: N, p, and n
> Parameter ranges: N = positive integers
> $0 \leq p \leq 1$
> $n = 0, 1, 2, \ldots, N$

The properties are written in terms of p and $q = 1 - p$.

- Mean: $\mu = np$

- Standard deviation: $\sigma = \left(npq \dfrac{N - n}{N - 1}\right)^{1/2}$

As the population size N increases, the factor $(N - n)/(N - 1)$ has a reduced effect on σ.

It is often necessary to estimate the parameter p from actual data. The best unbiased estimator of p is $\hat{p}$ (called "p hat"):

- $$\hat{p} = \frac{\text{number of successes observed}}{\text{population size}} = \frac{S}{N} \qquad (7.12)$$

Then we can compute the estimate $\hat{q} = 1 - \hat{p}$. Notice that in Eq. (7.12) the population size N is used, not the sample size n.

Example 7.7 Two government scientists were given a total of 15 specimens of drinking water from different reservoirs for a major city. If the parts per million (ppm) count for a given chemical pollutant exceeds a stated level, the water is termed *nonpotable*. Without the knowledge of the second person, the first scientist tested all specimens and five exceeded the limit. If the second person now tests four specimens drawn at random, what is (*a*) the fraction of the total specimens exceeding the limit, (*b*) the average number of nonpotable specimens in the random sample, (*c*) the number of nonpotable specimens that the second scientist must find to exceed 3 standard deviations above the mean?

SOLUTION (*a*) The fraction is $\hat{p}$ from Eq. (7.12), where a success is a sample which exceeds the limit:

$$\hat{p} = \frac{\text{samples above the limit}}{\text{total samples}}$$

$$= \frac{5}{15} = 0.33$$

(*b*) In the expression for the mean use $\hat{p}$ to estimate p:

$$\mu = np = 4(0.33) = 1.33 \text{ specimens}$$

In repeated samples of size 4 from similar populations of size 15, the scientist should observe an average of 1.33 nonpotable specimens.

(*c*) In the expression for σ we use $\hat{p} = \frac{1}{3}$ and $\hat{q} = \frac{2}{3}$.

$$\sigma = \sqrt{4\left(\frac{1}{3}\right)\left(\frac{2}{3}\right)\left(\frac{15-4}{15-1}\right)} = 0.836 \text{ samples}$$

The 3σ limit above the mean is $1.33 + 3(0.836) = 3.840$ specimens. The scientist must record all four specimens as nonpotable.

COMMENT The histogram for $h(x; 4, 5, 15)$ is shown in Fig. 7.5. The variable X is the number of specimens exceeding the ppm limit.

Problems P7.30–P7.34

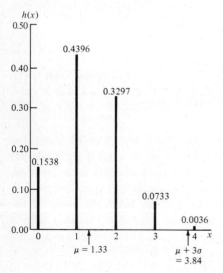

Figure 7.5 Histogram of $h(x; 4, 5, 15)$, Example 7.7.

7.8 Solving Problems with the Hypergeometric

The true test of knowledge of this or any distribution is the ability to set up and solve problems using it. Below are some simple steps to do many of the necessary computations.

1. Check assumptions. Be sure all distributional assumptions are reasonable.
2. Determine the variable and parameters. Define the random variable and compute the numerical value of all parameters. Write the descriptor for the pdf.

You will need to do at least one of the following to solve the problem:

3. Obtain a single probability. Use Eq. (7.10) to compute a single value for the hypergeometric.
4. Obtain the sum of several probability values. If several values are involved, use Eq. (7.10) successively and add. If the cdf can be used, apply Eq. (7.11).
5. Determine μ, σ, or σ^2. Use the formulas for these properties.
6. Plot the pdf or cdf. These require the pdf values for the entire variable range.

Example 7.8 A special close-tolerance retaining ring was manufactured for the space shuttle spacecraft. The machine which produced the ring has a defective rate of 8 percent. If 25 rings were made, determine (a) the chance of observing less than 2 defectives in a sample of 8 and (b) μ and σ of the number of defectives in samples of size 8 and 12.

SOLUTION Use the problem-solving steps above.
(a) Steps 1, 2, and 4 apply here.

1. The hypergeometric assumptions from Sec. 7.5 are correct because the population is finite, each ring is either defective (a success in the sample) or acceptable (a failure), and total defectives in the population can be determined.
2. Let X be the number of defective rings. The parameters are $N = 25$, $p = 0.08$, $n = 8$. The descriptor is $h(x; 8, 0.08, 25)$.
4. We want

$$P(x < 2) = \sum_{x=0}^{1} P(X = x)$$

There are $S = Np = 25(0.08) = 2$ defectives in the entire population; therefore we compute $P(x < 2)$ using Eq. (7.10) and add terms.

Since a defective is a success, $S = 2$ and $F = 23$.

$$P(x = 0) = \frac{{}_2C_0 \cdot {}_{23}C_8}{{}_{25}C_2} = \cancel{0.8433} \ 0.4533$$

$$P(x = 1) = \frac{{}_2C_1 \cdot {}_{23}C_7}{{}_{25}C_2} = \cancel{0.1533} \ 0.4533$$

$$P(x < 2) = \cancel{0.8433 + 0.1533 = 0.9966}$$
$$\phantom{P(x<2) = } 0.4533 \ 0.4533 = 0.9066$$

There is virtual certainty of observing less than two defectives in a sample of eight rings.

(b) Step 5 is used for $n = 8$ and $n = 12$. For $n = 8$

$$\mu = np = 8(0.08) = 0.64 \text{ defective ring}$$

$$\sigma = \left(npq \frac{N - n}{N - 1} \right)^{1/2} = \left[8(0.08)(0.92) \left(\frac{25 - 8}{25 - 1} \right) \right]^{1/2}$$
$$= 0.65 \text{ defective ring}$$

For $n = 12$

$$\mu = 12(0.08) = 0.96 \text{ ring}$$

$$\sigma = \left[12(0.08)(0.92) \left(\frac{25 - 12}{25 - 1} \right) \right]^{1/2} = 0.69 \text{ ring}$$

Example 7.12

Problems P7.35–P7.40

*7.9 Derivation of the Hypergeometric Mean

Derivation of the mean is found by using expected values (Sec. 6.5) or the moment generating function (Sec. 6.10). Determination of μ by $E[X]$ uses Eq. (6.4):

$$\mu = E[X] = \sum_{x=0}^{n} x h(x; n, S, N)$$

$$= \sum_{x=0}^{n} \frac{x \cdot {}_S C_x \cdot {}_F C_{n-x}}{{}_N C_n}$$

Factor out nS/N and simplify:

$$\mu = \frac{nS}{N} \sum_{x=1}^{n} \frac{(S-1)!}{(x-1)![S-1-(x-1)]!} \frac{F!}{(n-x)!(F-n+x)!}$$

$$\times \frac{(n-1)![N-1-(n-1)]!}{(N-1)!}$$

Let $z = x - 1$ to clarify the summation limits. Then

If $x = 1$, $z = 0$
If $x = n$, $z = n - 1$
If $x = x$, $n - x = n - (z + 1) = n - 1 - z$

The expression for μ is now

$$\mu = \frac{nS}{N} \underbrace{\sum_{z=0}^{n-1} \frac{{}_{S-1}C_{x-1} \cdot {}_{F}C_{n-1-z}}{{}_{N-1}C_{n-1}}}_{1}$$

$$= \frac{nS}{N} = np$$

The summation is equal to 1 since it is the total sum of the hypergeometric $h(z; n - 1, S - 1, N - 1)$. Also $\mu = np$ because $p = S/N$. You should try to compute σ^2 using $E[X]$ notation.

Problem P7.41

SOLVED PROBLEMS

Example 7.9 Determine the value of $48!/(42!6!)$ using (*a*) factorial cancellation and (*b*) Stirling's approximation.

SOLUTION (*a*) Cancel the appropriate terms in 48! and 42!:

$$\frac{48!}{42!6!} = \frac{48 \cdot 47 \cdot 46 \cdot 45 \cdot 44 \cdot 43}{6 \cdot 5 \cdot 4 \cdot 3 \cdot 2 \cdot 1} = 12{,}271{,}512$$

(*b*) Use Eq. (7.3) on each factorial term:

$$48! \simeq 48^{48} e^{-48} \sqrt{2(3.1416)(48)} = 1.2392 \times 10^{61}$$

$$42! \simeq 42^{42} e^{-42} \sqrt{2(3.1416)(42)} = 1.4022 \times 10^{51}$$

$$6! \simeq 6^{6} e^{-6} \sqrt{2(3.1416)(6)} = 7.1008 \times 10^{2}$$

Then $\dfrac{48!}{42!6!} \simeq 1.2446 \times 10^7 = 12{,}446{,}000$

This approximation is a 1.4 percent overestimate of the correct answer in (*a*).

COMMENT A calculator with an $n!$ key is very useful for small n values.

Section 7.2

Example 7.10 A total of 15 items have been classified as follows: good, 10; minor defects, 2; and major defects, 3 items. Compute the total number of ways to observe the completion of these 15 items on the production line.

SOLUTION Since the items are not identified in each class, use Eq. (7.5) with

$$n = 15 \quad n_1 = 10 \quad n_2 = 2 \quad n_3 = 3$$

$$_{15}P_{10, 2, 3} = \frac{15!}{10!2!3!} = 30{,}030 \text{ ways}$$

COMMENT You should realize that $n_1 + n_2 + n_3 = n$. If this were not true, the computation would be incorrect.

Section 7.3

Example 7.11 A civil engineer (CE) with the Sim-Plastic Co. has to check the maximum tensile strength on 3 of 25 prestressed beams. Determine (*a*) how many ways there are to observe one above and two below the design strength, if 4 beams are below the design; (*b*) how many samples may be taken from the 25 beams if each beam is indistinguishable; and (*c*) the number of samples if each beam is uniquely numbered 1, 2, ..., 25 and the order of the items in the sample must be maintained. Compare the answer with (*b*).

SOLUTION (*a*) There are $N = 25$ beams with $n_1 = 4$ below and $n_2 = 21$ above design. To get a sample with $r_1 = 2$ below design and $r_2 = 1$ above design, multiply as in Eq. (7.8):

$$_4C_2 \cdot {}_{21}C_1 = 6(21) = 126 \text{ ways}$$

(*b*) From $n = 25$ we can take samples of $r = 3$ in

$$_{25}C_3 = \frac{25!}{3!22!} = 2300 \text{ ways}$$

(*c*) If beams are each numbered, the total permutations are

$$_{25}P_3 = \frac{25!}{22!} = 13{,}800 \text{ ways}$$

The permutations are $r! = 3!$ times the combinations.

Sections 7.3, 7.4

Example 7.12 A grocer received 10 crates of avocados and discovered two overripe in a sample of three drawn from different crates. Because of perishability, it is estimated that as much as 20 percent of the fruit is unsalable. The grocer rejected all 10 crates as defective.
(*a*) Was this a reasonable conclusion?
(*b*) How many more defectives could the grocer have observed?

SOLUTION (*a*) Summarizing results from steps 1, 2, and 3 in Sec. 7.8, we have a hypergeometric variable with $S = Np = 10(0.20) = 2$ defectives. Let D = number defective crates. We need $h(d; n, S, N) = h(2; 3, 2, 10)$ to compute $P(d = 2)$:

$$h(2; 3, 2, 10) = \frac{_2C_2 \cdot {_8C_1}}{_{10}C_3} = 0.0667$$

There is about a 7 percent chance of finding two overripe crates. Since 7 percent is not small, there is a relatively good chance (subjectively speaking) that this lot is no more than 20 percent defective, so the grocer *should not reject* the 10 crates. We will further explore this type of decision making with probabilities in the next chapter.

(b) There are a maximum of $S = 2$ defective crates, since $p = 0.20$. Therefore no more should be observed regardless of sample size. If more are actually observed, then the hypergeometric we used is incorrect and the parameter S must be redefined because $p > 0.20$.

Section 7.8

REFERENCE

(1) Lieberman, G. J., and D. B. Owen: *Tables of the Hypergeometric Probability Distribution*, Stanford University Press, Stanford, Calif., 1961.

ADDITIONAL MATERIAL

Permutations and combinations (Secs. 7.3 and 7.4): Gibra [8], pp. 19–23; Hoel [13], pp. 21–26; Neville and Kennedy [19], Chap. 5; Volk [24], Chap. 2; Walpole and Myers [25], pp. 9–14.

Hypergeometric distribution (Secs. 7.5 to 7.9): Brownlee [5], pp. 158–163; Hahn and Shapiro [10], pp. 151, 152, 163–166; Hoel [13], pp. 67–71; Newton [20], pp. 69–75; Walpole and Myers [25], pp. 91–98.

PROBLEMS

P7.1 Evaluate the following factorials using cancellation wherever possible.
   (a) $\dfrac{7!}{3!}$
   (b) $\dfrac{21!}{8!15!}$
   (c) $\dfrac{40!}{3!37!}$
   (d) $\dfrac{19!}{15!}$

P7.2 Use Stirling's approximation to evaluate
   (a) $25!$
   (b) $\dfrac{21!}{15!}$
   (c) $\dfrac{38!}{5!33!}$

P7.3 Compare the approximations in Prob. P7.2 with the correct answers. Compute the percentage error in each introduced by Stirling's approximation.

P7.4 Determine the number of ways that five people can line up in front of a ticket window.

P7.5 There are seven seats across one row of a passenger aircraft.
   (a) How many ways can a group of four people be seated in these seats?
   (b) Determine the number of ways if one window seat is already reserved by a person not in this group.

P7.6 Three brands of peanut butter are to be tested by 10 people. If there are 5 of brand A, 3 of B, and 2 of C, determine the number of ways to observe the tests.

P7.7 A sample of size 7 is to be taken from 15 items. How many permutations are possible (a) with no replacement and (b) with replacement after each sample item is selected?

P7.8 A production scheduler has five priority I jobs and three priority II jobs. If only two of priority I and one of priority II can be completed next week, and if each job is different, how many ways are there for the person to schedule the three jobs? Assume that all priority I jobs will be completed first.

P7.9 Five children are asked to form a line.
   (a) Determine the number of ways for them to be arranged if two of them are friends and insist on being next to each other.
   (b) Compute the number of arrangements if two children will not stand next to each other.

P7.10 How many four-letter "words" can be arranged from the word *statistics* (a) if repeated letters are distinguishable and (b) if repetitions of letters are not allowed?

P7.11 An engineer has a plant layout problem to locate three new lathes in the places where there were five lathes previously and four new milling machines in the places where five milling machines previously stood. Determine the total number of permutations possible in locating the seven new machines.

P7.12 A farm equipment saleswoman is asked to priority order 10 sales contracts so the shipping department can prepare the implements for delivery. If three priority levels are used and company policy is to place 30 percent of the sales in priority A, 50 percent in B, and 20 percent in priority C, determine the number of ways possible for the saleswoman to order the sales.

P7.13 How many ways can five dogs be selected and lined up for final judging if there are eight candidates?

P7.14 In how many ways can you arrange 16 horses from four breeds if it is not necessary to distinguish between horses of the same breed? Assume there are four horses which represent each breed.

P7.15 A manager must select 4 workers from 12 to form a work team for the weekend. How many possible teams can the manager form?

P7.16 A sample of six breakfast cereals is to be tested for sugar content. The sample is to be composed of two brands from each of three different manufacturers. If company A makes seven brands, B makes four brands, and C makes eight brands, determine how many different samples can be selected.

P7.17 Show that Eq. (7.7) may also be written as $_{n+1}C_{r+1}$ using the expanded form of combinations given in Eq. (7.6).

P7.18 How many ways are there for a landscaper to select three elm trees from five and four oak trees from seven when he goes to the wholesale nursery?

P7.19 An assembly line produces 30 television sets per hour with 10 percent of them defective. How many ways are there to select a sample of five sets which has (a) no defective sets, (b) at most two defective sets, (c) at least one defective set?

P7.20 Calculate the number of ways to combine three cars from a group of ten and three drivers from a group of six people.

P7.21 Two inspectors select samples of size 1 each from 12 cameras. If three of the cameras are defective how many total ways are there for the first inspector to select at most one defective camera, and the second to select at least one defective camera? Assume that the first camera is replaced before the second one is selected.

P7.22 In how many ways is it possible to toss eight coins and observe five heads and three tails?

P7.23 Box A contains two white and four black balls, and box B contains three white and one black. Determine the number of ways to select:
   (a) two balls from box A
   (b) two balls from box A and two from box B
   (c) two balls from box A, at least one being white
   (d) the same as (c) from box B
   (e) two balls from one box or one from each box and obtain at least one white ball

---

The following hypergeometric distributions are used in some of the following problems.

A: $\qquad h(x; n, S, N) = h(x; 5, 4, 20)$

Problems P7.24, P7.25, P7.27, P7.31

B: $\qquad h(y; n, S, N) = h(y; 3, 6, 20)$

Problems P7.24, P7.28, P7.33, P7.35

C: $\qquad h(r; 3, 0.3, 10) = \dfrac{{}_3C_r \cdot {}_7C_{3-r}}{{}_{10}C_3} \quad r = 0, 1, 2, 3$

Problems P7.30, P7.33, P7.36

P7.24 How many successes and failures are there in the lot in (a) distribution A and (b) distribution B?

P7.25 Write the variable range for distribution A. Explain why this range is correct for this distribution.

P7.26 Write the descriptor and complete pdf for the following situation. An electronic component must have a life of 100,000 cycles to be accepted. The components are produced in lots of 20 each with 15 percent of each lot not meeting the minimum cycle specification. A sample of four components is put on life test to see if they last at least 100,000 cycles.

P7.27 For distribution A:
   (a) Plot the pdf and cdf.
   (b) Compute the probability that $x = 1$ or 2.

P7.28 Compute the pdf values for distribution B. Use them to (a) write out the cdf values and (b) plot the pdf and cdf.

P7.29 A car dealer sold six new solarmobiles last month. He has since learned that half of all these cars have a bad steering defect. If he received a total of 10 solarmobiles, half of which had the defect, plot the pdf of X, the number of cars sold with defective steering.

P7.30 For distribution C:
   (a) Determine the number of successes in each lot.
   (b) Plot the pdf.
   (c) Compute the mean and standard deviation.

P7.31 (a) Calculate μ and σ for distribution A.
   (b) Determine the change in μ and σ if the population size were doubled, but n and p were held constant.

P7.32 Compute the mean and standard deviation of the hypergeometric pdf developed in Prob. P7.29.

P7.33 Compare the mean and standard deviations for distributions B and C.

P7.34 Samples of size 7 are taken from lots of 20 riding lawn mowers. Study has shown that two mowers in each lot have incorrectly adjusted carburetors.
    (*a*) Define the variable and state the parameter values for a hypergeometric distribution.
    (*b*) Compute its mean and standard deviation.

P7.35 Assume that distribution B is correct for the number of people who choose soap 1 over soap 2. Determine the probability that at least half the people sampled select soap 1. Use all the appropriate steps in Sec. 7.8 to solve this problem.

P7.36 Use the cdf of distribution C to compute $P(r \leq 2)$.

P7.37 A contractor has 16 different home styles: 10 single-story and 6 two-story designs. If four designs are selected at random, develop an appropriate hypergeometric distribution to determine (*a*) the probability that all designs are two-story and (*b*) at least two designs are single-story.

P7.38 (*a*) Use the pdf developed in Prob. P7.34 to compute the probability that at most one incorrectly adjusted carburetor is in a sample of size 7.
    (*b*) How much will this probability change if the sample size is increased to 10?

P7.39 If samples of 10 students are selected from classes of 36 each in California, determine the chance that one student is from New York if in the past 2 in 12 students have been from New York.

P7.40 Compute μ and σ for the distribution of Prob. P7.39.

*P7.41 Derive the expression for σ^2 for the hypergeometric pdf.
    Hint: Use Eq. (6.3) with $u(X) = X(X - 1)$ to first show that

$$E[X(X - 1)] = \frac{nS}{N} \frac{(n - 1)(S - 1)}{N - 1}$$

and then proceed to compute the variance using the identity

$$\sigma^2 = E[X^2] - \mu^2 = E[X(X - 1)] + \mu - \mu^2$$

CHAPTER
EIGHT

BASICS OF THE BINOMIAL DISTRIBUTION

This chapter is designed to give you basic information about the binomial distribution. You learn about the binomial pdf, how to graph and compute with it, and when to use it to explain probabilistic variation.

CRITERIA

To use binomial distribution computations correctly, you must be able to:

1. Derive the binomial pdf and state its assumptions using the Bernoulli trial approach.
2. Graph the pdf and cdf of any binomial distribution, given the numerical values of the sample size and probability.
3. State the parameters, and compute the mean and standard deviation for any binomial distribution, given a statement of the problem.
4. Determine the pdf and cdf values for the binomial using a binomial cdf table, given values for the sample size, probability, and the random variable.
5. Use the steps to set up a binomial problem and compute one or more probability and property values, given a written description of the situation.
6. Compute the binomial probability that an observed outcome should occur and conclude whether the probability is reasonably high or quite small, given the observed outcome and the binomial parameter values.
7. State the conditions under which the binomial gives a correct and an approximate probability, given a statement of the problem.

8. State the parameters and properties of the distribution of proportions and state its mathematical relationship to the binomial distribution.

*9. Derive the binomial pdf from the binomial expansion, and derive the mean and variance formulas using expected-value relations.

STUDY GUIDE

8.1 Formulation of the Binomial pdf

This section presents a logical derivation of the formula for the binomial pdf. We start with a trial which results in one of two outcomes. If the experiment is tossing a coin, the two outcomes are heads (H) and tails (T). One outcome is called success and the other failure. We call this single experiment a Bernoulli trial. Let the probability of one outcome be p ($0 \leq p \leq 1$) and the other be $q = 1 - p$. The probability p will be the same each time a Bernoulli trial is observed. The Bernoulli pdf for a variable X is

$$f(x) = \begin{cases} p & x = \text{one outcome} \\ q & x = \text{other outcome} \end{cases} \quad (8.1)$$

The *discrete* variable X can assume only two values. The assumptions of the Bernoulli distribution are:

1. There is a single trial with only two possible outcomes.
2. The probability of an outcome is constant for each trial.

Example 8.1 Write the pdf for the observation of a product which is defective 1 time in 10.

SOLUTION Since the product is defective or acceptable, the experiment is a Bernoulli trial. Possible values and probabilities for the outcome Y are given by the Bernoulli pdf:

$$f(y) = \begin{cases} 0.1 & y = \text{defective} = 0 \\ 0.9 & y = \text{acceptable} = 1 \end{cases}$$

It is common to give y the values 0 and 1 as shown in $f(y)$. The graph of $f(y)$ is given in Fig. 8.1.

For the binomial pdf we observe more than one Bernoulli trial in an experiment and assume that each trial is independent. The assumptions are:

1. *n independent* Bernoulli trials
2. Only *two possible outcomes* on each trial
3. *Constant probability* for each outcome on each trial

8.1 FORMULATION OF THE BINOMIAL PDF

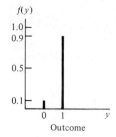

Figure 8.1 Graph of a Bernoulli pdf statement, Example 8.1.

These facts imply that we draw a *finite sample* of size n from an *infinite* population: so, the probabilities must remain constant.

To obtain the binomial pdf, we will consider the specific case for three trials ($n = 3$), using the toss of a fair coin. Then we will derive the pdf for any value of n. Suppose we toss a coin 3 times and record all possible outcomes. If X is the number of tails, we can record the possible outcomes $x = 0, 1, 2, 3$ tails using the probabilities $P(T) = P(H) = \frac{1}{2}$. Tabulation of the possible results gives us the following:

Number of tails x	Possible outcome	Probability of outcome
0	HHH	$1 \cdot P(T)^0 P(H)^3 = (\frac{1}{2})^0(\frac{1}{2})^3$
1	HHT HTH THH	$3 \cdot P(T)^1 P(H)^2 = 3(\frac{1}{2})^1(\frac{1}{2})^2$
2	TTH THT HTT	$3 \cdot P(T)^2 P(H)^1 = 3(\frac{1}{2})^2(\frac{1}{2})^1$
3	TTT	$1 \cdot P(T)^3 P(H)^0 = (\frac{1}{2})^3(\frac{1}{2})^0$

The coefficient 1 or 3 is the number of ways to combine a particular number of heads and tails. This makes the coefficient a *combination*, as learned in Sec. 7.4. If we write all four probabilities in general terms, the expressions below for $n = 3$ and $x = 0, 1, 2, 3$ are correct.

n	x	Probability of x tails
3	0	$_3C_0 P(T)^0 P(H)^3$
3	1	$_3C_1 P(T)^1 P(H)^2$
3	2	$_3C_2 P(T)^2 P(H)^1$
3	3	$_3C_3 P(T)^3 P(H)^0$

These may be summarized in the form

$$_3C_x P(T)^x P(H)^{3-x} \qquad x = 0, 1, 2, 3 \tag{8.2}$$

The maximum value of x is $n = 3$, since you cannot observe more tails than there are tails. The sum of the exponents is always equal to n.

Now, we can derive an expression similar to Eq. (8.2) for any n and x. If we have a coin and toss it n times, we can observe x tails in several ways. One of these ways is

$$\underbrace{\underbrace{TTT\ldots}_{x \text{ tails}} \underbrace{HHHH\ldots}_{n - x \text{ heads}}}_{n \text{ trials}}$$

This way has a probability of $P(T)^x P(H)^{n-x}$. A new combination occurs if, for example, one T is moved into the H series. So, there are $_nC_x$ ways to get x tails and $n - x$ heads. The entire probability can be stated as

$$_nC_x P(T)^x P(H)^{n-x}$$

The combination $_nC_x$ is called the *binomial coefficient* and is computed by using Eq. (7.6).

To obtain the complete binomial pdf, let $p = P(T)$ and $q = 1 - p = P(H)$, and write it in the form learned in Sec. 6.2. We use the binomial descriptor $b(x; n, p)$ to indicate that X is binomially distributed with specific n and p values:

$$b(x; n, p) = \begin{cases} _nC_x p^x q^{n-x} & x = 0, 1, 2, \ldots, n; \\ & 0 \leq p \leq 1; q = 1 - p \\ 0 & \text{elsewhere} \end{cases} \quad (8.3)$$

A definition of the binomial follows.

> The *binomial* is a *discrete* distribution which estimates the probability that a certain outcome will occur exactly x times in a finite sample of size n that is taken from an *infinite* population in which the probability of this outcome is a *constant p*.

Example 8.2 Write the complete pdf expression for the following: (*a*) S is binomially distributed with $n = 4$ and $p = 0.28$; (*b*) $b(t; 50, 0.025)$.

SOLUTION (*a*) The binomial descriptor is $b(s; 4, 0.28)$:

$$b(s; 4, 0.28) = \begin{cases} _4C_s (0.28)^s (0.72)^{4-s} & s = 0, 1, 2, 3, 4 \\ 0 & \text{elsewhere} \end{cases}$$

(*b*) The pdf for the variable T is

$$b(t; 50, 0.025) = \begin{cases} _{50}C_t (0.025)^t (0.975)^{50-t} & t = 0, 1, \ldots, 50 \\ 0 & \text{elsewhere} \end{cases}$$

COMMENT Once you know how to write the binomial pdf, you can drop the 0 and "elsewhere" since they are always the same.

Problems P8.1–P8.4

8.2 Graph of the Binomial pdf and cdf

To graph the pdf for $b(x; n, p)$, you need (1) the binomial pdf, Eq. (8.3), and (2) numerical values for n and p. For $x = 0, 1, 2, \ldots, n$, graph the probability values as a vertical line above each discrete x value. Once you have the binomial pdf values, the cumulative distribution function (cdf) is calculated by using the summation

$$F(x') = \sum_{x=0}^{x'} b(x; n, p) \qquad (8.4)$$

The binomial cdf may also be symbolized as $B(x'; n, p)$. To graph the cdf at x, you need (1) the values of the binomial pdf for $x = 0, 1, \ldots, x'$ and (2) the value of the binomial cdf from Eq. (8.4). This is a discrete cdf, so the graph is a nondecreasing step function.

Example 8.3 Graph the pdf and cdf for the following binomial distributions and comment on their shape: (a) $b(y; 7, 0.5)$; (b) $b(w; 8, 0.25)$.

SOLUTION (a) The pdf and cdf values for $b(y; 7, 0.5)$ are computed in Table 8.1 and graphed in Fig. 8.2 using the pdf formula

$$b(y; 7, 0.5) = {}_7C_y 0.5^y 0.5^{7-y} \qquad y = 0, 1, \ldots, 7$$

The pdf is symmetric because $p = q = 0.5$. The ordinates of the graphs are identified as $b(y)$ for the pdf and $B(y)$ for the cdf. Note that the exponents on p and q can be added only if $p = q = 0.5$.

Table 8.1 Computation of values for $b(y; 7, 0.5)$, Example 8.3

y (1)	$7-y$ (2)	${}_7C_y$ (3)	$(0.5)^y$ (4)	$(0.5)^{7-y}$ (5)	pdf (6) = (3)(4)(5)	cdf (7) = Σ(6)
0	7	1	1.0000	0.0078	0.0078	0.0078
1	6	7	0.5000	0.0156	0.0547	0.0625
2	5	21	0.2500	0.0313	0.1641	0.2266
3	4	35	0.1250	0.0625	0.2734	0.5000
4	3	35	0.0625	0.1250	0.2734	0.7734
5	2	21	0.0313	0.2500	0.1641	0.9375
6	1	7	0.0156	0.5000	0.0547	0.9922
7	0	1	0.0078	1.0000	0.0078	1.0000
					1.0000	

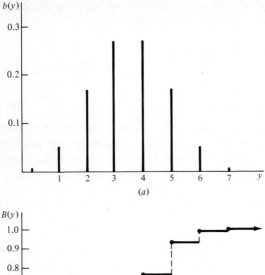
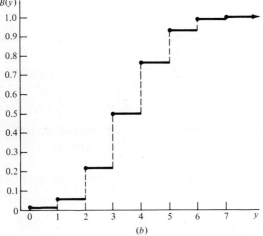

Figure 8.2 Graphs of the (a) pdf and (b) cdf for the $b(y; 7, 0.5)$ distribution.

(b) The pdf and cdf of $b(w; 8, 0.25)$ are graphed in Fig. 8.3. The pdf is skewed right.

COMMENT The following statements may be made about the relative shape of a binomial pdf.

$$p = q = 0.5 \quad \text{symmetric}$$
$$p < 0.5 \quad \text{skewed right}$$
$$p > 0.5 \quad \text{skewed left}$$

Example 8.11

Problems P8.5–P8.7

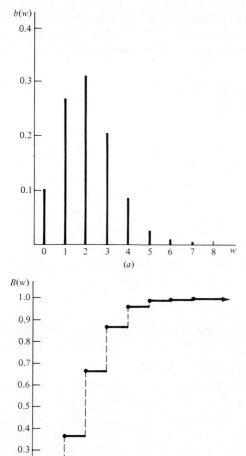

Figure 8.3 Graph of the (a) pdf and (b) cdf for b(w; 8, 0.25).

8.3 Parameters and Properties of the Binomial

The binomial is a commonly used distribution in engineering and other areas of statistics application. It is important to understand and know the following facts about this distribution.

        Two parameters: n and p
        Parameter ranges: n = positive integers
                                  $0 \leq p \leq 1$
        Mean: $\mu = np$
        Standard deviation: $\sigma = \sqrt{npq}$

The formulas for μ and σ are derived in Sec. 8.9. To estimate the parameter p, take a sample of size n and observe the number of outcomes. These outcomes are called *successes*, and the number of successes is the random variable X. The best estimate of p is $\hat{p}$:

$$\hat{p} = \frac{\text{number of successes}}{\text{total sample size}} = \frac{x}{n} \qquad (8.5)$$

Example 8.4 Mr. Tennyson is a student of fire technology. He had an assignment to check several business locations for obvious fire hazards in storage room areas. In one week he inspected 50 buildings and found 10 of them to be hazardous and 40 to be not hazardous. In the second week he checked 20 locations and found 6 to be hazardous. If he assumes that the number of hazardous buildings is binomially distributed and that the first week's observations are characteristic, how close to their expected value was he in the second week? Make the analysis in terms of (*a*) percentage deviation from the mean and (*b*) number of standard deviations from the mean.

SOLUTION All three assumptions of the binomial are correct. Therefore, for the first week let

X = number of hazardous buildings
p = probability that building is rated hazardous
$n = 50$
$\hat{p} = \frac{10}{50} = 0.20$

In the *second* week, if the binomial is used and $\hat{p} = 0.20$ is still correct, for $n = 20$

$$\mu = np = 20(0.20) = 4 \text{ buildings}$$

$$\sigma = \sqrt{npq} = \sqrt{4(0.80)} = 1.789 \text{ buildings}$$

(*a*) Mr. Tennyson observed $x = 6$ in the second week, but the binomial predicts an average of $x = 4$. Therefore, the actual observation of x is 50 percent greater than expected. This large difference should make him question the data from the first week.

(*b*) We know that $\sigma = 1.789$. The observed $x = 6$ is two buildings greater than the mean; so in terms of σ, it is $2/1.789 = 1.118\sigma$ above the mean. Figure 8.4 shows the pdf for $b(x; 20, 0.20)$ with $\mu + 1\sigma = 4 + 1.789 = 5.789$ marked. The value $x = 6$ deviates from the mean a little more than 1 standard deviation.

Problems P8.8–P8.13

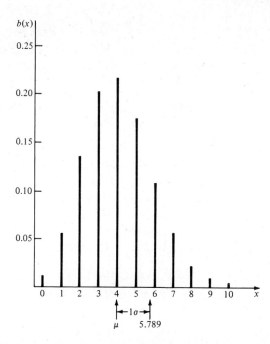

Figure 8.4 Graph of $b(x; 20, 0.20)$ showing $\mu + 1\sigma$, Example 8.4.

8.4 Use of a Binomial Distribution Table

You must be able to perform a binomial probability computation to correctly use the binomial distribution. The computation can be done most accurately by using Eq. (8.3) or (8.4) for the pdf or cdf, respectively; however, the pdf and cdf are tabulated for many different n and p values. Most tables have ranges of n from 1 to 20 and p from 0.05 to 0.50. Table B-1 gives the cdf for commonly used n values from 1 to 20 and p from 0.025 to 0.50. To use the cdf table, you need only know n, p, and $x'(x' = 0, 1, \ldots, n)$. The table entries are the cdf at x', that is, $F(x')$ which is also symbolized as $B(x'; n, p)$:

$$F(x') = B(x'; n, p) = \sum_{x=0}^{x'} b(x; n, p)$$

For example, if $n = 15$ and $p = 0.25$, the cdf for $x' = 6$ is $B(6; 15, 0.25) = 0.9434$. You can get other binomial values using these rules:

1. For a particular x' value the binomial pdf is
$$P(x') = b(x'; n, p) = B(x'; n, p) - B(x' - 1; n, p)$$
2. For the summation from x' to n use the complement cdf rule, that is,
$$\sum_{x=x'}^{n} b(x; n, p) = 1 - B(x' - 1; n, p) \qquad (8.6)$$

3. If $p > 0.5$, use the mirror-image relation for the binomial with $p' = 1 - p$. That is, enter the table for $x' = n - x - 1$ and $p' = 1 - p$ where n, x, and p are from the desired binomial. Thus

$$F(x') = B(x'; n, p') = 1 - B(n - x - 1; n, 1 - p)$$

Example 8.5 Determine the probability value for the following from Table B-1: (a) $B(5; 12, 0.30)$; (b) $b(3; 10, 0.25)$; (c) $b(5; 9, 0.15)$; (d) $\sum_{x=3}^{7} b(x; 7, 0.20)$; and (e) $B(4; 16, 0.60)$.

SOLUTION (a) The cdf value is read directly:

$$F(5) = B(5; 12, 0.30) = 0.8822$$

(b) The individual binomial value is obtained using rule 1:

$$P(x' = 3) = B(3; 10, 0.25) - B(2; 10, 0.25)$$
$$= 0.7759 - 0.5256 = 0.2503$$

You can check this by computing $_{10}C_3(0.25)^3(0.75)^7 = 0.2503$.

(c) Use rule 1 to compute $P(x' = 5)$:

$$b(5; 9, 0.15) = B(5; 9, 0.15) - B(4; 9, 0.15)$$

Since $p = 0.15$ is not in Table B-1, linear interpolation is used between $p = 0.10$ and $p = 0.20$:

$$b(5; 9, 0.10) = B(5; 9, 0.10) - B(4; 9, 0.10)$$
$$= 0.9999 - 0.9991 = 0.0008$$
$$b(5; 9, 0.20) = B(5; 9, 0.20) - B(4; 9, 0.20)$$
$$= 0.9969 - 0.9804 = 0.0165$$

By interpolation $b(5; 9, 0.15) = 0.0087$. The correct value by Eq. (8.3) is 0.0050. You can expect interpolation to be in error, especially for small probability values. As a rule, do not expect the probability to be accurate past the first decimal value.

(d) To obtain the sum from $x' = 3$ to $n = 7$, use rule 2. Applying Eq. (8.6) gives

$$\sum_{x=3}^{7} b(x; 7, 0.20) = 1.0 - B(2; 7, 0.20) = 1.0 - 0.8520 = 0.1480$$

There is a 14.8 percent chance that the binomial variable will be between 3 and 7. You should check this answer so you develop trust in the binomial tables. Use the same relation, but in formula form.

$$\sum_{x=3}^{7} b(x; 7, 0.20) = 1.0 - \sum_{x=0}^{2} {}_7C_x 0.20^x 0.80^{7-x}$$

(e) This summation is a simple cdf value to be looked up; however,

$p = 0.60$ is greater than 0.50. We use rule 3 and find the cdf value for $x' = n - x - 1 = 11$ and $p' = 1 - 0.60 = 0.40$.

$$F(4) = B(4; 16, 0.60) = 1 - B(11; 16, 0.40) = 1 - 0.9951 = 0.0049$$

If you do not trust this manipulation, compute $B(4; 16, 0.60) = 0.0049$.

COMMENT To correctly read and use distribution descriptions like $b(x; n, p)$, you must read them carefully. Pay attention to each symbol or number because it is very important to correct interpretation.

Problems P8.14–P8.17

8.5 Solving Problems with the Binomial

You must be able to take a real-world binomial situation and formulate a binomial statement from it. To set up and solve a binomial distribution problem, you can use the steps of Sec. 7.8, which may be summarized as:

1. Check the assumptions.
2. Determine the variable and parameters.
3. Obtain a single probability.
4. Obtain the sum of several probability values.
5. Determine μ, σ, or σ^2.
6. Plot the pdf or cdf.

If possible, you will want to use the cdf table to reduce computation time. The next two examples illustrate the solution of binomial problems.

Example 8.6 Rubber bands are produced with a designed breaking strength. Mr. Cleary took a sample of 50 from the manufacturing line, stretched them to the designed strength, and three bands broke.
(*a*) What is the probability of this observation?
(*b*) What is the probability that more than three rubber bands will break in a sample of 50?
(*c*) How many are expected to break in a test of 100 bands?

SOLUTION (*a*) Use steps 1 through 3 to determine the single probability.

1. On each of 50 trials, the rubber band will break or not break. The population is all manufactured bands (a very large number), so the probability of breaking is constant. Therefore, the binomial is appropriate.
2. Let X equal the number of broken bands in a sample of $n = 50$ bands. From Eq. (8.5) we estimate p as $\hat{p} = \frac{3}{50} = 0.06$.

3. The probability that $x = 3$ cannot be obtained from the table because $n = 50$ is not included. Using Eq. (8.3), we get

$$P(x = 3) = b(3; 50, 0.06) = {}_{50}C_3(0.06)^3(0.94)^{47} = \frac{50!}{3!47!}(0.06)^3(0.94)^{47}$$
$$= 19,600(0.0002)(0.0546) = 0.2311$$

There is a 23.11 percent chance, or about 1 in 4, that 3 bands will break if 50 are tested.

(b) To determine the chances for more than three breaks, use the results of step 4 to find $P(x \geq 4)$. We use the *complement* rule for the cdf, Eq. (8.6); that is, determine the probability for three or less and subtract from 1.0.

$$P(x \geq 4) = 1 - P(x \leq 3) = 1 - \sum_{x=0}^{3} b(x; 50, 0.06)$$
$$= 1 - \sum_{x=0}^{3} {}_{50}C_x 0.06^x 0.94^{50-x}$$
$$= 1 - (0.0453 + 0.1447 + 0.2262 + 0.2311)$$
$$= 0.3527$$

There is a quite large chance (35.27 percent) that more than 3 bands in 50 will break. You should actually work this one out to develop your computational skill.

(c) By using the value $\hat{p} = 0.06$, the expected value of the binomial $b(x; 100, 0.06)$ is the average number of breaks:

$$\mu = 100(0.06) = 6 \text{ bands}$$

Example 8.12

Problems P8.18–P8.20

8.6 Comparing Observed Data and Binomial Predictions

If observed data is known to follow the binomial pdf, the predicted outcome can be compared with the observed outcome to subjectively determine whether the observed value is as expected. This comparison can be performed for all possible values of a variable (Example 8.7) or for a single value. You should conclude that a single observed value is as expected if the binomial probability is *reasonably high*. If the binomial probability is quite small, you should suspect the observed value. Taking a larger sample is then in order, as is a reexamination of the binomial assumption. What is reasonably high? You will be better prepared to answer this with time and familiarity with particular engineering problems. However, a general guide is 5 percent or more. That is, if the binomial computation results in a 5 percent or more chance that the observed value will occur, trust its legitimacy. Example 8.8 illustrates this decision rule.

Example 8.7 An engineer who has often visited the tables in Las Vegas has decided to simulate four automatic turret lathes by tossing four dice. The number of even-numbered dice tossed will represent the number of idle lathes. The engineer plans to throw the four dice a total of 50 times. How well do the binomial frequencies compare with the following observed data?

Number of idle lathes	Frequency
0	4
1	13
2	16
3	13
4	4
	50

SOLUTION The sample size is $n = 4$ because each toss of four dice can be represented by a binomial. The probability of an even-numbered outcome is $p = 0.50$, and X is the number of idle machines ($x = 0, 1, 2, 3, 4$). Use successive subtraction in Table B-1 to get the individual probabilities for the binomial $b(x; 4, 0.5)$. Then use the formula

$$\text{Expected frequency} = 50(\text{binomial probability})$$

to determine the *expected* number of idle machines to be observed. Table 8.2 gives these results, and Fig. 8.5 is a graph of the two frequency curves. The closeness of the expected and observed frequencies shows that the binomial is a good distribution in this case.

Table 8.2 Expected frequency computations, Example 8.7

Idle lathes x	Binomial probability $P(x)$	Expected frequency $50P(x)$	Observed frequency	Absolute difference
0	0.0625	3.125	4.0	0.875
1	0.2500	12.500	13.0	0.500
2	0.3750	18.750	16.0	2.750
3	0.2500	12.500	13.0	0.500
4	0.0625	3.125	4.0	0.875
	1.0000	50.000	50.0	

174 BASICS OF THE BINOMIAL DISTRIBUTION

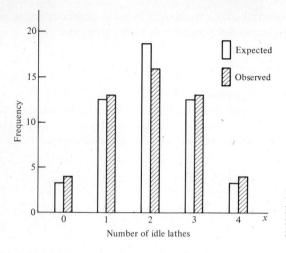

Figure 8.5 Comparison of expected and observed frequencies for a binomial variable, Example 8.7.

COMMENT You must judge when the observed and expected frequencies are different enough to reject the binomial as a good distribution. For example, the binomial would probably be rejected for this situation:

Expected frequency	Observed frequency	Absolute difference
3.125	3	0.125
12.500	9	3.500
18.750	15	3.750
12.500	16	3.500
3.125	7	3.875
50.000	50	

Example 8.8 A chemical mixing process has a 10 percent chance of mixing a closely toleranced drug incorrectly. In a sample of eight, two batches were incorrectly mixed. The inspector rejected all the batches on the basis of this sample. Was this decision correct?

SOLUTION To check the correctness of the reject decision, we must use the binomial to determine the chance that in $n = 8$ samples, $x = 2$ will be defective if $p = 0.10$. It would have been possible to observe three or more defective samples, and the decision to reject would still have been made. Therefore, we must compute the probability of *two or more* defectives, not just the probability of two. From Table B-1

$$P(x \geq 2) = 1 - B(1; 8, 0.10)$$
$$= 1 - 0.8131 = 0.1869$$

With an 18.7 percent chance of 2 or more defectives, the observation of two defectives is not unlikely. The batch should *not* have been rejected.

COMMENT Here is the logic to use in the above analysis:

I observed two defective samples.
The binomial probability of observing this outcome is reasonably high (18.7 percent).
An exceptionally bad batch was probably not observed.
I can believe the two defectives: accept the batch.

If the binomial result had been small, for example, 0.025, the logic would be as follows:

I observed two defective samples.
The binomial probability of observing this is quite small (2.5 percent).
An exceptionally bad batch may have been observed.
I think observing two defectives is too large: reject the batch.

Example 8.13

Problems P8.21–P8.24

8.7 Exact and Approximate Uses for the Binomial

As you learned in the assumptions of Sec. 8.1, the binomial distribution is theoretically correct only when p is constant. This is true when the population is infinite in size, which never happens in the real, finite world. Further, you learned in Sec. 7.5 that the hypergeometric distribution is theoretically correct for all finite population situations. The binomial is the limiting distribution for the hypergeometric when population size N grows and p remains constant. This is symbolically stated as

● $$\lim_{\substack{N \to \infty \\ p = \text{const.}}} h(x; n, p, N) = b(x; n, p)$$

Use the binomial as an approximating distribution under two separate situations:

1. If the *population size N is very large*. Assume the binomial is *correct* because, for all practical purposes, p is constant from trial to trial.
2. If the *sample size n is small in comparison to the population size N*. Use the approximation if $n/N \leq 0.10$, that is, for a 10 percent or less sample. The smaller the value of n/N, the smaller the error introduced by the approximation. The properties of the approximating binomial are $\mu = np$ and $\sigma^2 = npq$. Figure 8.6 summarizes the uses of the binomial.

176 BASICS OF THE BINOMIAL DISTRIBUTION

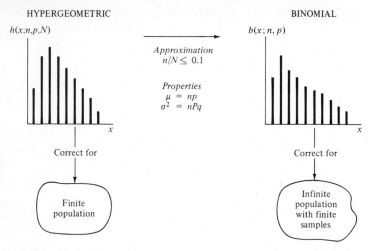

Figure 8.6 Summary of correct and approximation uses of the binomial distribution.

Example 8.9 The Bearss County Jail has a small, but regular, need for specially designed and formed steel bars. The bars are delivered in bundles or lots of 50. The county engineer takes a sample of four bars upon receipt. If two or more bars are unacceptable, the shipment is returned. If 10 percent of the bars are usually defective, what is the (*a*) correct and (*b*) approximate probability of lot rejection? (*c*) Answer (*a*) and (*b*) for lots of size 20.

SOLUTION (*a*) The hypergeometric is correct because of the small lot size. We have

$$X = \text{number of defective bars per lot}$$
$$p = 0.10 \quad \text{and} \quad S = (0.1)(50) = 5 \text{ bars}$$
$$n = 4 \text{ bars}$$
$$N = 50 \text{ bars}$$

Compute $P(x \geq 2)$ using the form of Eq. (7.11):

$$1 - P(x \leq 1) = 1 - H(1; 4, 0.10, 50)$$
$$= 1 - 0.9550 = 0.0450$$

There is a 4.5 percent chance of rejecting a lot.

(*b*) The hypergeometric can be approximated by the binomial because $n/N = \frac{4}{50} = 0.08 < 0.1$. Let

$$X = \text{number of defective bars per lot}$$
$$n = 4 \text{ bars}$$
$$p = 0.10$$

Compute $P(x \geq 2)$ using Eq. (8.6) and Table B-1:

$$\sum_{x=2}^{4} b(x; 4, 0.10) = 1 - B(1; 4, 0.10) = 1 - 0.9477 = 0.0523$$

The binomial probability (0.0523) overestimates the true value (0.0450) by 16 percent. This is considered a small error, especially for a small p like 0.10.

(c) If $N = 20$ for the hypergeometric, $S = 0.1(20) = 2$ defective bars. With only 2 defectives in 20, $P(x \geq 2)$ is the same as $P(x = 2)$ since $x = 3$ and 4 are not observable.

$$h(2; 2, 0.10, 20) = 0.0053$$

Since $n/N = \frac{4}{20} = 0.2 > 0.1$, approximation by the binomial is improper.

COMMENT If the binomial is used in (c), the answer is the same as in (b), that is, 0.0523, because for the binomial we approximate using the infinite-population assumption. This incorrect approximation is about 10 times the correct probability value of 0.0053.

Problems P8.25–P8.28

8.8 Distribution of Proportions

The binomial variable X is used to determine the *number* of times an outcome is observed. A closely related and equally useful variable is $Y = X/n$, which is the *proportion* of the time an outcome is observed. The variable range of Y is

$$y = 0, \frac{1}{n}, \frac{2}{n}, \ldots, 1$$

The binomial pdf for X is used to compute a probability value; that is,

- $P(Y = y) = P\left(\dfrac{X}{n} = \dfrac{x}{n}\right) = {}_nC_x \, p^x q^{n-x}$ $x = 0, 1, 2, \ldots, n; \; 0 \leq p \leq 1$

The pdf graph for $f(y)$ is the same as a binomial pdf graph, for example, Fig. 8.4, except that the abscissa scale is $0, 1/n, 2/n, \ldots, 1$ instead of $0, 1, 2, \ldots, n$.

The parameters of Y are n and p, and the properties are simply the binomial properties divided by n:

- Expected value: $\mu = p$

  Standard deviation: $\sigma = \sqrt{\dfrac{pq}{n}}$

This distribution is applied in quality control, where it is commonly called the fraction defective distribution, and work sampling studies. In fact, any application where the fraction, rather than the number, of occurrences is of interest can use this distribution.

Example 8.10 Management has asked you to determine if the fraction of warranty claims for a specific product has increased substantially compared to 10 years ago. You have taken samples of 200 claims for each year and observed the following number of claims for the product: 10, 8, 12, 15, 19, 8, 20, 12, 22, 25.

(a) Determine if the fraction has increased by checking if all values are within $\pm 3\sigma$ of the average fraction of claims for the product.

(b) State the distribution for the number of claims and show how the fraction of claims distribution is related to it.

SOLUTION (a) The variable Y is the fraction of claims with parameters $n = 200$ and p, which is the average of the yearly fractions for the 10 years. Table 8.3 gives the yearly fractions which have the average $p = 0.0755$. The mean and standard deviation of this distribution of fractions are

$$\mu = p = 0.0755$$

$$\sigma = \left[\frac{0.0755(0.9245)}{200} \right]^{1/2} = 0.0187$$

Three σ on each side of the mean gives the limits

$$\mu \pm 3\sigma = 0.0755 \pm 3(0.0187)$$
$$= 0.0194 \text{ and } 0.1316$$

From Table 8.3 all fraction values are within these limits, so the fraction has not substantially increased.

In quality control a chart set up in this fashion is called a p-chart or a fraction defective chart.

Table 8.3 Fraction of warranty claims, Example 8.10

Year	Claims in $n = 200$	Fraction of claims
1	10	0.050
2	8	0.040
3	12	0.060
4	15	0.075
5	19	0.095
6	8	0.040
7	20	0.100
8	12	0.060
9	22	0.110
10	25	0.125

(b) The number of claims X has a binomial distribution with parameters $n = 200$ and $\hat{p} = 0.0755$ estimated from the data. The properties are $\mu = 15.1$ and $\sigma = 3.74$. The fraction of claims is $Y = X/n$, which has the binomial properties divided by n:

$$\mu = \frac{15.1}{200} = 0.0755$$

$$\sigma = \frac{3.74}{200} = 0.0187$$

Problems P8.29–P8.33

*8.9 Derivation of the Binomial pdf and Its Mean

The binomial pdf formula may be easily derived by using the common binomial expansion, for given values of a and b:

$$(a + b)^n = {}_nC_0 a^0 b^n + {}_nC_1 a^1 b^{n-1} + \cdots + {}_nC_n a^n b^0$$

$$= \sum_{x=0}^{n} {}_nC_x a^x b^{n-x} \tag{8.7}$$

The combination terms ${}_nC_x$ are called the binomial coefficients. If $a = p$ and $b = q$ in the binomial expansion such that $p + q = 1$, we have the binomial pdf. Furthermore, in this form it is easy to see that the sum of the pdf is 1 because

$$(a + b)^n = (p + q)^n = 1^n = 1$$

The expressions for μ and σ of the binomial pdf are found by either of two methods: expected value or moment generating function. The expected value is used here, and the mgf is used in Solved Problems to find the mean. From Eq. (6.4) the mean is

$$\mu = E[X] = \sum_{x=0}^{n} x b(x; n, p) = \sum_{x=0}^{n} x \frac{n!}{x!(n-x)!} p^x q^{n-x}$$

Factor out np and show that the remaining terms are the sum of a binomial $b(z; n-1, p)$, where the variable $Z = X - 1$:

$$\mu = np \sum_{x=1}^{n} \frac{(n-1)!}{(x-1)!(n-x)!} p^{x-1} q^{n-x}$$

$$= np \sum_{z=0}^{n-1} \frac{(n-1)!}{z!(n-1-z)!} p^z q^{n-1-z}$$

$$= np \underbrace{\sum_{z=0}^{n-1} {}_{n-1}C_z p^z q^{n-1-z}}_{1}$$

$$= np \tag{8.8}$$

In Eq. (8.8), the index values are obtained by using $z = x - 1$ as in Sec. 7.9. For the variance relation, use Eq. (6.6):

$$\sigma^2 = V[X] = E[X^2] - \mu^2$$

and the fact that $E[X^2] = E[X(X - 1)] + E[X]$ to show that $\sigma^2 = npq$. See Prob. P8.34.

Example 8.14

Problems P8.34–P8.36

SOLVED PROBLEMS

Example 8.11 (*a*) In Example 8.3 the pdf $b(w; 8, 0.25)$ was plotted. Plot the pdf for $b(x; 8, 0.75)$ and compare the graphs.
(*b*) Determine the value of t for which 0.50 of the total probability is accumulated for $b(t; 20, 0.05)$.

SOLUTION (*a*) For $b(x; 8, 0.75)$ we have $p > 0.5$, which indicates that the pdf should be skewed left. Figure 8.7 shows that the pdf is the exact mirror image of the pdf for $b(w; 8, 0.25)$ in Fig. 8.3a. This relation is always true when the p in one binomial is $1 - p = q$ in another binomial and n is the same.
(*b*) The cdf for $b(t; 20, 0.05)$ needs to be computed to determine when at least 0.50 of the probability is accumulated. We do not need to construct the complete cdf because $p \ll 0.5$ and the pdf will be extremely skewed right.

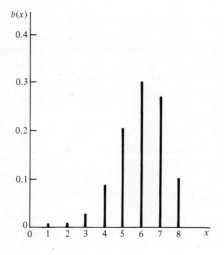

Figure 8.7 Graph of the pdf for $b(x; 8; 0.75)$, which is the mirror image of $b(w; 8, 0.25)$ in Fig. 8.3a.

t	pdf $b(t; 20, 0.05)$	cdf $B(t; 20, 0.05)$
0	0.3585	0.3585
1	0.3773	0.7358

When $t = 1$, over 0.50 of the probability is accumulated.

COMMENT In (b) we saved time by realizing that a highly skewed pdf accumulates probability very rapidly in the tails. In fact, for $b(t; 20, 0.05)$ the probability sums to 1 by the value $t = 6$. Of course, there is some inconsequential probability above $t = 6$. Therefore, general knowledge about the pdf can help solve problems. General rule: *Think before you compute*!

Section 8.2

Example 8.12 One production line produces 1800 integrated circuits (ICs) of type A.B2-4 each day. On a particular day a 1 percent sample was taken. The probability that an IC is defective is 2.5 percent.
(a) Determine the probability that the sample is no more than 10 percent defective.
(b) Compute the number of defectives necessary to exceed a control limit set at 3σ above the mean.

SOLUTION Use the first two steps of Sec. 8.5 to set up the correct binomial distribution. The sample is from a large population of ICs which are defective or acceptable when tested. Let

$$X = \text{number defective in the sample}$$
$$n = 1800(0.01) = 18$$
$$p = 0.025$$

(a) Determine the probability that no more than $18(0.1) = 1.8$ ICs are defective. Since the binomial is discrete, we must compute the chances that $x = 0$ or 1. From Table B-1

$$P(x \leq 1) = B(1; 18, 0.025) = 0.9266$$

There is an excellent chance that the sample will be 10 percent or less defective.

(b) The properties and control limit for this binomial are

$$\mu = 18(0.025) = 0.45 \text{ defective}$$
$$\sigma = \sqrt{18(0.025)(0.975)} = 0.66 \text{ defective}$$
$$\mu + 3\sigma = 0.45 + 3(0.66) = 2.43 \text{ defectives}$$

It would take three or more defectives in $n = 18$ to exceed the limit.

Section 8.5

182 BASICS OF THE BINOMIAL DISTRIBUTION

Example 8.13 A civil engineer (CE) with the state highway department is making a routine, end-of-year check of concrete specimen results on two different cement companies which furnish concrete for interstate bridges. The CE collected the following data by taking a random sample of all tests during the year.

	Company A	Company B
Sample size	20	40
Unacceptable specimens	2	4

There is evidence that on the average any company violates state standards 5 percent of the time. When a violation is discovered, the company may not bid on state projects for 1 year. Should either of the companies be considered for suspension on the basis of these results?

SOLUTION Use the binomial to compute the probability that at least the observed number of unacceptable specimens will occur for each company. Let X be the number of unacceptable specimens.

Company A: $\qquad n = 20 \qquad p = 0.05$

$$P(x \geq 2) = 1 - B(1; 20, 0.05)$$
$$= 1 - 0.7358 = 0.2642$$

Company B: $\qquad n = 40 \qquad p = 0.05$

$$P(x \geq 4) = 1 - B(3; 40, 0.05)$$
$$= 1 - 0.8619 = 0.1381$$

The chance of at least the observed number of unacceptable specimens is reasonably high for both companies. *Therefore, what the CE has observed in the sample is not unlikely.* The companies should probably not be considered for suspension based on these results.

COMMENT Notice that the p values are the same, but the larger sample size for company B decreases the probability that more than 5 percent of n is unacceptable.

Be sure to realize that in examples like this a summation of binomial terms is required. It would be incorrect to compute the probability for only 5 percent of n, since this would give $P(x = 1)$ for company A and $P(x = 2)$ for B. The sum is needed because the problem stated no more than 5 percent.

Section 8.6

Example 8.14 Determine (a) the moment generating function for the binomial $b(x; n, p)$ and (b) the mean using the mgf.

SOLUTION (a) From Eq. (6.21), the mgf $M(t)$ is

$$M(t) = E[e^{tx}] = \Sigma e^{tx} {}_nC_x p^x q^{n-x}$$
$$= \Sigma_n C_x (pe^t)^x q^{n-x}$$
$$= (pe^t + q)^n$$

The last step is possible because of the binomial expansion Eq. (8.7) with $a = pe^t$ and $b = q$.

(b) Using Eq. (6.24), the value of μ is obtained by using $M'(t = 0)$ and $p + q = 1$:

$$M'(t) = npe^t(pe^t + q)^{n-1}$$
$$M'(t = 0) = np(p + q)^{n-1}$$
$$= np$$

Section 8.9

ADDITIONAL MATERIAL

Belz [2], pp. 74–81; Hahn and Shapiro [10], pp. 138–151, 170, 171; Hoel [13], pp. 58–63; Kirkpatrick [15], pp. 92–101; Miller and Freund [18], pp. 54–60; Neville and Kennedy [19], Chap. 7.

PROBLEMS

The following distributions are used by the indicated problems in this chapter.

A: Samples of size 10 are taken from a processing system for ceramic tile. Since tiles have been historically 2.5 percent defective, the number X of defective tiles in each sample follows a binomial pdf of the form $b(x; 10, 0.025)$.

Problems P8.2, P8.10, P8.14, P8.21

B: $b(y; 4, 0.20)$ $y = 0, 1, 2, 3, 4$

Problems P8.5, P8.12, P8.15, P8.26

C: An estimated 75 percent of all house plants are overwatered by their owners. In repeated samples of size 5, the number W of overwatered plants is distributed $b(w; 5, 0.75)$.

Problems P8.11, P8.16, P8.23, P8.30

P8.1 State the assumptions for the Bernoulli distribution and the binomial distribution. Explain the difference(s) between them.

P8.2 Write the complete pdf expression for one of the Bernoulli trials in distribution A.

P8.3 In a particular state it has been estimated that one-fourth of the applicants for professional engineer pass the examination on the first try. Write the pdf for (a) the variable P, which is passing on the first attempt, and (b) the variable X, which is the number passing on the first attempt as observed in a sample of 50 applicants.

(c) Define the term *trial* in the context of this problem.

P8.4 Summarize the formulation of the binomial distribution and state its definition.

P8.5 Plot the pdf and cdf for distribution B.

P8.6 Let $p = 0.5$ for the binomial distribution. Plot the pdfs for (a) an even value of n and (b) an odd value of n.

(c) About what value of the variable are these pdf's symmetric?

P8.7 Use the binomial distribution tabulated in Example 8.3 to determine (a) $P(y \leq 4)$ and (b) $P(1 < y \leq 6)$.

P8.8 Samples of 15 items are taken each hour to determine if the production line should continue to run or if adjustments are necessary. If one or more items in the sample are defective, the line is stopped.

(a) Determine the values of μ and σ if the percentage defective is 2 percent.

(b) Compute the probability that the line will be stopped based on the results from one sample.

P8.9 The probability that an animal can be cured of a certain disease is 0.30. For a sample of 50 animals with the disease:

(a) Determine the parameters of the appropriate binomial distribution.

(b) Compute its mean, variance, and standard deviation.

(c) Calculate the chances that no more than 5 percent of the animals in the sample are cured.

P8.10 Determine the mean and standard deviation for distribution A.

P8.11 Compute μ and σ for distribution C and locate the mean on a graph of the pdf.

P8.12 Use distribution B to determine the probability that the number of defectives is no more than 2 standard deviations from the mean.

P8.13 Mr. Fennell has observed the number Y_i of unsealed bread bags that come out of the packaging machine for the last 10 days ($i = 1, 2, \ldots, 10$). The data below gives n_i, the number of bags checked each day, and the value of Y_i.

Day i	1	2	3	4	5	6	7	8	9	10
n_i	150	75	310	80	100	28	75	130	95	250
y_i	13	9	26	11	5	1	6	16	7	30

Compute the one best estimate of the fraction defective p for the packaging machine using the 10 days of data. Be sure to consider the fact that sample size n_i is different each day.

P8.14 Use Table B-1 to answer the following binomial probability questions for distribution A, which is $b(x; 10, 0.025)$.

(a) $P(x \leq 3)$

(b) The probability of observing more than one defective tile

(c) $B(2; 10, 0.025)$

(d) $P(1 < x \leq 3)$

P8.15 For distribution B, use Table B-1 to determine (a) $P(y = 2)$; (b) $\sum_{y=0}^{2} b(y; 4, 0.2)$; (c) $P(y \geq 2)$; and (d) the total probability between $\mu \pm 1\sigma$.

P8.16 For distribution C, which is $b(w; 5, 0.75)$, use Table B-1 to determine (a) $P(w \leq 3)$; (b) $P(w > 1)$; (c) $P(2 \leq w \leq 4)$; and (d) $P(w = 1)$.

P8.17 If Y is a binomial variable with $\mu = 2$ and $\sigma^2 = 1.6$, (a) determine the values of the parameters and (b) use Table B-1 to determine the cdf value at $y = 5$ and the probability that $2 \leq y \leq 8$.

P8.18 A team of geologists are taking seismic readings in 18 different offshore areas. Historically, there is a 20 percent chance that the readings will indicate oil is below the surface of the water. Only two readings indicate oil is present, whereas the geologists expected at least eight positive indications. Set up an appropriate distribution and use the problem-solving steps to determine (a) the probability of observing exactly two positive readings, (b) the expected number of positive readings in the 18 areas, and (c) the probability of at least eight positive indications. (d) If all conditions remain the same, how many different areas must be tested to have an expected value of eight positive readings?

P8.19 Use the problem-solving procedure of Sec. 8.5 to answer Prob. P8.9.

P8.20 Each day the production engineer of the Monocolor Paint Company checks the noise level in the plant. In the past, 2 out of every 10 days the Occupational Safety and Health Act (OSHA) limit has been exceeded in the can-closing area. A government inspector is in the plant this week (5 days).

   (a) Determine the chance that the inspector will not observe a noise level violation during the week.

   (b) What is the probability that the noise level will be exceeded for 2 or 3 days of the week?

   (c) What is the expected number of days that the level will be exceeded?

P8.21 In a sample of 10 ceramic tiles taken from the process described by distribution A, there was one defective tile. The new quality control engineer feels that the process system should be shut down and checked because 1 defective in 10 is too many. Is the suggestion to stop processing a reasonable one?

P8.22 If seismic readings are expected to give positive indications of oil 20 percent of the time, and if 8 out of 18 tests give positive results, should the method used to interpret the data be questioned for legitimacy?

P8.23 The person who concluded that 75 percent of all houseplant owners overwater their plants, as explained by distribution C, also believes it is not unusual in a sample of 5 to find that all 5 owners overwater. Is there a reasonably high probability that this can happen?

P8.24 The probability that any one of five sensitive circuit breakers is thrown by any stray voltage is estimated to be 0.5. In 75 tests the number X of breakers thrown was recorded as follows:

Breakers thrown X	Frequency of X
0	15
1	22
2	25
3	10
4	2
5	1

Do the observed frequencies tend to verify the statement that the number of thrown breakers will follow a binomial pdf?

P8.25 Shipments of 150 barrels of a particular chemical are sent out each week to the Eckwald Chemical Company. Ten of them are tested for chemical potency. Governmental tests in the past have shown that 12 percent of these barrels exceed potency limits.

   (a) Approximate the probability that of the 10 barrels tested today 3 will exceed the limit.

   (b) State exactly how the correct probability would be computed to answer (a).

P8.26 Assume that distribution B is an approximating pdf for samples of size 4 taken from populations of $N = 90$ items.

(a) Compare the μ and σ values of the hypergeometric and this approximating binomial distribution.

(b) How small can N get before you would start to suspect the results of the approximation?

(c) Compare the σ values at this value of N.

P8.27 (a) Solve Prob. P7.39 by using the binomial approximation.

(b) Compare the two answers and comment on the appropriateness of using the binomial approximation for this problem.

P8.28 Explain how the assumptions of the binomial distribution are used to approximate the hypergeometric distribution assumptions.

P8.29 The probability that the length of a metal sheet exceeds the upper specification limit is 0.10.

(a) State the complete forms of the binomial and proportion pdf's for this problem if $n = 5$.

(b) Compute the probability that at most 40 percent of a sample of five sheets exceeds the limit.

(c) What is the average proportion of a sample, regardless of size, that will exceed the limit?

P8.30 (a) Write the distribution of proportions for distribution C.

(b) Locate its mean on a graph of the pdf. Refer to Prob. P8.11 where (b) was answered for the binomial form of distribution C.

P8.31 Rework Prob. P8.9 using the pdf for proportions rather than the binomial pdf.

P8.32 If X has a $b(x; n, p)$ distribution, use the mean and standard deviation formulas for X to derive the corresponding formulas for the fraction defective distribution.

P8.33 (a) Compute the pdf values for the distribution of proportions with $p = 0.20$ and $n = 10$.

(b) Prepare the pdf graph and locate the mean μ and the points $\mu \pm 1\sigma$ on this graph.

*P8.34 Use expected value notation to derive the variance of the binomial distribution. See the hint at the end of Sec. 8.9.

*P8.35 Derive the variance of the binomial distribution by using its moment generating function.

*P8.36 The mgf for the distribution of proportions discussed in Sec. 8.8 is

$$M(t) = (pe^{t/n} + q)^n$$

Use it to derive the mean and variance for this distribution.

CHAPTER
NINE

BASICS OF THE POISSON DISTRIBUTION

This chapter covers the basic properties and uses of the Poisson distribution. You learn to make probability computations with the Poisson when it is the correct and approximation distribution.

CRITERIA

To correctly use the Poisson distribution, you must be able to:

1. State the assumptions of and the procedure used to formulate the Poisson pdf.
2. Graph the pdf and cdf of any Poisson distribution, given the average rate of occurrence.
3. State the parameter; state and compute the mean, standard deviation, and parameter estimate for any Poisson distribution, given a statement of the problem.
4. Determine the pdf and cdf value for any Poisson using a table of cdf values, given values of the parameter and the variable.
5. Use the problem-solving procedure to compute Poisson probability and property values, given a written statement of the problem.
6. Compute the Poisson probability that an observed outcome should occur and conclude whether the probability is reasonably high or quite small, given the observed outcome and the Poisson parameter value.

7. State the conditions under which the Poisson correctly approximates the binomial; and compute the approximating probability, given the binomial parameter values.
*8. Derive the Poisson pdf as the limit of the binomial, and derive the mean and variance formulas using expected-value relations or the moment generating function.

STUDY GUIDE

9.1 Formulation of the Poisson Distribution

We will use a specific example to derive the Poisson pdf. If you look closely at a photograph in a newspaper, you see that the image is composed of a large number of dots arranged in rows and columns. Each dot is printed larger for a dark area and smaller for light areas. Any one of the dots is printed incorrectly if it is too light, too dark, missing, or misplaced. Define

h = area represented by one dot; $h > 0$ but very small

μ = constant representing average number of times dot is printed incorrectly in h; $\mu > 0$

H = total area, divided into n equal areas, each containing a dot

The area $h = H/n$ gets smaller and smaller as n becomes countably infinite ($n = 0, 1, 2, \ldots$, no upper limit). See Fig. 9.1.

We postulate that the probability of an incorrectly printed dot is proportional to the size of h; that is, as the area h increases, the probability of an incorrect dot increases. Therefore

μh = probability of one incorrectly printed dot in h

$1 - \mu h$ = probability of a correctly printed dot in h

Since $\mu h + (1 - \mu h) = 1$, there is no chance of more than one mistake in any h. The small value μh is called a *rare event probability*, and it is used to derive the Poisson pdf from the binomial by substituting μh for p:

$$p = \mu h = \frac{\mu H}{n}$$

Take the limit as $n \to \infty$, as $p \to 0$, and with np constant (call its value λ), and

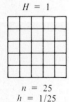
$n = 25$
$h = 1/25$

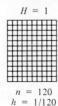
$n = 120$
$h = 1/120$

n is very large
h is very small

Figure 9.1 Derivation of the Poisson pdf requires the computation of the area h for an increasing number of areas n.

the Poisson pdf results (see Sec. 9.8):

$$\lim_{\substack{n\to\infty\\p\to 0\\np=\lambda}} b(x; n, p) = \frac{e^{-\lambda}\lambda^x}{x!} \quad (9.1)$$

The right side of Eq. (9.1) is the Poisson pdf, and the descriptor used is $P(x; \lambda)$ to show that X is Poisson with a particular value of λ. The complete Poisson pdf is

$$P(x; \lambda) = \begin{cases} e^{-\lambda}\lambda^x/x! & x = 0, 1, 2, \ldots; \lambda > 0 \\ 0 & \text{elsewhere} \end{cases} \quad (9.2)$$

Since x is a positive integer, the Poisson distribution is *discrete*. The value $e = 2.71828$ is the sum of the sequence

$$\sum_{r=0}^{\infty} \frac{1}{r!} = e$$

Example 9.1 Write the complete pdf expression for (a) $P(t; 3.5)$ and (b) W which is distributed as a Poisson with a constant of 0.5.

SOLUTION (a) The pdf for T is

$$P(t; 3.5) = \begin{cases} e^{-3.5}(3.5)^t/t! & t = 0, 1, 2, \ldots \\ 0 & \text{elsewhere} \end{cases}$$

(b) The descriptor is $P(w; 0.5)$, and the pdf is

$$P(w; 0.5) = e^{-0.5}(0.5)^w/w! \quad w = 0, 1, 2, \ldots$$

The "0 elsewhere" is omitted because it is always correct.

The Poisson is not much different from the binomial except that the Poisson probability is very small and the sample size is not actually known. Assumptions of the Poisson distribution are:

1. There are n independent trials where n is very large.
2. Only one outcome is of interest on each trial.
3. There is a constant probability of occurrence on each trial.
4. The probability of more than one occurrence per trial is negligible.

Some applications are number of defects per standard unit (a photograph, plane wing, sheet of glass, piece of material), radioactive decay of materials, and deaths per automobile accident. A definition is:

The *Poisson* is a *discrete* distribution which estimates the probability that a specified outcome will occur exactly x times in a standardized unit when the average rate of occurrence per unit is a constant λ.

Problems P9.1–P9.3

9.2 Graph of the Poisson pdf and cdf

To graph the pdf for $P(x; \lambda)$, you need the Poisson pdf, Eq. (9.2), and the average rate of occurrence λ. Plot the probability value on a vertical line above each discrete value $x = 0, 1, 2, \ldots$. The pdf is always *skewed right*.

The pdf values are used to compute the cdf, which can be symbolized by $P'(x'; \lambda)$, by the relation

$$F(x') = \sum_{x=0}^{x'} P(x; \lambda) \qquad (9.3)$$

Example 9.2 The variable R, which is the number of defects per page in the Funsity telephone directory, averages 2.3.
(*a*) Plot the pdf and cdf.
(*b*) Determine what value of R represents 50 percent of the mistakes per page.

SOLUTION (*a*) Computations and graphs for the pdf $P(r; \lambda)$ and the cdf $P'(r'; \lambda)$ are shown in Table 9.1 and Fig. 9.2, respectively. The skewed-right pdf uses the formula

$$P(r; 2.3) = \frac{e^{-2.3} 2.3^r}{r!} \qquad r = 0, 1, 2, \ldots$$

$$= \frac{0.1003(2.3)^r}{r!}$$

(*b*) From the cdf 59.6% of the pages include 2 or fewer mistakes. Therefore $r = 2$ mistakes is the answer.

Table 9.1 Computations for $P(r; 2.3)$, Example 9.2

r (1)	$r!$ (2)	2.3^r (3)	pdf $P(r; \lambda)$ (4) = 0.1003(3)/(2)	cdf $P'(r'; \lambda)$ (5) = Σ(4)
0	1	1.00	0.1003	0.1003
1	1	2.30	0.2306	0.3309
2	2	5.29	0.2652	0.5961
3	6	12.17	0.2033	0.7994
4	24	27.98	0.1169	0.9163
5	120	64.36	0.0538	0.9701
6	720	148.04	0.0206	0.9907
7	5,040	340.48	0.0068	0.9975
8	40,320	783.11	0.0019	0.9994
9	362,880	1,801.15	0.0005	0.9999
10	3,628,800	4,142.65	0.0001	1.0000

9.3 PARAMETER AND PROPERTIES OF THE POISSON

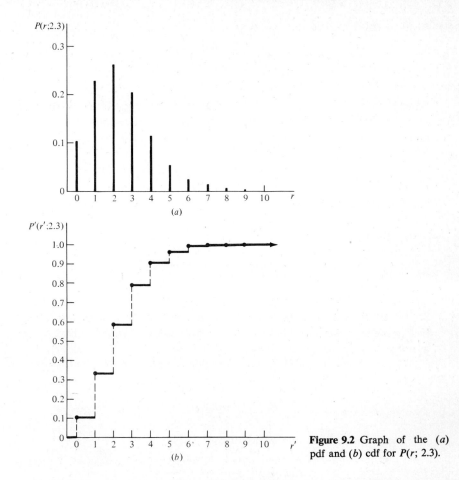

Figure 9.2 Graph of the (a) pdf and (b) cdf for $P(r; 2.3)$.

COMMENT You can see in Fig. 9.2a that the mode is at $r = 2$. The Poisson mode is at 0 for $\lambda \leq 1.0$ and greater than 0 for $\lambda > 1.0$.

Problems P9.4, P9.5

9.3 Parameter and Properties of the Poisson

The Poisson has one parameter which changes the shape of the pdf.

- Shape parameter: λ
- Parameter range: $\lambda > 0$
- Expected value: $\mu = \lambda$
- Standard deviation: $\sigma = \sqrt{\lambda}$

The expressions for μ and σ are derived in Sec. 9.8. To estimate the value of the parameter, simply compute the *average* $\overline{X}$ of the observed data for a sample of

BASICS OF THE POISSON DISTRIBUTION

size n:

$$\hat{\lambda} = \bar{X} = \frac{\Sigma f_i X_i}{n} \tag{9.4}$$

Example 9.3 Dolly and Darrel are engineers with a government-sponsored safety agency. They have determined the number of serious accidents each month at the construction sites of new government office buildings to be Poisson-distributed.
(a) Based on the data below, they predict a good chance of one serious accident each month. Are they correct?
(b) What are the 1σ limits above and below the mean of this data?

Accidents per month	0	1	2	3	4	5
Frequency	27	12	8	2	1	0

SOLUTION Before we answer the questions, be sure we understand the data. If X represents the number of serious accidents in 1 month, there were 27 months with no accidents, that is, $x = 0$.
(a) Compute the probability of one accident using the $P(x; \lambda)$ distribution. From Eq. (9.4), the average number of accidents is

$$\hat{\lambda} = \frac{27(0) + 12(1) + 8(2) + 2(3) + 1(4)}{27 + 12 + 8 + 2 + 1} = \frac{38}{50} = 0.76$$

The probability of one accident is

$$P(1; 0.76) = \frac{e^{-0.76} 0.76^1}{1!} = 0.3554$$

A chance of 35.54 percent is quite high, but there is an even larger chance that $x = 0$:

$$P(0; 0.76) = e^{-0.76} = 0.4677$$

Therefore, the couple should reconsider their estimate to include $x = 0$.
(b) For $P(x; 0.76)$ the properties are $\mu = \lambda = 0.76$ and $\sigma = \sqrt{\lambda} = 0.87$ accident. The 1σ limits are

$$\mu \pm 1\sigma = 0.76 \pm 0.87 = 1.63 \quad \text{and} \quad 0.0$$

The lower limit is 0 because $x < 0$ is not observable.

COMMENT Why use the Poisson here and not the binomial? Because the standardized time interval of one month, the small chance of an accident, and observed frequencies for $x = 0$ through $x = 4$ lead to the Poisson. For the binomial we have $n = 50$ months, but what is p? If some x value is called a "success" in order to compute $\hat{p}$, we have different p values for each x value. Since this violates the binomial assumptions, the Poisson is the appropriate distribution.

Problems P9.6–P9.9

9.4 Use of the Poisson Distribution Table

Equations (9.2) and (9.3) can be used for exact computations of the pdf and cdf, respectively. Table B-2 gives the Poisson cdf value at x' for λ values from 0.02 to 20.00. Individual pdf values can be obtained by subtraction:

$$P(x; \lambda) = P'(x'; \lambda) - P'(x' - 1; \lambda) \tag{9.5}$$

where $P'(x'; \lambda)$ is the cdf value at x'. If the sum from x' to ∞ is desired, use

$$\sum_{x=x'}^{\infty} P(x; \lambda) = 1 - P'(x' - 1; \lambda) \tag{9.6}$$

Example 9.4 Use Table B-2 to determine the following:
(a) $P'(12; 6.60)$
(b) $P(5; 12.0)$
(c) $P(4; 1.33)$
(d) $\sum_{x=2}^{\infty} P(x; 1.80)$

SOLUTION (a) The cdf value for $x' = 12$ is $F(12) = P'(12; 6.60) = 0.982$.
(b) The probability for $x = 5$ is found by using Eq. (9.5):

$$P(5; 12.0) = P'(5; 12.0) - P'(4; 12.0)$$
$$= 0.020 - 0.008 = 0.012$$

(c) Linear interpolation between $\lambda = 1.35$ and $\lambda = 1.30$ is necessary. Using Eq. (9.5), we get

$$P(4; 1.35) = 0.988 - 0.952 = 0.036$$
$$P(4; 1.30) = 0.989 - 0.957 = 0.032$$

By interpolation $P(4; 1.33) = 0.034$. The Poisson pdf, Eq. (9.2), for $\lambda = 1.33$ gives the same answer.
(d) To get the sum from $x = 2$ to $x = \infty$, use Eq. (9.6):

$$\sum_{x=2}^{\infty} P(x; 1.80) = 1 - P'(1; 1.80) = 1 - 0.463 = 0.537$$

Problems P9.10–P9.12

9.5 Solving Problems Using the Poisson

To set up and solve a Poisson problem, the problem-solving steps of Sec. 7.8 are used:

1. Check assumptions. The most important condition is that the probability of occurrence of the desired event is constant for each standardized unit.
2. Determine the variable and value of the parameter λ.
3. Obtain a single Poisson probability.
4. Obtain the sum of several Poisson probability values.
5. Determine μ and σ for the Poisson.
6. Plot the pdf or cdf.

Example 9.5 An ME with Flataz A. Pancake Rolling Mills is studying the defect pattern in cold-rolled steel. The engineer inspects 10-m sections to determine the number of defects D. The results for 50 sections are:

Defects per section D	Number of sections
0	35
1	8
2	3
3	2
4	1
6	1
	50

(a) Determine the probability of being on or above the specification limit of three defects per section.

(b) Compute the 3σ limits and explain them.

SOLUTION From steps 1 and 2 the Poisson is applicable for a 10-m section, and the variable is D with a parameter λ estimated by Eq. (9.4):

$$\hat{\lambda} = \bar{d} = \frac{(\text{defects per section})(\text{number of sections})}{\text{total number of sections}}$$

$$= \frac{0 + 8 + \cdots + 6}{50} = 0.6 \text{ defect per section}$$

(a) This is step 4. Use Eq. (9.6) and Table B-2:

$$\sum_{d=3}^{\infty} P(d; 0.6) = 1 - P'(2; 0.6) = 1 - 0.977 = 0.023$$

There is a 2.3 percent chance of being on or exceeding the specification limit of three defects.

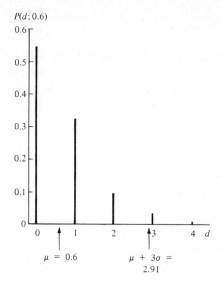

Figure 9.3 Plot of $P(d; 0.6)$ for defects in cold-rolled steel, Example 9.5.

(b) Step 5 is used to compute the limits of $\mu \pm 3\sigma$:

$$\mu = \lambda = 0.6 \text{ defect per section}$$

$$\sigma = \sqrt{\lambda} = \sqrt{0.6} = 0.77 \text{ defect per section}$$

$$\mu \pm 3\sigma = 0.6 \pm 3(0.77) = 2.91, -1.71$$

The lower limit is actually 0 because -1.71 defects is impossible. In Fig. 9.3 the pdf for $P(d; 0.6)$ is plotted with $\mu + 3\sigma = 2.91$ indicated. There is a chance that defects will exceed 2.91. In fact, in part (a) we showed the chances to be 2.3 percent for three or more defects.

Problems P9.13–P9.15

9.6 Comparing Observed Data and Poisson Probabilities

The Poisson distribution can be checked against observed data in one of two ways.

1. Graphically compare the entire Poisson pdf or frequency values with observed outcomes. The technique is the same as that used in Example 8.7 for a binomial variable. This is a subjective comparison at this point, but more quantitative techniques are presented later.
2. One particular outcome may be checked by computing the Poisson probability that the outcome would be observed. If the probability is reasonably high, conclude that the observation is as expected. If the chance is small, suspect the observation as exceptional. This technique is illustrated below.

Example 9.6 A chemical engineer, Gerald, has been asked to solve a sulfur content problem in the manufacture of tire rubber. Gerald found that today alone the 4 percent per tire limit was exceeded 5 times for process line A and 4 times for line B. Past records for 100 days on both lines indicate violations as follows:

Number of violations per day	0	1	2	3	4	5	6
Number of days	33	44	10	5	5	2	1

Which line, if either, should Gerald concentrate on first to solve the sulfur problem?

SOLUTION The Poisson is the correct distribution because each tire (the standardized unit) has a small probability of exceeding the 4 percent sulfur limit. Gerald should define the variable X as the number of violations per day and compute $\hat{\lambda} = \bar{X} = 1.15$ violations per day.

Use the Poisson $P(x; 1.15)$ to compute the probabilities of five (line A) or four (line B) *or more* violations. From Table B-2

Line A: $\quad \sum_{x=5}^{\infty} P(x; 1.15) = 1 - P'(4; 1.15) = 1 - 0.993 = 0.007$

Line B: $\quad \sum_{x=4}^{\infty} P(x; 1.15) = 1 - P'(3; 1.15) = 1 - 0.970 = 0.030$

The chances of the observed violations are very low for both lines, so they should both be given immediate attention.

COMMENT The results here are heavily biased since one $\hat{\lambda}$ value was used for both lines. Data collected separately for each line could result in different λ and probability values.

As a point of discussion, if the probability values are 0.007 (line A) and 0.25 (line B), line B is judged okay since 25 percent is a reasonably high chance of observing four or more violations.

Problems P9.16–P9.19

9.7 Exact and Approximate Uses of the Poisson

The Poisson is the *correct* distribution when there are a very large number of standardized units and an event has a constant, very small probability of

9.7 EXACT AND APPROXIMATE USES OF THE POISSON

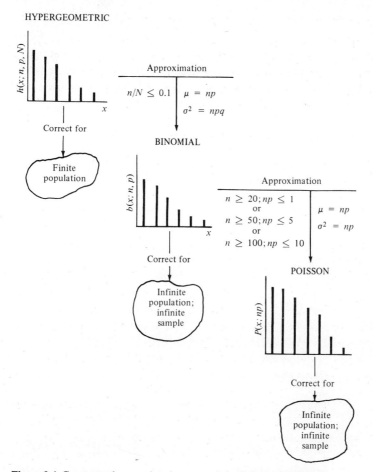

Figure 9.4 Correct and approximation uses of the Poisson distribution.

occurrence. Another equally useful application is that of approximating the binomial. As summarized in Sec. 9.1, the binomial approaches the Poisson with parameter $\lambda = np$ under the conditions given in Eq. (9.1). When used as an approximation to the binomial, the properties of the Poisson are $\mu = np$ and $\sigma = \sqrt{np}$. The approximation is correctly applied if $n \geq 20$ and $p \leq 0.05$ ($np \leq 1$). However, as n increases, the value of p can increase and the approximation is still close. If $n \geq 50$ and $np \leq 5$ or if $n \geq 100$ and $np \leq 10$, good results are obtained. A summary of correct and approximating uses of the Poisson and binomial are given in Fig. 9.4. A benefit of the Poisson approximation to the binomial, as illustrated below, is the reduction in computations.

Example 9.7 An IE working for the Random-Lines Telephone Company has collected data on X, the number of operater-aided local calls. In a sample of 25 one-minute intervals, the IE found a probability of 0.03 of such a call occurring. Compute the number of minutes that three or more calls should be observed.

SOLUTION Because the population is large (all calls), the sample is finite, and only two outcomes are possible for each call (local-aided or local regular), the binomial distribution is correct. Since $n = 25$ and $np = 0.75$, which is less than 1.0, the Poisson approximation can be used with $\lambda = 0.75$. The engineer should observe three or more calls 4.1 percent of the minutes, as shown here:

$$\sum_{x=3}^{\infty} P(x; 0.75) = 1 - P'(2; 0.75) = 1 - 0.959 = 0.041$$

In 25 min a total of $0.041(25) = 1.025$, or 1 min, should have three or more aided calls.

COMMENT If the binomial had been used, the probability would have been 0.0380:

$$\sum_{x=3}^{25} b(x; 25, 0.03) = 1 - \sum_{x=0}^{2} {}_{25}C_x(0.03^x)(0.97^{25-x})$$
$$= 1 - (0.4669 + 0.3611 + 0.1340)$$
$$= 0.0380$$

With rounding to two decimals both answers are 0.04; however, the binomial took more work.

The Poisson approximation to the binomial is accurate when the guidelines are followed. We will use the Poisson approximation *whenever correct* to reduce computations. We hope you do also!

Example 9.8

Problems P9.20–P9.23

*9.8 Derivation of the Poisson pdf, Mean, and Variance

The Poisson pdf is obtained from the binomial if the limit is taken as $n \to \infty$, $p \to 0$ and np is constant. In Sec. 9.1 we expressed the binomial probability as $p = \mu H/n$. The area $h = H/n$ will be reduced in size to make p approach 0. Substituting p into the binomial pdf, Eq. (8.3), we get the following expressions:

$$_nC_x p^x q^{n-x} = \frac{n!}{x!(n-x)!} \left(\frac{\mu H}{n}\right)^x \left(1 - \frac{\mu H}{n}\right)^{n-x}$$

$$= \frac{1}{x!} \left[\frac{n(n-1)(n-2)\cdots(n-x+1)}{n^x}\right] (\mu H)^x \left(1 - \frac{\mu H}{n}\right)^{n-x}$$

Since there are x terms in the numerator of the bracket,

$$_nC_x p^x q^{n-x} = \frac{1}{x!}\left[\left(\frac{n}{n} \cdot \frac{n-1}{n} \cdot \frac{n-2}{n} \cdots \frac{n-x+1}{n}\right)\left(1 - \frac{\mu H}{n}\right)^{-x}\right]$$
$$\times (\mu H)^x \left(1 - \frac{\mu H}{n}\right)^n$$
$$= \frac{1}{x!}(\mu H)^x \left[1 \cdot \left(1 - \frac{1}{n}\right) \cdot \left(1 - \frac{2}{n}\right) \cdots \left(1 - \frac{x-1}{n}\right)\right]$$
$$\times \left(1 - \frac{\mu H}{n}\right)^{-x}\left(1 - \frac{\mu H}{n}\right)^n$$

In the limit as $n \to \infty$, the term in brackets goes to 1, and $(1 - \mu H/n)^n$ goes to $e^{-\mu H}$ because of the general limit

$$\lim_{k \to \infty}\left(1 - \frac{a}{k}\right)^k = e^{-a}$$

Now we can state that

$$\lim_{\substack{n \to \infty \\ p \to 0 \\ np = \text{const.}}} {}_nC_x p^x q^{n-x} = \frac{1}{x!}(\mu H)^x (1) e^{-\mu H}$$

If $\lambda = \mu H$, we have the Poisson pdf.

The expressions for the mean μ and variance σ^2 are found by using expected values. From Eq. (6.4)

$$\mu = E[X] = \sum_{x=0}^{\infty} x P(x; \lambda) = \sum_{x=0}^{\infty} \frac{x e^{-\lambda} \lambda^x}{x!}$$

Factor out $\lambda e^{-\lambda}$, and the sum of the remaining series is e^λ because of the relation

$$\sum_{j=0}^{\infty} \frac{a^j}{j!} = e^a \tag{9.7}$$

Then

$$\mu = \lambda e^{-\lambda} \sum_{x=0}^{\infty} \frac{\lambda^{x-1}}{(x-1)!} = \lambda e^{-\lambda} e^\lambda = \lambda$$

For the variance use the fact that $X^2 = X(X-1) + X$. In expected-value terms

$$\sigma^2 = V[X] = E[X^2] - \mu^2$$
$$= E[X(X-1)] + E[X] - \mu^2$$

Expansion of these expected values and use of Eq. (9.7) will result in the variance $\sigma^2 = \lambda$.

Example 9.9

Problems P9.24, P9.25

SOLVED PROBLEMS

Example 9.8 The number of occurrences of thermal noise in a sample of circuit boards follows a binomial distribution. Plot the pdf for the binomial and the approximating Poisson for the following: (*a*) sample size of 40 and two noisy boards, and (*b*) sample size of 20 and two noisy boards.

SOLUTION (*a*) Let X be the number of occurrences of thermal noise. For $n = 40$ and two noisy boards, p is estimated as $\hat{p} = \frac{2}{40} = 0.05$ and $\lambda = np = 2$. The binomial and Poisson pdf values are listed and plotted in Fig. 9.5. The pdf values are very close for all x values.
(*b*) For the binomial $n = 20$ and $\hat{p} = \frac{2}{20} = 0.10$, and for the Poisson $\lambda = np = 2$. Figure 9.6 indicates close agreement of the values even though the approximation guideline (Fig. 9.4) is violated since $np = 2 > 1$.

COMMENT The Poisson can be used to compute binomial probabilities for a wide range of n and p values. Because the Poisson is the limit of the

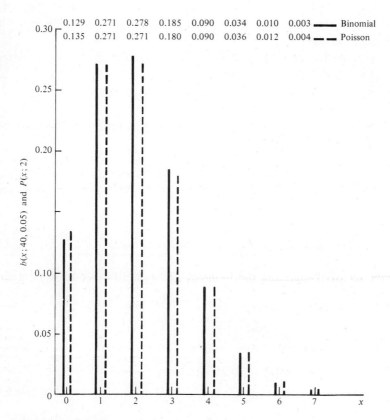

Figure 9.5 Comparison of binomial and Poisson pdf's $b(x; 40, 0.05)$ and $P(x; 2)$, Example 9.8.

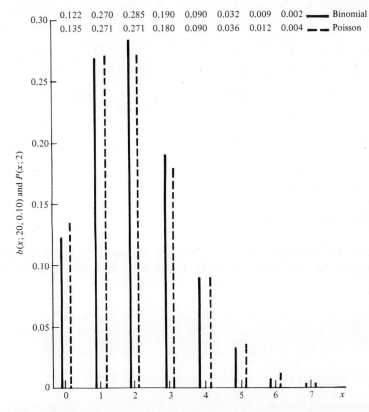

Figure 9.6 Comparison of binomial and Poisson pdf's $b(x; 20, 0.10)$ and $P(x; 2)$, Example 9.8.

binomial via an exponential expansion (Sec. 9.8), when E. C. Molina first published the Poisson pdf table, he called it the exponential binomial limit distribution.

Section 9.7

Example 9.9 (*a*) Determine the moment generating function $M(t)$ for the Poisson.
(*b*) Use it to obtain the mean.

SOLUTION (*a*) From Eq. (6.21), $M(t)$ is

$$M(t) = E[e^{tx}] = \sum \frac{e^{tx} e^{-\lambda} \lambda^x}{x!}$$

$$= e^{-\lambda} \sum_{x=0}^{\infty} \frac{(\lambda e^t)^x}{x!}$$

Using Eq. (9.7)

$$M(t) = e^{-\lambda}e^{\lambda e^t} = \exp[\lambda(e^t - 1)] \tag{9.8}$$

(b) By taking the derivative $M'(t)$ and finding the value at $t = 0$, we have the mean μ:

$$M'(t) = \lambda e^t \exp[\lambda(e^t - 1)]$$

$$\mu = M'(t = 0) = \lambda \exp[\lambda \cdot 0] = \lambda$$

Section 9.8

ADDITIONAL MATERIAL

Benjamin and Cornell [3], pp. 236–249; Brownlee [5], pp. 166–174; Hahn and Shapiro [10], pp. 154–162, 172–174; Hines and Montgomery [12], pp. 137–141; Neville and Kennedy [19], Chap. 8; Newton [20], pp. 78–84; Volk [24], pp. 46–53.

PROBLEMS

The following distributions are used by the indicated problems in this chapter.

A: The number of customers X served by a supermarket checkout averages 0.8 per minute and follows the Poisson distribution $P(x; 0.8)$.

Problems P9.1, P9.4, P9.10, P9.19

B: Sections of 10 m^2 of aluminum sheeting are checked each hour, and the number of defects D is counted. From past experience D follows a $P(d; 2.1)$ distribution.

Problems P9.1, P9.5, P9.11, P9.15

P9.1 Write the complete pdf expression of distributions A and B.

P9.2 State the conditions under which you would use the Poisson and the binomial distributions. How do these conditions differ?

P9.3 If the printed page of this book is inspected and the number of typesetting errors is counted, explain how the Poisson distribution might be applied in this inspection process.

P9.4 (a) Graph the pdf for distribution A.
(b) Determine the probability that $x = 1, 2,$ or 3.

P9.5 Plot the pdf and cdf for the Poisson described by distribution B.

P9.6 The number of orders X processed per day by an employee in the purchasing department is given below.
(a) Estimate the parameter of a Poisson distribution for X.
(b) Compute the mean and standard deviation.

Orders per day X	3	4	5	6	7	8	9	10	11	12	13
Frequency	4	8	11	14	12	15	13	8	3	3	2

P9.7 (a) Graph the pdf for the Poisson $P(y; 4.9)$ and locate μ on it.
(b) Compute the probability that Y is in the range $\mu \pm 1\sigma$.

P9.8 The number of telephone calls coming through a switchboard averages 8.5 per 15-min interval. What is the probability that the number of calls in 15 min will exceed the mean?

P9.9 Use the computations given in Table 9.1 to show that the mean is $\mu = 2.3$ for the Poisson distribution $P(r; 2.3)$.

P9.10 Use Table B-2 to determine the following probabilities for distribution A, which is $P(x; 0.8)$:
(a) $P(x = 1)$
(b) $P(x < 4)$
(c) $P(2 < x \le 5)$
(d) $P(x > 2)$

P9.11 For distribution B, what are the chances of two or less defects per 10 m^2? Use Table B-2.

P9.12 Use Table B-2 to determine the value of x' that places, as closely as possible, 5 percent of the area under the pdf curve in the right tail of the $P(x; 2.4)$ distribution.

P9.13 The average number of chocolate chips per cookie is supposed to be seven according to the general manager. If less than 3 or more than 10 chips are present, the process is readjusted. Develop a Poisson distribution for this situation, and compute the probability that the specification limits set by the manager can be met.

P9.14 The demand per day for a new type of tricycle at a department store has been observed for the past 32 days.

Demand per day	0	1	2	3	4	5
Number of days	10	12	6	2	1	1

Determine the number of tricycles to stock so there is at least a 90 percent chance of meeting demand on any day. Assume that the demand for such items has been Poisson-distributed in the past. Detail the problem-solving steps in Sec. 9.5 when solving this problem.

P9.15 A quality control warning limit is to be calculated 2 standard deviations above the mean for the inspection described by distribution B. What value should this limit have?

P9.16 Rework Example 9.6 with the data below which separates the violations per day for processing lines A and B. Is the answer the same as in the example?

Violations per day		0	1	2	3	4	5	6
Number of days	Line A	15	26	4	3	2	1	0
	Line B	18	18	6	2	3	1	1

P9.17 A frequently quoted example of the Poisson distribution is the number of deaths due to horse kicks in 20 years of observing 10 Prussian army corps in the days of the cavalry. The number of deaths X due to horse kicks is given below using a unit of one corp per year:

Number of deaths	0	1	2	3	4
Number of corps per year	109	65	22	3	1

Compare the observed and expected frequencies for the Poisson distribution.

P9.18 The number of defects in glass bottles follows a Poisson pdf with $\lambda = 1.75$. An inspector has just inspected a bottle and found three defects, so she wants to scrap the bottle.
 (a) Is this a correct decision?
 (b) What number of defects should be observed to change the answer you determined in part (a)?

P9.19 Use distribution A to critique the following logic. A customer service analyst has consistently observed three or more service completions per minute on several checkout lanes. Therefore, it is clear that the average is greater than 0.8 per minute, and a study should be conducted to determine the new average.

P9.20 Two technicians have taken data at different assembly plants to determine X, the number of employees late for work. In a sample of size 55 at plant A, the technician approximated the probability that an employee is late at 0.12. A similar study of 120 people at plant B resulted in a probability of 0.075.
 (a) For both plants determine the values of the parameters and properties (μ and σ) of the correct distribution for X. The total number of employees at plant A is 600 and at B is 1500.
 (b) What approximating distributions can be reliably used in computing probabilities for the variable X at each plant? What are the parameter values and properties for these distributions?

P9.21 (a) Rework Prob. P8.9 using a Poisson approximation. Compare the answers with those using the binomial.
 (b) Does this approximation follow the guides given in Fig. 9.4?

P9.22 Can a Poisson approximation be used to work Prob. P8.24? Why or why not?

P9.23 Summarize the conditions under which both the binomial and the Poisson can be used to approximate the hypergeometric distribution.

*P9.24 Derive the expression for the standard deviation of the Poisson distribution.

*P9.25 The additivity property of the Poisson states that if X_1 and X_2 are both Poisson with parameters λ_1 and λ_2, respectively, then $Y = X_1 + X_2$ is $P(y; \lambda_1 + \lambda_2)$. Use this property to write the pdf for and compute the properties of the distribution for the number of telephone calls in a 30-min interval if the number of calls per 15-min interval is Poisson with a parameter of 8.5 calls.

CHAPTER
TEN

OTHER DISCRETE DISTRIBUTIONS

This chapter covers the fundamentals of three discrete distributions—geometric, Pascal, and discrete uniform. The summary information here helps determine whether the distribution may be useful in a specific situation.

CRITERIA

To complete this chapter, you must be able to do the following for the *geometric* distribution, the *Pascal* distribution, and the *discrete uniform* distribution, respectively.

1. State and graph the pdf, and state the definition of the distribution (Secs. 10.1, 10.4, 10.7).
2. State the parameters, and state the formulas for and compute the expected value and standard deviation of the distribution (Secs. 10.2, 10.5, 10.8).
3. Compute probability values and all properties for an application of the distribution, given a written description of the situation (Secs. 10.3, 10.6, 10.9).

In addition you should be able to:

4. Determine random numbers from a discrete uniform distribution, given a table of random numbers and a description of the problem.

STUDY GUIDE

10.1 Geometric Distribution pdf and Graph

In a Bernoulli trial (Sec. 8.1) we observe one of two events with a constant probability p for success and $q = 1 - p$ for failure. If we continue to sample until the *first* success is observed, we observe x trials to get $x - 1$ failures and one success. Then

$$P(\text{observing } x - 1 \text{ failures}) = q^{x-1}$$
$$P(\text{observing one success}) = p$$

The variable X is the number of Bernoulli trials necessary to observe the first success. The pdf is the product of the two probabilities above:

$$f(x) = \begin{cases} pq^{x-1} & x = 1, 2, \ldots \, ; 0 \leq p \leq 1 \\ 0 & \text{elsewhere} \end{cases} \quad (10.1)$$

This is the *geometric distribution*. The variable is *discrete* and cannot equal 0 because at least one trial is needed for the first success. The geometric, unlike the binomial, does not use a fixed sample size n, because sampling continues until the first success is observed on the xth sample value, which is necessarily preceded by $x - 1$ failures. By definition:

> The *geometric* is a *discrete* distribution which estimates the probability that a certain outcome will occur for the *first* time on the xth sample value taken from an *infinite* population in which the probability of this outcome is a *constant* p.

The geometric pdf's for $p = 0.05$ and $p = 0.80$ are graphed in Fig. 10.1. The pdf is very spread out for a small p and is highly skewed right for a large p. The

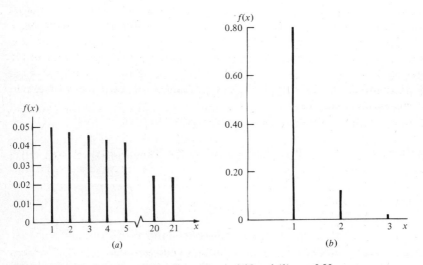

Figure 10.1 Plot of the geometric pdf for (a) $p = 0.05$ and (b) $p = 0.80$.

mode is always at $x = 1$ since $f(1) = p$, while for $x = 2, 3, \ldots f(x)$ decreases because q^{x-1} decreases.

10.2 Geometric Parameter and Properties

The geometric is a one-parameter distribution.

- One parameter: p
- Parameter range: $0 \leq p \leq 1$
- Expected value: $\mu = \dfrac{1}{p}$
- Standard deviation: $\sigma = \dfrac{\sqrt{q}}{p}$

The best estimate for p is the p used for the binomial:

$$\hat{p} = \frac{\text{number of successes}}{\text{sample size}} \qquad (10.2)$$

The sample size is fixed to determine $\hat{p}$, but the variable in the geometric pdf is for all $x = 1, 2, \ldots$

10.3 Geometric Distribution Application

The geometric distribution can be used whenever a sample is taken until the event of interest occurs the *first time*. Example 10.1 interprets the resulting probability values.

Example 10.1 A grocery products company markets 175 different products. Data indicate there is a 20 percent chance that in any one year legal proceedings will be initiated by a consumer or distributor against the company concerning any one of these products. A new product is to be introduced.
(a) Compute the probability of the first lawsuit in the third year of marketing. In the sixth year.
(b) Compute and interpret μ and σ for the distribution.

SOLUTION (a) If X is the year in which the first legal action is observed for the new product, the geometric distribution can be used to compute the chances that $x = 1, 2, \ldots$ The best estimate of p is 0.2. Using Eq. (10.1)

$$P(x = 3) = f(3) = 0.2(0.8)^{3-1} = 0.128$$
$$P(x = 6) = f(6) = 0.2(0.8)^{6-1} = 0.066$$

There is a 12.8 percent chance that the first suit occurs in year 3. The first suit could occur in the sixth year with about half this chance (6.6 percent).

(b) The distribution properties are

$$\mu = \frac{1}{0.20} = 5.0 \text{ years}$$

$$\sigma = \frac{\sqrt{0.8}}{0.2} = 4.5 \text{ years}$$

On the average the company can expect a new product to be marketed for 5 years before any legal involvement. The standard deviation around this 5-year figure is 4.5 years.

COMMENT The probability of 1σ from the mean is $P(\mu - 1\sigma \leq x \leq \mu + 1\sigma) = P(0.5 \leq x \leq 9.5) = 0.866$. Therefore, approximately 87 percent of the products will have legal action initiated between 6 months and 9.5 years after introduction.

Problems P10.1–P10.5

10.4 Pascal Distribution pdf and Graph

If we observe the outcome of successive Bernoulli trials as in the geometric distribution, but are interested in the probability of r successes after x failures, we have the *Pascal* distribution, also called the *negative binomial*. For example, the Pascal will compute the probability of observing $r = 3$ successes after $x = 5$ failures. In the case $r = 1$, the Pascal reduces to the geometric distribution.

If p is the constant probability of success on each trial and $q = 1 - p$, in a total of $x + r$ trials

$$P(\text{observing } x \text{ failures}) = q^x$$

$$P(\text{observing } r \text{ successes}) = p^r$$

In the $x + r$ trials, we observe x failures and $r - 1$ successes from the first $x + r - 1$ trials in any combination, with the rth success on trial $x + r$. So

$$\text{Number of combinations} = {}_{x+r-1}C_x$$

The Pascal pdf is the product of these three values:

● $$f(x) = \begin{cases} {}_{x+r-1}C_x \, p^r q^x & x = 0, 1, 2 \ldots \\ & 0 \leq p \leq 1; r = 1, 2, \ldots \\ 0 & \text{elsewhere} \end{cases} \quad (10.3)$$

Remember, X is the *number of failures*, not the total number of trials. By definition:

> The *Pascal* is a discrete distribution which estimates the probability that a certain outcome will occur for the rth time after x nonoccurrences in a sample taken from an *infinite* population in which the probability of the outcome occurring is a *constant p*.

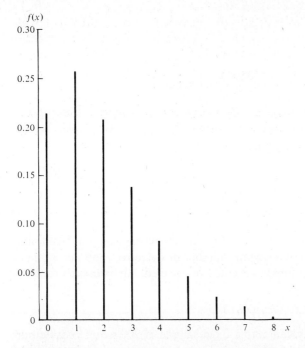

Figure 10.2 Pascal pdf for $p = 0.60$ and $r = 3$.

The pdf for $p = 0.6$ and $r = 3$ (Fig. 10.2) is skewed right. If the p value is decreased, the probability values become smaller, because it takes more failures to observe r successes. This makes the pdf wider and flatter.

10.5 Pascal Parameters and Properties

The Pascal is a two-parameter distribution:

    Two parameters: p and r
    Parameter ranges: $0 \leq p \leq 1$
                                 $r = 1, 2, 3, \ldots$
    Expected value: $\mu = \dfrac{rq}{p}$
    Standard deviation: $\sigma = \dfrac{\sqrt{rq}}{p}$

The best estimate of p is found by Eq. (10.2). The value of r is usually set by design specifications.

Example 10.2 In a discussion with the quality control manager a technician states that if a process line makes 5 percent defective product, and a product lot is rejected when 5 defectives are found, an average of 100 items must be inspected prior to rejection. Is the technician correct?

SOLUTION It is possible to translate the statement into a Pascal pdf with $p = 0.05$ and $r = 5$ successes, which are defective products. The expected value is

$$\mu = \frac{5(0.95)}{0.05} = 95 \text{ items}$$

The mean value is the average number of acceptable items. Including the $r = 5$ defective items, the technician is correct, because total sample size will average $x + r = 100$ items.

10.6 Pascal Distribution Application

The Pascal is applied in areas such as quality control when sequential sampling is used. Samples are taken until a certain number of defectives and acceptances are observed; then all the product is accepted or rejected. An example probability computation is shown here.

Example 10.3 A company can take a fixed sample of 200 items to accept or reject the product. Alternatively, a sequential sampling plan is available which requires rejection when five items are defective. In one day the sequential plan resulted in 5 defective items in 15 tests on a 3 percent defective line, so the day's product was rejected.
(a) Should the product be rejected?
(b) Is the average sample size prior to rejection smaller or larger than the fixed sample size of 200?

SOLUTION Use the problem-solving steps of Sec. 7.8.

(a) Step 1. Check assumptions. The Pascal pdf is appropriate because a sample is taken from an infinite population having a constant defective rate.
Step 2. Determine the variables and parameters. Let

X = number of acceptable items before r defectives are
 = observed
$p = 0.03$
$r = 5$ defective items

Step 4. Find the sum of several probabilities. To determine if the rejection is warranted, compute the probability that $r = 5$ and $x = 0, 1, \ldots, 10$ using Eq. (10.3).

$$P(x = 0 \text{ to } 10) = \sum_{x=0}^{10} {}_{x+4}C_x 0.03^5 0.97^x = 0.0001$$

Since the chances of observing 5 defectives and 10 or less acceptances are extremely small, this outcome is unusual, so rejection is necessary. The parameter $p = 0.03$ and test results for each item should be checked.

(b) Step 5. Compute the mean μ. The expected number of acceptable product prior to five defectives is the expected value of X.

$$\mu = \frac{rq}{p} = \frac{5(0.97)}{0.03} = 161.7 \text{ items}$$

Including the five defectives, the average sample size is $161.7 + 5 = 166.7$ items. This is less than the 200 fixed sample-size plan.

COMMENT As discussed in Sec. 8.6, the computation in part (a) for $P(x = 10)$ alone would be incorrect because five defectives could be observed after $0, 1, \ldots, 10$ acceptances and the product would still be rejected. The conclusion must be based on the sum of the probabilities for all possible outcomes, not only the one observed in this instance.

Problems P10.6–P10.9

10.7 Discrete Uniform Distribution pdf and Graph

If the probability of occurrence for a discrete variable is equal for all values, the *discrete uniform* is the correct distribution. The toss of a die is an example. All values 1 through 6 have a $\frac{1}{6}$ chance of appearing. The pdf for any discrete uniform variable is

- $$f(x) = \begin{cases} \dfrac{1}{b - a + 1} & x = a, a + 1, \ldots, b - 1, b \\ 0 & \text{elsewhere} \end{cases} \quad (10.4)$$

The value of $f(x)$ is the same for all x values between a and b. Figure 10.3a is a graph of $f(x) = \frac{1}{8}$ between $x = 0$ and $x = 7$, and Fig. 10.3b graphs $f(x) = \frac{1}{5}$ between 10 and 14. Note that a does not have to be 0. There are always as many vertical lines on the graph as there are x values. In fact, it is possible to state $f(x)$ as

- $$f(x) = \begin{cases} \dfrac{1}{k} & x = k \text{ different values} \\ 0 & \text{elsewhere} \end{cases} \quad (10.5)$$

A graph for nonequally spaced, noninteger x values can be expressed as a discrete uniform pdf, as shown in Fig. 10.3c. There are $k = 5$ different values, so $f(x) = \frac{1}{5}$. By definition:

> The *discrete uniform* pdf estimates the probability of observing a particular discrete value of a variable when all values have an equal chance of occurring.

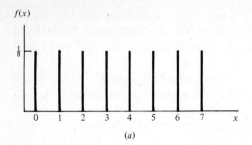
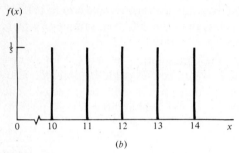
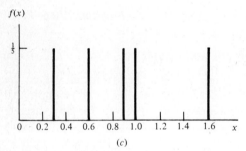

Figure 10.3 Graphs of the discrete uniform pdf.

10.8 Discrete Uniform Parameters and Properties

The pdf in Eq. (10.4) is the commonly used form. It has two parameters:

Two parameters: a and b
Parameter values: a and b = integers (positive or negative)

- Expected value: $\mu = \dfrac{a+b}{2}$

Standard deviation: $\sigma = \left[\dfrac{(b-a+1)^2 - 1}{12}\right]^{1/2}$

10.9 Discrete Uniform Distribution Application

Any situation which involves a discrete variable and equal probability for all values is an application of the discrete uniform.

Example 10.4 In a study on a certain type of missile, the number of electronic components with substandard reliability values was found to be evenly distributed between 10 and 35.
(a) Compute the average number of substandard components.
(b) Compute the chance that at least 20 components are substandard, thus indicating the need for a missile overhaul.

SOLUTION (a) The average is the expected value of X, the number of substandard components:

$$\mu = \frac{10 + 35}{2} = 22.5 \text{ components}$$

(b) We want $P(x \geq 20) = 1 - P(x \leq 19)$. The cdf can be used to compute

$$P(x \leq 19) = \sum_{x=10}^{19} f(x)$$

Using Eq. (10.4), the pdf is

$$f(x) = \frac{1}{35 - 10 + 1} = \frac{1}{26} \qquad x = 10, 11, \ldots, 35$$

Therefore $\quad P(x \leq 19) = \frac{10}{26}$

$$P(x \geq 20) = 1 - \frac{10}{26} = \frac{16}{26} = 0.615$$

There is a 61.5 percent chance that a missile will have to be overhauled when checked.

Problems P10.10–P10.14

10.10 Discrete Uniform and the Use of Random Numbers

The pdf for the integers $x = 0, 1, \ldots, 9$ is $f(x) = \frac{1}{10}$. Repeated random sampling from this $f(x)$ generates a table of one-digit random numbers which may be combined into random numbers of any size for use in experiments, simulations, etc. Tables of random numbers are easily obtained (1). A short sample is included in Table 10.1. Digit grouping, which is arbitrary, is by 5s for easy reading.

A random number table is correctly used with the following step-by-step procedure.

1. Determine the largest and smallest values to be selected from the table. The number of digits in each random number equals the number of digits in the largest value.

Table 10.1 Example of a random number table

48867	33971	29678	13151	56644	49193	93469	43252	14006	47173
32267	69746	00113	51336	36551	56310	85793	53453	09744	64346
27435	03196	33877	35032	98054	48358	21788	98862	67491	42221
55753	05256	51557	90419	40716	64589	90398	37070	78318	02918
93142	50675	04507	44001	06365	77897	84566	99600	67985	49133
98658	86583	97433	10733	80495	62709	61357	66903	76730	79355
68216	94830	41248	50712	46878	87317	80545	31484	03195	14755
17901	30815	78360	78260	67866	42304	07293	61290	61301	04815
88124	21868	14942	25893	72695	56231	18918	72534	86737	77792
83464	36749	22336	50443	83576	19238	91730	39507	22717	94719
91310	99003	25704	55581	00729	22024	61319	66162	20933	67713
32739	38352	91256	77744	75080	01492	90984	63090	53087	41301
07751	66724	03290	56386	06070	67105	64219	48192	70478	84722
55228	64156	90480	97774	08055	04435	26999	42039	16589	06757
89013	51781	81116	24383	95569	97247	44437	36293	29967	16088
51828	81819	81038	89146	39192	89470	76331	56420	14527	34828
59783	85454	93327	06078	64924	07271	77563	92710	42183	12380
80267	47103	90556	16128	41490	07996	78454	47929	81586	67024
82919	44210	61607	93001	26314	26865	26714	43793	94937	28439
77019	77417	19466	14967	75521	49967	74065	09746	27881	01070
66225	61832	06242	40093	40800	76849	29929	18988	10888	40344
98534	12777	84601	56336	00034	85939	32438	09549	01855	40550
63175	70789	51345	43723	06995	11186	38615	56646	54320	39632
92362	73011	09115	78303	38901	58107	95366	17226	74626	78208
61831	44794	65079	97130	94289	73502	04857	68855	47045	06309

Reproduced with permission of the Rand Corporation, from *A Million Random Numbers with 100,000 Normal Deviates*. Copyright © The Rand Corporation, Free Press, Glencoe, Ill., 1955.

2. Determine how many random numbers are needed.
3. Devise the scheme you will use in the tables. For example, start at the bottom of the left column, take the last two digits, and go up; then down the next column, etc. The scheme is arbitrary, but it should be followed for all numbers needed at one time.
4. Read the random numbers from the table using the limits from step 1. Disregard values outside the limits, but do *not* discard repeats.

Example 10.5 The time between truck arrivals at a single tollbooth for a freeway varies uniformly between 5 and 75 s. Develop the random numbers that a civil engineer might use to simulate 15 min of arrivals.

SOLUTION Use the previous steps:

1. All values must be between 05 and 75 and are therefore two digits long.

Table 10.2 Random number generation, example 10.5

Random number	Elapsed time (s)	Random number		Elapsed time (s)
48	48	64		529
73	121	44		573
39	160	46		619
71	231	32		651
29	260	52		703
67	327	14		717
31	358	64		781
51	409	71	STOP	852
56	465	73		925

2. The exact number of values needed is not known. We must keep track of total elapsed time and stop after 15(60) = 900 s.
3. The arbitrary scheme starts with two digits from the upper left and continues to the right around the perimeter. Proceed inward by two digits with the newly defined perimeter.
4. Table 10.2 presents only the numbers within the 05 and 75 limits. A total of 17 random values was needed before 900 s was reached, but not exceeded (indicated by STOP).

COMMENT Since the numbers are random, the next set may have more or less than 17 values to simulate 15 min of elapsed time. If the limits had been 75 to 125 s, each number would be three digits long, in which case the digits 089 represent 89 s. The leading 0 is needed to give all outcomes an equal chance of appearing, because two digits cannot include values from 100 to 125.

Problems P10.15–P10.18

REFERENCE

(1) The Rand Corporation, *A Million Random Numbers with 100,000 Normal Deviates*, Free Press, Glencoe, Ill., 1955.

ADDITIONAL MATERIAL

Geometric distribution (Secs. 10.1 to 10.3): Benjamin and Cornell [3], pp. 228–230; Gibra [8], pp. 144–145; Hahn and Shapiro [10], pp. 152–154; Miller and Freund [18], pp. 83–86.

Pascal distribution (Secs. 10.4 to 10.6): Benjamin and Cornell [3], pp. 230–235; Gibra [8], pp. 145–148; Hahn and Shapiro [10], pp. 154, 158, 159, 167–169; Walpole and Myers [25], pp. 103–105.

Random number generation (Sec. 10.10): Hahn and Shapiro [10], pp. 236–251, 257–259.

PROBLEMS

P10.1 A relay has a probability of working correctly of 0.995.
  (*a*) What is the probability that the switch will fail for the first time on trial number 11?
  (*b*) What is the expected number of trials to failure?

P10.2 A sequential testing scheme is being used to control insects on cotton patches. There is a 0.3 probability that the chemical used will destroy the insects. Write the pdf and properties for the variable Y, the number of fields treated before the first failure of the chemical is observed.

P10.3 The percentage defective of product lots has been 5 percent for a long time. A sampling plan has required a sample of 30 items and rejection if one or more defectives were found. Improvement to the process has reduced the defective rate to 4 percent. The new sampling plan calls for inspection until the first defective is found, followed by lot rejection.
  (*a*) Is the average number of items inspected prior to rejection for the new plan less than or more than the constant sample of 30 items in the old plan?
  (*b*) What is the probability that at most five items must be inspected prior to rejection?

P10.4 (*a*) Show that the cdf of the geometric distribution is $1 - q^x$ for $x = 1, 2, \ldots$ You will need to use the following summation relation with $q = 1 - p$:

$$\sum_{j=1}^{x} p(q)^{j-1} = p\frac{q^x - 1}{q - 1}$$

  (*b*) Use this cdf to determine the probability that the first defective item in Prob. P10.3 would be observed some time after the inspection of 30 items.

P10.5 The probability that a person wins any talent contest is 0.2.
  (*a*) What is the expected number of contests that must be entered before the person wins for the first time?
  (*b*) What is the probability that the person wins on some try prior to the expected value computed above?

P10.6 The probability that a particular transformer is ordered from a warehouse on any order is 0.2. If the stock level is 3, what is the probability that in exactly 10 orders (*a*) the demand will exactly equal the stock level and (*b*) the demand will not exceed the stock level?

P10.7 An inspector checks batches of clocks for acceptance and rejection. The sample size is 35; and as soon as 2 defectives are found, the batch is scrapped. If all 35 are inspected and less than 2 defectives found, the batch is approved.
  (*a*) If the fraction defective has been 0.08, what are the average number X of clocks checked prior to batch rejection?
  (*b*) Compute the standard deviation of X.

P10.8 John plans to toss a die until he observes two even numbers.
  (*a*) Compute the expected number of tosses required to observe the two even numbers.
  (*b*) What is the chance that the second even number will appear on the fifth toss? The tenth toss?

P10.9 Show that if $r = 1$, the Pascal and the geometric distributions are the same.

P10.10 The number of calls N received per week by a fire station varies uniformly between 20 and 30.

(a) Write the pdf for N using the forms of Eqs. (10.4) and (10.5).
(b) Locate μ and $\mu \pm 0.1\sigma$ on the pdf graph.
(c) Compute the probability that at least 28 calls are received in any one week.

P10.11 Write and graph the complete cdf relation for the discrete uniform distribution with parameters $a = 0$ and $b = 8$.

P10.12 Analysis has shown that one round of a particular dice game will return the player $5, $2, $0, or −$5, all with equal probability.
(a) Compute the mean and standard deviation of the return R.
(b) Plot the pdf for R and locate the mean.

P10.13 The response time T of an automatic underwater circuit activator occurs in 0.01-s units. Depending on the water temperature, the activation time for 100 tests varied from 0.01 to 0.06 s with these frequencies:

Response time (s)	0.01	0.02	0.03	0.04	0.05	0.06
Frequency	18	18	15	14	19	16

Graph the expected frequencies using the discrete uniform distribution and compare with the observed frequencies. Does this seem to you to be an appropriate pdf for the response time T?

P10.14 The size of back orders X for refrigerators at a central supply house has varied between 0 and 20 for the last few months. If all sizes of back orders have an equal chance of occurring, (a) compute the mean of X and (b) determine the one standard deviation limits above and below the mean.
(c) What is the probability that X is within these limits?

P10.15 Approximately 500 air conditioner coil systems are produced each day by the Air Cool Company. A random sample of 10 coils is to be checked for pressure as they come off the process line. Use the table of random digits (Table 10.1) to determine which of the 500 coils should be checked.

P10.16 A work sampling study requires that 15 random observations per day be made of an assembly operation to estimate the percentage of time spent working and the percentage spent idle. The operation is performed from 8 A.M. to 12 noon and from 1 P.M. to 5 P.M. Develop a technique and use it to obtain the 15 observation times in hours and minutes during the 8 working hours.

P10.17 Numbers between 00 and 59 are generated using a table of random numbers.
(a) What distribution is used to select the random number, and what are its parameter values?
(b) Write the pdf for these two-digit values.

P10.18 An IE technician wants to perform a study of file clerks in an insurance office. There are three clerks on floor 1, four on floor 2, and five on floor 3. If a total of 12 observations is needed on one day, starting at 8 A.M. and going straight through to 5 P.M., develop a technique and use it to determine the random times for observation and clerks to be observed. Allow at least 10 min between each observation for travel time. Also, at each observation time, determine a backup clerk on the same floor who will be observed in case the primary clerk is not available.

CHAPTER ELEVEN

BASICS OF THE NORMAL DISTRIBUTION

This chapter will help you understand the most commonly used continuous distribution—the normal. Emphasis is on computations and correct understanding of the properties of this distribution. This chapter includes a case study.

CRITERIA

To correctly use the normal distribution, you must be able to:

*1. State the pdf equation and descriptor.
2. State and compute the two parameters and properties of the normal.
3. Derive the *standard normal distribution* (SND) from the normal, and state the properties of the SND.
4. Construct the pdf graph for the SND; be able to compute, and draw on the graph, the area under the curve for the values ± 1, ± 2, and ± 3 for any normal distribution, given the mean and the standard deviation.
5. Use the SND table to determine the probability that a variable is between certain limits, given the mean and standard deviation of a normally distributed variable.
6. Solve any normal-curve problem by finding probability values or variable values given a statement of the situation and the parameters of the normal.
7. Fit a normal distribution to observed data using the steps outlined, given the observed frequency data.
8. State the conditions under which the normal can approximate the binomial distribution, and compute probabilities with it, given the binomial parameters.
*9. State the mgf of the normal distribution, and compute the mean and variance using it.

11.1 DEFINITION OF THE NORMAL PDF

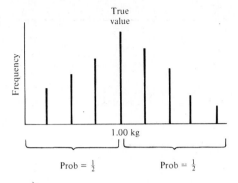

Figure 11.1 Plot of central tendency of observed weights around a true value of 1.00 kg.

STUDY GUIDE

11.1 Definition of the Normal pdf

If 1000 people use the same scale to weigh a package that actually weighs 1.00 kg, there will be values above and below 1.00 kg. If the probability of an error on either side of the true value is 0.5, a frequency plot of observed weights will have a strong central tendency around 1.00 kg (Fig. 11.1). The *error* about the true value may be defined as a random variable X which is continuous over the range $-\infty$ to $+\infty$. Derivation of a pdf $f(x)$ to describe the deviation from the true value μ results in the normal pdf:

$$f(x) = \begin{cases} \dfrac{1}{\sigma\sqrt{2\pi}} e^{-(x-\mu)^2/(2\sigma^2)} & -\infty \leq x \leq \infty;\, \sigma > 0;\, -\infty \leq \mu \leq \infty \\ 0 & \text{elsewhere} \end{cases} \quad (11.1)$$

Since the normal is so commonly used, we define the descriptor $X \sim N(\mu, \sigma^2)$, which is read "X is normally distributed with μ and σ^2." We will discuss μ and σ^2 later in this chapter. The normal is also referred to as the Gaussian, bell-shaped, or error distribution.

The normal is assumed to be the underlying population for statistical tests which evaluate hypotheses developed as a result of observing actual data. It is used to explain the distribution of IQ scores; errors around a manufacturing specification of size, weight, volume, etc.; and physical characteristics of animals, plant life, and people. It is, indeed, the most important distribution you will learn. By definition:

> The *normal* is a *continuous* distribution used to represent the randomly occurring error of a measurable variable that is observed in a sample taken from an infinite population.

Example 11.1 (a) What is the descriptor of

$$f(w) = \frac{1}{7.52} \exp\left[\frac{-(w-5.5)^2}{18}\right] \quad -\infty \leq w \leq \infty$$

(b) Write the pdf for the normal variable U using a true value of 7.5 cm and $\sigma = 0.9$ cm.

SOLUTION (a) From the exponential the true value of W is $\mu = 5.5$ and $18 = 2\sigma^2$ or $\sigma^2 = 9$. The descriptor is $W \sim N(5.5, 9)$.
(b) If $\mu = 7.5$ and $\sigma^2 = (0.9)^2 = 0.81$, the descriptor is $U \sim N(7.5, 0.81)$ and the pdf is

$$f(u) = 0.443 \exp\left[\frac{-(u-7.5)^2}{1.62}\right] \quad -\infty \leq u \leq \infty$$

COMMENT Remember that σ^2 is in the descriptor, not σ. The incorrect use of σ in $N(\mu, \sigma^2)$ will result in the wrong pdf and incorrect probability values.

Problems P11.1–P11.3

11.2 Parameters and Properties of the Normal

The normal is a continuous, two-parameter distribution:

- Two parameters: Location is μ and scale is σ
- Location parameter range: $-\infty \leq \mu \leq \infty$
- Scale parameter range: $\sigma > 0$

The location parameter moves the pdf back and forth along the real line. The scale parameter gives the pdf more or less spread. Expected-value formulas are used to obtain the properties

- Expected value: μ
- Standard deviation: σ

The parameters and the properties are the same. The best estimators for the parameters are:

$$\hat{\mu} = \text{sample mean} = \overline{X} \qquad (11.2)$$

$$\hat{\sigma} = \text{sample standard deviation} = s \qquad (11.3)$$

Equation (3.8) may be used to compute the biased s. The estimate is slightly biased by using n rather than $n - 1$, but sufficiently large samples ($n \geq 30$) substantially reduce this bias.

Table 11.1 Average mileage data for tires, Example 11.2

Mileage (+1000)	Midpoint (+1000)	Frequency f
20–22	21	2
22–24	23	11
24–26	25	20
26–28	27	35
28–30	29	22
30–32	31	9
32–34	33	1

Example 11.2 The average mileage M for 100 sets of tires made of a new synthetic rubber is given in Table 11.1. A normal distribution is to be fit to this data. Estimate the location and scale parameters and write the descriptor for M.

SOLUTION The estimators, Eqs. (11.2) and (11.3), are $\hat{\mu} = \overline{X} = 26{,}900$ mi and $\hat{\sigma} = s = 2440$ mi. The variable $M \sim N[26{,}900, (2440)^2]$, where $(2440)^2 = \sigma^2$.

COMMENT If Eq. (3.10) is used to estimate σ, the unbiased s value is 2450 mi, only 10 mi more. The difference is too small to worry about, especially when the data is already approximated by midpoints!

Problems P11.4–P11.7

11.3 Derivation of the Standard Normal Distribution (SND)

Rather than computing and tabulating $f(x)$ for many μ and σ values, the normal is transformed to the *standard normal distribution* (SND). The variable Z is defined as

$$Z = \frac{\text{deviation from mean}}{\text{standard deviation}} = \frac{x - \mu}{\sigma} \qquad (11.4)$$

Figure 11.2 shows how the x scale is transformed into z. When x is at the mean, $z = 0$; and when x is $\pm 1\sigma$ from μ, $z = \pm 1$. For example, if $\mu = 5$ and $\sigma = 0.3$, at $x = 5 + 0.3 = 5.3$ the SND value is $z = (5.3 - 5.0)/0.3 = +1$. Therefore, the SND always has $\mu = 0$ and $\sigma = 1$. The descriptor for Z is either $Z \sim N(0, 1)$ or $Z \sim \text{SND}$. Z has the same range as X, $-\infty$ to $+\infty$, and the pdf of Z is

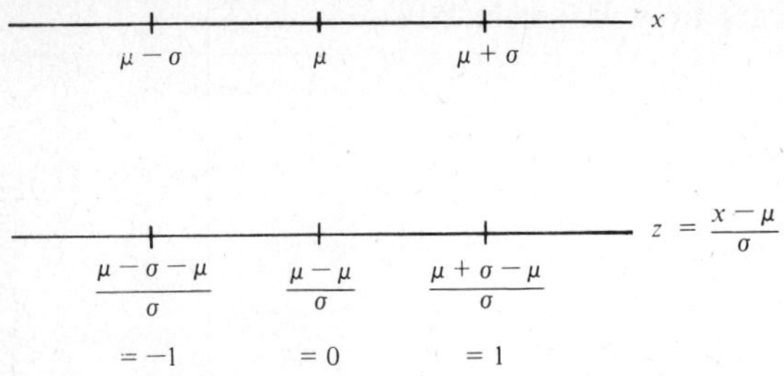

222 BASICS OF THE NORMAL DISTRIBUTION

$$\begin{array}{c|c|c|c}
\hline
 & & & x \\
\mu - \sigma & \mu & \mu + \sigma &
\end{array}$$

$$z = \frac{x - \mu}{\sigma}$$

$$\begin{array}{c|c|c}
\frac{\mu - \sigma - \mu}{\sigma} & \frac{\mu - \mu}{\sigma} & \frac{\mu + \sigma - \mu}{\sigma} \\
= -1 & = 0 & = 1
\end{array}$$

Figure 11.2 Relationship between the normal scale x and the SND scale $z = (x - \mu)/\sigma$.

obtained by using Eqs. (11.1) and (11.4).

$$f(z) = \begin{cases} \dfrac{1}{\sqrt{2\pi}} e^{-z^2/2} & -\infty \leq z \leq \infty \\ 0 & \text{elsewhere} \end{cases} \quad (11.5)$$

The parameters are μ and σ and properties for the SND are always the same:

Expected value: $\mu = 0$
Standard deviation: $\sigma = 1$

The coefficients of skewness a_3 and peakedness a_4 for the normal distribution are used as norm values in working with any other distribution or collected data. The norm values are $a_3 = 0$ and $a_4 = 3$ for the SND or any normal distribution.

Example 11.3 Julie is working with a variable $X \sim N(50, 9)$ while Jim is using $Y \sim N(47, 10.24)$. The distributions are for similar processes at two different plants. How can Jim and Julie discuss their variables without constantly accounting for different μ and σ values?

SOLUTION The two variables are different, so it would be easier to transform to the SND. Use Z_1 as the SND for Julie and Z_2 for Jim;

$$Z_1 \text{ and } Z_2 \sim \text{SND}$$

The difference in μ and σ will be accounted for in the probability statements. For example, using Eq. (11.4), we have

$$P(x \leq 52) = P\left(z_1 \leq \frac{52 - 50}{3}\right) = P(z_1 \leq 0.667)$$

$$P(y \leq 52) = P\left(z_2 \leq \frac{52 - 47}{3.2}\right) = P(z_2 \leq 1.563)$$

Problems P11.8–P11.11

11.4 Graph and Areas of the SND

You will find it helpful to draw a picture of what you want to find on the normal pdf before you solve the problem. Go slowly in this section and study the appropriate figure as you read the material.

11.4 GRAPH AND AREAS OF THE SND

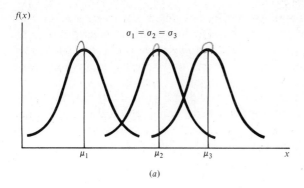

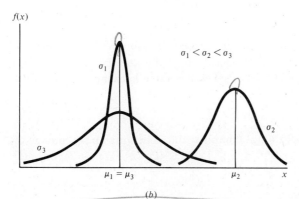

Figure 11.3 The normal pdf for different values of (a) μ and (b) μ and σ.

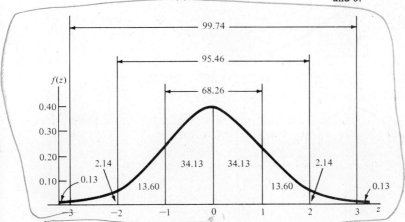

Figure 11.4 Percent areas under the SND curve.

224 BASICS OF THE NORMAL DISTRIBUTION

The characteristic bell shape of the normal pdf is shown in Fig. 11.3 for different values of the parameters μ and σ. The location parameter μ moves the curve along the x axis while the dispersion changes with σ, the scale parameter. All normal curves are symmetric about μ. Areas under the SND curve represent the probability that Z is between any two specific values. In this section we will concentrate on the area between ± 1, ± 2 and ± 3 (Fig. 11.4). On the corresponding graph for $N(\mu, \sigma^2)$ these areas are the same as $\mu \pm 1\sigma$, $\mu \pm 2\sigma$, and $\mu \pm 3\sigma$, respectively. In Fig. 11.4, for example, the area between ± 2 is 0.9546, which is the value of the integral below for Eq. (11.5):

$$\int_{-2}^{2} f(z)\, dz = \int_{-2}^{2} \frac{1}{\sqrt{2\pi}} \exp\left[-\frac{z^2}{2}\right] dz = 0.9546$$

This result is the same for $X \sim N(\mu, \sigma^2)$ in Eq. (11.1). Graphically, we always work with the SND. Since the pdf curve is symmetric around $\mu = 0$, it is possible to determine the area on one side of 0 and multiply by 2 to get the area for both sides. Symmetry also requires that 0.50 of the area be above and 0.50 below 0. So the area above (or below) a specific z value is found by subtraction from 0.50.

Example 11.4 Use Fig. 11.4 to compute the following areas under the normal curve described: (a) $P(0 \leq z \leq 1)$ for SND; (b) $P(|z| \geq 2)$ for SND; and (c) $P(21 \leq x \leq 33)$ for $N(25, 16)$.

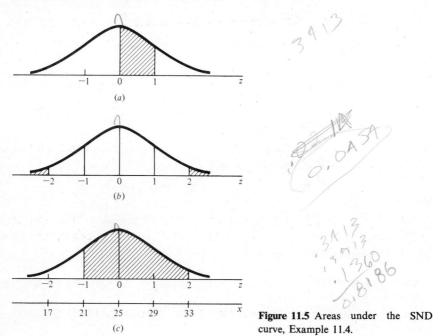

Figure 11.5 Areas under the SND curve, Example 11.4.

SOLUTION (*a*) See Fig. 11.5*a*. The hatched area between $z = 0$ and $z = 1$ is 0.3413 of the total area under the SND curve, as shown in Fig. 11.4. So there is a 34.13 percent chance that $0 \leq z \leq 1$.

(*b*) See Fig. 11.5*b*. If $-2 \geq z \geq 2$, we need the area outside the $z = 2$ and $z = -2$ values. There are two ways to obtain the value:

1. Find $P(z \geq 2)$ and multiply by 2 because of symmetry:

$$P(z \geq 2) = \tfrac{1}{2} \text{ total area} - P(0 \leq z \leq 2)$$
$$= 0.5 - (0.3413 + 0.1360) = 0.0227$$
$$P(|z| \geq 2) = 2P(z \geq 2) = 0.0454$$

2. Find $P(-2 \leq z \leq 2)$ and subtract from the total area:

$$P(-2 \leq z \leq 2) = 0.9546$$
$$P(|z| \geq 2) = 1.0 - 0.9546 = 0.0454$$

(*c*) See Fig. 11.5*c*. Use the transformation Eq. (11.4) for each *x* value:

$$x = 21: \qquad z = \frac{x - \mu}{\sigma} = \frac{21 - 25}{4} = -1$$

$$x = 33: \qquad z = \frac{33 - 25}{4} = 2$$

Then $P(-1 \leq z \leq 2)$ is most easily computed as

$$P(-1 \leq z \leq 0) + P(0 \leq z \leq 2) = 0.3413 + (0.3413 + 0.1360)$$
$$= 0.8186$$

Problems P11.12–P11.16

11.5 Use of the Standard Normal Table

The cdf area under the SND curve from $-\infty$ to a particular value z is tabulated in Table B-3. The table gives the probability value

$$P(-\infty \leq Z \leq z) = \int_{-\infty}^{z} N(0, 1) \, dz$$

Values of z from -3.59 to 0.00 are given on page 627 and from 0.00 up to $+3.59$ on page 628. Be sure you are on the correct page when extracting a value. Use the following steps to find a probability value from Table B-3:

1. Write the desired probability statement in terms of the variable, for example, $P(5 \leq x \leq 10)$.
2. Convert to a probability statement for the SND using Eq. (11.4), for example, $P(-1 < z \leq 2)$.

3. If the statement in step 2 has two fixed boundaries or is in the form $P(z \geq$ number), rewrite it, remembering that the table contains the cdf value. For example,
$$P(-1 \leq z \leq 2) = P(z \leq 2) - P(z \leq -1)$$
4. Look up the probability values:
$$P(z \leq 2) = 0.9773$$
$$P(z \leq -1) = 0.1587$$
5. Perform the arithmetic of step 3:
$$P(-1 \leq z \leq 2) = 0.9773 - 0.1587 = 0.8186$$
6. State the answer in terms of the original variable:
$$P(5 \leq x \leq 10) = 0.8186$$

Example 11.5 further illustrates the use of Table B-3.

Example 11.5 Compute the probability that:
(a) Y exceeds 17.5 if $Y \sim N(10, 9)$.
(b) W is between 9 and 12 if $W \sim N(10.5, 1)$.
(c) X is less than 5 and exceeds 15 if $X \sim N(12, 16)$.

SOLUTION Try to construct the SND graph yourself first. Then look at Fig. 11.6a through c.
(a) Using the steps outlined above (Fig. 11.6a),

1. Find $P(y \geq 17.5)$.
2. Find $P(z \geq (17.5 - 10)/3 = +2.5)$.
3. Rewrite in the cdf form $P(z \geq +2.5) = 1 - P(z \leq +2.5)$.
4. From Table B-3, $P(z \leq +2.5) = 0.9938$.
5. $P(z \geq +2.5) = 1 - 0.9938 = 0.0062$.
6. In terms of Y, $P(y \geq 17.5) = 0.0062$.

(b) See Fig. 11.6b. Without detailing the steps,
$$P(9 \leq w \leq 12) = P(-1.5 \leq z \leq +1.5)$$
$$= P(z \leq +1.5) - P(z \leq -1.5)$$

$P(z \leq +1.5) = 0.9332$ and $P(z \leq -1.5) = 0.0668$

$$P(-1.5 \leq z \leq +1.5) = 0.9332 - 0.0668 = 0.8664$$
Therefore, $P(9 \leq w \leq 12) = 0.8664$

(c) See Fig. 11.6c. The required probability is
$$P(5 > x > 15) = P(-1.75 > z > 0.75)$$

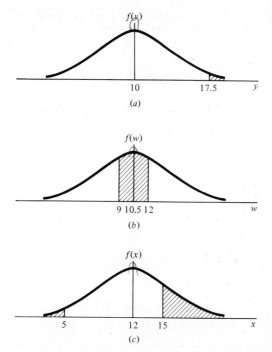

Figure 11.6 SND curves and required areas for Example 11.5.

This probability has two boundaries (step 3), but they are on the inside of the desired areas, so the two separate areas are added.

$$P(-1.75 > z > 0.75) = P(z < -1.75) + P(z > 0.75)$$
$$= 0.0401 + (1 - 0.7734) = 0.2667$$

Therefore, $\quad P(5 > x > 15) = 0.2667$

COMMENT You should rework part (b) using the symmetry property of the SND. The answer should still be 0.8664.

Table 11.2 is a summary of SND percentage areas for commonly used z values. For convenience, areas to one side of z or between two z values are given.

Problems P11.17–P11.21

11.6 Solving Problems with the Normal

The problem-solving steps of previous chapters are somewhat inappropriate for the normal. The Z variable relations are used to determine either probability values or values of the variable which satisfy specific probabilities. The latter

Table 11.2 Commonly used z values and percentage areas for the standard normal distribution

One-sided—outside			Two-sided—inside		
z	Percentage area (%)	Either graph	z	Percentage area (%)	Graph
3.0	0.135	Above z	3.0	99.73	
2.5	0.620		2.5	98.76	Between
2.326	1.0		2.326	98.00	+z and −z
2.0	2.28		2.0	95.44	
1.96	2.5	z	1.96	95.00	
1.645	5.0	Below −z	1.645	90.00	
1.5	6.68		1.5	86.64	
1.282	10.0		1.282	80.00	
1.0	15.87	−z	1.0	68.26	

application helps in the determination of design criteria to guarantee that some stated percentage of the product meets the criteria.

Example 11.6 A textile manufacturer is considering a new material-cutting machine. Tests indicate that cuts are normally distributed around the desired mean with $\sigma = 0.07$ cm.
(a) What symmetric tolerance limits are necessary to have no more than 2 percent defective cuts?
(b) What percentage will be acceptable if the limits are 12 ± 0.2 cm?

SOLUTION (a) The z values must be symmetrically located above and below 0 with 98 percent of the area under the SND curve (Fig. 11.7). Using z_L and z_U for the lower and upper limits, respectively, we want $P(z_L \leq z \leq z_U) = 0.98$. From Table 11.2, 98 percent of the area under the SND is between

$$z_U = 2.326 \quad \text{and} \quad z_L = -2.326$$

If X is the variable and μ is the center of the limits, we translate to the limits for X using the SND variable $Z = (X - \mu)/0.07$. For the lower limit

$$-2.326 = \frac{x_L - \mu}{0.07}$$

Solving for x_L, we get

$$x_L = \mu - 0.163$$

Similarly,

$$x_U = \mu + 0.163$$

Now, for every cut the desired mean is used to compute x_L and x_U.

11.7 FITTING THE NORMAL TO OBSERVED DATA

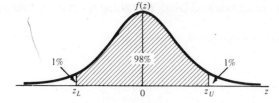

Figure 11.7 SND specifications which allow no more than 2 percent tive product, Example 11.6.

(b) The mean is 12.0, and the limits are set at $x_L = 11.98$ cm and $x_U = 12.02$ cm. Use the steps in Sec. 11.5 to find the area between these limits:

$$P(11.98 \leq x \leq 12.02) = P(-0.286 \leq z \leq 0.286)$$

By symmetry, we determine (Table B-3) that 22.5 percent of all cuts will meet this tolerance:

$$2P(0 \leq z \leq 0.286) = 2(0.1126) = 0.2252$$

Familiarity with all aspects of conclusion making on the normal curve is essential to good engineering statistics. As in previous chapters, one of the uses is to determine whether an observed value is reasonable or unexpected. You may review this procedure for the binomial (Sec. 8.6), since it does not change for different distributions. The case study in Solved Problem illustrates the technique for the normal.

Problems P11.22–P11.27

11.7 Fitting the Normal to Observed Data

A particular pdf is often fit to observed data to determine by observation or statistical test if the two compare favorably. The following procedure explains how to fit the $N(\mu, \sigma^2)$ curve to any data grouped into cells.

1. Estimate the parameters μ and σ^2 by using cell midpoints.
2. Compute the $z = (x - \mu)/\sigma$ for each cell, where x is the *upper* cell boundary.
3. Determine the area in each cell. Call this the cell probability P.
4. Compute the *expected* frequency E for each cell using

$$E = (\text{sample size})(\text{cell probability}) = nP$$

5. Compare the observed and expected cell frequencies numerically or graphically.

Later we discuss statistical methods to judge how well the normal fits. This section illustrates only the fitting of a normal pdf to observed data.

Table 11.3 Fitting the normal to observed data with $\hat{\mu} = 8.224$ and $\hat{\sigma} = 0.293$, Example 11.7

Cell midpoint (1)	Upper boundary x (2)	z (3)	Area left of z (4)	Cell area P (5)	Frequency Expected E (6)	Frequency Observed (7)
7.65	7.80	−1.45	0.0735	0.0735	5.88	5
7.95	8.10	−0.42	0.3372	0.2637	21.10	21
8.25	8.40	0.60	0.7257	0.3885	31.08	35
8.55	8.70	1.62	0.9474	0.2217	17.74	15
8.85	9.00	2.65	0.9960	0.0486	3.89	3
9.15	9.30	3.67	0.9999	0.0039	0.31	1
					80.00	80

Example 11.7 (*a*) Fit a normal distribution to the 80 daily flow rates collected on a computer-monitored process. The flows are midpoint values.

Flow (L/s)	Frequency
7.65	5
7.95	21
8.25	35
8.55	15
8.85	3
9.15	1

(*b*) Graph the expected and observed frequencies on one curve.

SOLUTION The steps to fit a normal curve are used. All column references are to Table 11.3.
1. The estimator Eqs. (11.2) and (11.3) give $\hat{\mu} = 8.224$ and $\hat{\sigma} = 0.293$.
2. z values (column 3) are computed using upper cell boundaries:

$$z = \frac{x - 8.224}{0.293}$$

3. Table B-3 is used to get the area from $-\infty$ to z for the SND (column 4). Successive area subtraction gives the probability P (column 5).
4. $E = 80P$ is the expected frequency (column 6).
5. Comparison indicates a close fit of the normal curve to observed flow rates (Fig. 11.8).

COMMENT Figure 11.8 shows only the right side of each histogram bar for easy comparison of observed and expected frequencies. You should not use

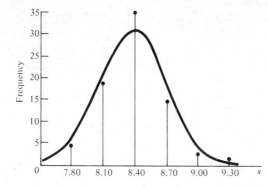

Figure 11.8 Comparison of observed frequency data and the normal distribution fit, Example 11.7.

midpoint values for z computation because this moves all cells one-half cell width to the left.

Problems P11.28–P11.30

11.8 Exact and Approximate Uses of the Normal

The normal distribution is an exact distribution for continuous data which can take on any value from $-\infty$ to $+\infty$. Since not many properties can assume all these values (especially not below 0), most uses are approximations to other discrete or continuous variables. The most common is the normal approximation to the discrete binomial. It can be shown (as done by Demoivre in 1733) that if $X \sim b(x; n, p)$ with $\mu = np$ and $\sigma = \sqrt{npq}$, the cdf of the variable

$$Z = \frac{X - np}{\sqrt{npq}} \tag{11.6}$$

has a limit of the SND as n increases. Thus,

$$\lim_{n \to \infty} F(z) = \int_{-\infty}^{x} N(0, 1) dx$$

This means that the binomial asymptotically approaches the standard normal distribution. Note that in Eq. (11.6) the variable Z uses the standard normal form $(X - \mu)/\sigma$.

The normal approximation to the binomial is very accurate when p is close to 0.5 because of the symmetry of the binomial pdf. As p deviates from 0.5, n must be larger for good results. As a guideline, compute np and nq. If $np \geq 5$ and $nq \geq 5$, the numerical results will be close to actual binomial computations.

Because the normal approximation to the binomial uses a continuous pdf to compute discrete probability values, a *continuity correction* of 0.5 is needed. If the probability that x is between two values a and b is desired, the approximation using the normal is between $a - 0.5$ and $b + 0.5$, that is, $P(a - 0.5 \leq x \leq b + 0.5)$. Figure 11.9 shows that for a cell width of 1 the correction is actually

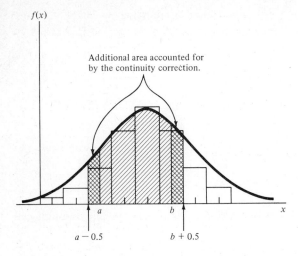

Figure 11.9 Effect of using the continuity correction when approximating the binomial with the normal distribution for a cell width of 1.

one-half cell, because the normal approximates the binomial with integer values as cell midpoints. The SND relation for $Z = (x - \mu)/\sigma$ is then used with $\mu = np$ and $\sigma = \sqrt{npq}$. The actual approximating computation would be

$$P\left(\frac{a - 0.5 - np}{\sqrt{npq}} \leq z \leq \frac{b + 0.5 - np}{\sqrt{npq}}\right) \qquad (11.7)$$

The continuity correction allows approximation of $P(x =$ one specific value $= c)$ using the normal by computing

$$P(x = c) = P(c - 0.5 \leq x \leq c + 0.5) \qquad (11.8)$$

Example 11.8 demonstrates the normal approximation and the continuity correction. If the normal is used to approximate any continuous variable, do not use the continuity correction.

Example 11.8 Jerri works as a consultant in agribusiness. In a particular project, studies showed that of 250 sacks of fertilizer, 32 were damaged during truck transport from the co-op mill to the wholesale outlet. If the number of damaged bags D is assumed to be binomially distributed with the historical damage rate of 10 percent, compute the chance that (*a*) the expected number of damaged bags is not exceeded and (*b*) exactly 32 bags are damaged.

SOLUTION (*a*) To compute the probability of at most $250(0.10) = 25$ damaged bags, using the $b(d; 250, 0.10)$ pdf, Jerri must determine the following binomial probability for D:

$$P(d \leq 25) = \sum_{d=0}^{25} {}_{250}C_d \, 0.10^d 0.90^{250-d}$$

This is a laborious task and tables up to $n = 250$ are not readily available.

We can apply the normal approximation because $np = 25 > 5$ and $nq = 225 \gg 5$. The values used with the continuity correction in Eq. (11.7) are $a = 0$, $b = 25$, $\mu = np = 25$, and $\sigma = \sqrt{npq} = 4.74$. Then

$$P(0 \le d \le 25) = P\left(\frac{0 - 0.5 - 25}{4.74} \le z \le \frac{25 + 0.5 - 25}{4.74}\right)$$
$$= P(-5.38 \le z \le 0.11) = P(z \le 0.11) - P(z \le -5.38)$$
$$= 0.5438$$

There is a 54.4 percent chance that no more than 10 percent of the 250 bags are damaged.

(b) Again, use the normal approximation. Since only a single value of 32 is involved, Eq. (11.8) applies with $c = 32$:

$$P(31.5 \le d \le 32.5) = P(1.37 \le z \le 1.58) = 0.0282$$

Jerri can expect that about 3 percent of the truckloads will have 32 damaged bags out of 250.

COMMENT How is the accuracy of the approximation in part (b)? If you have a hand calculator with factorial (!) and power/root (y^x) functions, and some time, you can compute the discrete binomial pdf value at $d = 32$:

$$b(32; 250, 0.10) = \frac{250!}{32! \, 218!} 0.1^{32} 0.9^{218} = 0.0274$$

As you can see, 2.74 percent is pretty close to the normal approximation of 2.82 percent, but it takes a lot longer to obtain.

A final comment on the continuity correction of ± 0.5. Its effect in (a) is small because n is large. Without it the answer is $P(0 \le d \le 25) = 0.50$ rather than 0.5438. As n decreases and as p gets closer to 0.5, the correction makes a slightly larger difference.

Since there is an asymptotic relation between the binomial and Poisson distributions and between the binomial and normal distributions, there is one between the Poisson and normal. If X is Poisson with $\mu = \sigma^2 = \lambda$, the cdf of $Z = (X - \lambda)/\sqrt{\lambda}$ is the SND. This may be stated symbolically as

$$\lim_{\lambda \to \infty} F(z) = \int_{-\infty}^{x} N(0, 1) dx$$

Because the Poisson has only one parameter, tables are easy to construct. Therefore, a computational approximation using the standard normal is not necessary. Figure 11.10 gives a complete summary of all approximating relations from the hypergeometric to the standard normal.

Example 11.9 (Case Study)

Problems P11.31–P11.38

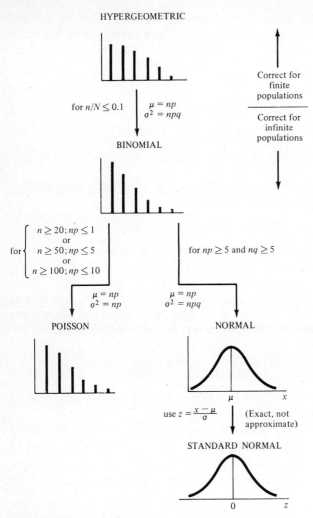

Figure 11.10 Complete series of distribution approximations from hypergeometric through standard normal.

*11.9 Moments of the Normal

Using Eq. (6.23) and some hefty integral work, the mgf of $X \sim N(\mu, \sigma^2)$ can be determined:
$$M(t) = \exp\left(\mu t + \tfrac{1}{2}t^2\sigma^2\right)$$
To obtain the expected value, compute $M'(t = 0)$:
$$M'(t) = \exp\left(\mu t + \tfrac{1}{2}t^2\sigma^2\right)(\mu + t\sigma^2)$$
$$E[X] = M'(t = 0) = \exp(0)(\mu + 0) = \mu$$
Just as easily it is possible to obtain $V[X] = \sigma^2$.

Problems P11.39–P11.42

SOLVED PROBLEM (Case Study)

Example 11.9 An inventory management consultant has been asked to help set safety stocks for printed wiring cards. One year of weekly demand D is given in Table 11.4. Help the consultant answer the following.
(a) Does the weekly demand pattern seem to fit a normal distribution?
(b) How does the mean demand compare with the median demand?
(c) What is the probability that demand will exceed 60 units per week for two consecutive weeks?
(d) What safety stock should be maintained to cover the demand 95 percent of the time?
(e) By what percentage will inventory investment increase if a 99 percent safety stock, rather than 95 percent, is maintained? Assume each card is worth $5.
(f) Compute the safety stock for 95 percent coverage if demand variance increases by 20 percent. Compare to the value obtained in part (d).

SOLUTION (a) The parameters of a normal pdf obtained from Eqs. (11.2) and (11.3), using cell midpoints, are $\hat{\mu} = 54.0$ and $\hat{\sigma} = 15.4$ units per week. The results of fitting the normal pdf to upper cell boundary values are shown in Table 11.5 and plotted with the demand histogram in Fig. 11.11. The fit is not too good, especially for the cells 40 to 50 and 50 to 60. However, we will use the normal to help manage the printed card stock level.
(b) The median cell must include the frequency value between 26 and 27, so that 50 percent of demand is less and 50 percent more than the median. Summing of observed frequencies determines the median cell as 50 to 60 units per week. The mean is also included in this cell.
(c) Use the SND and the $\hat{\mu}$ and $\hat{\sigma}$ estimates to compute the probability that demand will exceed 60 units per week. The continuity correction is

Table 11.4 Weekly demand of printed wiring cards, Example 11.9

Demand (units/wk)	Number of weeks
10–20	2
20–30	1
30–40	6
40–50	8
50–60	17
60–70	12
70–80	5
80–90	0
90–100	1
	52

Table 11.5 Normal distribution fit to demand data for $\hat{\mu} = 54.0$, $\hat{\sigma} = 15.4$, Example 11.9

Upper boundary	SND z	Area left of z	Cell area	Frequency Expected	Frequency Observed
20	−2.21	0.0136	0.0136	0.7	2
30	−1.56	0.0594	0.0458	2.4	1
40	−0.91	0.1814	0.1220	6.3	6
50	−0.26	0.3974	0.2160	11.2	8
60	0.39	0.6517	0.2543	13.2	17
70	1.04	0.8508	0.1991	10.4	12
80	1.69	0.9545	0.1037	5.4	5
90	2.34	0.9904	0.0359	1.9	0
100	2.99	0.9986	0.0082	0.4	1
				51.9	52

added to 60 to ensure that 60 is not included.

$$P(d > 60) = P\left(z \geq \frac{60.5 - 54.0}{15.4}\right) = P(z \geq 0.42) = 0.3372$$

There is about a 33 percent chance of exceeding 60 units per week every week. For two consecutive weeks the probability is $(0.3372)^2 = 0.1137$.

(d) To cover the demand 95 percent of the time, compute a value, call it d', which has 95 percent of the area to its left using the SND relation:

$$P(d \leq d') = 0.95$$

$$P\left(z \leq \frac{d' - 54.0}{15.4}\right) = 0.95$$

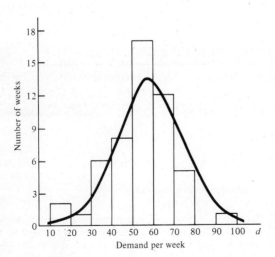

Figure 11.11 Normal curve fit to printed wiring card demand, Example 11.9.

From Table 11.2, $z = 1.645$. (Construct a graph of the SND if you are confused.) Now solve for d':

$$1.645 = \frac{d' - 54.0}{15.4}$$

$$d' = 25.3 + 54.0 = 79.3 \text{ or } 80 \text{ units per week}$$

Since the mean is 54.0, the required safety stock is 26 units per week.

(e) We repeat the computations in part (d) with $z = 2.326$ for 99 percent assurance:

$$P(d \leq d') = P\left(z \leq \frac{d' - 54.0}{15.4}\right) = 0.99$$

$$2.326 = \frac{d' - 54.0}{15.4}$$

$$d' = 35.8 + 54.0 = 90 \text{ units per week}$$

For 95 percent coverage, the total investment is $80(5) = \$400$ per week, and for 99 percent coverage, it is $90(5) = \$450$ per week. This is a 12.5 percent increase in investment to improve coverage from 95 percent to 99 percent, which is a 4.2 percent increase.

(f) A *variance* increase of 20 percent puts the new standard deviation at

$$\sigma = \sqrt{(15.4)^2(1.20)} = \sqrt{284.6} = 16.9 \text{ units per week}$$

The 95 percent safety stock level will be greater than the 26 units in part (d).

$$1.645 = \frac{d' - 54.0}{16.9}$$

$$d' = 27.8 \pm 54 = 82 \text{ units per week}$$

Safety stock is now 28 units per week.

ADDITIONAL MATERIAL

Belz [2], pp. 101–120; Bowker and Lieberman [4], Chap. 3; Hines and Montgomery [12], Chap. 7; Hoel [13], pp. 75–86, 232–234; Kirkpatrick [15], pp. 120–132; Miller and Freund [18], pp. 105–116; Neville and Kennedy [19], Chaps. 9, 10; Newton [20], Chap. 6; Walpole and Myers [25], pp. 110–127.

PROBLEMS

The following situations are used by the indicated problems in this chapter.

A: The current method of grading, handling, packaging, and transporting eggs used by the All Thumbs Co-op is capable in their estimation of breaking a minimum number of eggs in the processing from chicken to store shelf. Study has shown that the number of cracked or broken shells per 1000 eggs is best estimated by a normal distribution with a mean of 5 and a standard deviation of 0.5.

Problems P11.2, P11.8, P11.12, P11.17, P11.24

B: A certain dimension Y on a machined part has a normal distribution with the descriptor $Y \sim N(25 \text{ cm}, 0.09 \text{ cm}^2)$.

Problems P11.4, P11.10, P11.13, P11.17, P11.22

C: There is a maximum water content allowance of 10 percent for a certain meat product. In the past the actual water content from a company has been normal with an average of 8.5 percent and a variance of 1 (percent)^2.

Problems P11.2, P11.4, P11.15, P11.17, P11.25

D: In a nationwide sample of 5000 tires from one manufacturer, it has been found that 15 percent of the tires were returned for an adjustment prior to wearing out. Mileage readings have shown that the recorded miles at return time follow a normal pdf with $\mu = 18{,}500$ and $\sigma = 2800$ mi.

Problems P11.7, P11.10, P11.16, P11.18, P11.27

P11.1 A low-tar cigarette is advertised to contain 9 milligrams (mg) of tar and 0.8 mg of nicotine. In actual tests of 350 cigarettes it was found that the amounts of tar X and nicotine Y are normally distributed as follows: $X \sim N(9.01, 0.21)$ and $Y \sim N(0.76, 0.0025)$. Write the complete pdf expressions for X and Y.

P11.2 Write the pdf expression and the descriptor for (*a*) A and (*b*) C.

P11.3 Use the normal pdf expression in Eq. (11.1) to determine the value of $f(x)$ for the following situations:
  (*a*) $\mu = 0$, $\sigma = 1$, and $x = 0.5$
  (*b*) $\mu = -5$, $\sigma = 2$, and $x = -8$
  (*c*) $\mu = 2050$, $\sigma = 158$, and $x = 2130$

P11.4 State the mean and standard deviation for the situations described in (*a*) B and (*b*) C.

P11.5 A maintenance engineer plans to fit a normal distribution to the downtime data below collected from a sample of 30 milling machines. Determine the estimators for the mean and standard deviation.

Downtime (h/month)	Frequency	Downtime (h/month)	Frequency
0–1	2	6– 7	15
1–2	5	7– 8	8
2–3	4	8– 9	10
3–4	8	9–10	6
4–5	9	10–11	0
5–6	10	11–12	1

P11.6 State at least three facts that are always correct for the normal distribution but not so for any other distribution studied thus far.

P11.7 Construct a to-scale plot of the pdf described in D, and indicate the points μ and $\mu \pm 1\sigma$.

P11.8 Plot the general shape of the normal pdf described in A, and show the correspondence between the variable scale for the actual number of broken eggs and the SND scale for the transformation $z = (x - \mu)/\sigma$.

P11.9 Construct a to-scale plot of the SND.

P11.10 Compute the z values for:
 (a) $y = 25.8$ cm using the situation described in B
 (b) The situation in D if 12,500 mi was recorded for a returned tire

P11.11 If $z = -1.95$ was computed by an ME using a deviation from the mean of -28.95, determine the σ value for (a) the observed normal distribution and (b) the standard normal distribution.

P11.12 Determine the probability of the following for A:
 (a) At most 6 broken eggs per 1000
 (b) Between 4.5 and 5.5 broken eggs per 1000 on the average
 (c) More than 2 standard deviations from the mean number of broken eggs per 1000

P11.13 For the distribution in B determine the symmetric limits of Y with 0.26 percent of the area under the normal curve outside these limits.

P11.14 The average house in a particular city costs $51,000, and the sales price is normally distributed with $\sigma = \$5500$.
 (a) What percentage of the houses cost less than $40,000?
 (b) In a sample of 1000 houses how many will cost between $40,000 and $51,000?

P11.15 What is the probability that the water content in situation C will (a) exceed 9.5 percent and (b) not exceed the 3 standard deviation limits?

P11.16 Determine if each statement below is true or false for the distribution in D. If it is false, correct the statement.
 (a) Of the 750 tires returned, approximately 716 of them had been used between 15,700 and 24,100 mi.
 (b) It is quite likely that only one tire has gone less than 10,100 mi.
 (c) The probability that a tire was used between 15,700 and 18,500 mi is 0.136.
 (d) An estimated total of 84.13 percent of the tires were used at most 21,300 mi before return for adjustment.

P11.17 Use the steps in Sec. 11.5 to find:
 (a) The probability of at most 0.3 percent broken eggs for the distribution in A
 (b) The percentage of parts for situation B that measure between 24 and 26 cm
 (c) The observed water content for situation C that is exceeded 10 percent of the time

P11.18 For the distribution in D compute the number of miles x such that (a) $P(X \leq x) = 0.0495$, (b) $P(X \leq x) = 0.95$, and (c) $P(X \leq |x|) = 0.8413$ with the area equally distributed on each side of the mean.

P11.19 The area covered by a painter with 1 gal of paint is normally distributed with a mean of 400 ft^2 and a standard deviation of 60 ft^2. If the manufacturer specifies that 1 gal should cover between 375 and 450 ft^2, determine what fraction of the time the painter (a) exceeds the upper limit and (b) is within the manufacturer's limits.

P11.20 Use the SND Table B-3 to determine:
 (a) $P(z \leq -2.5)$
 (b) $P(1 \leq z \leq 3)$
 (c) $P(|z| \leq 1.65)$
 (d) $P(z > -0.70)$

P11.21 The number of grams in a 212-g can must be between 209 and 214 at least 95 percent of the time. If the dispensing machine fills cans with a normal distribution with mean 212.5 g and a variance of 0.8 g^2, is the 95 percent limit adhered to?

P11.22 For the distribution in situation B new tolerance limits of 25.5 ± 0.2 cm have been established. What is the probability of making a defective part?

P11.23 The length of a pencil is supposed to fit within the tolerance 7.25 ± 0.05 in. Past production has been normally distributed with $\mu = 7.25$ and $\sigma = 0.02$ in. However, due to wear on machinery, σ has increased to 0.025 in, but the mean has remained constant. What new tolerance limits must be used to produce the same percentage of scrap pencils as before the increase in σ?

P11.24 Determine a lower specification limit for the broken eggs in A that will be exceeded at least (*a*) 90 percent of the time and (*b*) 99 percent of the time.

P11.25 For the situation in C, an inspector found that the maximum limit was exceeded too often.

(*a*) What is the probability that this limit is violated?

(*b*) If the variance is fixed due to the type of injection equipment and the limit of 10 percent is fixed, determine the aimed-at mean necessary to violate the limit no more than 1 time in 1000.

P11.26 The management of a fast food restaurant exercises close control over the weight of an order of french fries. The recommended weight is 2.5 oz per order. In 85 trial orders the following weights were recorded.

Weight (oz)	Frequency
2.2–2.3	5
2.3–2.4	12
2.4–2.5	22
2.5–2.6	15
2.6–2.7	14
2.7–2.8	8
2.8–2.9	6
2.9–3.0	3

If the normal curve and data above are assumed useful in setting specification limits for the weight per order, what limits will be violated only 15 percent of the time? Set the limits so that the upper limit will be violated only one-half as often as the lower limit.

P11.27 For the distribution in D, determine the symmetric mileage limits that will include 98 percent of all tires to be returned.

P11.28 Fit a normal distribution to the data of Prob. P11.5.

P11.29 Fit a normal distribution to the data of Prob. P11.26.

P11.30 Fit a normal distribution to the machinability rating data given by data set D in Chap. 2. Does the normal seem to be a good fit for this data?

P11.31 The probability that an animal is cured of a certain disease is 0.30. For a sample of 50 animals, approximate the chances that no more than 5 percent of the animals are cured. Compare this answer with that obtained in Prob. P8.9c.

P11.32 If X is the number of successes in samples of size 200 and $p = 0.40$ is the probability of a success, estimate the probability that

(*a*) $x \leq 80$

(*b*) $55 \leq x \leq 65$

(*c*) $x = 90$

P11.33 In a computer simulation of an appliance store warehouse, the probability that any order includes a refrigerator is 0.25. From a total of 300 orders in one month a sample of 25 orders was drawn at random. Approximate the probability that a refrigerator is included on at most five orders using (*a*) the binomial and (*b*) the normal distribution.

P11.34 Approximate the probabilities for Prob. P8.18a and c using the normal distribution, and compare the answers.

P11.35 A die is tossed 30 times. Estimate the probability that (a) a 6 appears at least 10 times and (b) an even number appears exactly 15 times.

P11.36 In a manufacturing plant employing 1500 people, the probability that an employee is late for work is estimated to be 0.075 from a sample of 120 workers. Use the normal distribution to approximate the probability that at least 10 percent of the workers will be late on any given workday (a) in the sample of size 120 and (b) in the entire workforce of 1500.

(c) Compare the answer in (a) with the Poisson probability obtained for the sample of 120 employees.

P11.37 Summarize the conditions under which the normal is used to approximate the hypergeometric and binomial distributions.

P11.38 A sample of size 120 is taken from a processing line for cucumbers. The thickness of wax applied to the skin for appearance sake was too thick for only three of the cucumbers, so the line was not stopped to correct the thickness. If this same problem occurs an average of 5 percent of the time, use (a) the correct distribution and (b) a normal approximation to determine if the decision to not stop the line was reasonable.

*P11.39 Use the mgf of the normal distribution to show that the variance is σ^2.

*P11.40 The mgf of a normal distribution for the variable X is stated as $\exp(2.5t + 0.25t^2)$. Use this mgf to determine the mean and standard deviation of the variable.

*P11.41 The moment generating function of a variable W is $e^{30t + 3.5t^2}$. Determine $P(25 \leq w \leq 32)$.

*P11.42 The mgf for the variable X, which is the number of customer accounts for the I.M. Liquid Co. which have a balance due in excess of \$1500, is $(0.10e^t + 0.90)^{55}$. Approximate the probability using the appropriate distribution that at least 10 accounts have in excess of \$1500 due.

Hint: Check the form of the moment generating functions of the distributions that you have studied thus far.

CHAPTER
TWELVE

THE CENTRAL LIMIT THEOREM

Some of the most important properties which make much of statistical inference possible are expressed in the central limit theorem (CLT). This chapter discusses the meaning and implications of this theorem.

CRITERIA

To understand and correctly use the CLT, you must be able to:

1. State the CLT and compute the mean and standard deviation of the sample mean, given the population mean and standard deviation.
2. State the procedure used to show that the CLT actually works for any distribution.
3. State the approximate sample sizes that should be used to ensure correct application of the CLT.
4. Compute a normal distribution probability or a required sample size using the SND for sample means, given the parameters of the original population, the observed sample average, and sample size.

STUDY GUIDE

12.1 Purpose and Statement of the CLT

Most of the statistical inference and estimation techniques are based on the normal distribution. However, since the samples used in these techniques are taken from the real world, they may have a distribution far from normal. The

12.1 PURPOSE AND STATEMENT OF THE CLT

CLT allows us to use normal distribution theory to infer about the population from a nonnormal sampling distribution. To do this, we work with the mean of sample data, not the individual values.

The CLT may be stated as follows:

> The population may have any unknown distribution with a mean μ and a finite variance σ^2. Take samples of size n from the population. As the size of n increases, the distribution of sample means will approach a normal distribution with mean μ and variance σ^2/n.

Figure 12.1 will help explain the meaning of this theorem. If X has a distribution—any distribution—with a mean μ and a variance σ^2, that is, $X \sim \text{any}(\mu, \sigma^2)$, as increasingly larger samples are drawn, the means $\overline{X}$ of these samples will be normally distributed with mean μ, same as for X, and a variance σ^2/n, which we will call $\sigma_{\overline{X}}^2$. Thus, $\overline{X} \sim N(\mu, \sigma_{\overline{X}}^2)$. Symbolically,

$$X \sim \text{any}(\mu, \sigma^2) \xrightarrow[\text{increases}]{\text{as } n} \overline{X} \sim N(\mu, \sigma_{\overline{X}}^2)$$

Note that the sample size n—not necessarily the total number of observations, which is the sample size times the number of samples—increases.

The square root of $\sigma_{\overline{X}}^2$ is

$$\sigma_{\overline{X}} = \frac{\sigma}{\sqrt{n}}$$

This is called the *standard deviation of the mean*. As n increases, $\sigma/\sqrt{n}$ decreases. In the limit as n increases, $\sigma_{\overline{X}} = 0$, that is,

$$\lim_{n \to \infty} \sigma_{\overline{X}} = 0$$

This means that as n increases, more and more of the population is included in the makeup of the $\overline{X}$ distribution. If n is infinity, the sample is the same as the population and only one $\overline{X}$ value is present, so $\sigma_{\overline{X}}$ must be 0.

Remember, there are *two* distributions used in the CLT—one for X, the population, and one for $\overline{X}$, the sample means. The $\overline{X}$ distribution becomes normal as n increases. Our statistical tools will use the fact that $\overline{X} \sim$

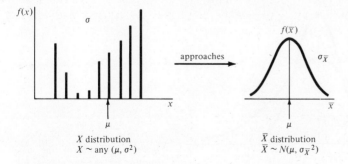

Figure 12.1 Graphical description of the central limit theorem.

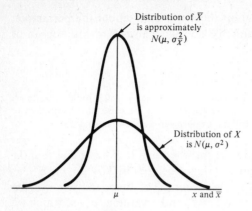

Figure 12.2 Relation between a population with an $N(u, \sigma^2)$ distribution and $\overline{X}$ for samples of size n taken from it.

$N(\mu, \sigma^2/n)$. By the way, X can also be $N(\mu, \sigma^2)$, and the CLT works just as well. Figure 12.2 shows that $\sigma_{\overline{X}}$ will be smaller than σ when sampling from X.

Example 12.1 Several children had bitten through teething rings from the Child's-Play Company and become sick from the liquid inside. In a lawsuit the crushing force F for a ring was stated to be 150 lb/in² with a variance of 121 (lb/in²)². Samples were taken with $n = 5$ and $n = 20$.
(a) Explain how the CLT applies in this case.
(b) Plot the distribution of sample means for both n values. State the relationship between the two $\sigma_{\overline{X}}$ values.

SOLUTION (a) We do not know the population distribution of F, but the CLT states that the sample means $\overline{F}$ will approach a normal distribution. For $n = 5$ the distribution of $\overline{F}$ will become normal with a mean of 150 lb/in² and a variance of $\frac{121}{5} = 24.2$ (lb/in²)². For $n = 20$, $\overline{F}$ is distributed approximately $N(150, 6.05)$.
(b) Figure 12.3 shows the normal distributions for sample means. For $n = 5$, $\sigma_{\overline{F}} = \sqrt{24.2} = 4.92$, and for $n = 20$, $\sigma_{\overline{F}} = \sqrt{6.05} = 2.46$ lb/in². The standard deviation for $n = 20$ is one-half that for $n = 5$.

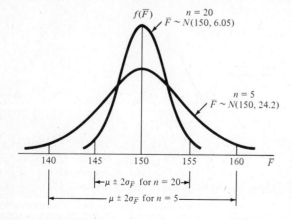

Figure 12.3 Graph of the distribution of sample means for $n = 5$ and $n = 20$, Example 12.1.

COMMENT To reduce $\sigma_{\bar{F}}$ to one-half of a specified value, n must be increased by a factor of 4. If the n above were $4(20) = 80$, then

$$\sigma_{\bar{F}} = \sqrt{\frac{121}{80}} = 1.23 \frac{\text{lb}}{\text{in}^2}$$

This value is one-half the $\sigma_{\bar{F}} = 2.46 \text{ lb/in}^2$ for $n = 20$.

Problems P12.1–P12.5

12.2 How the CLT Works

Assume there is a population X which has some distribution with mean μ and variance σ^2. The CLT may be illustrated by the following steps (Fig. 12.4).

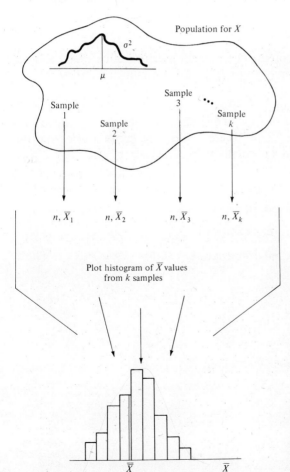

Figure 12.4 The CLT is illustrated by taking samples of size n and plotting sample means to observe the tendency toward the normal pdf.

1. Determine n.
2. Take a sample of size n and compute the sample mean $\bar{X}$.
3. Plot $\bar{X}$ on a histogram of $\bar{X}$ values.
4. Repeat steps 2 and 3 for k samples.
5. Compute the average and standard deviation of the $\bar{X}$ histogram. Call these $\bar{\bar{X}}$ (sample mean) and $s_{\bar{X}}$ (sample standard deviation of the mean).
6. Compare $\bar{\bar{X}}$ with μ and $s_{\bar{X}}$ with $\sigma/\sqrt{n}$.
7. Determine a larger n value and repeat steps 2 through 6.
8. Compare the shapes of the $\bar{X}$ histograms to notice the tendency toward a normal distribution.

These steps merely illustrate the CLT. The actual mathematical proof is quite involved. Regardless of the method of proof, the result is the same—a normal distribution may be used for conclusion making. Problem P12.6 asks you to use these steps to illustrate the CLT.

Problems P12.6, P12.7

12.3 Sample Sizes and the CLT

The CLT is a very powerful theorem. The value of n can be quite small, and the histogram of $\bar{X}$ values will rapidly converge to normality regardless of the underlying population. For example, a total of 400 random draws were taken from the bimodal distribution of Fig. 12.5. Samples of $n = 5$ were developed, and the 80 averages are plotted in Fig. 12.6a. The 400 values are grouped into $n = 10$ and are plotted in Fig. 12.6b. There is clearly a trend away from the original bimodal pdf and toward a normally shaped histogram.

The CLT formulas are:

Original distribution X: 	Expected value = μ
	Standard deviation = σ

Sampling distribution $\bar{X}$: 	Expected value = μ 	(12.1)

	Standard deviation = $\sigma_{\bar{X}} = \dfrac{\sigma}{\sqrt{n}}$ 	(12.2)

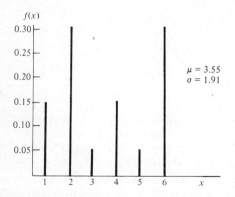

Figure 12.5 Bimodal distribution used to illustrate the central limit theorem.

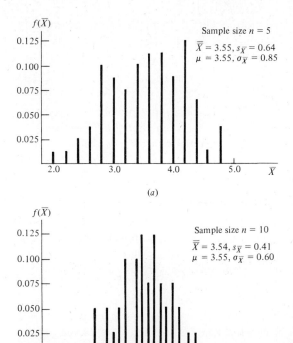

Figure 12.6 Relative frequency graphs of $\overline{X}$ values observed in samples from the distribution of Fig. 12.5 using (a) $n = 5$ and (b) $n = 10$.

Figure 12.6 gives the values of means and standard deviation for the X and $\overline{X}$ distributions for the example above using Eqs. (12.1) and (12.2). Comparison shows that the means are very close to the original value $\mu = 3.55$. The $s_{\overline{X}}$ values are slightly less than the predicted $\sigma_{\overline{X}} = \sigma/\sqrt{n}$ values. If larger n values were used, $s_{\overline{X}}$ would continue to decrease. Considering the nonnormal shape of the original pdf, agreement with the CLT is remarkable for sample sizes of 5 and 10.

To ensure rapid convergence to a normal, the sample size ranges below are recommended. You will not know the exact shape of the population pdf, but you should have a general idea of how it looks.

Original population	Sample size
Approximately normal	$n \geq 5$
Skewed or flattened	$10 \leq n \leq 30$
Highly skewed	$n > 30$
High tails and low center	$n > 100$

12.4 Common Uses of the CLT

The most important thing to remember when using the results of the CLT is that *you are working with the distribution of sample averages, namely $\overline{X}$, **not** the* *original X population*. The standard normal distribution (SND) transformation (Sec. 11.3) is used with $\mu = \overline{X}$ and $\sigma_{\overline{X}} = \sigma/\sqrt{n}$. The form is

$$Z = \frac{\overline{X} - \mu}{\sigma/\sqrt{n}} \qquad (12.3)$$

This relation may be used either to compute probabilities for observed average values or to determine the required sample size such that the observed $\overline{X}$ is within a specified range around the true population mean μ. Example 12.2 illustrates these techniques.

Example 12.2 The average lead time from vendors has increased recently because of the scarcity of resources. JB, an engineer with an oil products company, has collected historical data from 350 vendors and found an average lead time of 10 days with a standard deviation of 5 days. To determine the current average, JB sampled 28 vendors and computed a value of 12 days.
(*a*) What are the chances that the average lead time will exceed 12 days?
(*b*) If JB wants 98 percent assurance that the estimate of average lead time is no more than 1 day off, can the sample average of the 28 vendors be relied on?

SOLUTION Define the variable L as lead time and $\overline{L}$ as the average lead time from samples. Figure 12.7 shows these two distributions. From the CLT, $\overline{L} \sim N(10, 25/n)$.
(*a*) Since the average of 28 vendors was obtained, the $\overline{L}$ distribution is normal with $\mu = 10$ and $\sigma_{\overline{L}} = 5/\sqrt{28} = 0.945$ days. To compute $P(\overline{L} \geq 12$ days), use the SND from Eq. (12.3):

$$P(\overline{L} \geq 12) = P\left(z \geq \frac{12 - 10}{5/\sqrt{28}}\right) = P(z \geq 2.12) = 0.0170$$

There is a 1.7 percent chance that the *average* lead time will exceed 12 days *provided the sample size is 28*.
(*b*) In Fig. 12.8 you can see what JB wants to do. The observed sample average $\overline{L}$ should not be more than 1 day on either side of the true mean 98 percent of the time. Since by the CLT the true mean μ is the same as the sample mean, the symmetry of the normal pdf requires that the 0.98 area be equally divided. Equation (12.3) for the SND of sample means is used to compute a required *minimum n* value. We know that $\overline{L} - \mu$ must be between -1 and $+1$. Therefore,

$$P\left[-1 \leq (\overline{L} - \mu) \leq 1\right] = 0.98$$

If we divide by $\sigma/\sqrt{n}$, we have the SND relation:

$$P\left(\frac{-1}{\sigma/\sqrt{n}} \leq z \leq \frac{+1}{\sigma/\sqrt{n}}\right) = 0.98$$

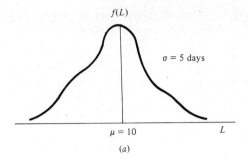

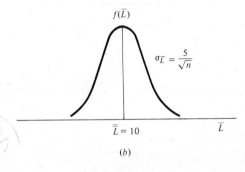

Figure 12.7 Distribution of (a) lead times and (b) sample averages of lead times according to the CLT, Example 12.2.

By symmetry

$$P\left(0 \leq z \leq \frac{1}{\sigma/\sqrt{n}}\right) = 0.49 \qquad (12.4)$$

From Table 11.2 the z value is 2.326. We solve Eq. (12.4) for n after substituting z and $\sigma = 5$ and removing the probability statement portion.

$$2.326 \leq \frac{1}{\sigma/\sqrt{n}}$$

$$\sqrt{n} \geq 5(2.326)$$

$$n \geq 135.3$$

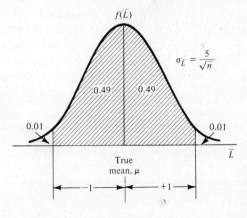

Figure 12.8 Distribution of average lead time with 98 percent of the area between $\bar{L} - \mu = -1$ and $\bar{L} + \mu = +1$, Example 12.2.

At least 136 vendors should be sampled. Since only 28 were checked, 108 more are needed to ensure that the average is no more than 1 day from the true mean.

COMMENT You must use the distribution of averages when samples are involved. In part (a) the use of σ rather than $\sigma_{\bar{L}}$ would give the following incorrect result:

$$P\left(z \geq \frac{12 - 10}{5}\right) = P(z \geq 0.4) = 0.3446$$

This indicates a 34 percent chance of exceeding 12 days, rather than the correct 1.7 percent.

Problems P12.8–P12.15

ADDITIONAL MATERIAL

Gibra [8], pp. 123–126; Hines and Montgomery [12], pp. 176–181; Hogg and Craig [14], pp. 196–199.

PROBLEMS

P12.1 The sugar content in a food product is normally distributed with an average of 10 percent and a standard deviation of 2 percent. Fifty samples of 25 packages each are taken from the production line. Determine the expected mean and standard deviation of the average sugar content computed from these samples.

P12.2 A total of 30 samples of 4 each were taken of structural concrete specimens. The sample means for drying time in hours to reach a specified strength are given below. From past experience under different weather conditions the time has averaged 5.8 h with a variance of 3.0. Compute the mean and variance of the sample means and compare them with those predicted by the CLT.

4.6	7.2	5.4	6.1	5.9	3.9
7.8	4.9	5.6	5.8	6.0	4.9
6.2	5.1	4.8	7.1	4.2	6.3
7.1	8.1	4.9	5.2	5.7	6.0
4.8	5.5	6.2	7.5	8.0	4.9

P12.3 Samples of 12 people were used to determine the average percentage of income spent on housing. The results showed an average of 20 percent with a standard deviation of the sample means of 2.6 percent.
(a) What are the estimated mean and standard deviation of the entire population?
(b) If the same results had been obtained for $n = 20$, how would the results above change?

P12.4 If samples are taken from a finite population of size N, rather than an infinite population, the standard deviation of the mean $\sigma_{\bar{X}}$ is computed as

$$\sigma_{\bar{X}} = \frac{\sigma}{\sqrt{n}} \sqrt{\frac{N-n}{N-1}}$$

If $N = 500$ and $n = 10$, use $\sigma = 0.50$ to compute $\sigma_{\bar{X}}$ using the finite and infinite population relations. What is the value of the correction factor for the finite population?

P12.5 The ages of five patients in a doctor's office are 18, 35, 41, 52, and 54. Develop all possible samples of size 3 and compute (a) the population mean and standard deviation and (b) the mean and standard deviation of the sample means.

(c) Compare the actual standard deviation of the mean value in (b) with that estimated using the finite population equation in Prob. P12.4.

P12.6 Use the steps in Sec. 12.2 to illustrate that the CLT does work for samples of size 2, 5, 10, and 20. Do the experiment for $k = 20$ samples using the uniform distribution

$$f(x) = 0.1 \qquad x = 0, 1, \ldots, 9$$

and the random digits in Table 10.1.

P12.7 Explain in your own words the difference in meaning between the two standard deviations σ and $\sigma_{\bar{X}}$.

P12.8 Two hundred samples of size 25 each are taken from an infinite population with $\mu = 10.5$ cm and $\sigma = 3.1$ cm. Determine the number of sample means that should be between $\bar{X} = 9$ and $\bar{X} = 11$ cm.

P12.9 The creep rate of a nonferrous alloy averages 10 percent per hour with a standard deviation of 0.2 percent per hour. If samples of size 5 are taken, determine a lower specification limit for the mean that will be exceeded in 95 percent of the samples.

P12.10 The time to complete an assembly operation has a standard of 75 s with a standard deviation of 10 s. Jeni computed a normal distribution probability of 0.067 of observing a time less than 60 s. However, she was quite surprised when, after observing a lot of assembly times grouped into 5 each, the overall mean was 72.8 s and the smallest sample mean value was 69.5 s.

(a) Verify the 0.067 probability value.

(b) Use probability analysis to determine why Jeni did not observe any sample means less than 60 s.

P12.11 The mileage figure for a new car is advertised as 48 km/gal in country driving. The standard deviation for this mean estimate is expected to be 5 km/gal. Samples are to be taken of actual driver mileage in several parts of the country to determine the symmetric 90 percent limits on the 48-km/gal estimate.

(a) What is the minimum size sample necessary to ensure that the mean of the sample means is no more than 3 km/gal from the true mean?

(b) If samples of the size you computed in (a) are taken and a mean of $\bar{X} = 45$ km/gal is found to be an accurate estimate of the mileage, what chance is there that in any particular sample the average quoted value of 48 km/gal will be exceeded?

P12.12 An engineer wants to determine the size of the sample needed to estimate the mean of a specific thickness within 0.0008 cm. Quality control records show that the last production run had a thickness distribution with $\mu = 0.05$ and $\sigma = 0.002$ cm. Use a 95 percent assurance level and determine the sample size for the engineer.

P12.13 A grocery store manager has been checking 10 different canned meat items to determine if the display area should be increased or decreased. In the past between 6 cases (144 cans) and 8 cases (192 cans) of each item have been displayed. The sales per week have averaged 150 cans with a standard deviation of 20 cans. What is the probability than the mean sales per week for the sample of 10 items is below or above the historical stocking level? Assume restocking takes place only once a week.

P12.14 A product has a prescribed mean dimension of 10.5 in. If the standard deviation is 1.5 in, what is the chance that the average dimension for 36 items exceeds this specification by more than 0.5 in?

P12.15 A finite population of 100 control devices with a stated mean life of 1500 h and a life standard deviation of 150 h was delivered by the vendor. The minimum specification limit for the devices is 1050 h. The purchaser took a sample of seven devices and used nondestructive, accelerated life testing to estimate a mean life of 1375 h.

(*a*) Use the normal distribution to determine the number of devices that, according to the manufacturer, should conform to the specification limit.

(*b*) Use the equation in Prob. P12.4 to estimate the percentage of samples which should have a mean life less than a value of 1300 h if samples of seven devices are tested and the observed mean of 1375 is correct.

(*c*) Provided 1500 is the correct mean, should a shipment be rejected if a sample of seven devices has a mean life of 1375 h? Why or why not?

(*d*) What mean life of the sample would you compare to 1375 h to reject this particular shipment 5 percent of the time?

CHAPTER THIRTEEN

BASICS OF THE χ^2 DISTRIBUTION

This chapter discusses the χ^2 (chi-square) distribution. You will learn how a χ^2 variable is formed, what the pdf graph looks like, and how to use a table of χ^2 values. Understanding this chapter is essential to the correct use of statistical methods which assume an underlying χ^2 distribution.

CRITERIA

To complete this chapter, you must be able to:

1. State the equation for and give an example of *degrees of freedom*.
2. State the steps necessary to generate a histogram which will approximate a χ^2 distribution, given a normal distribution and its variance.
3. Sketch the pdf, and state and compute the mean and standard deviation of any χ^2 distribution, given the degrees of freedom.
4. Use a χ^2 distribution table to determine a χ^2 value, given the degrees of freedom and the area above the desired value; or, determine the area above a stated χ^2 value, given the degrees of freedom.
5. Compute the probability of observing a specific sample variance value, given the variance for a sample from a normal population which has a known variance.
*6. Summarize the derivation of the χ^2 distribution from a normal population, and state the mgf and additivity property for χ^2.

STUDY GUIDE

13.1 Concept of Degrees of Freedom

The *degrees of freedom* is a parameter used in several continuous distributions. It is an integer value equal to the sample size n minus the number of population parameters k to be estimated from the sample. The symbol is ν (nu). Mathematically,

$$\nu = n - k \tag{13.1}$$

If we observe n values and compute s^2 as an estimate of the population variance σ^2, the formula for s^2 will have $n - 1$ degrees of freedom. This is why the unbiased formula for s^2 has $n - 1$, not n, in the denominator.

The term *degrees of freedom* is commonly used in engineering mechanics and other fields in a manner analogous to its use in statistics. For example, in mechanics a rigid body has a maximum of 6 degrees of freedom—movement along and rotation around each of the three axes. Every time movement is restricted, degrees of freedom are lost (Fig. 13.1). Example 13.1 discusses statistical degrees of freedom.

Example 13.1 Jannette, an engineering student, and Gloria, a mathematics student, will play the following two-stage game.
(*a*) Gloria has five numbers in mind. She gives Jannette four values and the average of all five numbers, and Jannette is to state the value of the fifth number.
(*b*) If stated correctly, Gloria then gives Jannette three of five numbers plus the average and variance of all five numbers. Help Jannette state the

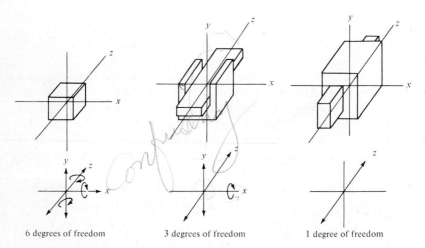

Figure 13.1 Examples of restricting the degrees of freedom in engineering mechanics.

missing numbers for the following situations:

  Numbers: 1, 2, 3, 4 Numbers: 1, 2, 3
  Average: 3   Average: 3
            Variance: 2.5

How many degrees of freedom are lost in each stage of the game?

SOLUTION (a) If y_i ($i = 1, 2, \ldots, 5$) are the values, Jannette must state y_5. The mean may be written

$$\bar{y} = \frac{1 + 2 + 3 + 4 + y_5}{5} = 3.0$$

Solution gives $y_5 = 5$. The mean of 3.0 removed 1 degree of freedom, thereby requiring y_5 to be 5.

(b) We can find y_4 and y_5 by writing the expression for s^2:

$$s^2 = \frac{4 + 1 + 0 + (y_4 - 3)^2 + (y_5 - 3)^2}{4} = 2.5$$

Then

$$(y_4 - 3)^2 + (y_5 - 3)^2 = 5 \qquad (13.2)$$

From the mean of 3, we have

$$y_4 + y_5 = 9 \qquad (13.3)$$

Simultaneous solution of Eqs. (13.2) and (13.3) yields $y_4 = 4$ and $y_5 = 5$. Here 2 degrees of freedom were lost by fixing $\bar{y}$ and s^2, so two y values are predetermined.

13.2 Formulation of the χ^2 pdf

The χ^2 distribution is commonly used to test statistical hypotheses. Since testing is a primary engineering application of statistics, you should understand how the χ^2 pdf can be generated. Start with an $N(\mu, \sigma^2)$ population. Follow Fig. 13.2 through the following procedure:

1. Determine the sample size n.
2. Take one sample of size n and compute the sample variance s^2.
3. Calculate the value of the statistic χ^2 for this sample, where

$$\chi^2 = \frac{(n-1)s^2}{\sigma^2} \qquad (13.4)$$

4. Plot χ^2 on a histogram of χ^2 values.
5. Repeat steps 2 through 4 for k samples.

The histogram (Fig. 13.2) will approximate a χ^2 pdf with $\nu = n - 1$ degrees of freedom.

256 BASICS OF THE χ^2 DISTRIBUTION

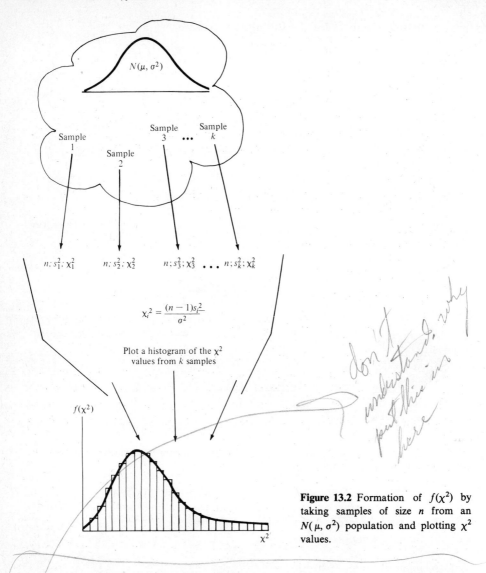

Figure 13.2 Formation of $f(\chi^2)$ by taking samples of size n from an $N(\mu, \sigma^2)$ population and plotting χ^2 values.

If a variable Y has a χ^2 distribution with ν degrees of freedom, the descriptor is $Y \sim \chi^2(\nu)$. The pdf relation for $f(y)$ is

$$f(y) = \left[2^{\nu/2}\Gamma(\nu/2)\right]^{-1} y^{(\nu-2)/2} \exp(-y/2) \qquad y \geq 0$$

The $\Gamma(\nu/2)$ is the gamma function for $\nu/2$. If the argument of Γ is k, a positive integer exceeding 1, then $\Gamma(k) = (k-1)!$. If $k > 1$, in general, $\Gamma(k) = (k-1)\Gamma(k-1)$. Also, $\Gamma(1) = 1$ and $\Gamma(\frac{1}{2}) = \sqrt{\pi}$.

From the pdf relation you can see the χ^2 distribution is not simple. Therefore, we will use tabulated values for the χ^2 pdf in future sections.

Problems P13.1, P13.2

13.3 GRAPHS, PARAMETER, AND PROPERTIES OF THE χ^2 PDF

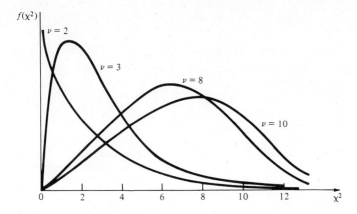

Figure 13.3 Graphs of the χ^2 pdf for different values of the degrees of freedom parameter ν.

13.3 Graphs, Parameter, and Properties of the χ^2 pdf

Once the sample size is determined, ν is known and the χ^2 pdf can be graphed. Figure 13.3 shows $f(\chi^2)$ for several values of ν. The range of χ^2 is always 0 to ∞. As ν increases, $f(\chi^2)$ becomes symmetric. For values above $\nu = 30$, the statistic $\sqrt{2\chi^2}$ has a pdf that is approximately normal with the mean $\sqrt{2\nu - 1}$ and a variance of 1. Symbolically

$$f(\sqrt{2\chi^2}) \sim N(\sqrt{2\nu - 1}, 1) \quad \text{for } \nu > 30$$

The coefficient of skewness, which is $\alpha_3 = 2\sqrt{2/\nu}$, shows that for smaller ν values, the pdf is skewed right.

Chi square is a continuous distribution with one parameter and properties as follows.

- Shape parameter: ν (degrees of freedom)
- Parameter range: ν = positive integers
- Best estimate of ν: $\hat{\nu} = n - 1$
- Expected value: $\mu = \nu$
- Standard deviation: $\sigma = \sqrt{2\nu}$

Example 13.2 A production engineer has studied the errors above design tolerances produced by numerically controlled (N/C) machines on successive punch-drill-ream operations. The error W, in thousandths of centimeters, has a χ^2 distribution with 4 degrees of freedom.
(a) Write the descriptor for W.
(b) Compute the properties.
(c) Sketch the pdf and indicate the properties.

258 BASICS OF THE χ^2 DISTRIBUTION

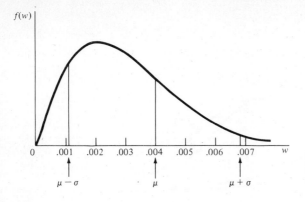

Figure 13.4 Graph of $\chi^2(4)$ for manufacturing errors in 0.001 cm, Example 13.2.

SOLUTION (a) Setting $\nu = 4$, the descriptor is $W \sim \chi^2(4)$.
(b) The properties are computed using $\nu = 4$.

$$\mu = 4$$

$$\sigma = \sqrt{8} = 2.8$$

In 0.001 cm units, the properties are

$$\mu = 0.0040 \text{ cm}$$

$$\sigma = 0.0028 \text{ cm}$$

(c) Figure 13.4 is a sketch of $\chi^2(4)$ with μ and $\mu \pm 1\sigma$ indicated. Most of the area is included within 1σ of the mean.

COMMENT To get μ and σ, $\nu = 4$ was used and then converted to 0.001 cm. If σ is computed using $\nu = 0.004$, an incorrect value of $\sqrt{2(0.004)} = 0.089$ is obtained.

Problems P13.3–P13.7

13.4 Use of a χ^2 Distribution Table

The χ^2 pdf is tabulated in appendix Table B-4 for degrees of freedom $\nu = 1$ to 100. The table entries are specific values χ_0^2 with a stated probability α of being exceeded; that is,

$$P(\chi^2 \geq \chi_0^2) = \int_{\chi_0^2}^{\infty} f(\chi^2) \, d\chi^2 = \alpha$$

For example, if $Y \sim \chi^2(10)$, there is a probability of 0.05 that χ^2 will exceed $\chi_0^2 = 18.31$ (Table B-4). It is possible to get a χ_0^2 value for any α from 0.995 to 0.005 (Fig. 13.5).

The value α is actually $1 - F(\chi_0^2)$, where $F(\chi_0^2)$ is the cdf at χ_0^2. This fact is useful if you want the pdf value with α to the left of χ_0^2. For example, if $\nu = 5$, then $\chi_0^2 = 0.55$ has 0.01 of the area to its left and 0.99 to its right.

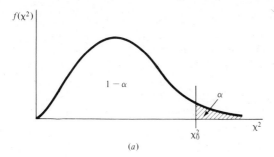

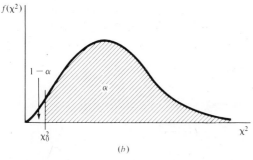

Figure 13.5 The χ^2 pdf with (a) a small portion and (b) a large portion of the area to the right of χ_0^2.

There are two ways to use a χ^2 distribution table, both illustrated in Example 13.3:

1. Given the values of ν and α, determine χ_0^2.
2. Given the values of ν and χ_0^2, determine α.

Example 13.3 Tony is the engineer at a nuclear-powered electric generation plant. He is about to perform two statistical tests. Test 1 uses the standard deviation of the temperature of rejected cooling water, and test 2 involves the standard deviation of power outage time if the reactor fails. Both tests use the $\chi^2(20)$ pdf.
(a) For test 1 determine a χ^2 value with 10 percent of the area to its right.
(b) For test 2, find two χ^2 values with 90 percent of the area between them and the remainder symmetrically distributed outside them.
(c) For test 2, if the resulting χ^2 value is 20.05, what area is to its right?

SOLUTION (a) For $\nu = 20$ there is 10 percent of the area above $\chi_0^2 = 28.41$ (Table B-4). This may be written

$$P[\chi^2(20) \geq 28.41] = 0.10$$

(b) See Fig. 13.6a. Of the area 5 percent is outside each of the values χ_L^2 (lower) and χ_U^2 (upper). In addition, we know that 95 percent of the area

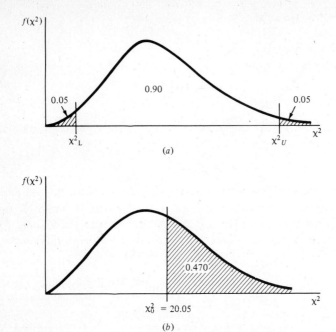

Figure 13.6 χ^2 areas (a) in both tails and (b) above a specific χ_0^2 value, Example 13.3.

is to the right of χ_L^2. For $\nu = 20$, Table B-4 gives $\chi_L^2 = 10.85$ and $\chi_U^2 = 31.41$. This is called a two-tail, symmetric area.

(c) The computations for test 2 result in $\chi_0^2 = 20.05$ for a $\chi^2(20)$ distribution. The area to its right is found by entering Table B-4 at $\nu = 20$ and going to the column with an entry just less than 20.05. At $\alpha = 0.50$, $\chi_0^2 = 19.38$. By linear interpolation 47.0 percent of the area is above 20.05 (Fig. 13.6b). The corresponding probability statement is

$$P[\chi^2(20) \geq 20.05] = 0.470$$

Problems P13.8–P13.12

13.5 Solving Problems with the χ^2 Distribution

In Sec. 13.2 we learned that the statistic $(n-1)s^2/\sigma^2$ is distributed as χ^2 with $\nu = n - 1$ degrees of freedom. We use this fact to decide if an observed s^2 value from a small sample ($n < 30$) is significantly different from a previous σ^2 value. The assumptions are (1) that the sample is from a normal population with variance σ^2 and (2) that an estimate of σ^2 from prior data is available. To perform the analysis, the procedure is:

1. Take a sample of size n and compute s^2.
2. Determine the value of the statistic

$$\chi_0^2 = \frac{(n-1)s^2}{\sigma^2} \tag{13.5}$$

3. Calculate $\nu = n - 1$ and look up the value of χ_0^2. Determine the probability of observing a value of χ_0^2 or greater; that is, obtain α in the relation

$$P(\chi^2 \geq \chi_0^2) = \alpha \tag{13.6}$$

4. If the probability in step 3 is small, conclude that the variance has changed because the chance of observing the value you observed is small. If the probability is relatively large and the sample is from a normal population, the variance is unchanged because the value you observed had a good chance of being observed. This is the same logic used in Sec. 8.6.

If the sample is larger than 30, use the CLT and the normal distribution to study σ^2 (Chap. 12).

Example 13.4 For some time the critical value of the Reynolds number for laminar flow in a certain pipe has averaged 2020 with a standard deviation of 75. A sample of 25 Reynolds values was taken because scale buildup might have been sufficient to alter the variance. If the sample's s was 85, does it seem probable that the variance has increased?

SOLUTION Use the steps outlined earlier:

1. $n = 25$ and $s^2 = (85)^2$.
2. $\chi_0^2 = \dfrac{24(85)^2}{(75)^2} = 30.827$.
3. $\nu = 25 - 1 = 24$. From Table B-4, after interpolation, $P(\chi^2 \geq 30.827) = 0.196$.
4. There is a 19.6 percent chance that a value of $s \geq 85$ will be observed. This is a large probability, so we conclude that the variance has *not* increased.

COMMENT The probability value used to decide if s^2 is significantly different is always in question. Commonly, 0.05 is used. Then if $P(\chi^2 \geq \chi_0^2) < 0.05$, the conclusion is that variance has increased in a statistical sense. Experience and familiarity with a process make an analyst better able to select the best α value.

It is easy to use Eqs. (13.5) and (13.6) to determine the minimum observed variance s^2 required to decide that population variance has changed statistically. See the Solved Problem section.

Example 13.5

Problems P13.13–P13.17

*13.6 The pdf Derivation, mgf, and Additivity Property of χ^2

The steps in Sec. 13.2 described how the statistic $(n-1)s^2/\sigma^2$ will form a $\chi^2(n-1)$ pdf if samples are from an $N(\mu, \sigma^2)$ population. In addition, it is possible to prove the following derivation of the χ^2 pdf. Take k samples, each of size n, from $N(\mu, \sigma^2)$. Compute

$$s_j^2 = \frac{\sum_{i=1}^{n}(y_{ij} - \bar{y}_j)^2}{n-1}$$

where there are $i = 1, 2, \ldots, n$ values of y per sample for $j = 1, 2, \ldots, k$ samples. Multiplication by $n - 1$ and division by σ^2 gives χ^2 in the form $(n-1)s^2/\sigma^2$.

$$\chi^2 = (n-1)\frac{\sum_{i=1}^{n}(y_{ij} - \bar{y}_j)^2}{(n-1)\sigma^2}$$

$$= \sum_{i=1}^{n}\left(\frac{y_{ij} - \bar{y}_j}{\sigma}\right)^2$$

The value in the parentheses follows the standard normal transformation:

$$Z = \frac{\text{value} - \text{mean}}{\text{standard deviation}}$$

Therefore, χ^2 with $n - 1$ degrees of freedom is the sum of the squares of n standard normal variables. Symbolically, if $Z_i \sim N(0, 1)$ for $i = 1, 2, \ldots, n$ and $Y = Z_1^2 + Z_2^2 + \cdots + Z_n^2$, the variable $Y \sim \chi^2(n - 1)$.

Equation (6.23) gives the moment generating function for the χ^2 pdf as

$$M(t) = (1 - 2t)^{-\nu/2}$$

It is easy to get the derivatives of $M(t)$ to show that $\mu = \nu$ and $\sigma^2 = 2\nu$ for the χ^2 pdf.

It is possible to use $M(t)$ for several χ^2 distributions to prove the *additivity property* of χ^2 variables, which is stated as follows (Fig. 13.7):

If $\quad W \sim \chi^2(\nu_1) \quad$ and $\quad Y \sim \chi^2(\nu_2) \quad$ and $\quad Z = W + Y$

then $\quad Z \sim \chi^2(\nu_1 + \nu_2)$

The sum of two independent χ^2 variables also has a χ^2 distribution with degrees

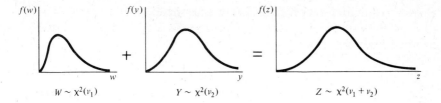

Figure 13.7 The additivity property of the χ^2 distribution.

of freedom equal to the sum of the two degrees of freedom. This property is true for the sum of as many χ^2 variables as you like.

Problems P13.18–P13.23

SOLVED PROBLEM

Example 13.5 Use the data of Example 13.4 to determine the minimum value of s to decide that the variability of the Reynolds values has actually increased. The critical probability is 0.05.

SOLUTION The value of $\chi_0^2 = (n-1)s^2/\sigma^2$ must exceed the χ^2 distribution value for $\nu = n-1$. Substitute Eq. (13.5) into Eq. (13.6) to obtain the required probability statement:

$$P\left[\chi^2 \geq \frac{(n-1)s^2}{\sigma^2}\right] = \alpha$$

In this case we have $n = 25$ and $\sigma^2 = (75)^2$:

$$P\left[\chi^2 \geq \frac{24s^2}{(75)^2}\right] = 0.05 \qquad (13.7)$$

From Table B-4, $\chi_0^2 = 36.42$ has a 5 percent chance of being exceeded. Solve for s using

$$\chi_0^2 = \frac{24s^2}{(75)^2} = 36.42$$

$$s = 92.4$$

For all values of s above 92.4 there is less than a 5 percent chance that $\chi_0^2 \geq 36.42$. If $s \leq 92.4$, then α exceeds 0.05 and we conclude that no increase occurred.

COMMENT The preceding analysis is only for a sample size $n = 25$. Similar computations for several values of n give the engineer a good idea of what values of s to look for. Some results are given here:

Sample size n	Critical s value
10	102.8
15	97.5
20	94.5
25	92.4
30	90.9

You should confirm at least one of these s values by using Eq. (13.7) to determine that the probability is less than 0.05 for any s values above critical.

Section 13.5

ADDITIONAL MATERIAL

Degrees of freedom (Sec. 13.1): Kirkpatrick [15], pp. 156–158; Lipson and Sheth [17], pp. 16, 17; Neville and Kennedy [19], pp. 130–131; Volk [24], p. 76.

χ^2 *pdf basics* (Secs. 13.3 to 13.6): Belz [2], pp. 143–145; Bowker and Lieberman [4], pp. 107–114; Brownlee [5], pp. 82–84; Hogg and Craig [14], pp. 94, 101, 139, 140; Kirkpatrick [15], pp. 158–163; Lipson and Sheth [17], pp. 68–70; Walpole and Myers [25], pp. 172–174.

PROBLEMS

P13.1 Use the χ^2 statistic to compute the following:
(a) The χ^2 value for a sample of size 25 taken from a normal population with a standard deviation of 15 and a sample variance of 150
(b) The minimum sample size required to obtain a value $\chi^2 \geq 24.5$ if a previous sample from a normal population with $\sigma = 10$ resulted in $s = 11.7$
(c) The degrees-of-freedom value for the χ^2 pdf if samples of size 12 are repeatedly taken from a normal population with $\sigma^2 = 5.75$. With $\sigma^2 = 6.50$

P13.2 Compute the value of the χ^2 pdf for (a) $\nu = 4$ and $y = 0.50$, (b) $\nu = 10$ and $y = 1.5$, and (c) $\nu = 2$ and $y = 2$.

P13.3 Sketch the χ^2 pdf for $\nu = 4$ and indicate the points μ and $\mu \pm 1\sigma$ on the graph.

P13.4 Compute the values of μ, σ, and a_3 for (a) $Y \sim \chi^2(3)$ and (b) $Y \sim \chi^2(25)$.

P13.5 Samples of size 10 are taken from a normal population with $\sigma^2 = 15$. In one sample $s^2 = 8.43$ and the analyst wants to use the χ^2 pdf to perform a test on the sample variance.
(a) Plot the χ^2 pdf to scale using the pdf relation.
(b) Hatch-mark the area under the pdf curve corresponding to sample variance values of $s^2 \geq 8.43$.

P13.6 Samples of size $n = 51$ are taken from a normal population. Approximate the probability that (a) $\chi^2 \geq 65$ and (b) $40 \leq \chi^2 \leq 65$.

P13.7 John is taking samples from the processing line in a paint factory. A sample of size 75 in the last weeks has had a variance of $s^2 = 0.3$ (percent)^2 in the water content of the latex paints. Past data indicates the water content has followed a normal distribution with $\sigma^2 = 0.25$ (percent)^2. Use the normal approximation to the χ^2 pdf to determine the probability that the sample variance will exceed 0.3 (percent)^2.

P13.8 Determine the α value for the area above $\chi^2 = 15.99$ for $n = 11$.

P13.9 Maria is taking samples from two machines. For machine A the sample is 15, and for machine B the size is 22. Use the χ^2 pdf tables to determine (a) the area outside the χ^2 values of 7.79 and 23.68 for machine A and (b) the area between the χ^2 values of 12.0 and 34.0 for machine B.

P13.10 For $\nu = 15$ find the χ^2 values with (a) 5 percent of the area above it and (b) 2 percent of the area below it.

P13.11 A machinist wants to be sure that the variance of a product does not violate specified limits. A sample of $n = 24$ items is taken, and the χ^2 distribution is used to test the variance. Determine the χ^2 values which have 90 percent of the area between them and 5 percent outside each value.

P13.12 Use the normal distribution approximation to compute two χ^2 values for $\nu = 75$ with 5 percent of the area outside each value.

P13.13 Samples of size 25 are taken each day from a nozzle production line. If the nozzle diameter follows an $N(\mu, 0.20 \times 10^{-4} \text{ mm}^2)$ distribution and $s^2 = 0.10 \times 10^{-4}$ is observed, use the steps in Sec. 13.5 to determine if the variance has decreased.

P13.14 In Prob. P13.13 what s^2 value can you expect to exceed 5 percent of the time?

P13.15 A quality control specialist wants to set up the 98 percent control limits for sample variance on a normal population having $\sigma^2 = 0.012$. It has been decided that samples of size 10 are to be taken, and the χ^2 distribution is used to set the limits on s^2.

(a) Compute the values so that there is a 1 percent chance of being outside each limit.

(b) Explain why these limits are not equally spaced around the average sample variance s^2 that will be observed.

P13.16 The time during which an insect repellent is considered effective by users varies between 1 and 10 h with the sample variance of 2.4 exceeded by 5 of 100 samples. The users were grouped into samples of size 15. Use these sample results to estimate the true variance of the assumed normal population from which these samples were taken.

P13.17 Mr. Kelly takes three samples each day from finished product. In the past the data has been normally distributed with a widely fluctuating variance. Results of the three samples for one day are given below. Use this data to compute three estimates of the probability that the sample variance will exceed 40.0 grams.

Sample	Sample data, grams				
1	47.2	49.6	52.8	61.4	51.0
2	62.9	51.4	53.1	50.2	61.3
	59.4	65.4	61.4		
3	50.4	48.6	52.9	55.8	58.4
	49.4				

*__P13.18__ Use the moment generating function for the χ^2 pdf to determine the μ and σ^2 relations.

*__P13.19__ The variable X has the mgf $(1 - 2t)^{-4}$ for $t < \frac{1}{2}$, and Y has the mgf $(1 - 2t)^{-3}$. Compute the mean and variance of the variable $W = X + Y$.

*__P13.20__ Suppose that samples are taken from 10 standard normal distributions $X_1, X_2, \ldots, X_{10}$ and the results squared and summed to form the variable $Y = \sum_{i=1}^{10} X_i^2$. For the variable Y, (a) state the values of μ and σ and (b) compute $P(Y \geq 20.00)$.

*P13.21 If $X_1 \sim \chi_1^2(5)$ and $X_2 \sim \chi_2^2(10)$, find the probability that $X_1 + X_2$ does not exceed 7.50.

*P13.22 If Y_1 has a χ^2 distribution with $\nu_1 = 20$ and Y_2 has a χ^2 distribution with $\nu_2 = 30$, use (a) Table B-4 and (b) the normal approximation to determine $P(z \leq 67.5)$ where $Z = Y_1 + Y_2$.

*P13.23 Experimenters at agricultural station A have been working with different irrigation levels for a certain strain of seed. Yield has been normally distributed with $\mu = 10$ tons per acre and $\sigma^2 = 4.84$ (tons per acre)^2. Researchers at station B have used different fertilizer levels on the same seed with yields distributed as $N(8, 2.86)$. Mutual experiments using samples of size $n_A = 15$ and $n_B = 10$ have shown that sample variances seem to differ between the methods of irrigation and fertilization.

(a) What distribution and parameter value should be used to investigate the total variance of the two yield-improvement methods?

(b) What is the probability that the value of a statistic from the distribution determined in (a) will exceed 30? 40?

CHAPTER FOURTEEN

BASICS OF THE t DISTRIBUTION

This chapter introduces the t distribution. You will learn the shape and properties of a t distribution and how to use a table of t values, which is necessary to perform any statistical test utilizing the t distribution.

CRITERIA

To complete this chapter, you must be able to:

1. State the steps necessary to generate a histogram which will approximate a t distribution, given a normal distribution and its mean.
2. Sketch the pdf, and state and compute the mean and standard deviation of any t distribution, given the degrees of freedom.
3. Use a t distribution table to determine a t value, given the degrees of freedom and the one- or two-tail area above or below the desired value; or, determine the area above or below a stated t value, given the degrees of freedom.
4. Compute the probability of observing a specific sample mean value, given the mean and variance computed for a sample of size n from a normal population with a known mean and an unknown variance.
*5. Summarize the derivation of the t distribution from the standard normal and χ^2 distribution.

268 BASICS OF THE t DISTRIBUTION

STUDY GUIDE

14.1 Formulation of the t pdf

The t distribution is used to perform tests on sample means when the actual population variance is not known but can be estimated from the sample by s^2. The sampling procedure to obtain the t pdf assumes you have a normal population $X \sim N(\mu, \sigma^2)$ and you know μ but have no idea of the correct value

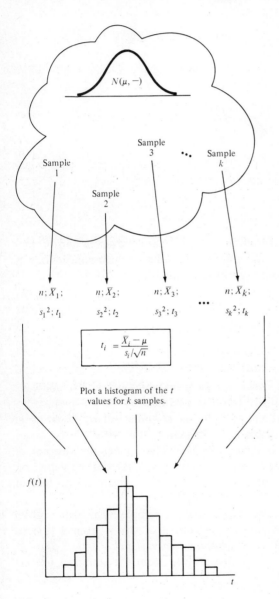

Figure 14.1 Formation of $f(t)$ by taking samples of size n from a normal population and plotting t values.

of σ^2. The steps are (Fig. 14.1):

1. Determine the sample size n.
2. Take a sample of size n and compute $\bar{X}$ and s^2.
3. Calculate the statistic t for the sample

$$t = \frac{\bar{X} - \mu}{s/\sqrt{n}} \tag{14.1}$$

4. Plot t on a histogram of t values.
5. Repeat steps 2 through 4 for k samples.

The histogram will approximate the t distribution with $\nu = n - 1$ degrees of freedom.

The descriptor for a variable T which is distributed as t with $\nu = n - 1$ is $T \sim t(n - 1)$. The expression for the pdf is

$$f(t) = \frac{1}{(\pi\nu)^{1/2}} \frac{\Gamma[(\nu+1)/2]}{\Gamma(\nu/2)} \left(1 + \frac{t^2}{\nu}\right)^{-(\nu+1)/2} \quad -\infty < t < \infty$$

where Γ is the gamma function.

The t pdf uses the central limit theorem (CLT) because it deals with *sample means*, and Eq. (14.1) uses the standard normal form:

$$\frac{\text{sample mean} - \text{population mean}}{\text{standard deviation of sample means}}$$

The steps to form the χ^2 or t pdf from a normal are similar, but the assumptions are different. The statistic (step 3) is different, and the eventual applications are definitely not the same.

Problems P14.1–P14.4

14.2 Graphs, Parameter, and Properties of the t pdf

Once you know the sample size, you can compute $\nu = n - 1$ and plot the t pdf. Several examples are given in Fig. 14.2. All t distributions are symmetric around 0 and range from $-\infty$ to $+\infty$. There is a larger area in the tails of the t than for a corresponding normal pdf. But, as ν increases toward infinity, t approaches an $N(0, 1)$ pdf. The t is a small sample distribution; therefore, if $n \geq 30$, the SND should be used because there is virtually no difference between the two.

The t is a *continuous* distribution with one parameter.

Shape parameter: ν
Parameter range: positive integers
Best estimate of ν: $\hat{\nu} = n - 1$
Expected value: $\mu = 0$
Standard deviation: $\sigma = \sqrt{\dfrac{\nu}{\nu - 2}}$ for $\nu > 2$

270 BASICS OF THE t DISTRIBUTION

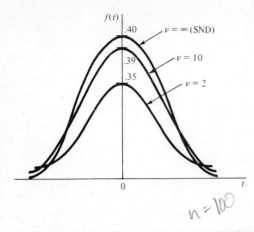

Figure 14.2 The t pdf for $\nu = 2$ and 10 degrees of freedom, and the standard normal distribution which is the shape of $f(t)$ as $\nu \to \infty$.

Example 14.1 In a sample of 100 clothes dryer manufacturers, 28 percent of the machines failed prior to the warranty expiration. Historically the failure rate has been 24 percent.
(a) How can the t distribution be used to study this increase?
(b) What are the μ and σ values for this distribution?

SOLUTION (a) The t distribution can help study the difference between the population mean (24 percent) and the observed sample mean (28 percent) if the population can be assumed to be normal with $\mu = 24$ percent and the standard deviation of the sample is a good estimate of the population σ. From Eq. (14.1) we have a t distribution with $\nu = 10 - 1 = 9$. The t statistic to use is

$$t = \frac{28 - 24}{s/\sqrt{10}} = \frac{12.649}{s} \qquad (14.2)$$

(b) The expected value is $\mu = 0$; and using $\nu = n - 1 = 9$, the standard deviation is $\sigma = \sqrt{9/7} = 1.134$ percent. This means that the expected difference between historical and sample failure percentages is 0, and the standard deviation *of the t distribution* is 1.134 percent. Remember that the σ of the assumed normal population is estimated by the sample s value.

COMMENT Since the t has only one parameter ν, the $\bar{X}$, μ, and s in Eq. (14.2) give us the values of the $t(9)$ pdf. If sample size changes to $n = 12$, the new t distribution is $t(11)$.

Problems P14.5–P14.8

14.3 Use of a t Distribution Table

Appendix Table B-5 gives t values for ν degrees of freedom. Since the area in one or both tails of the pdf is needed, you can use the table for either case. Turn to Table B-5 to follow this explanation:

1. *One-tail α.* The table gives a t value with α of the area to its right; that is, given ν and α, the entries are t_0 values such that

$$P(t \geq t_0) = \int_{t_0}^{\infty} f(t)\, dt = \alpha$$

For example, if $\nu = 13$ and the one-tail $\alpha = 0.05$, then $t_0 = 1.771$ has 5 percent of the area above it (Fig. 14.3a). The pdf is symmetric around 0, so the area below $-t_0$ is also α. Therefore, 5 percent of the area under $f(t)$ is below $t_0 = -1.771$ (Fig. 14.3a).
2. *Two-tail α.* If the total α is *symmetrically* divided between the tails, enter with α in the two-tail α row. The entry is $|t_0|$ that has $\alpha/2$ below and $\alpha/2$ above it. In probabilistic terms,

$$P(t \geq t_0) = P(t \leq -t_0) = \frac{\alpha}{2}$$

For $\nu = 15$ and a two-tail $\alpha = 0.05$, the Table B-5 entry is 2.131. There is 2.5 percent of the area above $t_0 = 2.131$ and 2.5 percent below $-t_0 = -2.131$ (Fig. 14.3b).

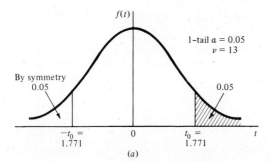

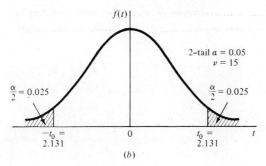

Figure 14.3 (*a*) One-tail and (*b*) two-tail areas of the t distribution.

There are two ways to use the t distribution table:

1. Given ν and α, determine the t_0 value(s).
2. Given ν and the t_0 value(s), determine α.

Since t approaches the SND as n increases, the entries in the $\nu = \infty$ row of Table B-5 are the same as those in the SND Table B-3 for the corresponding probability values.

Example 14.2 (a) Find symmetric t values with 80 percent of the area between them if $\nu = 25$.
(b) Find t values with 10 percent of the area above and 5 percent below for $\nu = 5$.
(c) A statistical test computation results in $\nu = 10$ and $t_0 = 2.5$. What are the one-tail and symmetric two-tail α levels?

SOLUTION (a) Use Table B-5 for $\nu = 25$ and a symmetric two-tail $\alpha = 0.20$ so that 20 percent of the area is outside the $|t_0|$ values. This results in $t_0 = 1.316$ and $-t_0 = -1.316$. You should construct the t distribution graph now and label the t_0 values and the $\alpha/2$ areas.
(b) See Fig. 14.4. The symbols t_1 and t_2 replace t_0 to avoid confusion for nonsymmetric limits. Two one-tail α values must be obtained. For the upper tail $\alpha = 0.10$ and $t_1 = 1.476$. For the lower tail $\alpha = 0.05$ and $t_2 = -2.015$.
(c) For $\nu = 10$, $t_0 = 2.5$ must be linearly interpolated to give a one-tail $\alpha = 0.017$. The two-tail α is 0.034 because it is always twice the one-tail value, thus indicating that 1.7 percent of the area is below and 1.7 percent above $|t_0| = 2.5$.

Problems P14.9–P14.12

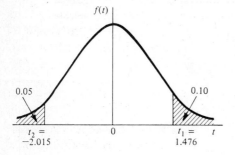

Figure 14.4 Unsymmetric, two-tail areas on the t distribution, Example 14.2b.

PROBLEMS

P14.1 Compute the values of the t statistic for the following situations:

(a) A sample of germination rates from several different cotton fields resulted in the following percentages: 82, 78, 89, 75, 72, and 84. Experience has shown that the rate is normally distributed with an average of 76 percent.

(b) In samples of $n = 22$ items, the sample mean and normal population mean have never been separated by more than 10.5. The sample standard deviation has remained constant at 15.4.

P14.2 A trial sample of net weights was taken from a new can-filling machine for soft drinks with the results $\bar{X} = 11.89$ oz and $s = 0.10$ oz.

(a) If the machine is expected to fill cans according to a normal distribution with $\mu = 11.98$ oz, determine the n value required to observe any t value between $+2.5$ and -2.5. Assume that the $\bar{X}$ and s values are reliable estimates of machine capacity.

(b) Is this n value a maximum or minimum required to stay in the given range of t?

P14.3 A sample of 20 is observed, and the values $\bar{X} = 150.7$ and $s = 19.6$ are computed. Use these values to estimate the value of μ that the parent normal population is supposed to have. A similar sample last week resulted in $t = 2.9$.

P14.4 Compute the t pdf value for (a) $\nu = 4$ and $t = 1.5$ and (b) $\nu = 10$ and $t = 0$. See Sec. 13.2 for helpful comments on gamma functions.

P14.5 Plot the pdf's for the t distribution with $\nu = 4$ and the SND to scale on the same graph. Compare the shape of the pdf's in the tail areas.

P14.6 Compute the means and standard deviations for the t distributions described in Prob. P14.1.

P14.7 Why is it that the t statistic has $s/\sqrt{n}$ as the standard deviation and the z statistic uses s as the estimate of the standard deviation?

P14.8 Compute and compare the μ and σ values of the t distributions for the following ν values: 5, 10, 100, 1000, and 10,000.

P14.9 Plot the general shape of a t distribution and indicate the area on the graph and compute the probability using Table B-5 that (a) $t > 2.228$ for $\nu = 10$, (b) $|t| < 2.131$ for $\nu = 15$, and (c) $|t| > 2.0$ for $\nu = 25$.

P14.10 Samples of size 12 are taken from a normal population with mean 17.8 g. In a particular sample $s = 1.8$ g. Compute the probability of exceeding a sample mean value of 19.0 g using Table B-5.

P14.11 Determine the symmetric t values with 20 percent of the area outside them for 8 degrees of freedom.

P14.12 Carroll has taken six readings from the monthly sales ledger and obtained the following values: \$121,532; \$194,312; \$152,923; \$138,692; \$142,107; \$176,917. If sales have followed a normal distribution in the past with $\mu = \$130,000$, compute the one-tail and symmetric two-tail α values from this sample using Table B-5.

P14.13 Julie is an aerospace engineer with the Fly-at-Night Company. She has taken a sample of 10 pressure readings in a test chamber and computed $\bar{X} = 350$ lb/in^2 and $s = 52$ lb/in^2. Is it reasonable to assume that this sample is from an $N(300, \sigma^2)$ population?

P14.14 Set up the symmetric 95 percent control limits for the sample mean if samples of size 25 are taken from a normal population with $\mu = 5.25$ and the value $s^2 = 1.89$ is a reliable estimator of population variance.

P14.15 Thorndike is in charge of incoming material inspection. An inspector tested eight items and observed a sample mean of 29.1 and a standard deviation of 3.5 from a lot that is assumed to follow a normal distribution with a mean of 27.0. Thorndike accepted the lot because the sample mean did not exceed the single specification limit of 30.0 which must be met by each individual item, not necessarily the sample mean.

(a) Is it reasonable to state that the items are from an $N(27.0, \sigma^2)$ population?

(b) Critique the logic used to accept the lot and determine if this was the correct decision.

P14.16 Compute the probability that the sample mean does not exceed 292 if in repeated samples of $n = 10$ from an $N(300, \sigma^2)$ population the sample variance has been stable at 52.

P14.17 A sample of 25 has been taken with the results $\bar{X} = 2.78$ and $\Sigma(X_i - \bar{X})^2 = 10.71$. If the sample is from a normal population with $\mu = 3.00$, use a one-tail α to determine if it is reasonable to reject the lot using (a) the t distribution and (b) the normal distribution. Comment on the results.

*P14.18 On different days samples are taken from an $N(10, 0.8)$ population to determine process mean and variance. On the first day the results were $n = 25$ and $\bar{X} = 10.5$. On the second day $n = 10$ and $s^2 = 1.1$.

(a) Use the statistic in Eq. (14.1) to compute the t statistic value.

(b) Use the definition of the T variable, $T = Z/\sqrt{Y/\nu}$, as discussed in Sec. 14.5 to determine the values of Z, Y, and t. The two t values should be the same.

CHAPTER FIFTEEN

BASICS OF THE *F* DISTRIBUTION

This chapter teaches you the basic facts about the F distribution. This distribution is used to perform tests on the variances of two samples when the true population variances are not known.

CRITERIA

To complete this chapter, you must be able to:

1. State the steps necessary to generate a histogram which approximates the F pdf, given two normal populations with equal but unknown variances.
2. Sketch the pdf, and state the formula for and compute the mean and standard deviation of an F distribution, given the degrees of freedom values.
3. Use an F distribution table to determine an F value, given the degrees of freedom and the area above the desired value.
4. Compute the probability of observing a specific F value, given the sample sizes and observed sample variance.
*5. Summarize the derivation of the F distribution from two normal populations.

STUDY GUIDE

15.1 Formulation of the *F* pdf

The primary use of the F distribution is to test the ratio of two sample variances when it is reasonable to assume that (1) the population variances are equal and (2) the samples are from normal populations. To generate the F distribution

278 BASICS OF THE F DISTRIBUTION

sample from two normal populations with equal variances $N(\mu_1, \sigma^2)$ and $N(\mu_2, \sigma^2)$, use the following steps (Fig. 15.1):

1. Determine the sample sizes n_1 and n_2.
2. Take the samples from each population and compute s_1^2 and s_2^2.
3. Compute the F statistic for the two samples:

$$1 \le F = \frac{s_1^2}{s_2^2} \quad (15.1)$$

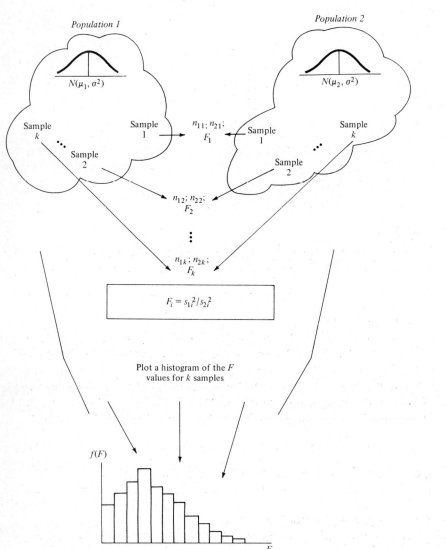

Figure 15.1 Formation of $f(F)$ by taking sample pairs n_1 and n_2 from normal populations and plotting F values.

4. Plot F on a histogram.
5. Repeat steps 2 through 4 for k samples.

The histogram will approximate the pdf for F with $\nu_1 = n_1 - 1$ and $\nu_2 = n_2 - 1$ degrees of freedom.

If Z has an F distribution with ν_1 and ν_2 degrees of freedom, the descriptor is $Z \sim F(\nu_1, \nu_2)$. The pdf expression for $F \geq 0$ is

$$f(F) = \frac{\Gamma[(\nu_1 + \nu_2)/2]}{\Gamma(\nu_1/2)\Gamma(\nu_2/2)} \left(\frac{\nu_1}{\nu_2}\right)^{\nu_1/2} \frac{F^{(\nu_1/2)-1}}{(1 + \nu_1 F/\nu_2)^{(\nu_1+\nu_2)/2}} \qquad F \geq 0$$

Like the χ^2 and t, the F pdf is tabulated so it is not necessary to compute $f(F)$ values.

Problems P15.1, P15.2

15.2 Graphs, Parameters, and Properties of the F pdf

The pdf $f(F)$, which is shown in Fig. 15.2 for several (ν_1, ν_2) values, becomes less skewed right as the values of ν_1 and ν_2 increase. The F distribution has two shape parameters called degrees of freedom:

Two shape parameters: ν_1 and ν_2

Parameter ranges: positive integers

- Best estimates of ν_1 and ν_2: $\hat{\nu}_1 = n_1 - 1$ and $\hat{\nu}_2 = n_2 - 1$

Expected value: $\mu = \dfrac{\nu_2}{\nu_2 - 2}$ for $\nu_2 > 2$

Standard deviation: $\sigma = \left[\dfrac{2\nu_2^2(\nu_1 + \nu_2 - 2)}{\nu_1(\nu_2 - 4)(\nu_2 - 2)^2}\right]^{1/2}$ for $\nu_2 > 4$

Example 15.1 Engineers sampled the crude flowing in two oil field pipelines. Variations in viscosity are studied using the sample variances. If sample sizes are $n_1 = 26$ and $n_2 = 11$, determine and interpret the μ and σ values of the distribution of sample variance ratios.

SOLUTION Assuming normal populations for the two flows, the variance ratio has an F distribution which can be used to study relative changes between viscosity variances. In this case the $F(25, 10)$ distribution applies, and the μ and σ values are

$$\mu = \frac{10}{10 - 2} = 1.25$$

$$\sigma = \left[\frac{2(100)(33)}{25(6)(64)}\right]^{1/2} = (0.6875)^{1/2} = 0.83$$

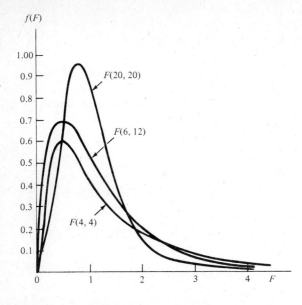

Figure 15.2 Plot of the F distribution for different (ν_1, ν_2) pairs.

These values are dimensionless because they represent the ratio s_1^2/s_2^2, where each s_i^2 is expressed in $(\text{dyn} \cdot \text{s}/\text{cm}^2)^2$ for viscosity. The expected value of 1.25 means that the sample 1 variance is 1.25 times the sample 2 variance. The standard deviation represents spread around the mean of the ratio of variances.

Problems P15.3–P15.5

15.3 Use of an F Distribution Table

Table B-6 gives the F_0 value for ν_1 and ν_2 degrees of freedom having $\alpha = 0.10$, 0.05, and 0.01 of the area to its right; that is,

$$\int_{F_0}^{\infty} F(\nu_1, \nu_2)\, dF = \alpha$$

Because F has two parameters and because it does not rapidly approximate another distribution as ν_1 and ν_2 increase, the F table can get extremely large if many α values are included. The three α values in Table B-6 are sufficient for all common uses of the F distribution. Table B-6 is used as follows. Find the correct ν_1 value across the top and ν_2 and α down the side. Go across the α row until the ν_1 column is reached. Read the F value. For example, if $\nu_1 = 6$ and $\nu_2 = 10$, then $F_0 = 3.22$ has 5 percent of the area to its right and $F_0 = 5.39$ has 1 percent of the area above it.

Since Table B-6 gives F values for the right tail only, an F value with α of the area in the left tail is found by the following procedure:

1. Switch the order of the degrees of freedom.
2. Look up the $F(\nu_2, \nu_1)$ value in Table B-6 with α in the right tail.
3. Take the reciprocal.

In symbols, if F_0' is a left-tail value of $F(\nu_1, \nu_2)$, then

$$F_0' = \frac{1}{F(\nu_2, \nu_1)} \qquad (15.2)$$

Figure 15.3 indicates that for $\nu_1 = 8$ and $\nu_2 = 14$, 5 percent of the area is above $F(8, 14) = 2.70$ and that by using Eq. (15.2) 5 percent is below F_0'.

$$F_0' = \frac{1}{F(14, 8)} = \frac{1}{3.24} = 0.31$$

All values in Table B-6 are greater than 1.00, even though the F pdf is defined for $F \geq 0$. An $F \geq 1$ is ensured by always placing the larger sample variance in the numerator of Eq. (15.1) and referring to the associated data as sample 1.

Example 15.2 Determine the following.
(a) Two F values such that 5 percent is in each tail for $\nu_1 = 8$ and $\nu_2 = 12$
(b) The area below $F = 0.2114$ if $\nu_1 = 9$ and $\nu_2 = 24$

SOLUTION (a) Table B-6 shows that 5 percent of the area is above $F(8, 12) = 2.85$. The F value with 5 percent of the area below it is found by using the procedure above.

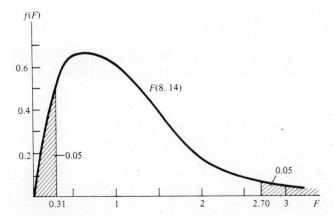

Figure 15.3 F distribution for $\nu_1 = 8$ and $\nu_2 = 14$ showing 5 percent of the area above $F = 2.70$ and 5 percent below $F = 0.31$.

1. Switch the degrees of freedom so that $\nu_1 = 12$ and $\nu_2 = 8$.
2. Find $F(12, 8) = 3.28$.
3. Use Eq. (15.2) to compute

$$F'_0 = \frac{1}{3.28} = 0.30$$

We can now state

$$P(0.30 \leq F \leq 2.85) = 0.90$$

(b) In Eq. (15.2), $F'_0 = 0.2114$ because the desired area is below the given F value. Solve for $F(24, 9)$ and determine α from Table B-6.

$$F(24, 9) = \frac{1}{F'_0} = \frac{1}{0.2114} = 4.73 \qquad (15.3)$$

Therefore, $\alpha = 0.01$. Interpolation to find the α value is necessary if the exact F is not tabulated.

Notice that the ν_1 and ν_2 values are switched in Eq. (15.3) so that the correct F value is found in Table B-6.

Problems P15.6–P15.11

15.4 A Simple Application of the F pdf

The F relation, Eq. (15.1), may be used to determine the probability that the variance of one sample will equal or exceed some specified multiple of the second-sample variance. Table B-6 is used to obtain this probability value once F is calculated or specified. If α is reasonably high (greater than 0.5) so that an observed F will occur, conclude that the sample variances come from populations with equal variances. The assumptions are that (1) the samples are from normal populations with variances σ_1^2 and σ_2^2 and (2) the variances are equal and unknown but can be reliably estimated by sample variances s_1^2 and s_2^2. In these computations, sample 1 should have the larger variance to ensure that $F \geq 1$.

Example 15.3 A consumer product is made at two different plants using similar machines. Dimensions are sampled at both sites with the following results.

Plant	A	B
Sample size	15	8
Sample variance, cm²	0.100	0.115

(a) Are the plants making product with equal variances?
(b) How much larger or smaller does s_B^2 have to be to alter the decision in (a)?

SOLUTION Use s_B^2 as the numerator in Eq. (15.1) since $s_B^2 > s_A^2$.
(a) Compute the F statistic:

$$F = \frac{s_B^2}{s_A^2} = \frac{0.115}{0.100} = 1.15$$

With $\nu_1 = n_B - 1 = 7$ and $\nu_2 = n_A - 1 = 14$, Table B-6 gives $F(7, 14) = 2.76$ for $\alpha = 0.05$. Therefore, the probability $P(F \geq 1.15)$ is larger than 5 percent. We conclude that the samples are from populations with no significant difference in variances.

(b) Since $P(F \geq 2.76) = 0.05$, any F larger than 2.76 has a small chance of occurring. If $F = s_B^2/s_A^2$ exceeds 2.76, conclude that the population variances are unequal. To observe $F > 2.76$, s_B^2 must be at least 2.76 times as large as s_A^2. For example, if s_A^2 remains constant at 0.100, an $s_B^2 \geq 0.276$ is required to have unequal population variances.

COMMENT Notice that no conclusions are drawn about the sample means using the F distribution. Also, if only one variance is to be studied, the χ^2 distribution, not the F, must be used.

Problems P15.12–P15.14

*15.5 Derivation of the F Distribution

Mathematically the F distribution is derived by defining a variable F which is the ratio of two χ^2 distributions each divided by its degree of freedom $\nu_i = n_i - 1$:

$$F = \frac{\chi_1^2/\nu_1}{\chi_2^2/\nu_2} \tag{15.4}$$

Equation (15.1) is then obtained from two $N(\mu, \sigma^2)$ populations. Since the variable $(n - 1)s^2/\sigma^2$ has a $\chi^2(n - 1)$ distribution, substitution into Eq. (15.4) gives

$$F = \frac{[(n_1 - 1)s_1^2]/[\sigma_1^2(n_1 - 1)]}{[(n_2 - 1)s_2^2]/[\sigma_2^2(n_2 - 1)]} = \frac{s_1^2/\sigma_1^2}{s_2^2/\sigma_2^2}$$

Under the assumption that the normals have variances which are not significantly different, $\sigma_1^2 = \sigma_2^2$ and $F = s_1^2/s_2^2$ results.

There is a direct mathematical relation between the F and t distributions. For a given ν and α value, $F(1, \nu) = [t(\nu)]^2$, where α is the two-tail α for the t variable. For example, if $\nu = 10$ and $\alpha = 0.01$, $F(1, 10) = 10.04$ and $[t(10)]^2 = (3.169)^2 = 10.04$.

Problems P15.15, P15.16

ADDITIONAL MATERIAL

Belz [2], pp. 145–147; Bowker and Lieberman [4], pp. 120–123; Brownlee [5], pp. 285–288; Lipson and Sheth [17], pp. 106–110; Miller and Freund [18], pp. 175–178; Volk [24], pp. 159–166; Walpole and Myers [25], pp. 179–183.

PROBLEMS

P15.1 Compute the value of the F statistic for the following situations.
    (a) Samples of size 35 are taken from two vendors' products which follow normal distributions, and the standard deviations are computed. The results are $s_1 = 58.2$ and $s_2 = 71.9$.
    (b) The sample 1 and sample 2 data collected by Mr. Kelly and presented in Prob. P13.17

P15.2 Compute the F pdf values for (a) $\nu_1 = \nu_2 = 4$ at the value $F = 1.5$ and (b) $n_1 = 5$ and $n_2 = 7$ at $F = 3$.

P15.3 Jerry took two samples of size 12 each at one plant and Jeannie took two samples of sizes $n_1 = 20$ and $n_2 = 12$ at another plant. Determine and compare the μ and σ values for the F distributions that should be used to study sample variances.

P15.4 Construct the pdf graph for the $F(6, 12)$ distribution. Indicate (a) the point μ and (b) the area outside the points $\mu \pm \sigma$ on this graph.

P15.5 Determine a simplified expression for the standard deviation of the F distribution if $\nu_1 = \nu_2 = \nu$ and $\nu > 4$.

P15.6 An F distribution has $\nu_1 = 10$ and $\nu_2 = 6$. Find (a) the F value with 5 percent of the area to its right and (b) the area to the right of the value $F = 2.94$.

P15.7 If samples of size $n_1 = n_2 = 9$ have been taken to study sample variances, determine the F values with 1 percent of the area in each tail.

P15.8 The values $F = 0.50$ and $F = 2.86$ were computed from two samples with $\nu_1 = 24$ and $\nu_2 = 20$. Determine the area under the F distribution graph between these two F values.

P15.9 The following variance data was computed at the Ur-a-Stinker Cosmetic Company.

Sample size	Sample variance
8	1.49
15	6.08

    (a) What are the F value and the degree-of-freedom values which result in an $F > 1$?
    (b) What is the probability that the F value computed in (a) is not exceeded?

P15.10 If the area below the value $F = 0.2907$ is exactly 0.05 and $\nu_1 = \nu_2$, determine (a) the values of ν_1 and ν_2 and (b) the F value with 0.025 of the area to its left.

P15.11 (a) Determine the F value with 5 percent of the area in the left tail if $\nu_1 = 100$ and $\nu_2 = 500$.
    (b) Does this value decrease or increase if ν_1 and ν_2 are interchanged?

P15.12 A new product is being marketed in two different foreign market areas. In a sample of 31 days' sales in one area, the standard deviation is $8500 per day and for 20 days in the other the standard deviation is $7200 per day. Is it reasonable to assume that the samples come from populations with statistically equal variances?

P15.13 Samples of size $n_1 = n_2 = 15$ are taken from normal populations. After the first sample is taken, the value $s_1^2 = 18.7$ is computed. Determine the probability that s_2^2 will be at least 3 times s_1^2.

P15.14 John Harris is a chemical engineer in a refining plant. Sample flow rates in liters per second have been taken during two different weeks. Use the values shown to determine if it is reasonable to state that the population variance has decreased, increased, or remained constant.

Week 1:	178.3	104.9	219.6	154.2	201.5
Week 2:	162.4	189.4	142.7	152.8	192.0
	151.6	174.8	180.1	149.6	

*P15.15 Verify that the $F(1, 10)$ value is the square of the $t(10)$ value for $\alpha = 0.05$.

*P15.16 The variable T was defined in Sec. 14.5 as the ratio of an $N(0, 1)$ variable Z divided by the square root of a χ^2 variable Y divided by its degrees of freedom ν; that is,

$$T = \frac{Z}{\sqrt{Y/\nu}}$$

Show that the variable T^2 has the distribution $F(1, \nu)$ as stated in Eq. (15.4).

CHAPTER
SIXTEEN

OTHER CONTINUOUS DISTRIBUTIONS

This chapter covers the basics of the uniform, exponential, and gamma distributions. You will learn how they are commonly applied and how they are related to distributions already discussed.

CRITERIA

To correctly understand and apply these distributions, you must be able to:

1. State and plot the *uniform* pdf and cdf, state its parameters and properties, and compute probability values, given a statement of the problem.

(Criteria 2 to 5 are for the exponential distribution.)

2. State and plot the *exponential* pdf and cdf, and give a definition of the exponential probability law.
3. State and calculate the parameter and properties of the exponential.
4. Compute the probability of failure and the reliability for an exponentially failing unit, given values of the parameter and time.
*5. State the relation between the Poisson and the exponential distributions.
6. Plot the *gamma* pdf for different parameter values, and state and compute the properties, given a written description of the situation.
*7. Write the gamma distribution parameter values which generate the exponential, χ^2, and Erlang pdfs. (*Note*: The beta, Weibull, log normal, and Rayleigh distributions are summarized in this section.)

STUDY GUIDE

16.1 The Uniform Distribution

The *uniform* (or rectangular) distribution is a *continuous* pdf used when an event occurs with equal probability between two extreme values. For example, weight might be uniformly distributed between two kilogram values. If X is uniform between α and β, the descriptor is $X \sim U(\alpha, \beta)$ and the pdf is

$$f(x) = \begin{cases} \dfrac{1}{\beta - \alpha} & \alpha \leq x \leq \beta \\ 0 & \text{elsewhere} \end{cases} \qquad (16.1)$$

Figure 16.1a shows that $f(x)$ is constant at $1/(\beta - \alpha)$ in the range $\alpha \leq x \leq \beta$. The parameters and properties are:

Two scale parameters: α and β

Parameter ranges: $-\infty < \alpha < \infty; \; -\infty < \beta < \infty$

Best estimates: $\hat{\alpha} = X_{\min}$

$\hat{\beta} = X_{\max}$

Expected value: $\mu = \dfrac{\alpha + \beta}{2}$

Standard deviation: $\sigma = \dfrac{\beta - \alpha}{\sqrt{12}}$

The best estimates of the parameters are the smallest and largest observations in a sample, if no better information about the population is available.

The cdf for the uniform is often used. If $f(x)$ is integrated from α to x, the cdf is

$$F(x) = \begin{cases} 0 & x < \alpha \\ \dfrac{x - \alpha}{\beta - \alpha} & \alpha \leq x \leq \beta \\ 1 & x > \beta \end{cases} \qquad (16.2)$$

Figure 16.1b of $F(x)$ shows a constant gradient between α and β.

Example 16.1 The noise level N in a machine room is uniformly distributed between 80 and 95 decibels (db). Government standards maintain that 90 db is a maximum safe level for extended exposure.
(a) Write the pdf for N and compute the chance of exceeding the 90-db standard.
(b) Determine μ and σ and find the probability that the level is within $\pm 1\sigma$ of the mean.

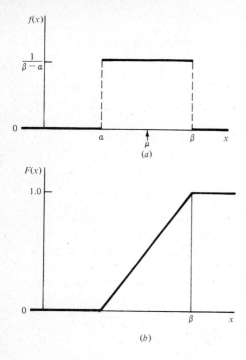

Figure 16.1 Plot of the uniform distribution (*a*) pdf and (*b*) cdf.

SOLUTION (*a*) $N \sim U(80, 95)$ is the descriptor. From Eq. (16.1) the pdf is

$$f(n) = \tfrac{1}{15} \qquad 80 \leq n \leq 95$$

The cdf Eq. (16.2) can be used to compute $P(n > 90)$ by taking

$$P(n > 90) = 1 - P(n \leq 90) = 1 - F(90)$$
$$= 1 - \frac{90 - 80}{95 - 80} = 0.333$$

At any point in time there is a $\tfrac{1}{3}$ chance of exceeding the maximum safe level.

(*b*) The properties μ and σ for $U(80, 95)$ are $\mu = 87.50$ db and $\sigma = 4.33$ db. One standard deviation from μ is 83.17 and 91.83 db. The probability of being in these limits is 0.577, which is the integral of $f(n)$ over the range 83.17 to 91.83:

$$\int_{83.17}^{91.83} \frac{1}{15} \, dn = \frac{n}{15} \bigg|_{83.17}^{91.83} = 0.577$$

Example 16.5

Problems P16.1–P16.6

16.2 Formulation of the Exponential Distribution

The probability of failure of an item prior to some specified time t is often of interest. Commonly it is postulated that failure depends only on a constant rate λ multiplied by the time t. In this case the chance of *no failure in time t* is called *reliability R(t)* and can be written

$$R(t) = e^{-\lambda t} \tag{16.3}$$

However, $R(t)$ is the complement of the cumulative probability of a failure in time t. This is 1 minus the cdf of failure at time t:

$$R(t) = 1 - F(t) = e^{-\lambda t}$$

According to Eq. (6.18) we get the failure pdf $f(t)$ by differentiating $F(t)$:

$$F(t) = 1 - e^{-\lambda t} \tag{16.4}$$

$$f(t) = \frac{dF(t)}{dt} = \lambda e^{-\lambda t}$$

This is the exponential distribution. The complete pdf relation is

●
$$f(t) = \begin{cases} \lambda e^{-\lambda t} & t > 0;\ \lambda > 0 \\ 0 & \text{elsewhere} \end{cases} \tag{16.5}$$

The descriptor of the exponential distribution is $T \sim \exp(\lambda)$. The pdf and cdf are plotted in Fig. 16.2. You can see that $f(t) = 0$ at $t = 0$ and equals λ immediately after $t = 0$. $F(t)$ asymptotically approaches 1.

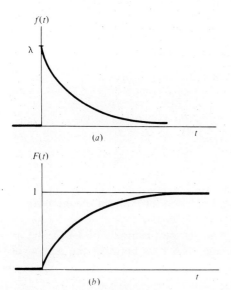

Figure 16.2 Plot of the exponential (*a*) pdf and (*b*) cdf.

The exponential is a continuous distribution defined by the following exponential probability law.

> If the instantaneous occurrence rate of an event is a constant λ, the probability of occurrence will depend only on λ and the time interval t and will follow the exponential distribution.

For example, if the event is the failure of an item, the chance of failure in a set time t will be the same regardless of where t occurs in the item's life. The exponential is commonly used in reliability and queuing analysis. Later examples illustrate some of these uses.

Problems P16.7–P16.9

16.3 Parameters and Properties of the Exponential Distribution

The exponential has one parameter, and its mean equals its standard deviation.

$$\text{One parameter:} \quad \lambda$$
$$\text{Parameter range:} \quad \lambda > 0$$
$$\text{Best estimate of } \lambda: \quad \hat{\lambda} = \frac{1}{\bar{X}}$$
$$\text{Expected value:} \quad \mu = \frac{1}{\lambda}$$
$$\text{Standard deviation:} \quad \sigma = \frac{1}{\lambda}$$

The dimension of the data is usually time, so λ is an occurrence per time unit, for example, failure per hour or arrival per minute.

Example 16.2 The time T between the arrival of orders at a regional warehouse is recorded for 24 orders:

17	19	25	34	35	35
37	39	40	40	41	41
42	42	44	46	51	52
52	56	71	72	80	93

(*a*) Estimate and interpret the parameter λ.
(*b*) Assume arrivals follow an exponential law. Plot μ and σ on the pdf.

SOLUTION (*a*) The mean is $\bar{t} = 46$ h, and the best estimate of λ is $\hat{\lambda} = \frac{1}{46} = 0.0217$ arrivals per hour. On the average 46 h pass between the arrival of orders.

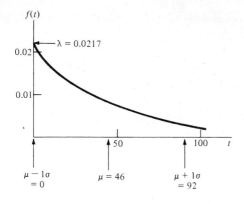

Figure 16.3 Plot of the exponential pdf with $\lambda = 0.0217$ arrivals per hour, Example 16.2.

(b) The pdf (Fig. 16.3) is obtained from Eq. (16.5) and has the properties $\mu = \sigma = 46$ hours per arrival. One standard deviation from the mean is 0 and 92 h.

Problems P16.10–P16.12

16.4 Solving Reliability Problems with the Exponential

The exponential is commonly applied in reliability problems where it is hypothesized that the failure rate λ is constant over the entire life of an item. The pdf gives the instantaneous probability of failure at any given time $t > 0$. Then the cdf $F(t) = 1 - e^{-\lambda t}$ is used to compute the probability of failure *prior to t*. Similarly, the reliability $R(t)$ is the probability of successful operation for *at least* a time t, and $R(t) = 1 - F(t)$, as mentioned in Sec. 16.2. In reliability the reciprocal of λ is called θ and is termed the *mean time between failures (MTBF)*, or average life.

$$\text{MTBF:} \quad \theta = \frac{1}{\lambda}$$

$$\text{Failure rate:} \quad \lambda = \frac{1}{\theta}$$

Example 16.3 demonstrates an application of the exponential.

Example 16.3 Engineers have collected data from 100 compressors on natural gas pipelines and found that the average life is 5.75 years and that failures follow the exponential distribution.
(a) Compute the probability of failure during the first year after installation. During the first 3 months.
(b) Compute the probability of failure prior to the average life.

(c) Compute the probability of operating at least 10 years.
(d) Plot the reliability curve and compare it with the pdf curve.

SOLUTION The MTBF is $\theta = 5.75$, so $\lambda = 0.174$ is the parameter for the exponential pdf.

$$f(t) = 0.174 e^{-0.174t} \quad t > 0$$

where T is the life of a compressor. Use $f(t)$ to answer the questions.

(a) Failure at any time t follows $f(t)$, so failure during the first year is the cdf at $t = 1$. From Eq. (16.4)

$$F(t = 1) = 1 - e^{-0.174(1)} = 1 - 0.84 = 0.16$$

There is a 16 percent chance of failure in the first year. Similarly, you can calculate a 4 percent chance of failure in the first 3 months using $t = 0.25$.

(b) Failure prior to the average life of $\theta = 5.75$ years equals the cdf at $t = 5.75$:

$$F(t = 5.75) = 1 - e^{-0.174(5.75)} = 1 - e^{-1} = 0.632$$

The chance of failure by the average life is *always* 63.2 percent for any exponential, because at $t = \theta = 1/\lambda$ the area to the left of $1/\lambda$ is 0.632:

$$\int_0^{1/\lambda} \lambda e^{-\lambda t} \, dt = 0.632$$

(c) This is the exponential reliability at $t = 10$. By Eq. (16.3)

$$R(t = 10) = e^{-0.174(10)} = 0.176$$

There is a 17.6 percent chance that the compressor will operate successfully for at least 10 years.

(d) The exponential reliability, Eq. (16.3), and the $f(t)$ curve are shown in Fig. 16.4. Every point on $R(t)$ gives the probability of operating for at least t years, and $f(t)$ gives the instantaneous probability of failure at t years. $R(t)$ always starts at 1.00, and $f(t)$ always starts at λ.

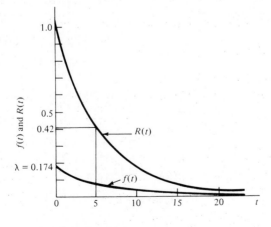

Figure 16.4 Plot of the reliability curve and the failure pdf for exponential failures with $\lambda = 0.174$, Example 16.3.

Remember, the exponential is appropriate only when the occurrence rate of an event is constant over a given period, regardless of when the period occurs. Solved Problems shows how to fit the exponential to observed data.

Example 16.6

Problems P16.13–P16.21

*16.5 Relation between the Exponential and Poisson Distributions

In Chap. 9 the Poisson parameter λ was defined as the constant rate of occurrence per specified unit. If time is the unit, the Poisson can have a parameter λt for t time units. The Poisson probability of no occurrences prior to t, by Eq. (9.2), is $P(0; \lambda t) = e^{-\lambda t}$. This means the first occurrence takes place at or after time t, because it did not happen prior to t. That is,

$$e^{-\lambda t} = P(\text{no occurrence before time } t)$$
$$= P(\text{first occurrence at or after time } t)$$

The last probability statement is 1 minus the cdf at t, so

$$e^{-\lambda t} = 1 - F(t)$$

or
$$F(t) = 1 - e^{-\lambda t} \qquad (16.6)$$

Equation (16.6) is the same as Eq. (16.4), which was used to obtain the exponential pdf. The following statement relates the Poisson and exponential distributions.

> If the number of events X follows the discrete Poisson with parameter λ per time unit, the number of times units T between occurrences of each event follows the continuous exponential with parameter λ.

This direct connection between a discrete and a continuous distribution is important in the analysis of queuing (waiting line) systems. For example, if the number of cars arriving at a tollbooth is Poisson, the time between each arrival is exponential. Since time starts anew at 0 after each arrival, the rate of arrival is constant and the exponential is appropriate.

Problems P16.22, P16.23

16.6 The Gamma Distribution

If events occur at a constant rate λ (as for the Poisson and exponential) but we are interested in the time T to observe n events, then the variable T has a gamma pdf. The descriptor is $T \sim G(t; n, \lambda)$, and the pdf relation is

$$f(t) = \begin{cases} \dfrac{\lambda^n}{\Gamma(n)} t^{n-1} e^{-\lambda t} & t > 0 \\ 0 & \text{elsewhere} \end{cases}$$

where $\Gamma(n)$ is the gamma function defined by the integral

$$\Gamma(n) = \int_0^\infty x^{n-1} e^{-x} \, dx \qquad (16.7)$$

The gamma pdf is continuous and has two parameters.

    Shape parameter: n

    Scale parameter: λ

    Parameter ranges: $n > 0; \lambda > 0$

    Parameter estimates: $\hat{\lambda} = \dfrac{\bar{X}}{s^2}$

                             $\hat{n} = \bar{X}\hat{\lambda}$ or known

    Expected value: $\mu = \dfrac{n}{\lambda}$

    Standard deviation: $\sigma = \dfrac{\sqrt{n}}{\lambda}$

The estimates $\hat{\lambda}$ and $\hat{n}$ are not the best estimators, as discussed in Sec. 6.9; however, they are accurate enough for most applications.

The gamma pdf is used often because its shape is so flexible. Different values of the shape parameter n cause the following. The pdf graphs are shown in the next example.

Range of n	pdf shape	Location of mode
$n \leq 1$	Reverse J	0
$n > 1$	Unimodal	$\dfrac{n-1}{\lambda}$

As implied by its name, the scale parameter λ adjusts only scale, never shape.

Example 16.4 A sensitive microswitch, which has a constant failure rate of 2 per year, is duplicated 4 times to ensure a sufficiently high reliability of a tracking radar unit. All switches are active during operation to reduce deterioration; however, it has been found that when three switches fail, the radar unit must be shut down and the switches replaced.
(a) State the parameters and plot the pdf with its mean and mode for the time to shutdown.
(b) Plot the pdf if the failure rate is reduced to 1 per year and replacement occurs after two switches fail.
(c) Give an example of a gamma variable unrelated to the one above with the parameter $n < 1$.

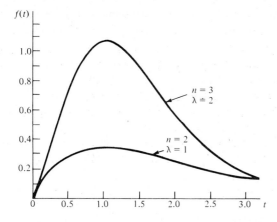

Figure 16.5 Plot of the gamma pdf with $n > 1$, Example 16.4.

SOLUTION (a) The distribution of the variable T, time to radar unit shutdown, is gamma with parameters $\lambda = 2$ per year and $n = 3$ failures; that is, $T \sim G(t; 3, 2)$. The pdf is shown in Fig. 16.5 with mean and mode

$$\mu = \frac{n}{\lambda} = \frac{3}{2} = 1.5 \text{ years}$$

$$m = \frac{n-1}{\lambda} = \frac{2}{2} = 1.0 \text{ year}$$

(b) If the failure rate is $\lambda = 1$ per year and $n = 2$ failures, the pdf (Fig. 16.5) has a much lower profile. It is still unimodel because $n > 1$ with $m = 1$ year but $\mu = 2$ years before unit shutdown.

(c) Assume that orders for tons of grain arrive at a bulk feed warehouse at a constant rate per week for a particular type of cattle feed. The warehouse receives its supply in very large lots by rail from distant processors. If computations for time to reorder are based on fractions of rail cars remaining, the time T between reordering will have $T \sim G(t; n, \lambda)$, with n the number of carloads in stock at reorder time and λ the number of orders per week. If less than 1 rail car is in stock at

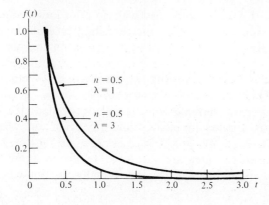

Figure 16.6 Plot of the gamma pdf with $n < 1$, Example 16.4.

reorder time, the gamma pdf will be reverse *J* in shape. Figure 16.6 is a plot for $n = 0.5$ with $\lambda = 1$ and 3 orders per week.

COMMENT The gamma is a versatile distribution which is useful when shape is clearly altered by one parameter (called *n* here). It is used in reliability, inventory, and queuing analyses.

Problems P16.24–P16.27

*16.7 Relationship of Gamma to Other Distributions

The gamma is directly transformed into the exponential or χ^2 distribution by fixing the value of one or both parameters. You should perform these substitutions to show that the pdf relations and properties are correct.

	Gamma parameter value	
Distribution	n	λ
Exponential	1	No change
χ^2	$\nu/2$	$\frac{1}{2}$
Erlang	Positive integer	No change

The Erlang is commonly used instead of the gamma because *n* is usually a positive integer. For the gamma, $\Gamma(n)$ is found by Eq. (16.7), but in the Erlang $\Gamma(n) = (n-1)!$. The pdf for Erlang is identical to the gamma, as is the interpretation.

For your information, some other useful continuous distributions are summarized in Table 16.1 with pdf shape, parameters, and properties. Major areas of application are summarized here:

Beta. If the probability of occurrence is established with no prior knowledge or experience, the beta is assumed. It is used in scheduling projects by the Program Evaluation and Review Technique (PERT) and in quality control theory.

Weibull. If the time to failure is not constant (not exponential), the Weibull is often used for reliability analysis.

Log Normal. If the logarithm of a variable is normally distributed, the log normal can be used. It is applied in biology and financial and wealth distribution analysis.

Rayleigh. If errors from a circle's center are measured in distance on two perpendicular, planar radials and the errors are normally distributed, the

Rayleigh may be used. It is applied in bombing and artillery sighting and communication theory.

Problems P16.28–P16.32

SOLVED PROBLEMS

Example 16.5 Joel has a table of random numbers generated from a continuous $U(0, 1)$ distribution. He wants to generate uniform values from $U(10, 25)$ using the transformation

$$U(\alpha, \beta) = (\beta - \alpha)u + \alpha \tag{16.8}$$

where u is from $U(0, 1)$. Use these four values of u to compute the required numbers.

0.92406 0.76522
0.58342 0.08341

SOLUTION For $U(10, 25)$ use $\alpha = 10$ and $\beta = 25$ in Eq. (16.8). The transformed values are:

23.86 21.48
18.75 11.25

COMMENT Table 10.1 is from a discrete uniform distribution over the range 0 to 9. By placing a decimal before any number from the table, a random number from a $U(0, 1)$ distribution can be transformed into any $U(\alpha, \beta)$ using Eq. (16.8).

Section 16.1

Example 16.6 The hours between breakdowns (8 A.M. to 5 P.M.) of a delicate machine were recorded for 20 breakdowns (Table 16.2). Do breakdowns seem to occur at a constant rate per hour regardless of previous history?

SOLUTION Table 16.3 gives the observed frequency for the 20 breakdowns using the data of Table 16.2. The observed cumulative distribution value is

$$\text{Observed } F(t) = \frac{\text{cumulative frequency}}{\text{sample size}} = \frac{F}{20}$$

If the breakdown rate per hour is independent of operating time, the exponential is applicable with the parameter

$$\hat{\lambda} = \frac{1}{\text{average time between breakdowns}} = \frac{1}{\bar{t}}$$

$$= \frac{1}{2.95} = 0.339 \frac{\text{hour}}{\text{breakdown}}$$

Table 16.1 Summary of some continuous distributions

Name and variable range	Parameters and estimates	Properties of μ and σ^2	General pdf shapes	Comments
Beta $0 \leq x \leq 1$	Shape: $\beta > 0$ Shape: $\alpha > 0$ $\hat{\beta} = \dfrac{1-\overline{X}}{s^2}\left[\overline{X}(1-\overline{X}) - s^2\right]$ $\hat{\alpha} = \dfrac{\overline{X}\hat{\beta}}{1-\overline{X}}$	$\mu = \dfrac{\alpha}{\alpha+\beta}$ $\sigma^2 = \dfrac{\alpha\beta}{(\alpha+\beta)^2(\alpha+\beta+1)}$		Collapses to uniform when $\alpha = \beta = 1$

Distribution		Parameters	Moments	Graphs	Notes
Weibull $x \geq 0$		Shape: $\beta > 0$ Scale: $\theta > 0$ $\hat{\beta}$ and $\hat{\theta}$ are graphical	$\mu = \theta \Gamma(\frac{1}{\beta} + 1)$ $\sigma^2 = \theta^2 [\Gamma(\frac{2}{\beta} + 1)] - \theta^2 [\Gamma(\frac{1}{\beta} + 1)]^2$		Collapses to exponential when $\beta = 1$ and $\lambda = 1/\theta$
Log normal $x \geq 0$		Shape: $\sigma > 0$ Scale: $-\infty \leq \mu \leq \infty$ $\hat{\mu}$ and $\hat{\sigma}$ are graphical	$E[X] = \exp(\mu + \frac{1}{2}\sigma^2)$ $V[X] = \exp(2\mu + \sigma^2)(e^{\mu^2} - 1)$		Collapses to Rayleigh when $\beta = 2$
Rayleigh $x \geq 0$		Scale: $\theta > 0$ $\hat{\theta}$ is graphical	$\mu = \theta \sqrt{\pi/2}$ $\sigma^2 = 0.429 \theta^2$		

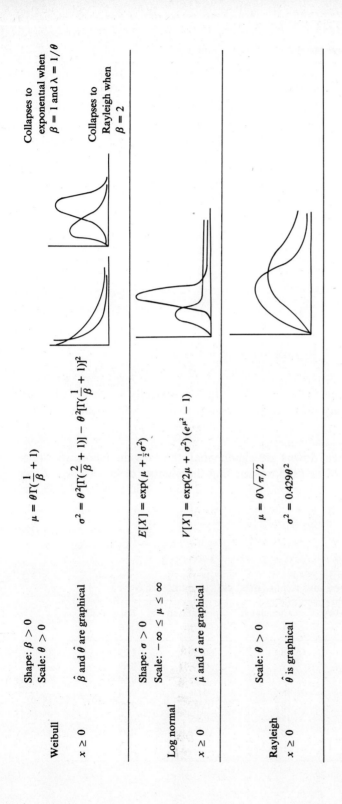

Table 16.2 Observed breakdown times, Example 16.6

Day	Breakdown hour	Hours between breakdowns t	Day	Breakdown hour	Hours between breakdown t
1	8 A.M.	Start study	4	9 A.M.	2
	9	1		10	1
	2 P.M.	5		2 P.M.	4
	5	3		5	3
2	9 A.M.	1	5	9 A.M.	1
	11	2		2 P.M.	5
	3 P.M.	4	6	9 A.M.	4
3	9 A.M.	3		11	2
	3 P.M.	6		12	1
	4	1		3 P.M.	3
			7	1 P.M.	7

Using the exponential cdf, Eq. (16.4),

$$\text{Exponential } F(t) = 1 - e^{-0.339t}$$

Table 16.3 gives the value, and Fig. 16.7 graphically compares the two curves. They are very close, so we conclude that breakdowns are exponential.

COMMENT Since breakdowns are exponential, the Poisson-exponential relation (Sec. 16.5) could be used to state that the number of breakdowns B per time unit t is $P(b; \hat{\lambda}t)$. For example, if $t = 1$ day $= 9$ h, then $\hat{\lambda}t = 3.05$ and $B \sim P(b; 3.05)$.

Section 16.4

Table 16.3 Observed and exponential cdf, Example 16.6

Hours t	Frequency f(t)	Cumulative F	Observed F(t)	Exponential F(t)
1	6	6	0.30	0.29
2	3	9	0.45	0.49
3	4	13	0.65	0.64
4	3	16	0.80	0.74
5	2	18	0.90	0.82
6	1	19	0.95	0.87
7	1	20	1.00	0.91
	$\overline{20}$			

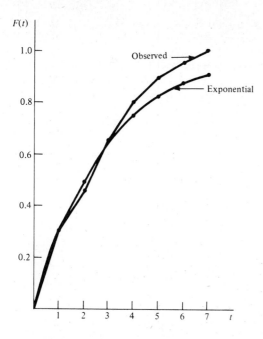

Figure 16.7 Observed and exponential cdf graphed for comparison, Example 16.6.

ADDITIONAL MATERIAL

Uniform distribution (Sec. 16.1): Hines and Montgomery [12], pp. 148–151.
Exponential distribution and reliability (Secs. 16.2 to 16.5): Gibra [8], pp. 155–157; Grant and Leavenworth [9], Chap. 18; Hahn and Shapiro [10], pp. 90, 105–107, 122–124, 159, 162; Hines and Montgomery [12], pp. 151–155, 425–430; Lipson and Sheth [17], pp. 44–48; Miller and Freund [18], pp. 119, 452–469.
Gamma distribution (Secs. 16.6 and 16.7): Gibra [8], pp. 158–160; Hahn and Shapiro [10], pp. 83–91, 118–120; Hines and Montgomery [12], pp. 155–159; Wilks [26], pp. 170–173.

PROBLEMS

P16.1 The time T between arrivals at a self-serve gas pump island varies uniformly between 1 and 5 min.
    (*a*) Write the complete pdf for T.
    (*b*) Locate the expected value on the graph of the pdf.
    (*c*) Compute the percentage of the time that $t \leq 1.5$ min.

P16.2 The response time R for the five departments in the city of Smokey the Bear is uniformly distributed between 3 and 10 min.
    (*a*) In 1000 calls how many will be answered in less than the average response time?
    (*b*) In 1000 calls how many response times will be between 8 and 10 min?

P16.3 The variable X has the distribution $U(150, 225)$. (a) Write and plot the cdf for X. Use the cdf to determine (b) $P(150 \leq x \leq 200)$ and (c) $P(\mu - \sigma \leq x \leq \mu + \sigma)$.

P16.4 (a) Derive the moment generating function for the variable $X \sim U(\alpha, \beta)$.
(b) Use expected value formulas to obtain the general expressions for $E[X]$ and $V[X]$.

P16.5 The time to failure for a nuclear reactor component has been observed to be between 1500 and 3500 h. If the time is uniformly distributed, compute the probability that the component operates in excess of 2800 h.

P16.6 For the component in Prob. P16.5, symmetric 90 percent control limits need to be established on the time to failure. Determine the values for these upper and lower control limits.

P16.7 Show that the exponential distribution possesses the three properties of a true pdf for a continuous variable.

P16.8 Plot the exponential pdf for $\lambda = 0.15$.

P16.9 Use the exponential pdf expression in Eq. (16.5) to derive the exponential cdf given by Eq. (16.4).

P16.10 The time to breakdown in integer hours for numerically controlled drill presses in a large machining department has been observed for 150 different breakdowns. If this data is assumed to follow an exponential distribution, estimate the parameter, mean, and standard deviation of the time between breakdowns.

Hours to breakdown	Number of breakdowns	Hours to breakdowns	Number of breakdowns
1	30	7	8
2	28	8	9
3	17	9	5
4	21	10	4
5	14	12	1
6	12	15	1

P16.11 A particular gear is expected to fail according to the exponential distribution at a rate of 0.0015 failures per hour of operation.
(a) Compute the mean life for a gear.
(b) Determine the number of gears out of 50 which can be expected to be replaced prior to 500 h.

P16.12 The service time T that it takes to fill an order at a drugstore prescription window is being computer-simulated. The analyst, who has a choice of the uniform or the exponential service model, has decided to use either a $U(0, 20)$ or an $\exp(0.10)$ distribution, because the mean service time is the same for both.
(a) Verify that the two means are equal.
(b) The analyst is concerned that the exponential model may place too few service times outside the range $\mu \pm 1\sigma$ when compared with the times outside this range for the uniform. Determine the percentage of service times that will exceed these limits for both models.

P16.13 An electronic calculator has a constant failure rate of 0.08 per 100 h. Determine the probability that the calculator will (a) fail during the first 1000 h, (b) operate for at least 2000 h, and (c) fail sometime between 100 and 500 h of operation.

P16.14 The following failure times have been observed during the test of 10 digital wristwatches which fail according to the exponential distribution:

8,804	9,435	10,890	7,281	11,385
7,652	12,209	8,934	9,962	10,073

(a) Estimate the MTBF.
(b) Determine the reliability of the watch for 10,000 h.

P16.15 Use the best estimate of λ for the data in Prob. P16.14 to plot the exponential pdf and the associated reliability curve.

P16.16 The shelf life for a highly perishable food item follows the exponential failure model. If no preservative is used, the mean time to spoilage is 100 h after processing; the addition of a preservative increases the mean time to 500 h but makes the item unusable for the first 250 h.
(a) Compute an upper 95 percent control limit on the spoilage time of an unpreserved item.
(b) If items are selected at random from the preserved items, what percentage of the selected items will be unusable?
(c) What percentage of the items selected in (b) will spoil in a time exceeding that of the 95 percent control limit for the unpreserved items?

P16.17 A sealed gear that was tested under varying conditions is marketed as having a constant failure rate of 6.94×10^{-3} per month, or a mean life of 12 years. Suppose 500 gears were installed in motors which receive virtually no maintenance checks.
(a) How many gears will be operating successfully after 6 months? 12 years?
(b) After 8 months a total of 15 gears had failed, so the company owning the motors removed all the gears because they were unreliable. Is this a reasonable decision?

P16.18 Batches of 10 space capsule items are delivered. Destructive testing of one item indicated an anticipated life of 500 flight hours. The mean life is stated as 5000 h with failures from an exponential distribution.
(a) Should the batch of 10 items be returned as not meeting the mean life specification? Use 10 percent as a critical chance of observing a particular failure time.
(b) What test result would change the decision made in (a)?

P16.19 The exponential failure model is used when the failure rate is constant throughout the life of the item. This is interpreted as a memoryless failure model, because regardless of how long the item has already lasted, the probability that it lasts a time r longer is computed as the reliability for the time r only. Thus, if the item has lasted a time $T = t$, the probability of lasting at least $T = t + r$ is

$$P(T > t + r / T > t) = P(T > r)$$

Use conditional probability to show that this relation is correct for the exponential distribution.

P16.20 Fit an exponential distribution to the data in Prob. P16.10.

P16.21 Select a real-world queuing situation and observe 100 arrivals to the serving station. Fit an exponential distribution to the frequency data in a manner similar to Example 16.6 and compare the observed and expected cdf values.

*P16.22 The time between arrivals for buses at a downtown bus stop follows an exponential distribution with a mean of 0.667 minute per arrival.
(a) What is the average number of buses that will arrive in a 15-min period?
(b) What is the probability that at least 10 buses will arrive in an 8-min interval?

*P16.23 The number N of quality errors per hour for a particular inspector follows a Poisson pdf with $\lambda = 0.08$ per minute.
(a) What is the average number of errors per hour?
(b) Compute the probability that the time between observed errors will be at most 12.5 min. At least 30 min.

P16.24 A machine attendant must refill a machine at a constant rate of 0.25 times per hour. Compute the mean and standard deviation of the time T it takes to observe the need to refill a total of five machines.

P16.25 An assembly operator uses four subassembly units to complete one circuit board. The subassemblies are produced and arrive at the operator's workplace at a constant rate of 2 per minute.
(a) Plot the pdf for T, the time to observe the arrival of four subassemblies at the work station.
(b) Locate the mean and mode on the graph.

P16.26 The time between machine breakdowns is assumed to be exponentially distributed. If the data in Prob. P16.10 is a summary of 150 observed times, estimate
   (a) the parameters and
   (b) the properties of the gamma distribution that may be used to describe the time before three breakdowns are observed.

P16.27 An unstaffed radar station operates correctly provided no more than five sensors fail. If all sensors fail at a constant rate of 0.005 per hour, determine the μ and σ values for the time T that the station is expected to operate correctly.

*P16.28 (a) Rework Prob. P16.27 under the condition that only one sensor need fail for the radar station to become inoperable.
   (b) What is the relation between the pdf used in (a) and the one used in Prob. P16.27?

*P16.29 Demonstrate that the gamma distribution mean and standard deviation are the same as those of χ^2 for $n = \nu/2$ and $\lambda = \frac{1}{2}$.

*P16.30 What is the difference between the gamma and the Erlang distributions?

*P16.31 If $X \sim G(x; n, \frac{1}{2})$, determine the corresponding value of ν for the χ^2 distributions when (a) $n = 4$ and (b) $n = 0.5$.

*P16.32 Develop an example of how the (a) beta and (b) Weibull distributions may be used.
   (c) Compute the values of μ and σ for these illustrations and explain their meaning in terms of your example.

CHAPTER SEVENTEEN

SAMPLE SIZE DETERMINATION AND INTERVAL ESTIMATION

In this chapter you will learn how to compute the required size of a sample to meet a stated statistical objective. You will also learn to compute intervals which include the population mean, variance, or proportion with a stated probability. This chapter includes a case study.

CRITERIA

To complete this chapter, you must be able to:

1. Define the terms *confidence level* and *significance level* and indicate them on a pdf graph.
2. State and define the two aspects of accuracy. Compute the *sample size* to estimate the population mean or the binomial proportion, given the confidence and precision requirements.
3. Define, interpret, and give examples of a *point estimate* and an *interval estimate*.

For criteria 4 to 6 state the relation for and compute a one-sided or two-sided interval estimate for the specified population parameter:

4. Population *mean*, given the confidence level, sample mean, and standard deviation.

5. Population *standard deviation*, given the confidence level and sample variance.
6. Population *proportion* for a large sample, given the confidence level and sample proportion.

STUDY GUIDE

17.1 Confidence and Significance Levels

The words *confidence* and *significance* are common in statistics. The first deals with our "faith" in the statistical conclusion and the second with our "mistrust" of it. By definition,

> *Confidence level* is the degree of assurance that a particular statistical statement is correct, under specified conditions.

> *Significance level* is the degree of uncertainty about the statistical statement under the same conditions used to determine the confidence level.

Significance levels are symbolized by α ($0 \leq \alpha \leq 1$) and confidence levels by $1 - \alpha$. Mathematically,

$$\text{Confidence level} + \text{significance level} = 1$$
$$(1 - \alpha) + \alpha = 1$$

If the significance is $\alpha = 0.05$, confidence is $1 - 0.05 = 0.95$. In percentage terms confidence is $100(1 - \alpha)$ percent $= 100(0.95) = 95$ percent. In this case the statement or conclusion is expected to be wrong about 5 in 100 times. If we run the same test 100 times and compare the results to reality, on the average, 5 of the results will be incorrect and 95 correct. The smaller the significance level α, the less chance we have of being incorrect.

Significance and confidence are easily pictured on a pdf graph. There are one-tail and two-tail significance areas for a significance level. A one-tail significance (Fig. 17.1a) places the entire α area in one tail, lower or upper, and the remainder is the confidence $1 - \alpha$. A two-tail area (Fig. 17.1b) places $\alpha/2$ in each tail and the confidence $1 - \alpha$ between these two areas. This is also called a two-tail, symmetric confidence area. In Fig. 17.1a the integral of $f(x)$ from x_0 to ∞ equals α, and in Fig. 17.1b the integral from x_L to x_U equals $1 - \alpha$.

Problems P17.1–P17.4

17.2 Sample Size Determination

When a population parameter is estimated, the sample size n should be computed, rather than guessed, so that the parameter value is accurate. Statistical

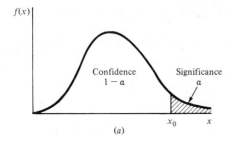
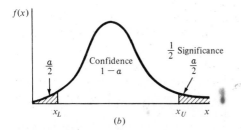

Figure 17.1 (a) One-tail and (b) two-tail significance areas on a general pdf f(x).

accuracy has two aspects:

1. *Confidence level* $1 - \alpha$. For instance, 95 percent is the $100(1 - \alpha)$ percent confidence area in Fig. 17.1.
2. *Precision requirement* h. The amount of deviation from the true value in actual units or percent that is allowed.

An accuracy statement to estimate the mean stress might be: "I want to be 98 percent confident that my estimate μ is no more than 10 lb/in² from the true stress." Here the precision requirement is an actual value. Equivalently, this accuracy statement may be: "I want to be 98 percent confident that μ is no more than 3 percent from the true value." This second method requires a preliminary idea of μ so that 3 percent of it can be calculated. Precision statements in actual units are preferred.

Figure 17.2 pictures a population and its unknown true mean. The precision band of width $2h$ around the true value should, on the average, contain $100(1 - \alpha)$ percent of the estimates. That is, if 98 percent confidence is required and 100 estimates are made, 98 will be inside the band. The two outside represent the 2 percent significance.

The central limit theorem for sample means and the standard normal distribution (SND)

$$z = \frac{x - \mu}{\sigma_p} \qquad (17.1)$$

are used to compute the required sample size. The variable z in Eq. (17.1) is the SND value for the two-tail significance. The value $h = x - \mu$ is the precision

308 SAMPLE SIZE DETERMINATION AND INTERVAL ESTIMATION

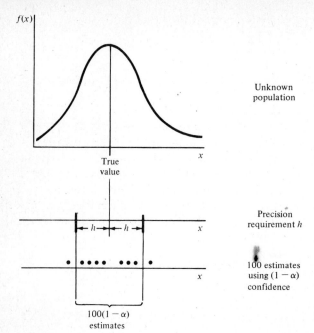

Figure 17.2 Relationship among the true value, the precision, and the confidence in estimating a parameter.

requirement in actual units, and σ_p is the *standard deviation of the parameter estimated*. We present formulas to estimate the population mean μ and the binomial parameter p.

Equation (17.1) is used to estimate the required sample size for the population mean by substituting the precision requirement h for $x - \mu$ and the standard deviation of the mean $\sigma_{\bar{x}} = \sigma/\sqrt{n}$ for σ_p.

$$z = \frac{h\sqrt{n}}{\sigma}$$

● $$n = \left(\frac{z\sigma}{h}\right)^2 \qquad (17.2)$$

An estimate of σ is obtained from either past data or a small trial sample. Any sample size which exceeds n will yield more confidence and precision.

The binomial parameter p is the fraction of the population possessing a certain attribute. If π is the true population value we want to estimate, the precision is $h = p - \pi$. Since the binomial variable represents the number of outcomes with the attribute and because we want the corresponding proportion, divide the binomial variable and its standard deviation by the sample size, as done in Sec. 8.8, to obtain the distribution of proportions. This results in s_p, the best estimate of the standard deviation of the proportion:

$$s_p = \frac{\sqrt{np(1-p)}}{n} = \sqrt{\frac{p(1-p)}{n}}$$

17.2 SAMPLE SIZE DETERMINATION

Substituting h and s_p into Eq. (17.1) allows solution for n:

$$n = \frac{z^2 p(1-p)}{h^2} \qquad (17.3)$$

Example 17.1 shows how to use these sample size determination formulas.

Example 17.1 A person working as an engineering management specialist for a fast service doughnut company must estimate the annual nationwide revenue loss caused by the incorrect use of a newly installed cash register system which automatically registers an item's cost when the item-name button is pressed. Because there is no historical data, the engineer plans to take two samples—one to estimate π, the fraction of all entries that are incorrect, and another to estimate μ_e, the average error size per incorrect entry e. Then the mean revenue loss μ will be

μ = (number of entries)(fraction incorrect)(error per entry)
   = (number of entries) (estimate of π)(estimate of μ_e)

The confidence required is 95 percent, and the precisions in actual differences are 0.2 percent for π and 1¢ for μ_e.

Determine the sample sizes n_π and n_e if in a preliminary sample of 500 entries, 5 were incorrect and the standard deviation of the errors was 4.7¢.

SOLUTION The z value for a two-tail significance of 5 percent is 1.96 (Table 11.2). To compute the required n_π, substitute $h = 0.002$ and $p = \frac{5}{500} = 0.01$ into Eq. (17.3):

$$n_\pi = \frac{1.96^2(0.01)(0.99)}{0.002^2} = 9508$$

Observe 9508 entries and compute the fraction incorrectly entered. This is the π estimate.

To compute n_e, substitute $h = 1¢$ and $\sigma = 4.7¢$ into Eq. (17.2):

$$n_e = \left(\frac{1.96(4.7)}{1.0}\right)^2 = 85$$

The 85 errors can be obtained from incorrect entries found in the n_π sample. In fact, the 500 entries initially observed can be included in the samples n_π and n_e.

If the resulting estimates are $\pi = 0.012$ and $\mu_e = 6.2¢$ per entry, there is a 95 percent confidence that the expected annual dollar loss for an estimated total of 548 million entries is

$$\mu = (548 \times 10^6)(0.012)(0.062) = \$407{,}712 \text{ per year}$$

Problems P17.5–P17.11

17.3 Point Estimates and Interval Estimates

A *point estimate* is a single-valued estimate of a population parameter made from a sample. For example, $\overline{X} = 51.5$ kg is a point estimate of the population mean μ. There is no probability connected with this type of estimate, and the accuracy is not readily known. We have made point estimates thus far by using the best estimator formulas.

An *interval estimate* is a probability statement that a population parameter is between two computed values. An example interval estimate is: We are 95 percent certain that μ is between 18.5 and 26.3 kg. In symbols, $P(18.5 \leq \mu \leq 26.3) = 0.95$. This is a symmetric, two-sided or two-tail interval estimate with a 95 percent confidence. The general form of a two-sided interval with a $1 - \alpha$ confidence level is

● $\quad\quad\quad P(\text{lower limit} \leq \text{parameter} \leq \text{upper limit}) = 1 - \alpha \quad\quad\quad (17.4)$

If the estimate is stated as

$$P(\text{parameter} \leq \text{upper limit}) = 1 - \alpha$$

it is an upper, one-sided 95 percent confidence interval. The symbol $CI_{1-\alpha}$ will be used to indicate an interval with confidence $1 - \alpha$. A 98 percent confidence interval estimate of the mean, for example, is written as $CI_{0.98}$ for μ.

To understand a $CI_{1-\alpha}$, first recall that in the sample size determination section we computed n after setting $1 - \alpha$ and the precision band $2h$. Then, once n is computed and taken, we knew that $100(1 - \alpha)$ of our estimates were no more than h from the true value. Here we set $1 - \alpha$ and n and then compute the precision band or interval. An interval estimate means that $100(1 - \alpha)$ of the true parameter values will fall between the upper and lower limits. The limits, if actually computed many times, will vary in value and width. Figure 17.3 shows this using a true mean value μ_0 and several intervals. In the long run, a total of $100(1 - \alpha)$ of these intervals will include μ_0, and 100α of them will not include μ_0.

Problem P17.12

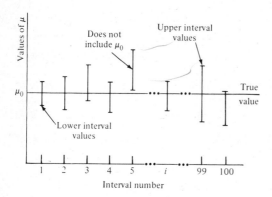

Figure 17.3 Limits of several confidence interval computations for the population mean μ.

17.4 Interval Estimates for Population Mean

An interval estimate for μ is computed by using the statistic

$$t = \frac{\bar{X} - \mu}{s/\sqrt{n}} \tag{17.5}$$

which has a t distribution with $\nu = n - 1$ degrees of freedom (Chap. 14). The confidence that t is between two symmetrically placed t values is $1 - \alpha$; that is, $P(-t_\alpha \leq t \leq +t_\alpha) = 1 - \alpha$, where $\pm t_\alpha$ are the two-tail α values. With the substitution of Eq. (17.5) into this probability statement and some algebraic simplification we have

$$P\left(\bar{X} - \frac{st_\alpha}{\sqrt{n}} \leq \mu \leq \bar{X} + \frac{st_\alpha}{\sqrt{n}}\right) = 1 - \alpha$$

The expression in parentheses is the two-sided $CI_{1-\alpha}$ for μ. A sample of size n is taken, and the unbiased values of $\bar{X}$ and s are computed. It is assumed that the sample is taken from a normal population and that $\bar{X}$ and s are reliable estimates of population parameters. The interval limits are summarized in Table 17.1 and graphically explained in Fig. 17.4a.

If a one-sided upper or lower interval estimate is desired, t_α is the one-tail α value in Table B-5. Figure 17.4b shows that the upper interval estimate is bounded above by $\bar{X} + st_\alpha/\sqrt{n}$, but unbounded below, while the lower interval is unbounded above. These limits are also summarized in Table 17.1. The one-sided limit is closer to $\bar{X}$ than the two-sided limits (Fig. 17.4) because the one-tail t_α value is smaller (closer to the mean) than the two-tail value.

The procedure to set a $CI_{1-\alpha}$ on μ is:

1. Determine if the interval is one-sided or two-sided, symmetric. Set the confidence level $1 - \alpha$.
2. Determine the sample size, take the sample, and compute $\bar{X}$ and s.
3. Find the correct t_α value(s) in Table B-5.
4. Compute the interval limits (Table 17.1).
5. Write the complete interval statement, for example, "I know 95 out of 100 times that μ is computed it will be between the two limits."

● **Table 17.1 Limits for one-sided and two-sided interval estimates on population mean**

Type of interval	t_α value	Lower limit	Upper limit
Two-sided, symmetric	Two-tail α	$\bar{X} - \frac{st_\alpha}{\sqrt{n}}$	$\bar{X} + \frac{st_\alpha}{\sqrt{n}}$
One-sided, upper	One-tail α	$-\infty$	$\bar{X} + \frac{st_\alpha}{\sqrt{n}}$
One-sided, lower	One-tail α	$\bar{X} - \frac{st_\alpha}{\sqrt{n}}$	$+\infty$

312 SAMPLE SIZE DETERMINATION AND INTERVAL ESTIMATION

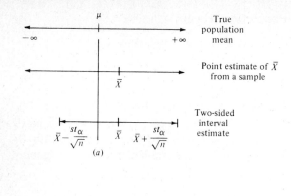

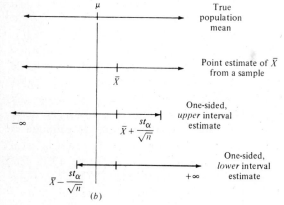

Figure 17.4 Relation between (a) two-sided and (b) one-sided interval estimates of population mean μ.

As the sample size increases, t approaches the SND (see the last row of Table B-5 for $\nu = \infty$). Therefore, as a guide, if $n \geq 30$, substitute z for t in Table 17.1. See Example 17.2 for an illustration.

If needed, it is also possible to compute an interval estimate for the sum or difference of two population means μ_1 and μ_2. The interval is set on the statistic $\mu_1 + \mu_2$ or $\mu_1 - \mu_2$. For the difference, the interval statement takes the form

$$P(\text{lower limit} \leq \mu_1 - \mu_2 \leq \text{upper limit}) = 1 - \alpha$$

Example 17.2 A chemist wishes to determine an interval estimate on the mean percentage of molybdenum (Mo) in a common steel. A sample of 15 specimens had $\overline{X} = 0.4$ percent Mo with $s = 0.06$ percent Mo. Compute (a) the two-sided $CI_{0.98}$; (b) the one-sided, upper $CI_{0.98}$; and (c) the two-sided $CI_{0.98}$ for $n = 500$ with the same $\overline{X}$ and s. Compare with the interval in (a).

SOLUTION (a) The chemist should use the steps above:

1. Two-sided symmetric with $1 - \alpha = 0.98$.
2. $n = 15$, $\overline{X} = 0.4$ percent, and $s = 0.06$ percent.

3. From Table B-5 for $\alpha = 0.02$ and $\nu = 14$, $t_{0.02} = 2.624$.
4. From Table 17.1,

Lower limit: $\quad 0.4 - \dfrac{0.06(2.624)}{\sqrt{15}} = 0.360\%$ Mo

Upper limit: $\quad 0.4 + \dfrac{0.06(2.624)}{\sqrt{15}} = 0.440\%$ Mo

5. The chemist can state that 98 out of 100 times the mean Mo content will be between 0.360 and 0.440 percent. In equality form

$$0.360\% \text{ Mo} \leq \mu \leq 0.440\% \text{ Mo}$$

(b) For the one-sided, upper interval, t decreases to $t_{0.02} = 2.305$. From Table 17.1

Lower limit: $\quad -\infty$

Upper limit: $\quad 0.4 + \dfrac{0.06(2.305)}{\sqrt{15}} = 0.436\%$ Mo

There is 98 percent confidence that μ does not exceed 0.436 percent.

(c) If $n = 500$, the t value is from the SND curve, so it is a z value. For a two-tail $\alpha = 0.02$, $z = 2.326$ (Table 11.2). The limits are (step 4)

$$0.4 \pm \dfrac{0.06(2.326)}{\sqrt{500}} = 0.4 \pm 0.006 = 0.394 \text{ to } 0.406\% \text{ Mo}$$

The interval is narrower by 0.068 percent than the one in part (a). Use of the t pdf will make a difference when n is small, especially if $n \leq 10$.

COMMENT A reminder: The 98 percent confidence means that 98 out of 100 intervals computed under these same conditions contain μ. It is *incorrect* to state that with a *probability* of 0.98 the population mean is in the interval. The 0.98 probability is that an interval contains μ, not that μ is in this particular interval. There is a difference since the intervals will change from one sample to the next, but μ is an unknown constant.

A case study involving the statistical computation of n and the relation to an interval estimate of μ is in the Solved Problem.

Example 17.4 (Case Study)

Problems P17.13–P17.19

17.5 Interval Estimates for Population Standard Deviation

The procedure to set a confidence interval on the population variance or standard deviation is similar to that for the mean. We assume the sample of size n is from a normal distribution and use the χ^2 pdf to compute the limits. From

Chap. 13 the statistic

$$\chi^2 = \frac{(n-1)s^2}{\sigma^2} \qquad (17.6)$$

has a χ^2 distribution with $\nu = n - 1$ degrees of freedom. Since the χ^2 pdf is not symmetric, a total of $1 - \alpha$ of the area is between the two values χ_L^2 (lower) and χ_U^2 (upper) obtained from Table B-4. In probability form

$$P(\chi_L^2 \leq \chi^2 \leq \chi_U^2) = 1 - \alpha$$

See Fig. 13.6a for an example.

By substituting Eq. (17.6) into the probability statement above and solving for σ, we have

$$P\left(\chi_L^2 \leq \frac{(n-1)s^2}{\sigma^2} \leq \chi_U^2\right) = 1 - \alpha$$

$$P\left[\sqrt{\frac{(n-1)s^2}{\chi_U^2}} \leq \sigma \leq \sqrt{\frac{(n-1)s^2}{\chi_L^2}}\right] = 1 - \alpha \qquad (17.7)$$

The limits in parentheses form a two-sided $CI_{1-\alpha}$ on σ. The squared values will give the $CI_{1-\alpha}$ for σ^2. See Fig. 17.5a for a graphical summary. The one-sided intervals on σ are bounded by the appropriate limit from Eq. (17.7) with α in one tail of the χ^2 pdf. The interval is unbounded above for the lower $CI_{1-\alpha}$ and bounded below by 0 for the upper $CI_{1-\alpha}$ because $\sigma > 0$ in all situations (Fig. 17.5b). Table 17.2 summarizes all interval limits.

Be sure to recognize that the lower chi-square value χ_L^2 is used to compute the upper limit, and the upper chi-square value χ_U^2 is included in the lower-limit computation.

The procedure to set a $CI_{1-\alpha}$ on σ is:

1. Determine if the interval is one-sided or two-sided. Set the confidence level $1 - \alpha$.
2. Determine the sample size, take the sample, and compute the sample variance s^2.
3. Find the correct χ_L^2 and χ_U^2 values in Table B-4.
4. Compute the interval limits (Table 17.2).
5. Write the complete interval statement.

If samples are large enough to exceed Table B-4 ($n > 100$), use the normal approximation to χ^2 explained in Sec. 13.3. See Example 17.3.

It is useful at times to compute an interval estimate on the ratio of two population variances σ_1^2/σ_2^2. The method uses the F distribution and is discussed in several of the bibliography texts.

17.5 INTERVAL ESTIMATES FOR POPULATION STANDARD DEVIATION

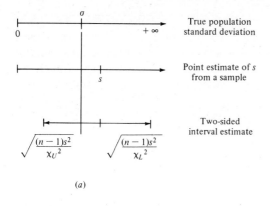

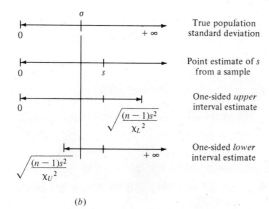

Figure 17.5 Relation between (a) two-sided and (b) one-sided interval estimates of the population standard deviation σ.

● **Table 17.2 Limits for one-sided and two-sided interval estimates on population standard deviation**

Type of interval	χ^2 values	Lower limit	Upper limit
Two-sided	χ_L^2, χ_U^2	$\sqrt{\dfrac{(n-1)s^2}{\chi_U^2}}$	$\sqrt{\dfrac{(n-1)s^2}{\chi_L^2}}$
One-sided, upper	χ_L^2	0	$\sqrt{\dfrac{(n-1)s^2}{\chi_L^2}}$
One-sided, lower	χ_U^2	$\sqrt{\dfrac{(n-1)s^2}{\chi_U^2}}$	$+\infty$

Example 17.3 Traffic research engineers are studying the variation in speed in kilometers per hour for speeding and nonspeeding interstate drivers.
(a) In a sample of 30 speeders the standard deviation was 15.5 km/h. Set the 95 percent two-sided and one-sided, upper confidence intervals on σ.
(b) In a sample of 250 nonspeeders, $s = 15.4$ km/h. Compute the two-sided $CI_{0.95}$ on σ and determine the effect of the larger sample on the interval's size.

SOLUTION (a) Use the procedure outlined above.

1. Two-sided with $1 - \alpha = 0.95$.
2. $n = 30$, $s^2 = (15.5 \text{ km/h})^2 = 240.25 \text{ km}^2/\text{h}^2$.
3. For $\nu = 29$ the χ^2 values are $\chi_L^2 = 16.05$ and $\chi_U^2 = 45.72$.
4. From Table 17.2 the limits on σ are

Lower limit: $\sqrt{\dfrac{(29)(240.25)}{45.72}} = \sqrt{152.39} = 12.34 \dfrac{\text{km}}{\text{h}}$

Upper limit: $\sqrt{\dfrac{(29)(240.25)}{16.05}} = \sqrt{434.10} = 20.83 \dfrac{\text{km}}{\text{h}}$

5. In 95 out of 100 two-sided intervals, the standard deviation will be contained in the interval 12.34 to 20.83 km/h. Or, with 95 percent confidence

$$12.34 \text{ km/h} \leq \sigma \leq 20.83 \text{ km/h}$$

For the upper, one-sided $CI_{0.95}$ use $\chi_L^2 = 17.71$, and the limits in step 4 are 0 and 19.83 km/h.

(b) Using the steps above, the traffic engineer would have:
1. Two-sided with $1 - \alpha = 0.95$.
2. $n = 250$, $s^2 = (15.4 \text{ km/h})^2 = 237.16 \text{ km}^2/\text{h}^2$.
3. For the large sample $n = 250$, the χ_L^2 and χ_U^2 values are obtained by using the fact that as n increases, the variable $\sqrt{2\chi^2}$ becomes normal with $\mu = \sqrt{2\nu - 1}$ and $\sigma = 1$ (Sec. 13.3). In this case $\mu = \sqrt{2(249) - 1} = 22.29$. From the SND Table 11.2, $z = \pm 1.96$ have $\alpha/2 = 0.025$ of the area outside them. Use the SND transformation to get the χ^2 values.

$$z = \dfrac{\text{variable value} - \text{mean}}{\text{standard deviation}}$$

for χ_L^2: $\quad \dfrac{\sqrt{2\chi_L^2} - 22.29}{1} = -1.96 \quad$ and $\quad \chi_L^2 = 206.65$

for χ_U^2: $\quad \dfrac{\sqrt{2\chi_U^2} - 22.29}{1} = +1.96 \quad$ and $\quad \chi_U^2 = 294.03$

4. From Table 17.2

Lower limit: $\sqrt{\dfrac{(249)(237.16)}{294.03}} = 14.17 \dfrac{\text{km}}{\text{h}}$

Upper limit: $\sqrt{\dfrac{(249)(237.16)}{206.65}} = 16.90 \dfrac{\text{km}}{\text{h}}$

The two sample s values are virtually identical, but the interval width has significantly decreased from 8.49 to 2.73 km/h, because of the larger sample. Unfortunately, we do not often have large samples upon which to base conclusions.

Problems P17.20–P17.26

17.6 Interval Estimates for Population Proportion

The proportion of a population which possesses a particular characteristic can be interval-estimated by using the distribution of proportions (Sec. 8.8). Examples are the fraction of exposed persons who will contact a disease or the fraction of a product lot which is defective.

The true proportion is π, and its sample estimate is

$$\hat{p} = \dfrac{\text{number of successes}}{\text{sample size}} = \dfrac{x}{n}$$

In Sec. 17.2 we learned that the standard deviation estimate is s_p, which is a function of p:

$$s_p = \sqrt{\dfrac{p(1-p)}{n}} \qquad (17.8)$$

This presents a problem when an interval is set on π because an estimate of p is necessary to get s_p, which changes every time p is altered. A similar problem does not exist for intervals on μ and σ because the standard deviation is not a function of the parameter to be estimated. We circumvent the above problem for large samples ($n \geq 30$) by using the normal approximation to the binomial. For a symmetric, two-sided interval the form of Eq. (17.1) is used to make the SND probability statement.

$$P\left(-z \leq \dfrac{p - \pi}{\sigma_p} \leq z\right) = 1 - \alpha$$

Algebraic simplification gives the interval limits.

$$p - z\sigma_p \leq \pi \leq p + z\sigma_p$$

where p is from a trial sample and σ_p is estimated from Eq. (17.8). Naturally, the larger n, the more accurate p, thus improving the interval for π. One-sided limits are similarly computed using a one-tail α value for z. The limits are summarized

● **Table 17.3 Limits for one-sided and two-sided interval estimates for population proportion**

Type of interval	z value	Lower limit	Upper limit
Two-sided, symmetric	Two-tail α	$p - zs_p$	$p + zs_p$
One-sided, upper	One-tail α	0	$p + zs_p$
One-sided, lower	One-tail α	$p - zs_p$	1

in Table 17.3 and computed by using the steps below ($n \geq 30$):

1. Determine if a one-sided or two-sided interval is desired. Set the confidence level $1 - \alpha$.
2. Determine the sample size, take the sample, count the number possessing the desired characteristic, and compute p and s_p.
3. Find the correct SND value(s) in Table 11.2 or B-3.
4. Compute the interval limits (Table 17.3).
5. Write the complete interval statement.

For small samples ($n < 30$) several approximate interval estimates can be made. The best method is graphical using the correct form of the discrete binomial pdf to find, as closely as possible, two values x_1 and x_2 with $1 - \alpha$ between them for several different π values; that is,

$$\sum_{x=x_1}^{x_2} b(x; n, \pi) = 1 - \alpha$$

This method is thoroughly covered in Miller and Freund (see Additional Material).

Example 17.4

Problems P17.27–P17.32

SOLVED PROBLEM (CASE STUDY)

Example 17.4 Assume you are an engineering management graduate working for a consulting firm which is studying trends in apartment rental fees. You want to estimate the average monthly rental μ in dollars and the fraction π of renters spending over $300 per month on rent. In a preliminary sample of 10 renters you compute $\overline{X} = \$250$, $s = \$47$ per month, and the fraction $p = 0.20$. Using a 90 percent confidence and an actual precision of $10 per month for μ and 4 percent for π, compute the required sample sizes and two-sided interval estimates of (a) μ and (b) π. Compare the attained precision with that required for μ and π. (c) In addition, compute the interval for μ using only the preliminary sample data and compare it to the interval in (a).

SOLUTION (a) To estimate the average montly rental, use Eq. (17.2) to compute the required n value with $z = 1.645$ for 90 percent confidence. The precision is $h = \$10$, and $s = \$47$ estimates σ.

$$n = \left[\frac{1.645(47)}{10}\right]^2 = 60 \text{ renters}$$

Assume 50 more renters are sampled and the results are $\overline{X} = \$250$ and $s = \$54$ per month. The two-sided interval on μ (Table 17.1) uses the SND $z = 1.645$ because $n > 30$.

Limits for μ: $250 \pm \dfrac{54(1.645)}{\sqrt{60}} = \$238.53$ to $\$261.47$ per month

You are 90 percent confident that monthly rent averages between $238.53 and $261.47. The actual precision is $54(1.645)/\sqrt{60} = \$11.47$, which exceeds the desired $10. This occurred because s rose from $47 to $54 when the required $n = 60$ was taken. Had s remained at $47, $h = \$10$ would have resulted. Substitution of $s = \$47$ into the limits computation will give $240 \leq \mu \leq 260$.

(b) For the proportion of renters spending over $300 per month, Eq. (17.3) gives the required sample size. Using $z = 1.645$, $p = 0.20$, and a precision of 4 percent, we have

$$n = \frac{(1.645)^2(0.20)(0.80)}{(0.04)^2} = 271 \text{ renters}$$

Assume additional renters are sampled and that $p = 0.15$. From Table 17.3

Limits on π: $0.15 \pm 1.645\sqrt{\dfrac{(0.15)(0.85)}{271}} = 0.114$ to 0.186 per month

You are 90 percent confident that the percentage is between 11.4 and 18.6 percent of the renters. Unlike the interval on μ, the observed precision is 3.6, which is less than the desired 4 percent, because p decreased when $n = 271$ observations were taken.

(c) From the preliminary sample $n = 10$, $\overline{X} = \$250$, and $s = \$47$. Use the Table 17.1 limits with $t_{0.10} = 1.833$.

Limits on μ: $250 \pm \dfrac{47(1.833)}{\sqrt{10}} = \$222.76$ to $\$277.24$ per month

These limits are much wider than those in (a). The precision is $27.24 because a larger sample of size 60 is required to obtain the desired precision of $10.

320 SAMPLE SIZE DETERMINATION AND INTERVAL ESTIMATION

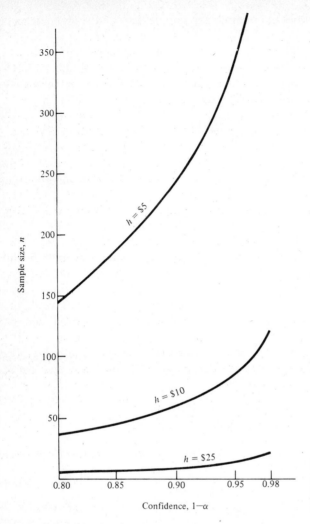

Figure 17.6 Plot of required sample sizes for several values of confidence $1 - \alpha$ and precision h, Example 17.4.

COMMENT The sample size computations in Eqs. (17.2) and (17.3) are quite sensitive to the confidence and precision levels. Figure 17.6 is a plot of the sample size required by Eq. (17.2) to estimate μ at various confidence and precision levels. As $1 - \alpha$ and h become restrictive, the n value skyrockets. A healthy balance of h, $1 - \alpha$, and n is always necessary to keep the costs of estimation in control. In all cases, the larger the preliminary sample, the more accurate estimates such as $\overline{X}$, s, and p become. This accuracy will result in a smaller required n value to obtain the desired precision and confidence.

Sections 17.4 and 17.6

ADDITIONAL MATERIAL

Sample size determination (Secs. 17.1 and 17.2): Miller and Freund [18], pp. 182–185, 245; Newton [20], pp. 195–199, 207, 208; Volk [24], pp. 137–146.

Confidence interval for means (Sec. 17.4): Beck and Arnold [1], pp. 102–108; Bowker and Lieberman [4], pp. 294–308; Duncan [7], pp. 510–516; Gibra [8], pp. 322–325; Hahn and Shapiro [10], pp. 74–77; Hines and Montgomery [12], pp. 234, 235, 276–278; Hoel [13], pp. 200–202, 261–265; Kirkpatrick [15], pp. 249–254; Miller and Freund [18], pp. 185–187; Newton [20], pp. 193–195.

Confidence interval for standard deviations (Sec. 17.5): Duncan [7], pp. 516, 517; Gibra [8], pp. 325–328; Hines and Montgomery [12], pp. 278–280; Hoel [13], pp. 254–257; Miller and Freund [18], pp. 230, 231; Newton [20], pp. 205–207, 208–210.

Confidence interval for proportions (Sec. 17.6): Duncan [7], pp. 517–524; Gibra [8], pp. 328–332; Hahn and Shapiro [10], pp. 146–148; Miller and Freund [18], Chap. 9; Newton [20], pp. 205–207.

PROBLEMS

P17.1 The control limits for a machined dimension which is normally distributed have been computed using the SND values $z = \pm 2.50$. What are the sizes of the significance level and the confidence level?

P17.2 The χ^2 distribution values of 7.96 and 23.54 are used by an engineer in determining if the variance has changed in samples of size 17.
  (a) Show the significance and confidence areas on the graph of a pdf.
  (b) Compute the significance level for the study.

P17.3 Study has shown that the number X of people under 25 years old who do not carry health insurance may be estimated by the $b(x; 10, 0.30)$ distribution.
  (a) Determine the values of X for an approximate 90 percent confidence level with one-half the significance in each tail.
  (b) What is α if an upper, one-tail value of $x = 7$ is used to limit the confidence area?

P17.4 If the significance level for a distribution is switched from one-tail, upper to two-tail, symmetric, but the α level remains the same, will the upper-tail variable value move to the right, move to the left, or remain unchanged? Give the reason for your answer.

P17.5 A metallurgical engineer gets some exercise by collecting aluminum cans along roadways. He wants to determine the sample size necessary to estimate with 95 percent confidence the proportion of cans made of aluminum and found on roadways. The estimate should be accurate within 3 percent of the actual proportion. A preliminary sample of size 50 contained 36 aluminum cans.

P17.6 The time to replace a burned-out traffic signal bulb was observed 25 times and resulted in $\bar{X} = 18$ min and $s = 6$ min. If replacement times are assumed to be normally distributed and if s is a reliable estimate of σ, use a 95 percent confidence level to:
  (a) Determine the maximum width of the precision band around the true mean time.
  (b) Compute the n value needed to cut the precision band to one-half the width computed above.

P17.7 An economist wants to have 90 percent assurance that an estimate of mean income is at most 1.5 percent from the true mean income. In the past the required sample sizes have been small with an average of $n = 26$, and the best parameter estimates are $\hat{\mu} = \$14{,}275$ per year and $\hat{\sigma} = \$2700$ per year. Do the following:

(a) Alter the sample size determination formula, Eq. (17.2), to be usable for small samples in which $n - 1$ is an estimate of the degrees of freedom. Use this equation to determine the required sample size.

(b) Determine the sample size if the normal distribution had been used.

P17.8 An audit team wants to estimate with 98 percent confidence the total dollar value of the error made in journal entries during the last year at the Running Scared Manufacturing Company. In a preliminary sample of 75 out of the 10,000 entries, a total of 3 had an error with a mean of \$232 and a standard deviation of \$42. The auditors want the precision to be \$10 for incorrect entries and 1 percent for the estimated percentage of incorrect entries.

(a) Write the equation you would use to estimate the total dollar error for the year.

(b) Compute the sample sizes required to use this equation.

P17.9 The fraction of the registered voters who voted in a party primary was estimated at 0.65 from a sample of 700 names.

(a) What is the error of this estimate on either side of the true fraction at a 95 percent confidence level?

(b) What confidence can be placed on this estimate if the required precision is ± 5 percent of the true fraction?

P17.10 Mr. I. M. Macked is a fluid mechanic analyst. He wants to estimate the mean temperature of a by-product in a smelting process within 1°C. A sample of 15 readings has a mean of 275°C and a standard deviation of 5.5°C. Compute the required sample size to achieve a 95 percent confidence assuming that the temperature is normally distributed.

P17.11 Use the results of the 15 readings in Prob. P17.10 and the normal distribution to determine the confidence level that Mr. Macked can place on the mean estimate of 275°C if the precision requirement of 1°C is maintained.

P17.12 Write the general form of the point estimate and 90 percent interval estimate statements for each of the following:

(a) In a sample of 1000 values the mean was 18.5. There were 50 values less than 13.8 and 50 greater than 23.4.

(b) The fraction defective p, which has averaged 0.075, has been at least 0.05 for 90 out of the last 100 samples. In addition, 90 of these 100 samples have a p value less than or equal to 0.08. Develop two one-sided interval estimates for π.

The following situations are used by the indicated problems in the remainder of this chapter.

A: A data processing department has collected sample data on the number of cathode-ray tube (CRT) terminals active at any one time. At present, 235 terminals are connected to the computer, and this study is designed to determine the capability of the hardware to handle more terminals. In a sample of 25 randomly selected times the number of active terminals were:

134	202	118	156	172
125	192	201	168	139
158	186	191	210	124
151	140	212	173	213
183	189	162	124	208

Problems P17.13, P17.20, P17.27

B. A supermarket wants to set up an express lane for customers who have less than a specific number of items to purchase. In a sample of 250 orders, the number of items averaged 21 with a standard deviation of 4 items. A total of 18 percent of the 250 orders included produce items which had to be weighed and priced at the checkout counter. This statistic is important because the express checkout would have no weighing capability.

Problems P17.14, P17.21, P17.29

C: The number of accidents per week was recorded for six representative weeks at the Safety Alarm Company. Each accident was classified as a head, body, or limb injury. The statistics are given in the table.

Week number	Number of accidents per week			
	Head	Body	Limb	Total
1	3	0	4	7
2	2	10	2	14
3	4	6	7	17
4	3	3	8	14
5	2	4	15	21
6	0	3	9	12

Problems P17.16, P17.17, P17.23, P17.24, P17.32

P17.13 Use the data in situation A and the procedure presented in Sec. 17.4 to set the following confidence intervals on the mean number of active terminals:
(a) Two-sided with $\alpha = 0.05$
(b) One-sided lower for a confidence of 0.90

P17.14 Determine a symmetric 98 percent confidence interval for the population mean μ if the variable X is the number of items per order described in situation B.

P17.15 A sample of 20 measurements with $\bar{X} = 15{,}000$ hertz (Hz) and $s = 900$ was taken in plant 1. In a similar sample of size 200 in plant 2 the same $\bar{X}$ and s were observed. Use a significance level of 20 percent and the t and normal distributions to determine the effect of the increased sample size on the width of the two-sided confidence interval for the population mean μ.

P17.16 Use the data of situation C to set the upper one-sided $CI_{0.95}$ on the number of accidents per week classified as limb injuries.

P17.17 (a) For situation C determine the width of a 90 percent confidence interval (precision band) for the mean number of accidents per week classified as head injuries.

(b) Determine if the sample data collected for six weeks is sufficient to estimate the true number of head injuries per week within one injury. It is known that the population is normal with $\sigma = 1.37$, which is the same as the s value of the collected data. Use $\alpha = 0.10$.

P17.18 The diameter of a machined bolt is measured on 12 finished products. If the bolt diameters are from a normal population and the sample statistics are $\bar{X} = 10.35$ mm and $s = 0.08$ mm, determine the percentage of measurements that you would expect to fall below the one-sided, lower confidence interval limit of 10.28 mm.

P17.19 If samples of size n are taken from a finite population of size N, the standard deviation of the mean $\sigma_{\bar{X}} = \sigma/\sqrt{n}$ is replaced with

$$\sigma_{\bar{X}} = \frac{\sigma}{\sqrt{n}} \sqrt{\frac{N-n}{N-1}}$$

(a) Use this fact to write the expression for the two-sided, symmetric confidence interval for the population mean μ if σ is estimated by the sample s value.

(b) Compute the 95 percent confidence interval on μ using the relation in (a) for a sample of $n = 15$ from $N = 150$ if $\bar{X} = 1275.3$ and $s = 73.5$.

P17.20 Use the procedure in Sec. 17.5 and the data in situation A to set a two-sided $CI_{0.98}$ on the population (a) standard deviation and (b) variance for the number of active terminals.

P17.21 For situation B and a 95 percent confidence level determine (a) the two-sided and (b) the one-sided, upper interval on the standard deviation of the number of items per order.

P17.22 Why is the lower limit of the one-sided, upper confidence interval for σ always 0, rather than $-\infty$?

P17.23 (a) Determine the sample size required to estimate the population mean for data taken from a normal distribution with $\sigma = 2.5$. Use a 95 percent confidence and a precision of 2.

(b) Assume that the sample is taken and the result is the total accidents per week given in situation C. Compute the two-sided $CI_{0.95}$ for both μ and σ.

(c) How does the width of the confidence interval for μ in (b) compare with the required precision level used in (a)?

P17.24 Calculate the limits of the one-sided, lower $CI_{0.92}$ for σ using the number of accidents per week classified as body injuries in situation C.

P17.25 The number of seconds that a driver must wait at a blinking-red traffic signal is assumed to follow a normal distribution with $\mu = 5$ and $\sigma = 1$ s. A sample of 10 times is taken with the result $s = 1.4$ s. Place a 98 percent, two-sided confidence interval on σ to determine if the assumed population parameter value is included.

P17.26 Rework Prob. P17.25 assuming that the sample size was 100 but s remained at 1.4 s.

P17.27 Use the 25 data values in situation A to compute the average proportion of the 235 terminals active at one time. Use this average as a value for the proportion p to compute (a) the two-sided and (b) the one-sided, lower $CI_{0.90}$ on the population proportion π.

P17.28 The proportion of time that the workers on a newly installed assembly line are idle is estimated at 0.132 from a sample of 550 work sampling observations. Place a symmetric $CI_{0.95}$ on π, the true fraction of idle time.

P17.29 Determine the limits for the one-sided, upper confidence interval using (a) $\alpha = 0.02$ and (b) $\alpha = 0.05$ for the proportion of orders in situation B that require the use of a scale.

P17.30 Returnable soft drink bottles are delivered to the Oops! Slipped Bottling Company for refilling. Historically the fraction of chipped or damaged bottles has averaged $\pi = 0.065$. In a sample of 375 bottles a total of 30 were unusable. Use the two-sided 90 percent confidence interval on π to determine if this fraction has increased.

P17.31 For the situation described in Prob. P17.30 assume that the fraction of unusable bottles remains at $30/375 = 0.08$ as the sample size changes. Use the relations for a two-sided $CI_{0.90}$ on π to determine the minimum sample size n to show that π has increased from the assumed value of $\pi = 0.065$.

P17.32 Determine the following interval estimates for situation C using a confidence of 98 percent. Base your computations on the total number of accidents in each injury category and the assumption that the number of accidents is normally distributed.

(a) Two-sided for the proportion of total accidents per week classified as head injuries

(b) One-sided, upper for the proportion of total accidents per week classified as body or limb injuries

CHAPTER
EIGHTEEN

DISTRIBUTIONS OF MORE THAN ONE VARIABLE AND THE TRANSFORMATION OF VARIABLES

This chapter has two primary objectives. In the first five sections you learn to formulate and work with a pdf for more than one variable. Simple bivariate distributions for two variables are emphasized. In the final five sections you apply the rules of expected values to functions of one, two, or more variables, and you learn how to find the distributions of functions of random variables.

CRITERIA

To understand a pdf of more than one variable, you must be able to:

1. Plot the possible bivariate values and the pdf, given the form of the pdf; and state the three properties of a discrete or continuous bivariate distribution.
2. Define and compute the *marginal distributions* using a certain procedure, given the value or pdf relation for a bivariate distribution.
3. State the definition of independence of two variables, and mathematically conclude independence or dependence, given probability values or pdf relations for the bivariate and marginal distributions.
4. Write the equations for and compute the *conditional distribution* and *conditional expectation*, given the form of the bivariate distribution.
5. State and define the meaning of the variables, parameters, and properties of the *multinomial* and *bivariate normal* distributions.

6. State the formulas for and calculate the *expected values* of different functions of one or more random variables, given the form of the functions.
7. Do the same as in 6 except for the *variance*.
*8. Derive the distribution of the function of a discrete random variable using the *transformation-of-variable* procedure, given the original distribution and a one-to-one transformation function.
*9. Do the same as in 8 except for a continuous variable.
*10. Derive the bivariate distribution of discrete or continuous variables using the transformation of variable procedure for two variables, given the original bivariate distribution and two one-to-one transformation functions.

STUDY GUIDE

18.1 Formation and Properties of a Bivariate pdf

In all previous chapters we have dealt with only one variable in the pdf. The range of a variable, for example, $x = 0, 1, \ldots$ is used to construct the x axis of the pdf graph. If we are simultaneously interested in two variables X and Y, we have a two-variable, or *bivariate, distribution*. This can also be called a *joint distribution*. Whether X and Y are discrete or continuous, you should first carefully define and graph the possible values for each variable. Figure 18.1 gives graphical examples for (a) $x = 1, 2, \ldots, 6$ and $y = 0, 1$; (b) $0 \le x \le 10$ and $0 \le y \le x$; (c) $x \ge 0$ and $y = 2, 4, 6$.

With one variable (the *univariate* case) the pdf $f(x)$ is plotted in two-dimensional space. The bivariate pdf is identified as $f(x, y)$ and is plotted on the third axis. The interpretation of $f(x, y)$ is simple; it is the probability that X equals x and Y equals y *simultaneously*, that is,

$$P(X = x, Y = y) = f(x, y)$$

Figure 18.2 is a plot of the bivariate pdf

$$f(x, y) = \begin{cases} 0.0156(_4C_x) & x = 0, 1, 2, 3, 4; y = 1, 2, 3, 4 \\ 0 & \text{elsewhere} \end{cases} \quad (18.1)$$

where $_4C_x$ is the combination of four things taken x at a time. Note that $f(x, y)$ is plotted as a vertical line above each discrete (x, y) grid intersection.

It is possible to extend theory and imagine $(k + 1)$-dimensional space for a *multivariate* pdf $f(x_1, x_2, \ldots, x_k)$ of k variables. The actual form of such distributions is usually sheer speculation, but the theory exists for these joint distributions and is used to develop correct statistical procedures, some of which we use later.

The properties for discrete and continuous distributions of more than one variable parallel those of the univariate case (Secs. 6.4 and 6.6), except that the summing and integrating take place over all variables. The bivariate properties are summarized here.

18.1 FORMATION AND PROPERTIES OF A BIVARIATE PDF

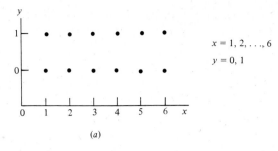

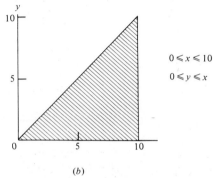

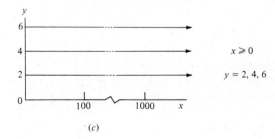

Figure 18.1 Possible values for two variables when (*a*) both are discrete; (*b*) both are continuous; (*c*) one is continuous and the other discrete.

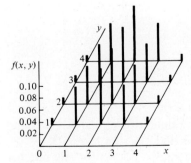

Figure 18.2 Plot of the bivariate pdf $f(x,y)$ given by Eq. (18.1).

Discrete pdf	Continuous pdf
1. $f(x,y) > 0$ at all defined (x,y)	1. $f(x,y) \geq 0$ for all $-\infty \leq x \leq \infty$ and $-\infty \leq y \leq \infty$
2. $P(a \leq x \leq b, c \leq y \leq d)$ $= \sum_{x=a}^{b} \sum_{y=c}^{d} f(x,y)$	2. $P(a \leq x \leq b, c \leq y \leq d)$ $= \int_{c}^{d} \int_{a}^{b} f(x,y)\, dx\, dy$
3. $\sum\sum f(x,y) = 1$, summed over all defined (x,y)	3. $\int_{-\infty}^{\infty} \int_{-\infty}^{\infty} f(x,y)\, dx\, dy = 1$

These properties are demonstrated in the same way as the univariate properties.

Problems P18.1–P18.11

18.2 Marginal Distributions

A *marginal distribution* is a pdf developed from a multivariate pdf by summing out (discrete) or integrating out (continuous) all variables no longer of interest. The name is meaningful: if you place all probability values for a bivariate pdf in a table and add down columns and across rows, the *margin* entries are the pdf values for the marginal distributions. This is illustrated in Example 18.1 for two discrete variables using the following procedure.

1. Tabulate all pdf values from $f(x,y)$.
2. Decide which marginal pdf is desired, $f(x)$ or $f(y)$.
3. Sum over all values of the *deleted* variable for each value of the remaining variable. Be careful to sum over the deleted variable.
4. Verify that the marginal distribution is a true pdf.
5. Write the values or the pdf relation for the marginal distribution.

In symbols, the marginal distributions of a continuous $f(x,y)$ are

- Marginal pdf of X: $\qquad f(x) = \int_{y} f(x,y)\, dy$

- Marginal pdf of Y: $\qquad f(y) = \int_{x} f(x,y)\, dx$

The integral is taken from $-\infty$ to $+\infty$. The continuous marginal pdf is found by using the steps above with the integral rather than the sum in step 3.

The logic above is extended to the pdf $f(x_1, x_2, x_3, \ldots, x_k)$ for any number of variables k. The marginal pdf for $f(x_1)$ is

$$f(x_1) = \int_{x_k} \cdots \int_{x_3} \int_{x_2} f(x_1, x_2, x_3, \ldots, x_k)\, dx_2\, dx_3 \cdots dx_k$$

Example 18.1 Many animals were selected for a biomechanical experiment and placed into one of four feeding areas. It has been hypothesized that a particular muscular disorder exists in these animals with a probability of 0.5. In one experiment a random sample of four animals was selected and observed for the disorder. Define the variables

$$X = \text{number of animals with the disorder}$$
$$Y = \text{number of the feeding area selected}$$

Assume that each feeding area is selected with equal probability on each trial and that $f(x, y)$ follows the distribution of Eq. (18.1). The individual probability values (Table 18.1) are graphed in Fig. 18.2. Develop and comment on the marginal distribution for (*a*) the number of animals with the disorder and (*b*) the feeding area selected on any one trial.

SOLUTION (*a*) The procedure to find a marginal distribution is used.

1. Table 18.1 tabulates $f(x, y)$.
2. The marginal pdf $f(x)$ is required.
3. Sum over all y values to obtain the column totals. These are the values of $f(x)$.

x	0	1	2	3	4
$f(x)$	0.060	0.252	0.376	0.252	0.060

4. Since $f(x)$ sums to 1.0 and possesses all three pdf properties (Sec. 6.4), it is a true pdf.
5. Because many animals are present and because each one should show symptoms of the disorder with $p = 0.5$, X has a binomial pdf. The $f(x)$ values above are from

$$b(x; 4, 0.5) = {}_4C_x 0.5^4 \qquad x = 0, 1, 2, 3, 4 \qquad (18.2)$$

Table 18.1 Tabulation of the pdf for two variables, Example 18.1

Area y	Number of animals x				
	0	1	2	3	4
1	0.015	0.063	0.094	0.063	0.015
2	0.015	0.063	0.094	0.063	0.015
3	0.015	0.063	0.094	0.063	0.015
4	0.015	0.063	0.094	0.063	0.015

(b) By following the same five steps and summing over values of $x = 0, 1, \ldots, 4$ for each y, the row totals in Table 18.1 give $f(y)$:

y	1	2	3	4
$f(y)$	0.25	0.25	0.25	0.25

Since a feeding area has an equal chance of selection on each trial, Y has a discrete uniform distribution (Sec. 10.7) with the pdf

$$f(y) = \tfrac{1}{4} = 0.25 \qquad y = 1, 2, 3, 4 \qquad (18.3)$$

COMMENT You will always obtain the marginal pdf's using the method above provided the entries in the table possess the three properties of $f(x, y)$ in Sec. 18.1. The values for $f(x)$ and $f(y)$ are best found simultaneously in Table 18.1 by appending a right column for $f(y)$ and a bottom row for $f(x)$.

Problems P18.12–P18.16

18.3 Independent Random Variables

By definition, for discrete or continuous variables

Two random variables X and Y are *independent* if it is true that for every (x, y) pair the product of the marginal pdf's is equal to the bivariate pdf; that is,

$$f(x, y) = f(x)f(y) \qquad (18.4)$$

If Eq. (18.4) is not true for all (x, y), the variables are called *dependent*. Independence may be shown in two ways.

1. Multiply the pdf relations $f(x)$ and $f(y)$ and compare with the $f(x, y)$ relation. This method requires that all pdf relations be known.
2. Multiply each individual probability value in $f(x)$ and $f(y)$ and compare to the corresponding $f(x, y)$ value. This method is practical if the variables are discrete and all individual probability values are known.

Example 18.2 We will use the distributions of Example 18.1 in a slightly different context. In tests on 4 of 1000 animals randomly selected from four feeding areas Y, the number X with a muscular disorder was recorded. The joint and marginal probability values are given in Table 18.2. Are X and Y independent?

SOLUTION The second method is used to determine the independence or dependence of X and Y. Each $f(x)f(y)$ gives a joint probability value. For

Table 18.2 Joint and marginal probability values, Example 18.2

y \ x	0	1	2	3	4	f(y)
1	0.015	0.063	0.094	0.063	0.015	0.250
2	0.015	0.063	0.094	0.063	0.015	0.250
3	0.015	0.063	0.094	0.063	0.015	0.250
4	0.015	0.063	0.094	0.063	0.015	0.250
f(x)	0.060	0.252	0.376	0.252	0.060	1.000

example,

$$f(x=1)f(y=2) = (0.252)(0.250) = 0.063 = f(x=1, y=2)$$

Similar computations for all (x, y) pairs produce a correct $f(x, y)$ value. We conclude that X and Y are independent.

Since in Example 18.1 we showed that X is binomial and Y is discrete uniform, we could use the first method above and multiply the pdf relations, Eqs. (18.2) and (18.3).

$$f(x)f(y) = f(x, y) = {}_4C_x(0.5)^4(0.25) = 0.0156({}_4C_x)$$

This is the $f(x, y)$ of Eq. (18.1) used to determine the table entries. Again, X and Y are independent.

The property of independence is very important in dealing with two or more variables. Often independence is assumed because it is inconvenient or difficult to demonstrate. Actually no two variables are ever completely independent, because some indirect, undetected connection probably exists. However, the analyst must frequently use independence to obtain "ball park" estimates.

Problems P18.17–P18.25

18.4 Conditional Distributions

If two random variables X and Y are not independent, it is possible to rewrite Eq. (18.4) using a conditional distribution of Y, given X or given Y:

$$f(x, y) = f(x)f(y/x) = f(y)f(x/y) \qquad (18.5)$$

By definition, the conditional distribution of Y for a given X, which is written $f(y/x)$, generates the pdf values for Y using a specific value $X = x$. There are as many conditional distributions $f(y/x)$ as there are values of x. Computationally, for discrete and continuous variables

$$f(y/x) = \frac{f(x, y)}{f(x)} \qquad (18.6)$$

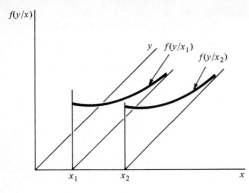

Figure 18.3 Conditional distributions of Y at two specific values of X.

where $f(x) > 0$. The pdf $f(x/y)$ is defined similarly with $f(y) > 0$. This definition is analagous to the conditional probability relation in Eq. (5.2). Figure 18.3 illustrates the fact that a pdf is present at each x value; therefore, each one has the three properties of a pdf discussed in Secs. 6.4 and 6.6. Specifically, for continuous variables Eq. (18.6) may be used to compute the total area under $f(y/x)$.

$$\int_y f(y/x)\, dy = \int_y \frac{f(x,y)}{f(x)}\, dy$$

$$= \frac{1}{f(x)} \int_y f(x,y)\, dy = \frac{1}{f(x)} f(x) = 1$$

The division by $f(x)$ makes the conditional pdf integrate to 1. The same result is obtainable for $f(x/y)$.

Equation (18.5) may be used to compute the values of the discrete joint distribution $f(x, y)$ or the form of the continuous joint distribution if the marginal pdf and the associated conditional pdf are known.

The distributions that have been studied thus far may be thought of as conditional pdf's given a specific value of the parameters. For example, if X has a binomial distribution, for a specific value of the parameter $p = p_1$, the conditional binomial distribution $f(x/p_1)$ is

$$f(x/p_1) = {}_nC_x p_1 x(1 - p_1)^{n-x}$$

If p is changed to p_2 where $p_2 \neq p_1$, a new conditional distribution $f(x/p_2)$ may be written. Since there are an infinite number of values for $0 \leq p \leq 1$, there are an infinite number of binomial distributions $f(x/p)$.

Example 18.3 illustrates a discrete conditional pdf, and Example 18.4 involves continuous variables.

Example 18.3 A consultant to an airline company has determined that the joint distribution for spare engine supply is

$$f(x, y) = \frac{x + y}{15} \qquad x = 0, 1, 2; y = 1, 2 \qquad (18.7)$$

18.4 CONDITIONAL DISTRIBUTIONS

Table 18.3 Values of the joint distribution in Eq. (18.7)

x \ y	1	2	f(x)
0	$\frac{1}{15}$	$\frac{2}{15}$	$\frac{3}{15}$
1	$\frac{2}{15}$	$\frac{3}{15}$	$\frac{5}{15}$
2	$\frac{3}{15}$	$\frac{4}{15}$	$\frac{7}{15}$

Table 18.4 Conditional pdf values $f(y/x)$, Example 18.3

x \ y	1	2
0	0.33	0.67
1	0.40	0.60
2	0.43	0.57

where X = number of engines required per week
Y = number of engines stocked per week

(a) Determine the probability values for and (b) plot the conditional distribution of the number of engines stocked for each demand value.

SOLUTION (a) Since $f(x, y)$ is discrete, the conditional probability of Y given X is computed for each value of y given x = 0, 1, 2. Equation (18.6) is used to compute $f(y/x)$, so the marginal distribution $f(x)$ is needed. Table 18.3 gives $f(x, y)$ from Eq. (18.7) and the marginal pdf $f(x)$, while Table 18.4 presents the probabilities in $f(y/x)$. For example, if the demand is $x = 0$ and the stock level is $y = 1$,

$$f(y = 1/x = 0) = \frac{f(0, 1)}{f(0)} = \frac{\frac{1}{15}}{\frac{3}{15}} = \frac{1}{3} = 0.33$$

Note that the sum of the probabilities across each row is 1; that is, $f(y/x)$ for each x is a true pdf.

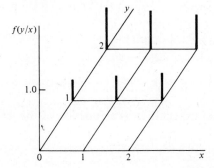

Figure 18.4 Plot of the conditional distribution computed in Example 18.3.

(b) Figure 18.4 is a plot of the probability values in Table 18.4. The conditional pdf is different for each value $x = 0, 1,$ and 2 engines.

The expected value of the conditional pdf $f(y/x)$ is written $E[Y/X]$ and is called the *conditional expectation* of Y given X. The formula is the same as in previous chapters, except the conditional distribution is used.

- Discrete:
$$E[Y/X] = \sum_y y f(y/x) \tag{18.8}$$

- Continuous:
$$E[Y/X] = \int_y y f(y/x) \, dy \tag{18.9}$$

If $E[X/Y]$ is required, the variables are merely switched.

Example 18.4 The variables X and Y have the uniform joint distribution
$$f(x, y) = 2 \qquad 0 < x < y < 1$$
Compute the conditional expectation of Y given X.

SOLUTION It is necessary to determine the conditional pdf $f(y/x)$ by Eq. (18.6), which means the marginal pdf $f(x)$ must first be obtained from $f(x, y)$.

$$f(x) = \int_y f(x, y) \, dy = \int_x^1 2 \, dy = 2y \Big|_x^1$$
$$= 2(1 - x) \qquad 0 < x < 1 \tag{18.10}$$

Now Eq. (18.10) is used to compute $f(y/x)$:

$$f(y/x) = \frac{f(x, y)}{f(x)} = \frac{2}{2(1 - x)}$$
$$= \frac{1}{1 - x} \qquad y > x; \, 0 < x < 1$$

By Eq. (18.9) the conditional expectation of Y given X is

$$E[Y/X] = \int_x^1 y \frac{1}{1-x} \, dy = \frac{1}{1-x} \left(\frac{y^2}{2} \right) \Big|_x^1$$
$$= \frac{1 - x^2}{2(1 - x)} = \frac{(1 - x)(1 + x)}{2(1 - x)}$$
$$= \frac{1 + x}{2} \qquad 0 < x < 1$$

As you might expect, the conditional expectation is a function of the given variable X.

COMMENT You should rework this problem now to find $E[X/Y] = y/2$ for $0 < y < 1$.

Problems P18.26–P18.31

18.5 Two Specific Distributions of More than One Variable

The *multinomial* distribution is discrete. It is used when three or more variables $X_1, X_2, X_3, \ldots, X_k$ are defined and used to categorize sample outcomes. The probability that each X_i occurs is p_i. This distribution is simply the k-dimensional equivalent of the binomial distribution. In the multinomial there are k possible outcomes rather than only two.

Using the symbol Π to represent multiplication for $i = 1, 2, \ldots, k$, the pdf is

$$f(x_1, x_2, \ldots, x_k) = \frac{n!}{\Pi x_i!} \Pi p_i^{x_i} \qquad (18.11)$$

where n is the sample size and each $x_i = 0, 1, 2, \ldots, n$. Also, it is necessary that $\Sigma x_i = n$ and $\Sigma p_i = 1$. The pdf computes the chance of x_i outcomes in category i.

Parameters and properties are defined for each variable.

$k + 1$ parameters:	$n, p_1, p_2, \ldots, p_k$
Parameter ranges:	$n = 1, 2, \ldots$
	$0 \leq$ each $p_i \leq 1$
Best estimate of p_i:	For each category i
	$\hat{p}_i = \dfrac{\text{number in category } i}{\text{sample size}} = \dfrac{x_i}{n}$
Expected value of X_i:	$\mu_i = np_i$
Standard deviation of X_i:	$\sigma_i = \sqrt{np_i(1 - p_i)}$

Example 18.5 An inspector classes mirrors as acceptable (X_1), flawed (X_2), or rejected (X_3). In a production run of 100 mirrors, $X_1 = 90$, $X_2 = 7$, and $X_3 = 3$.
(a) State the multinomial pdf for this run and compute the chance of this outcome.
(b) How can this distribution be reduced to a binomial?

SOLUTION (a) The parameters are $n = 100$ and

$$\hat{p}_1 = \tfrac{90}{100} = 0.90 \qquad \hat{p}_2 = \tfrac{7}{100} = 0.07 \qquad \hat{p}_3 = \tfrac{3}{100} = 0.03$$

The pdf relation is

$$f(x_1, x_2, x_3) = \frac{100!}{x_1! x_2! x_3!} 0.90^{x_1}(0.07^{x_2})(0.03^{x_3})$$

This pdf cannot be graphically visualized because it is four-dimensional. The probability of the observed values is 0.035.

$$f(x_1 = 90, x_2 = 7, x_3 = 3) = \frac{100!}{90!7!3!} 0.90^{90}(0.07^7)(0.03^3)$$
$$= 0.035$$

(b) If the two outcomes were acceptable (X_1) and nonacceptable ($Y = X_2 + X_3$), which combines flawed and rejected, the general pdf would be

$$f(x_1, y) = \frac{n!}{x_1! y!} p_1^{x} p_2^{y}$$

where $y = n - x_1$ and $p_2 = 1 - p_1$. This is a binomial pdf; so, the two-variable multinomial is a binomial.

The second distribution to be discussed is the *bivariate normal* (BVN), which is the two-variable form of the normal distribution. The pdf is graphed in Fig. 18.5. You can see that there is a normal distribution at each value of the variables X and Y. The BVN has five parameters, with the expected values and variances the same as four of these parameters.

- Five parameters: $\mu_x, \mu_y, \sigma_x^2, \sigma_y^2, \rho$
- Parameter ranges: $-\infty \le \mu_x$ and $\mu_y \le \infty$
  σ_x^2 and $\sigma_y^2 > 0$
  $-1 \le \rho \le 1$
- Expected values: $E[X] = \mu_x$ and $E[Y] = \mu_y$
- Variances: $V[X] = \sigma_x^2$ and $V[Y] = \sigma_y^2$

The parameter ρ is called the *coefficient of correlation* and measures the mathematical dependence between X and Y. A value $\rho = 0$ implies independent variables. We discuss correlation at length later in the book. For now, you should realize that $\rho = 0$ implies independence, in which case the BVN pdf is the product of two normals, that is,

- $\text{BVN} = N(\mu_x, \sigma_x^2) N(\mu_y, \sigma_y^2)$ if $\rho = 0$

We use the BVN as an underlying assumption in techniques covered later. When more than two variables are present, a multivariate normal is used. This expansion is similar to the above using μ and σ^2 for each variable and ρ for each pair of variables.

Problems P18.32–P18.35

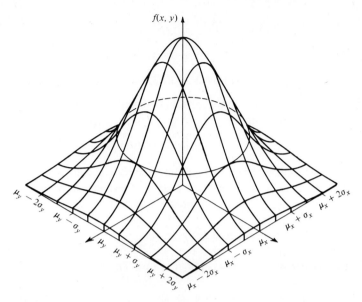

Figure 18.5 Graph of the bivariate normal (BVN) pdf. (Reproduced with the permission of the publisher from *Statistical Theory with Engineering Applications* by A. Hald, John Wiley and Sons, New York, 1952.)

18.6 Means of the Functions of Variables

It is correct to state that the function of a random variable is also a random variable. Therefore, we can use the expected-value relations for discrete and continuous variables of Secs. 6.5 and 6.7 to compute the means of functions of random variables. If $u(X)$ is any function of X, then

Discrete: $\quad E[u(X)] = \Sigma u(x)f(x)$

Continuous: $\quad E[u(X)] = \int u(x)f(x)\, dx$

There are several common rules for the expected value of functions of random variables. Three of these are important:

1. The expected value of a constant c is the constant:
   - $$E[c] = c$$

2. The expected value of any variable is μ, its mean. We know this as
   - $$E[X] = \mu$$

3. The expected value of a constant c times the variable is c times the expected value of the variable:
   - $$E[cX] = cE[X]$$

If two variables X and Y are involved, then the function is $u(X, Y)$, and two rules are important.

4. The expected value of the sum (difference) is the sum (difference) of expected values *under all conditions*.

• $$E[X + Y] = E[X] + E[Y]$$
$$E[X - Y] = E[X] - E[Y]$$

5. The expected value of the product is the product of expected values *if and only if the variables are independent*.

• $$E[XY] = E[X]E[Y] \quad \text{if independent}$$

The last two rules may be extended to three or more variables. Rule 5 is proved below, but you should be able to demonstrate that the rest are correct. For continuous variables the expected value of the product of two variables is

$$E[XY] = \int_x \int_y xyf(x, y)\, dx\, dy$$

Only if X and Y are independent does $f(x, y) = f(x)f(y)$. Then

$$E[XY] = \int_x \int_y xyf(x)f(y)\, dx\, dy$$
$$= \left[\int_x xf(x)\, dx\right]\left[\int_y yf(y)\, dy\right]$$
$$= E[X]E[Y]$$

Example 18.6 A linkage system with two independently operating bearings has been redesigned. The previous average allowable torque on each bearing was 200 kg · cm and 350 kg · cm, whereas the new single-bearing design has an average capacity of twice the sum of the old bearings. Compute the average allowed torque of the new design.

SOLUTION This problem is very simple, because only the summing of random variables is involved. Let X be the torque of the first bearing and Y be the torque of the second. The redesigned torque Z is equal to $2(X + Y)$. Finding the mean of Z applies rules 3 and 4 above with $c = 2$.

$$\begin{aligned} E[Z] &= E[2(X + Y)] = E[2X + 2Y] \\ &= E[2X] + E[2Y] \quad \text{rule 4} \\ &= 2E[X] + 2E[Y] \quad \text{rule 3} \\ &= 2(200) + 2(350) \\ &= 1100 \text{ kg} \cdot \text{cm} \end{aligned}$$

Problems P18.36–P18.42

18.7 Variances of Functions of Variables

Variances of functions of random variables are found using expected-value relations learned in Secs. 6.5 and 6.7. We use expected-value formulas to verify these rules for discrete or continuous variables. If $u(X)$ is any function of X and $u(X, Y)$ is any function of the variables X and Y,

$$V[u(X)] = E[(u(X))^2] - \{E[u(X)]\}^2 \qquad (18.12)$$

$$V[u(X, Y)] = E[(u(X, Y))^2] - \{E[u(X, Y)]\}^2$$

For single variables, three important rules are:

1. The variance of a constant c is always 0, because there is no variation.

$$V[c] = 0$$

2. The variance of any variable is σ^2. In Eq. (18.12), if $u(X) = X$, we have the familiar formula

$$V[X] = E[X^2] - \mu^2$$

3. The variance of a constant c times a variable is c^2 times the variance of the variable:

$$V[cX] = c^2 V[X]$$

For two variables the three rules are:

4. The variance of the sum is the sum of the variances *if and only if the variables are independent*.

$$V[X + Y] = V[X] + V[Y] \qquad \text{if independent}$$

5. The variance of the difference is the *sum* of the variances *if and only if the variables are independent*.

$$V[X - Y] = V[X] + V[Y] \qquad \text{if independent}$$

6. The variance of a linear combination of k variables each times a constant c_i is the sum of each c_i^2 times the variance of the variable.

$$V[\Sigma c_i X_i] = \Sigma c_i^2 V[X_i]$$

For the example of two variables

$$V[c_1 X_1 + c_2 X_2] = c_1^2 V[X_1] + c_2^2 V[X_2]$$

This is a combination of rules 3 and 4.

You should be able to verify all these rules. Rule 3 is proved using expected-value notation:

$$V[cX] = E[(cX)^2] - \{E[cX]\}^2$$
$$= E[c^2X^2] - \{cE[X]\}^2$$
$$= c^2E[X^2] - c^2\{E[X]\}^2$$
$$= c^2\{E[X^2] - \mu^2\}$$
$$= c^2V[X]$$

Example 18.7 Assume that in tests on the old two-bearing design of Example 18.6 the torque variances averaged 10 and 12 $(kg \cdot cm)^2$. What do you expect the variance to be when the average allowable torque of the new design is tested?

SOLUTION The variables are independent, so we can add the variances of X and Y. The best variance estimate comes from the same linear combination used in Example 18.6, that is, $Z = 2(X + Y)$. We can apply rule 6 to this function of X and Y:

$$V[Z] = V[2X + 2Y] = 4V[X] + 4V[Y]$$
$$= 4(10) + 4(12) = 88 \; (kg \cdot cm)^2$$

Remember to always *square the constant* when computing the variance of a constant times a variable. If the function Z had been $Z = 2(X - Y)$, the variance $V[Z]$ would still have the same value, because variances are always added (rules 4 and 5).

Problems P18.43–P18.48

*18.8 Transformation of a Single Discrete Variable

It is possible to obtain the complete pdf expression for a function of a random variable X, for example, X^2, $2X + 1$, or $\ln X$, by using the *transformation-of-variable* technique. This is the technique used to derive the pdf relations for distributions such as fraction defective, standard normal, χ^2, t, and F. We will discuss only *discrete* variable transformations in this section, using the following terminology.

X is a random variable with the pdf of $f(x)$.
Y is a function of X; that is,

$$Y = u(X) \tag{18.13}$$

18.8 TRANSFORMATION OF A SINGLE DISCRETE VARIABLE

The *inverse* function of Y is Eq. (18.13) solved for X, which is written as

$$X = w(Y) \tag{18.14}$$

The pdf of Y is $f(y)$, which is the complete pdf expression that is to be determined.

To apply the transformation-of-variable procedure discussed here, it is vital that there be a one-to-one correspondence between the values of X and Y; that is, a specific function $y = u(x)$ is a single-valued function of x, and its inverse, $x = w(y)$, can be used to determine a unique value of x. An example of a one-to-one transformation is $y = 3x + 1$ where $x = 0, 1, \ldots$ For each x there is only one value of y; and if the function is solved for $x = (y - 1)/3$, there is only one value of x for each y. Show yourself right now that two functions which do not give one-to-one correspondence are $y = x^2$ ($-\infty \leq x \leq \infty$) and $y = \cos x$ ($0 \leq x \leq 2\pi$ rad).

Once a one-to-one correspondence is guaranteed, the transformation-of-variable technique for a single discrete variable requires that (1) the range of the variable Y be determined and (2) the inverse function $x = w(y)$ be used to obtain the pdf $f(y)$ by direct substitution into $f(x)$. The following procedure may be used for single discrete variables only.

1. Ensure that $y = u(x)$ is a one-to-one transformation.
2. Convert the variable range on X into the corresponding range for Y using Eq. (18.13), $y = u(x)$.
3. Substitute the inverse function $w(y)$ determined by Eq. (18.14) into $f(x)$ each place that x appears to obtain the distribution $f(y)$.

$$f(y) = f[w(y)] \tag{18.15}$$

4. Write the complete $f(y)$ including the variable and parameter ranges. It is reasonable at this time, if desired, to check that $f(y)$ is a true pdf which possesses the properties in Sec. 6.4. Example 18.8 illustrates this procedure.

Example 18.8 (*a*) The probability that a missile component fails on its xth activation has followed a geometric distribution. Recent modifications have doubled the number of activations needed for failure, so the function $Y = 2X$ is now used for failure prediction. Determine the pdf of Y if the component failure characteristics are unchanged.
(*b*) Determine the pdf of the variable $Y = X^2$ if X has a geometric distribution.
(*c*) The number of cabs arriving per quarter hour at a taxi stand follows the $P(x; 6)$ distribution. What is the distribution of the function $Y = 3X + 1$?

SOLUTION (a) From Eq. (10.1) the geometric pdf is

$$f(x) = \begin{cases} pq^{x-1} & x = 1, 2, 3 \ldots ; 0 \le p \le 1 \\ 0 & \text{elsewhere} \end{cases} \quad (18.16)$$

The transformation-of-variable procedure is used.

1. The function $u(x)$ is $2x$. For $x = 1, 2, 3, \ldots$ there is a one-to-one correspondence between all x and y values in $y = 2x$.
2. The variable range for Y is $y = 2, 4, 6, \ldots$.
3. From Eq. (18.14) the inverse function is $x = y/2$, which is substituted into $f(y)$ according to Eq. (18.15):

$$f(y) = pq^{y/2-1}$$

4. The complete pdf for Y is

$$f(y) = \begin{cases} pq^{y/2-1} & y = 2, 4, 6, \ldots ; 0 \le p \le 1 \\ 0 & \text{elsewhere} \end{cases}$$

(b) The geometric distribution in Eq. (18.16) and the steps above are used.
1. In general, $y = x^2$ does not yield a one-to-one correspondence because $x = \pm \sqrt{y}$ has two solutions; but the range of x in this case has no negative values, so both x and y are single-valued.
2. The variable range for Y is $y = 1, 4, 9, \ldots$.
3. The inverse function $x = \sqrt{y}$ is used to obtain $f(y) = pq^{\sqrt{y}-1}$
4. The complete pdf for Y is

$$f(y) = \begin{cases} pq^{\sqrt{y}-1} & y = 1, 4, 9, \ldots ; 0 \le p \le 1 \\ 0 & \text{elsewhere} \end{cases}$$

(c) The procedure to determine the pdf of $Y = 3X + 1$ for a Poisson distribution, Eq. (9.2), is summarized. The function $y = 3x + 1$ is single-valued for $x = 0, 1, 2, \ldots$ and the range of Y is $y = 1, 4, 7, \ldots$ The inverse function is $x = (y-1)/3$ and by Eq. (18.15) the complete pdf of Y is

$$f(y) = \begin{cases} \dfrac{e^{-6} 6^{(y-1)/3}}{\left(\dfrac{y-1}{3}\right)!} & y = 1, 4, 7, \ldots \\ 0 & \text{elsewhere} \end{cases}$$

COMMENT If $Y = u(X)$ is not a one-to-one transformation, it is still possible to use the transformation-of-variable technique, but it is necessary to determine and add the probability values for all x's which correspond to each y value.

Problems P18.49–P18.55

*18.9 Transformation of a Single Continuous Variable

The transformation procedure for a single continuous variable is the same as for a discrete variable. The one-to-one transformation may now be interpreted as increasing or decreasing functions $Y = u(X)$ and $X = w(Y)$. In step 3 of the procedure of Sec. 18.8 the expression of $f(y)$ is computed as

$$f(y) = f[w(y)] \left| \frac{dx}{dy} \right| \qquad (18.17)$$

where $|dx/dy|$ is the absolute value of the first derivative of the inverse function $x = w(y)$. This derivative is assumed to be continuous and not vanish at all points in the variable range of Y. It is common to refer to dx/dy as J, the Jacobian of the transformation, so Eq. (18.17) may be written

$$f(y) = f[w(y)]|J| \qquad (18.18)$$

Multiplication by $|J|$ ensures that $f(y)$ integrates to 1, thus making it a true pdf.

The following examples explain the use of the transformation procedure for continuous variables.

Example 18.9 The time to failure for a self-actuating valve follows an exponential distribution with a mean of 100 h. Determine the pdf for the reciprocal of the time-to-failure variable.

SOLUTION If X is the time-to-failure variable, $X \sim \exp(0.01)$ because the parameter is $\lambda = 1/E[X] = \frac{1}{100}$. The variable X is continuous with the pdf

$$f(x) = 0.01 e^{-0.01x} \qquad x > 0 \qquad (18.19)$$

Use the transformation steps for a continuous variable to obtain the pdf of the reciprocal of X, that is, $Y = 1/X$.

1. The function $y = 1/x$ decreases for all $x > 0$, so it is single-valued.
2. The variable range of Y is $y > 0$.
3. The inverse function $x = w(y)$ is $1/y$, and the absolute value of the Jacobian is

$$|J| = \left| \frac{dx}{dy} \right| = \left| \frac{d}{dy} \frac{1}{y} \right| = \left| -\frac{1}{y^2} \right| = \frac{1}{y^2}$$

Therefore, the pdf for Y is found by substituting $x = 1/y$ into the exponential Eq. (18.19) according to Eq. (18.18):

$$f(y) = 0.01 e^{-0.01/y} \left(\frac{1}{y^2} \right)$$

4. The complete pdf for Y is

$$f(y) = \begin{cases} \dfrac{0.01}{y^2} e^{-0.01/y} & y > 0 \\ 0 & \text{elsewhere} \end{cases}$$

Example 18.10 If the variable X has a uniform distribution over the range $(0, 1)$, that is,

$$f(x) = 1 \quad 0 \le x \le 1$$

find the distribution of $Y = -2 \ln X$.

SOLUTION The transformation procedure may be summarized as follows. The function $y = -2 \ln x$ is increasing for $0 \le x \le 1$, and the range for Y is $y > 0$, because at $x = 1$, $y = -2 \ln(1) = 0$ and as x approaches 0, y approaches ∞. The inverse function $w(y)$ is $x = e^{-y/2}$, and the Jacobian to be used in Eq. (18.18) is

$$J = \frac{d}{dy} e^{-y/2} = -\frac{1}{2} e^{-y/2}$$

The pdf for Y may now be written as

$$f(y) = \tfrac{1}{2} e^{-y/2} \quad y > 0$$

Comparison of this $f(y)$ and the $f(y)$ in Sec. 13.2 for the χ^2 distribution indicates that this is the pdf of a χ^2 variable with 2 degrees of freedom.

Problems P18.56–P18.61

*18.10 Transformation of Two Variables

The procedures for single-variable transformations of the last two sections are easily extended to two or more variables. For two variables the bivariate distribution $f(x_1, x_2)$ for X_1 and X_2 is known and the distribution $f(y_1, y_2)$ is to be determined using the transformation functions Y_1 and Y_2, which are each functions of X_1 and X_2; that is,

$$Y_1 = u_1(X_1, X_2) \quad \text{and} \quad Y_2 = u_2(X_1, X_2)$$

It is necessary that Y_1 and Y_2 define a one-to-one transformation between (x_1, x_2) and (y_1, y_2) so that inverse relations can be uniquely solved.

The following procedure can be used for the transformation of two discrete or continuous variables. These steps explain how the Jacobian is extended for use with two continuous variables.

1. Ensure that the functions $y_1 = u_1(x_1, x_2)$ and $y_2 = u_2(x_1, x_2)$ are one-to-one transformations.
2. Convert the variable ranges on X_1 and X_2 into the corresponding ranges for Y_1 and Y_2.
3. Solve for the inverse functions

$$x_1 = w_1(y_1, y_2) \quad \text{and} \quad x_2 = w_2(y_1, y_2)$$

If $f(x_1, x_2)$ is a *discrete* pdf, obtain the joint distribution of Y_1 and Y_2 by direct substitution into $f(x_1, x_2)$:

$$f(y_1, y_2) = f[w_1(y_1, y_2), w_2(y_1, y_2)] \qquad (18.20)$$

If $f(x_1, x_2)$ is a *continuous* pdf, obtain the joint distribution of Y_1 and Y_2 by direct substitution and multiplication by the absolute value of the Jacobian for two variables:

$$f(y_1, y_2) = f[w_1(y_1, y_2), w_2(y_1, y_2)]|J| \qquad (18.21)$$

The Jacobian J is the value of the determinant of partial derivatives.

$$J = \begin{bmatrix} \dfrac{\partial x_1}{\partial y_1} & \dfrac{\partial x_1}{\partial y_2} \\ \dfrac{\partial x_2}{\partial y_1} & \dfrac{\partial x_2}{\partial y_2} \end{bmatrix} \qquad (18.22)$$

Note that the absolute value of J is used in Eq. (18.21).
4. Write the complete pdf $f(y_1, y_2)$ including all variable and parameter ranges.

If the marginal distribution of Y_1 or Y_2 is needed, one variable is summed or integrated out according to the logic in Sec. 18.2. Example 18.11 illustrates the procedure above for continuous variables.

Example 18.11 The variables X_1 and X_2 have the bivariate distribution

$$f(x_1, x_2) = e^{-(x_1 + x_2)} \qquad 0 \le x_1 \le \infty; \, 0 \le x_2 \le \infty$$

Determine the following:
(a) The distribution of $Y_1 = X_1 + X_2$ and $Y_2 = X_1/(X_1 + X_2)$
(b) The marginal distribution of Y_2

SOLUTION (a) The procedure above for two continuous variables is used to find $f(y_1, y_2)$.

1. The functions for Y_1 and Y_2 are single-valued.
2. The ranges for Y_1 and Y_2 are $0 \le y_1 \le \infty$ and $0 \le y_2 \le 1$.
3. The inverse functions are found by the solution of $y_1 = x_1 + x_2$ and $y_2 = x_1/(x_1 + x_2)$ for x_1 and x_2.

$$x_1 = y_1 y_2 \qquad x_2 = y_1(1 - y_2)$$

The Jacobian from Eq. (18.22) is

$$J = \begin{bmatrix} y_2 & y_1 \\ 1 - y_2 & -y_1 \end{bmatrix}$$

$$= -y_1 y_2 - y_1(1 - y_2) = -y_1$$

Using Eq. (18.21), the pdf of Y_1 and Y_2 is

$$f(y_1, y_2) = e^{-[y_1 y_2 + y_1(1-y_2)]} y_1$$
$$= y_1 e^{-y_1}$$

4. The complete pdf expression is

$$f(y_1, y_2) = \begin{cases} y_1 e^{-y_1} & 0 \leq y_1 \leq \infty;\ 0 \leq y_2 \leq 1 \\ 0 & \text{elsewhere} \end{cases}$$

(b) The marginal distribution of Y_2 is obtained by integrating out Y_1:

$$f(y_2) = \int_0^\infty f(y_1, y_2)\, dy_1$$
$$= \int_0^\infty y_1 e^{-y_1}\, dy_1$$

Reference to Eq. (16.7) shows that this integral is the gamma function for $n = 2$. Therefore

$$f(y_2) = \Gamma(2) = 1 \qquad 0 \leq y_2 \leq 1$$

COMMENT You should now find the marginal distribution for Y_1 from $f(y_1, y_2)$.

With knowledge of the transformation of two continuous variables it is now possible for you to derive distributions such as the t and F. For example, the derivation of the t distribution (Sec. 14.5) can be summarized as follows. Let $Z \sim N(0, 1)$, let $Y \sim \chi^2(\nu)$, and let Z and Y be independent random variables. Their joint pdf is

$$f(z, y) = \frac{1}{\sqrt{2\pi}} e^{-z^2/2} [2^{\nu/2} \Gamma(\nu/2)]^{-1} y^{(\nu/2)-1} e^{-y/2}$$

$$-\infty \leq z \leq \infty;\ 0 \leq y \leq \infty$$

If the transformation functions are

$$T = \frac{Z}{\sqrt{Y/\nu}} \qquad \text{and} \qquad U = Y$$

a one-to-one transformation is present. The joint distribution $f(t, u)$ can be formulated by Eq. (18.21), and the marginal distribution $f(t)$ has a t distribution with ν degrees of freedom. See Hogg and Craig for the details. The χ^2, F, and many other distributions are derived in a similar fashion.

Problems P18.62–P18.65

ADDITIONAL MATERIAL

Distribution of several variables (Secs. 18.1 to 18.4): Beck and Arnold [1], pp. 67–76; Bowker and Lieberman [4], pp. 52–66; Gibra [8], Chap. 3; Hahn and Shapiro [10], pp. 51–62; Hines and Montgomery [12], Chap. 4; Hogg and Craig [14], Chap. 2; Wilks [26], pp. 39–53.

BVN distribution (Sec. 18.5): Beck and Arnold [1], pp. 401–413; Hahn and Shapiro [10], pp. 79–83; Hoel [13], pp. 151–157; Hogg and Craig [14], Chap. 13; Wilks [26], pp. 158–170.

Properties of several variables (Secs. 18.6 and 18.7): Gibra [8], Chap. 5; Hahn and Shapiro [10], pp. 62–66; Hines and Montgomery [12], Chap. 3; Hoel [13], pp. 117–132; Wilks [26], pp. 77–87.

Transformation of variables (Secs. 18.8 to 18.10): Gibra [8], Chap. 4; Hahn and Shapiro [10], Chap. 5; Hines and Montgomery [12], pp. 63–69; Hoel [13], pp. 244–252; Hogg and Craig [14], Chap. 4; Walpole and Myers [25], pp. 135–151; Wilks [26], pp. 53–59.

PROBLEMS

The following distributions are used by the indicated problems in this chapter.

A: A solid-state component is produced by sequentially going through operations on two different processing lines: line A followed by line B. Two items on each component are given quality checks on line A, and three are checked on line B. The number X of defectives found on line A and the number Y found on line B follow the pdf

$$f(x, y) = \frac{x + y}{30} \qquad x = 0, 1, 2; y = 0, 1, 2, 3$$

Problems P18.4, P18.15, P18.17, P18.26, P18.39

B: Two pieces of equipment used in operating rooms fail independently according to the exponential failure law. The joint pdf for X, failure time in years of the first, and Y, failure time in years of the second, is

$$f(x, y) = e^{-(x+y)} \qquad x \text{ and } y > 0$$

Problems P18.5, P18.6, P18.18

C: A consultant doing a cash flow analysis for the City of Big Bucks has determined that in a 1-month period the city will receive the industrial tax revenues and payment to a municipal contractor will be due. Let

X = receipt time of the industrial taxes

Y = payment time to the municipal contractor

If X and Y are expressed in fractions of a month and if both follow a uniform distribution, the joint pdf is

$$f(x, y) = 1 \qquad 0 \le x \le 1; 0 \le y \le 1$$

Problems P18.10, P18.16, P18.19, P18.40, P18.45

P18.1 Locate on a two-dimensional plot all possible values of the variables whose ranges are described below.
  (a) $x \geq 0$ and $y \leq 10$
  (b) Two dice are tossed, and X is the number on the first die while Y is the outcome of the second toss such that
$$y = \begin{cases} 0 & \text{if number on second die exceeds number on first} \\ 1 & \text{otherwise} \end{cases}$$
  (c) $1 \leq x \leq 5$ and $x - 1 \leq y \leq x + 1$
  (d) $x = 0, 1, 2, 3$ and $y = 0, 1, 2$ with $0 \leq x + y \leq 3$

P18.2 A distribution of the form $f(x, y) = cx(1 + 2y^2)$, where c is a constant, has been suggested to describe the probability that $x = 0, 1, 2$ and $y = 0, 1$. Determine the value of c to ensure that $f(x, y)$ is a pdf.

P18.3 (a) Determine if each expression is a pdf for $0 \leq x \leq 2$ and $0 \leq y \leq 1$.
  (1) $f(x, y) = \tfrac{3}{10}x(1 + 2y^2)$ (2) $f(x, y) = \tfrac{3}{2}x(1 - 2y^2)$
  (b) If a sample of size 1000 is taken from population (1), determine how many in the sample have $x \leq 1$ and $y \geq \tfrac{1}{2}$.

P18.4 Determine the probability that the total number of defectives for situation A will equal (a) exactly two and (b) at most two.

P18.5 The cumulative distribution function $F(x', y')$ for a bivariate distribution is determined like the cdf for a single variable. If $f(x, y)$ is the pdf for two continuous variables, the cdf is defined as
$$F(x', y') = P(X \leq x', Y \leq y')$$
$$= \int_{-\infty}^{y'} \int_{-\infty}^{x'} f(x, y) \, dx \, dy$$
Use this definition to derive the cdf expression for the joint exponential pdf in situation B.

P18.6 For the operating room experiment in situation B, compute the probability that both pieces will fail within (a) the first 6 months of installation and (b) the first year of installation.

P18.7 A shipment of 10 electric room heaters is received, and 2 are damaged. For a sample of size 2, determine the complete joint pdf for the variables X and Y, where
$$x = \begin{cases} 0 & \text{if first heater is damaged} \\ 1 & \text{if first heater is not damaged} \end{cases}$$
$$y = \begin{cases} 0 & \text{if second heater is damaged} \\ 1 & \text{if second heater is not damaged} \end{cases}$$
Assume that the heater selected first is replaced before the second is chosen.

P18.8 Rework Prob. P18.7 but assume the first heater is not replaced prior to selecting the second heater.

P18.9 Determine the value of the constant c to make the expression $cx(6 - x - y)$ a true pdf. The variable ranges are $0 \leq x$ and $y \leq 1$.

P18.10 For the joint uniform pdf in situation C, first predict (using your knowledge of the probability distribution) and then compute the probability that (a) both the tax receipt and contractor payment will occur in the first half of the month and (b) the contractor payment must precede the tax receipt.

P18.11 For the joint pdf $f(x, y) = cx(1 - y)$, determine the value of the constant c to make certain that it integrates to 1 for the variable ranges (a) $0 \leq x \leq 2$ and $0 \leq y \leq 1$ and (b) $0 \leq x$ and $y \leq 1$.

P18.12 Let P_1 be the variable representing the probability that one event occurs and P_2 be the variable for the occurrence of another event. If the joint pdf is
$$f(p_1, p_2) = 4p_1(1 - p_2) \qquad 0 \leq p_1 \text{ and } p_2 \leq 1$$
  (a) Find the marginal distributions $f(p_1)$ and $f(p_2)$.

(b) Plot the marginal distributions.
(c) Determine the expected values of the marginal distributions.

P18.13 A probability table has been constructed for the two variables X and Y, which represent observable outcomes in a sample of 12 items.
  (a) Use the steps in Sec. 18.2 to determine the marginal distributions $f(x)$ and $f(y)$.
  (b) Determine if the following relation is true $f(x = 2, y = 2) = f(x = 2)f(y = 2)$.

x \ y	1	2	3
0	$\frac{3}{12}$	$\frac{2}{12}$	$\frac{1}{12}$
2	0	$\frac{3}{12}$	0
4	0	$\frac{1}{12}$	$\frac{2}{12}$

P18.14 The trivariate distribution

$$f(x, y, z) = 0.00015 \exp[-(0.05x + 0.03y + 0.10z)] \quad x, y, z > 0$$

is used to describe the failure model for a three-component subassembly. Determine the following marginal distributions:
  (a) $f(x, y)$
  (b) $f(y)$
  (c) $f(y, z)$

P18.15 Determine the probability that $x \leq 1$ defectives for situation A.

P18.16 Compute the chance that the contractor payment in situation C will be due in the last quarter of the month.

P18.17 Are the variables X and Y independent as described in situation A?

P18.18 (a) Use the bivariate distribution in situation B to show that the two exponential variables are independent.
  (b) What are the values of the exponential parameters λ_x and λ_y for the distributions of X and Y?

P18.19 Derive the marginal distributions for situation C and determine if the variables are independent.

P18.20 A chemical process is comprised of two reactions, both of which produce an amount of poisonous gas. Let X be the percentage of gas that is poisonous in reaction 1 and Y be the total percentage in both reactions. The upper limit on Y is 2 percent, and the joint pdf which best explains the percentage of gas generated is

$$f(x, y) = \tfrac{1}{2} \quad 0 < x < y < 2$$

  (a) Plot $f(x, y)$.
  (b) Find $P(x \leq 1, y \leq 1.5)$.
  (c) Find the marginal distributions for X and Y.
  (d) Determine if the two variables are independent.
  (e) Compute $P(x \leq 0.8)$.
  (f) Compute $P(y \geq 1.5)$.

P18.21 Use the marginal distributions determined in Prob. P18.13 to decide if the variables X and Y are dependent or independent.

P18.22 Answer the questions below for the relation

$$f(x, y) = cx(1 + 2y^2) \quad x \text{ and } y = 0, 1$$

  (a) What value of the constant c makes this a true pdf?
  (b) Are X and Y independent? Use the value for c from part (a).
  (c) What percentage of the time is $x = 1$?

P18.23 Use the pdf's that you derived in the following problems to determine whether the variables in each problem are independent:
(a) Prob. P18.7
(b) Prob. P18.8
(c) Is it possible to predict if the variables are independent without performing all these computations? Why or why not?

P18.24 The variables X and Y represent the time in minutes to perform a welding operation and a sanding operation, respectively. The completion times are independent, following uniform distributions.

$$f(x) = \tfrac{1}{3} \quad 1 \leq x \leq 4$$
$$f(y) = \tfrac{1}{5} \quad 5 \leq y \leq 10$$

(a) Determine the bivariate distribution of X and Y.
(b) Compute the probability that the welding time is less than 3 min.
(c) Use the joint pdf $f(x, y)$ to verify that

$$P(x \leq E[X], y \leq E[Y]) = 0.25$$

P18.25 Traffic flow off an expressway at two exit ramps has independent Poisson distributions with means 20 and 15 cars per minute. Jamie, a transportation engineer, wants to use the joint pdf for traffic flow at the two locations to determine the probability of observing 20 or fewer cars per minute on each ramp. Find the joint pdf expression and the probability for Jamie.

P18.26 (a) For situation A determine the conditional distribution of the number of defectives found on line B given the number on line A.
(b) Compute an expression for the conditional expectation of Y given X and use it to determine $E[Y/x = 1]$.

P18.27 Use the joint distribution

$$f(x, y) = \tfrac{3}{10} x(1 + 2y^2) \quad 0 \leq x \leq 2; 0 \leq y \leq 1$$

to determine:
(a) $f(x/y)$ and $E[X/Y]$
(b) $f(y/x)$ and $E[Y/X]$

P18.28 (a) Determine the conditional distribution of P_2 given P_1 for the pdf in Prob. P18.12.
(b) Use this conditional pdf to find $P(p_2 \geq 0.5/p_1 = 0.3)$ and $P(p_2 \geq 0.5/p_1 = 0.8)$.

P18.29 Use the probability table in Prob. P18.13 to obtain the conditional expectation of X given Y for (a) $y = 2$ and (b) $y = 4$.

P18.30 Susanne has decided to use the joint pdf in Prob. P18.20 for her study of a chemical process in a smelter.
(a) What is the conditional pdf of the total percentage of poisonous gas if the percentage from reaction 1 is known?
(b) Determine the probability that the total percentage exceeds 1.2 percent if the first reaction generates 1 percent poisonous gas.
(c) Compute $E[Y/x = 1]$.
(*Note*: You already determined the marginal pdf's in Prob. P18.20c.)

P18.31 The maximum pollutant output allowed is stated as 2 parts per 1000. Two processes generate the pollutant according to the pdf

$$f(x, y) = 2 - x - y \quad 0 \leq x \text{ and } y \leq 1$$

(a) Plot the conditional pdf $f(x/y)$ for $y = \tfrac{1}{2}$ and $y = 1$.
(b) Compute $P(x \leq \tfrac{1}{2} / y = \tfrac{1}{2})$ and hatch this area on the plot in (a).

P18.32 Michael can dial a number on his telephone and have access to a nationwide long distance (US WATS) line. For several reasons he often has to dial more than once to get a line. In a comprehensive study the following probabilities were determined.

Category	Number of dials to obtain a line	Probability
1	1	$p_1 = 0.10$
2	2 through 5	$p_2 = 0.30$
3	6 through 10	$p_3 = 0.40$
4	11 or greater	$p_4 = 0.20$

(a) Write the complete pdf for the variables X_i (i = 1, 2, 3, 4), where in a sample of n attempts there are x_i times that the dials necessary fall into category i.

(b) Michael has made 10 calls on the system today. He recorded the following dials per call: 3, 5, 12, 1, 9, 2, 7, 10, 6, 4. What is the probability of this outcome?

(c) For 10 calls how many should Michael expect to record in each category?

P18.33 In a sample of 40 months of sales records for the Go-for-Broke Glass Company, a total of 5 months had at most $500,000 in sales, 20 were in the range $500,000 to $800,000, and 15 were in the range of $800,000 to $1.2 million per month.

(a) If these results are used to estimate the probability of future monthly sales potential, what is the chance of meeting top management's sales goal—at least half of next year with million-dollar months and the remainder in the $500,000 to $800,000 range?

(b) Based on past sales, how many months during the next two years should be expected to fall into each sales category?

P18.34 The bivariate normal (BVN) distribution has the pdf form

$$f(x, y) = \left(2\pi\sigma_x\sigma_y\sqrt{1 - \rho^2}\right)^{-1} e^{-r/2}$$

where

$$r = \frac{1}{1 - \rho^2}\left[\left(\frac{x - \mu_x}{\sigma_x}\right)^2 - 2\rho\left(\frac{x - \mu_x}{\sigma_x}\right)\left(\frac{y - \mu_y}{\sigma_y}\right) + \left(\frac{y - \mu_y}{\sigma_y}\right)^2\right]$$

Show that if $\rho = 0$, the two variables are independent because $f(x, y) = f(x)f(y)$ where $X \sim N(\mu_x, \sigma_x^2)$ and $Y \sim N(\mu_y, \sigma_y^2)$.

P18.35 If $f(x, y)$ is the BVN pdf given in Prob. P18.34, it is possible to show that the conditional distributions of Y/X and X/Y are themselves normal with the parameters below.

Distribution	Expected value	Variance
$f(y/x)$	$\mu_y + \rho\dfrac{\sigma_y}{\sigma_x}(x - \mu_x)$	$\sigma_y^2(1 - \rho^2)$
$f(x/y)$	$\mu_x + \rho\dfrac{\sigma_x}{\sigma_y}(y - \mu_y)$	$\sigma_x^2(1 - \rho^2)$

(a) Determine the expected values and variances of the conditional distribution for a BVN with $\mu_x = 5$, $\mu_y = 10$, $\sigma_x = 1$, $\sigma_y = 1.5$, and $\rho = 0.8$.

(b) Use the parameter values in (a) to determine the conditional probability that $10 \le x \le 12$ given that $y = 20$.

P18.36 Show that the following expected-value rules are correct for continuous variables:

(a) Rule 1, $E[c] = c$

(b) Rule 4, $E[X + Y] = E[X] + E[Y]$

P18.37 The variables X and Y have the pdf's given for the welding and sanding times in Prob. P18.24. Compute the expected values for the functions (a) $X + Y$ and (b) $3X$.

P18.38 Use the marginal distributions determined in Prob. P18.13a to compute (a) $E[1.5Y]$ and (b) $E[X + 5Y]$.

P18.39 Find the marginal distributions for situation A and compute as many of the following expected values as possible:
 (a) $E[4X]$
 (b) $E[20X - 0.5Y]$
 (c) $E[XY]$

P18.40 The joint uniform variables X and Y in situation C are independent. Use this fact to verify that $E[XY] = E[X]E[Y]$.

P18.41 The time X in minutes to complete an assembly operation follows an $N(10.5, 1.2)$ distribution, and the time Y to inspect the product is exponential with a parameter of $\lambda = 2$ min. What is the expected time to assemble and inspect three products?

P18.42 Use the poisonous gas distribution of Prob. P18.20, which is

$$f(x,y) = \tfrac{1}{2} \qquad 0 < x < y < 2$$

to show that the $E[X]E[Y] \neq E[XY]$ when X and Y are dependent.

P18.43 If the distributions in Prob. P18.41 are independent, compute the variances for the functions (a) $2X$ and (b) $2X + 3Y$.

P18.44 Prove that the variance rule 5, $V[X - Y] = V[X] + V[Y]$, is correct for independent variables by using expected-value notation.

P18.45 For the distribution in situation C, use the marginal distributions (determined in Prob. P18.19) to compute the mean and variance of the variable $Z = 3X - 2Y$.

P18.46 The width X of a steel plate is normally distributed with $\mu_x = 3.5$ cm and $\sigma_x = 0.2$ cm. This plate is attached to a beam which has a thickness Y that is also normally distributed with $\mu_y = 10.5$ cm and $\sigma_y = 1.1$ cm. If the two measurements are independent, determine the mean and standard deviation of the total measurement when three plates are placed between two beams.

P18.47 If two variables X and Y are not independent and their dependence is measured by a correlation coefficient $\rho(-1 \leq \rho \leq 1)$, the expected value and variance of the linear combination $Z = c_1 X + c_2 Y$, where c_1 and c_2 are constants, may be written

$$E[Z] = c_1 E[X] + c_2 E[Y]$$

$$V[Z] = c_1^2 V[X] + c_2^2 V[Y] + 2c_1 c_2 \rho \sqrt{V[X]V[Y]}$$

Compute $E[Z]$ and $V[Z]$ for the assembly and inspection of three products in Prob. P18.41 if the best estimate of correlation is $\hat{\rho} = 0.2$.

P18.48 Two dependent variables are distributed according to the Poisson: $X \sim P(x; 10)$ and $Y \sim P(y; 3)$.
 (a) If $\rho = 0.7$, use the formulas for dependent variables in Prob. P18.47 and the variable $Z = X + Y$ to determine $E[Z]$ and $V[Z]$.
 (b) By how much will $V[Z]$ decrease if X and Y are independent?

*P18.49 If the number of defectives X in a sample of n items follows a binomial distribution, determine the complete pdf of Y, the proportion of the time that X defectives are observed (Sec. 8.8).

*P18.50 The time X in minutes to complete a welding operation follows the discrete uniform distribution

$$f(x) = \tfrac{1}{5} \qquad x = 6, 7, 8, 9, 10$$

Find the distribution of the variables (a) $Y = 1/X$ and (b) $Y = 2X^2$.

*P18.51 If $X \sim P(x; \lambda)$, write the pdf relation for (a) $Y = X^3$ and (b) $Y = X + 1$.

*P18.52 Factory workers are allowed to take off 1 h early one workday of the week. Study shows that they choose Monday ($x = 1$), Tuesday ($x = 2$), etc., according to the distribution

$$f(x) = \frac{x}{15} \qquad x = 1, 2, 3, 4, 5$$

In working with the data, the analyst needs the distribution for (a) $Y = \ln X$ and (b) $Y = 2(X + 1)$. Determine the pdf's for these two functions.

*P18.53 What is the pdf relation for $Y = X^2$ if X has a binomial distribution with parameters n and p?

*P18.54 If $X \sim b(x; n, p)$, do the following for the function $Y = 3X$.
 (a) Write the complete pdf expression.
 (b) Use the appropriate expected-value rule in Sec. 18.6 to determine $E[Y]$.
 (c) Derive the expected value for the distribution in (a) to show that the result is the same as $E[Y]$ in (b).

*P18.55 A gambler plans to toss one die and double the outcome.
 (a) Write the complete pdf expressions $f(x)$ for X, which is the actual outcome, and $f(y)$ for Y, which is the outcome doubled.
 (b) Compute $E[Y]$ and $V[Y]$ using the rules in Secs. 18.6 and 18.7.
 (c) Calculate $E[Y]$ and $V[Y]$ using the $f(y)$ from (a). Are they the same?

*P18.56 Let the variable X have a normal distribution with parametrs μ and σ, and show that the function $Z = (X - \mu)/\sigma$ has the standard normal distribution given by Eq. (11.5).

*P18.57 Tax payments are due in city hall the first workday of June, but citizens are given three extra months, with some penalty, to make late payments. Study has shown that the fraction of unpaid tax dollars during the 3-month late payment period follows the distribution

$$f(x) = \left(1 - \frac{x}{3}\right)^2 \qquad 0 \le x \le 3 \text{ months}$$

A new city council ruling now allows only 1 month for late payment.
 (a) Find the new distribution for the fraction of unpaid taxes using the transformation function $Y = X/3$.
 (b) Compute $E[X]$.
 (c) Determine $E[Y]$ by two methods: using first the transformation function and then $f(y)$ from (a). The results should be the same.

*P18.58 If X has the distribution

$$f(x) = 2x \qquad 0 \le x \le 1$$

find the distribution of (a) $4X^2$ and (c) $2X - 1$.

*P18.59 In Example 6.10 the distribution for computer processing time T is

$$f(t) = \frac{10 + t}{2250} \qquad 10 \le t \le 60 \text{ s}$$

New hardware has cut the processing time by 25 percent.
 (a) Find the processing time distribution for $Y = 3T/4$.
 (b) The properties of T were found to be $\mu = 39.63$ s and $\sigma = 13.67$ s. What are the corresponding properties for Y?

*P18.60 Find the distribution of $Y = -\ln X$ for the following.
 (a) $f(x) = 3x^2/8 \qquad 0 \le x \le 2$
 (b) $f(x) = 20x(1 - x)^3 \qquad 0 \le x \le 1$

*P18.61 The variable X follows an exponential distribution with $\lambda = 1$. Determine the distribution of the functions (a) $Y = 2X$ and (b) $Y = \ln X$.

*P18.62 For the bivariate distribution in Prob. P18.3 a(1) determine the joint distribution of $Z_1 = X$ and $Z_2 = 2Y$.

*P18.63 A discrete bivariate distribution has been derived. It takes the form

$$f(x_1, x_2) = \tfrac{1}{12} x_1(1 + 2x_2^2) \qquad x_1 = 0, 1, 2;\ x_2 = 0, 1$$

(a) Use the transformation-of-variable technique for discrete variables to determine the distribution of Y_1 and Y_2 where $Y_1 = X_1^2$ and $Y_2 = 2X_2$.

(b) Find the marginal distribution of Y_1. Of Y_2.

*P18.64 Use the bivariate $f(x_1, x_2)$ in Example 18.11 to find the joint distribution of $Y_1 = X_1 + X_2$ and $Y_2 = X_2/(X_1 + X_2)$.

*P18.65 Use the transformation-of-variable procedure to find the joint distribution of $Y_1 = X_1 + X_2$ and $Y_2 = X_1 - X_2$ for:

(a) $f(x_1, x_2) = 1 \qquad 0 \le x_1 \le 1;\ 0 \le x_2 \le 1$

(b) $f(x_1, x_2) = 4x_1 x_2 \qquad 0 \le x_1 \le 1;\ 0 \le x_2 \le 1$

The variable ranges for Y_1 and Y_2 are found by substituting the values 0 and 1 into the inverse functions $w_1(y_1, y_2)$ and $w_2(y_1, y_2)$ and plotting the resulting equations.

LEVEL THREE
STATISTICAL INFERENCE

The chapters in this level all discuss the testing of statistical hypotheses for the purpose of making inferences about population parameters or distributions. You will learn to test hypotheses about the population mean, variance, and proportion using the distributions of the previous level. In addition, statistical tests which do not make specific distributional assumptions are discussed in a chapter on nonparametric techniques.

It is often necessary to determine if a set of collected data fits a particular statistical distribution. This level covers both statistical testing procedures and graphical procedures for evaluating the fit of a hypothesized distribution. The first chapter (19) should be considered a guide for all the hypothesis testing procedures in this level. It is here that the basic rationale and steps of statistical inference are introduced.

The final chapter on Bayesian estimation gives you an introduction to another method of making parameter estimates quite different from the approach taken in previous chapters. It may be considered an optional chapter on first reading.

The flowchart shows the relations between the chapters. The dotted lines identify some prerequisite chapters from Level Two that might be reviewed before the chapter in this level is read.

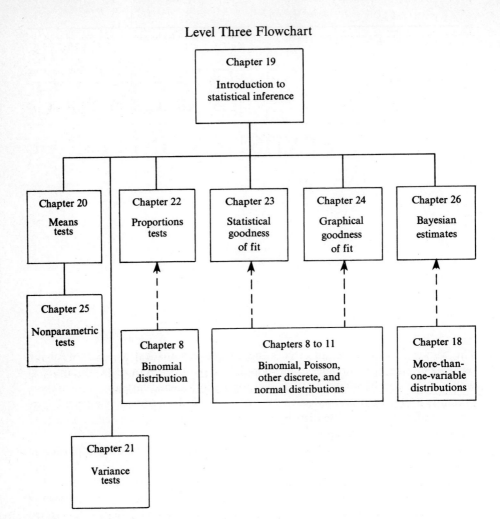

CHAPTER
NINETEEN

INTRODUCTION TO STATISTICAL INFERENCE

This chapter introduces the basic concepts and terminology used to make statistical inferences using the hypothesis testing procedure. This material is generally applicable to all statistical tests and is used in the following chapters to develop specific-purpose tests.

CRITERIA

To complete this chapter, you must be able to:

1. Define the terms *hypothesis testing, null hypothesis, alternative hypothesis, acceptance region*, and *critical region*.
2. State the meaning of a type I and type II error; and compute the probability of a type II error, given a description of the situation.
3. State the difference between a one-sided and two-sided hypothesis test, and write the acceptance region for each, given the significance level and the distribution.
4. State and demonstrate the use of the steps in the hypothesis testing procedure, given the hypothesis, significance level, and test statistic formula.
*5. Compute one of the following given the other two and a test statistic: probability of a type I error, probability of a type II error, or the sample size.
6. State the purpose of an operating characteristic (OC) curve, and read the value of sample size or type II error probability, given the OC curve and details of the hypothesis to be tested.

STUDY GUIDE

19.1 Rationale of Hypothesis Testing

Once a sample is taken, it is usually characterized by one or more sample statistics. The purpose of hypothesis testing is to use these statistics and our knowledge of statistical distributions to make inferences about the population from which the sample originates.

Inferences may be made about the pdf of the population or the value of population parameters. We discuss both uses. If the inference is about parameters, there are one-sample and two-sample tests which we will investigate.

A hypothesis is a statistical statement that is to be rejected or not rejected (accepted). Hypotheses can be formulated about means, variances, differences of means, or pdf forms. Some examples are:

Subject	Hypothesis description	Statistical statement
Mean	Population mean is 25	$\mu = 25$
Two means	Two populations have the same mean	$\mu_1 = \mu_2$
Proportion	Fraction of population that is defective is 0.05	$\pi = 0.05$
pdf	Population follows Poisson pdf with $\lambda = 0.5$	$P(x; 0.5)$ is correct
Two variances	Two populations have equal variances	$\sigma_1^2 = \sigma_2^2$

There are two hypotheses for any statistical test. The first and most important is H_0, the *null hypothesis*. The *alternative hypothesis* H_1 is automatically accepted if the test shows that H_0 should be rejected. A hypothesis is simple if it involves a single value, or composite if no specific value is given. A test of two simple hypotheses is $H_0 : \mu = 25$ and $H_1: \mu = 27$. A test of one simple and one composite is $H_0: \mu = 25$ and $H_1: \mu > 25$.

The usual desire is to not reject H_0. The terminology *not reject* is correct because it is always possible to reformulate the test to reject H_0. However, for ease of interpretation we will also use the term *accept* H_0 instead of *not reject* H_0. To accept H_0, the result of the statistical test must be some number which falls into the *acceptance region*. Any other value is in the *critical region*, as shown in Fig. 19.1, and requires the rejection of H_0. For example, if the true mean of a normal population is $\mu = 100$ and we hypothesize $H_0: \mu = 100$ or $H_1: \mu \neq 100$, two values x_1 and x_2 must be determined to separate the acceptance and critical regions (Fig. 19.2).

Problems P19.1, P19.2

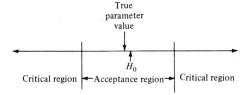

Figure 19.1 Acceptance and critical regions for a statistical test.

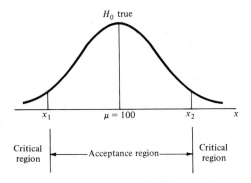

Figure 19.2 Acceptance and critical regions for testing the mean of a normal population.

19.2 Types of Errors and Their Probabilities

Since either H_0 or H_1 has got to be correct in reality, we can make two types of errors in our decision. These errors, called type I and type II, are explained here:

Reality	Decision	
	Reject H_0	Accept H_0
H_0 true	Type I error	Correct
H_0 false	Correct	Type II error

- Type I error: Reject a true H_0
- Type II error: Accept a false H_0

The probability of these errors is important because it determines how sensitive and powerful the test is in distinguishing between true and false hypotheses.

- $P(\text{type I error}) = \alpha$
- $P(\text{type II error}) = \beta$

Figure 19.3 shows H_0 and H_1. If H_0 is actually true but is rejected in favor of a false H_1 (type I error), the area outside c_1 and c_2 and under the H_0-true curve represents the probability α. For this same test and graph, if H_1 is actually true but the false H_0 is accepted (type II error), the area in the acceptance region but under the H_1 curve represents the probability β.

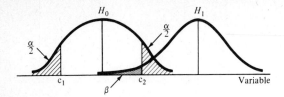

Figure 19.3 Areas of type I and type II errors.

Example 19.1 A petroleum engineer has taken a sample of 25 readings and hypothesized that the pressure in a vessel is 3.5 atmospheres (atm). The test is therefore between H_0: $\mu = 3.5$ and H_1: $\mu \neq 3.5$. The pressure is normally distributed with a standard deviation of 2.5 atm.
(a) Determine the critical region limits if a type I error is allowed 5 percent of the time.
(b) Compute the probability for a type II error if in reality $\mu = 4.8$ atm.

SOLUTION (a) The sample standard deviation is $2.5/\sqrt{25} = 0.5$ by the CLT (Sec. 12.1). The mean pressure is distributed $N(3.5, 0.25)$ if H_0 is true. To allow an $\alpha = 0.05$ probability for a type I error requires that $\alpha/2 = 0.025$ of the area be in each tail. The two corresponding values w_1 and w_2, where W is the sample average in atmospheres, on the normal curve define the critical region. Use the standard normal distribution (SND) variable $Z = (W - \mu)/\sigma$ to determine these limits. Figure 19.4 graphically summarizes the results.

$$P(z \leq z_1) = P\left(\frac{w_1 - 3.5}{0.5} \leq z_1\right) = 0.025$$

$$\frac{w_1 - 3.5}{0.5} = -1.96$$

$$w_1 = 2.52 \text{ atm}$$

and

$$P(z \geq z_2) = P\left(\frac{w_2 - 3.5}{0.5} \geq z_2\right) = 0.025$$

$$\frac{w_2 - 3.5}{0.5} = +1.96$$

$$w_2 = 4.48 \text{ atm}$$

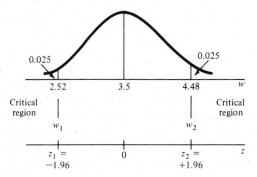

Figure 19.4 Critical region for a test H_0: $\mu = 3.5$ and H_1: $\mu \neq 3.5$, Example 19.1.

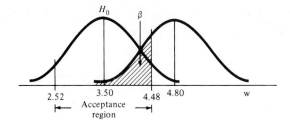

Figure 19.5 The probability β of a type II error for Example 19.1.

The critical region is defined by $w_1 = 2.52$ and $w_2 = 4.48$ atm. Any pressure value between these is in the acceptance region and requires that the engineer not reject (accept) H_0: $\mu = 3.5$ atm.

(b) If $\mu = 4.8$, β is the probability of incorrectly accepting H_0: $\mu = 3.5$, which is done if the sample mean w is between 2.52 and 4.48 atm. If we use the SND to compute $P(2.52 \leq w \leq 4.48)$, we have the area under the H_0 curve for the true $\mu = 4.8$. See Fig. 19.5.

$$P(w \leq 4.48) = P\left(z \leq \frac{4.48 - 4.80}{0.50}\right) = P(z \leq -0.64) = 0.2611$$

$$P(w \leq 2.52) = P\left(z \leq \frac{2.52 - 4.80}{0.50}\right) = P(z \leq -4.56) = 0$$

The sum 0.2611 is β, so there is a 26.11 percent chance of accepting H_0: $\mu = 3.5$ when the true value is 4.8 atm. The β value will be different for each true value. We explore β further in this chapter.

Problems P19.3–P19.7

19.3 One-Sided and Two-Sided Tests

The terms *significance* and *confidence area* used in interval estimation (Sec. 17.1) also apply to hypothesis testing. The significance is indicated by α, the type I error probability. Confidence in the test's conclusion is $1 - \alpha$. Therefore,

• Significance + confidence = $\alpha + (1 - \alpha) = 1$

The α level is usually set before the test is run. Examples are 0.10, 0.05, and 0.01, with 0.05 being most common. If $\alpha = 0.05$, there is a 5 percent level of significance, which means that on the average a type I error (reject a true H_0) will occur 5 in 100 times that H_0 and H_1 are tested. In addition, there is a 95 percent confidence level that the result is correct.

If H_1 involves a not-equal relation, for instance, H_0: $\mu = 3.5$ and H_1: $\mu \neq 3.5$, no direction is given, so the significance area is equally divided between the two tails of the testing distribution in a fashion similar to that shown in Fig. 17.1b. This is called a two-sided or two-tail test. If it is known that the parameter can go in only one direction, H_1: $\mu > 3.5$ or H_1: $\mu < 3.5$ may be used. This is a one-sided test, and the α area is in one tail of the distribution (Fig. 17.1a). Since the critical region will shift between a one-sided and a two-sided test, it is important that the appropriate H_0 and H_1 are determined before the test is performed.

Example 19.2 Determine the acceptance regions for the t distribution when $\alpha = 0.05$ and $\nu = 10$ degrees of freedom using a one-sided and a two-sided test.

SOLUTION From Table B-5 the acceptance region for a one-sided test must be determined for the upper and lower tails. Each tail has $\alpha/2$ for the two-sided test. The acceptance regions for the tests are:

$$\text{One-sided upper:} \quad -\infty \leq t \leq 1.812$$
$$\text{One-sided lower:} \quad -1.812 \leq t \leq \infty$$
$$\text{Two-sided:} \quad -2.228 \leq t \leq 2.228$$

Problems P19.8, P19.9

19.4 Steps of the Hypothesis Testing Procedure

To make a statistical inference, a standard procedure is followed once the data is collected. The steps are:

1. State the hypotheses. Develop the exact form of H_0 and H_1. The alternative hypothesis H_1 is used to make the test one-sided or two-sided.
2. Select the significance level. The α value will determine the probability of a type I error for the test.
3. Compute sample statistics and estimate parameters. One or more statistics may be needed to perform the test.
4. Compute the test statistic. The formula used to test H_0 is called the *test statistic*. It will result in a specific value on the distribution used to perform the test, such as normal, t, etc. This computation will require certain assumptions about the population.
5. Determine the acceptance and critical regions of the test statistic. Use the α level and estimated parameters as needed to find the value for a one- or two-sided test from the table of the appropriate distribution.
6. Reject or do not reject H_0. If the computed test statistic value (step 4) is in the acceptance region, do not reject H_0; if it is in the critical region, reject H_0.

We use these steps in the following chapters on hypothesis testing. Example 19.3 illustrates this general procedure using the t distribution.

Example 19.3 A meteorologist is studying the annual rainfall in a desert region. With all available records the average has been 7.8 cm/yr. For the past 5 yr rainfall has been 8.0, 6.2, 4.1, 6.9, and 5.6 cm/yr. The meteorologist wants to know if the decrease is statistically significant.

SOLUTION Use the hypothesis testing steps.

1. Since only a decrease is to be tested, use a one-sided test for the mean rainfall, with the historic $\mu = 7.8$ cm.

$$H_0: \mu = 7.8 \text{ cm versus } H_1: \mu < 7.8 \text{ cm}$$

2. Select an $\alpha = 0.05$ significance level.
3. The t distribution is used to test means for small samples. The next chapter discusses this in detail. The sample statistics $\overline{X}$ and s are needed for the rainfall data: $\overline{X} = 6.16$ cm and $s = 1.45$ cm. The degrees-of-freedom parameter is estimated by $\hat{\nu} = n - 1 = 4$.
4. The test statistic used here is the t statistic, Eq. (14.1):

$$t = \frac{\overline{X} - \mu}{s/\sqrt{n}} = \frac{6.16 - 7.80}{1.45/\sqrt{5}} = -2.53$$

The value $\mu = 7.80$ is used under the assumption that the mean is still the same, that is, H_0 is true.

5. Table B-5 gives $t_0 = -2.132$ as the lower, one-tail t value for $\nu = 4$ degrees of freedom. The acceptance region is $t \geq -2.132$, and the critical region is $t < -2.132$ (Fig. 19.6). There is no critical-region component in the right tail because of the one-sided hypothesis $H_1: \mu < 7.8$.
6. The test statistic $t = -2.53$ is in the critical region, so H_0 *is rejected* and the meteorologist should be told that the mean annual rainfall has decreased from 7.8 cm. The confidence in this result is $100(1 - \alpha)$ percent, or 95 percent.

COMMENT The α value of 0.05 is the probability that a true $H_0: \mu = 7.8$ cm has been rejected. If the sample average of 6.16 is actually correct and the $H_0: \mu = 7.8$ is not rejected, a type II error is present. β, the probability of this error, is computed using the t statistic. First, compute the value of $\overline{X}$ which corresponds to -2.132, the t value separating the acceptance and

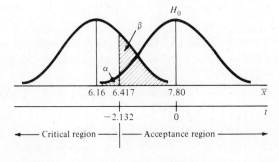

Figure 19.6 The α and β areas for a one-sided test of means, Example 19.3.

critical regions (Fig. 19.6).

$$\frac{\bar{X} - 7.8}{1.45/\sqrt{5}} = -2.132$$

$$\bar{X} = 6.417$$

The value of β is the area under the curve with $\mu = 6.16$ and $s = 1.45$ but in the acceptance region, that is, above 6.417.

$$t = \frac{6.417 - 6.16}{1.45/\sqrt{5}} = 0.396$$

$$\beta = P(t \geq 0.396) = 0.36$$

Therefore, there is a 36 percent chance of accepting a false $\mu = 7.8$ hypothesis if the true mean is 6.16. The β value will change as the correct mean shifts from 6.16.

Problems P19.10, P19.11

*19.5 Dependence between α, β, and Sample Size

There is a distinct relationship between the two probability values α and β and the sample size n for any hypothesis. The value of any one is found by using the test statistic and set values of the other two. Usually the α and n values are the most crucial, so they are established and the β value is not controlled (as in Example 19.1). The procedures to compute α, β, or n are given here and illustrated in Example 19.4.

1. Given α and n, determine β (Fig. 19.3). Find the area under the H_1 curve but in the acceptance region for H_0 using an H_1 value of the parameter in the test statistic.
2. Given β and n, determine α. Find the area under the H_0 curve that is in the critical region for H_0 using the H_0 value of the parameter in the test statistic.
3. Given α and β, determine n. Set up the test statistic for α and β with the H_0 value and an H_1 value of the parameter and two different n values. It is usually necessary to decide whether the selected α or β is more important, because they cannot both be obtained exactly since n must be an integer.

Example 19.4 Use the t statistic for the hypotheses H_0: $\mu = 5$ and H_1: $\mu = 6$ with $\sigma = 1$ to compute (a) β if $\alpha = 0.05$ and $n = 16$, (b) α if $\beta = 0.025$ and $n = 16$, and (c) n to obtain $\alpha = 0.05$ and $\beta = 0.025$ as closely as possible.

SOLUTION The t statistic is

$$t = \frac{(\bar{X} - \mu)\sqrt{n}}{\sigma} \qquad (19.1)$$

19.5 DEPENDENCE BETWEEN α, β, AND SAMPLE SIZE

(a) For $\alpha = 0.05$ and $\nu = 15$, the one-sided critical region is $t > 1.753$. The corresponding $\overline{X}$ value using $\mu = 5$ and Eq. (19.1) is

$$\frac{(\overline{X} - 5)\sqrt{16}}{1} = 1.753$$

$$\overline{X} = 5.438$$

According to procedure 1, find the area below $\overline{X}$ using $\mu = 6$ (Fig. 19.7a). From Eq. (19.1), $t = (5.438 - 6)\sqrt{16}/1 = -2.248$ and

$$\beta = P(t \leq -2.248) = 0.021$$

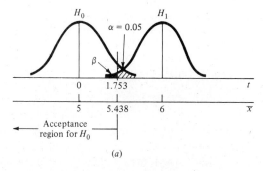

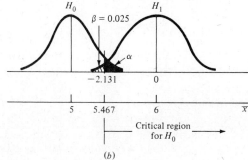

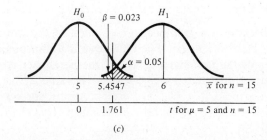

Figure 19.7 Graphs used to determine (a) β, (b) α, and (c) n, Example 19.4.

(b) For $\beta = 0.025$ and $\nu = 15$, the one-sided ciritical region for H_1 is $t < -2.131$. Use Eq. (19.1) to find the corresponding $\overline{X}$ if H_1: $\mu = 6$ is correct.

$$\frac{(\overline{X} - 6)\sqrt{16}}{1} = -2.131$$

$$\overline{X} = 5.467$$

According to procedure 2, find the area above $\overline{X} = 5.467$ using $\mu = 5$ (Fig. 19.7b). Computation in Eq. (19.1) gives $t = 1.868$ and

$$\alpha = P(t \geq 1.868) = 0.042$$

(c) Use Eq. (19.1) for $\alpha = 0.05$ with $\mu = 5$, and for $\beta = 0.025$ with $\mu = 6$ at $n = 15$ and $n = 16$. Solution gives the following $\overline{X}$ and actual α and β values:

n	μ	$\overline{X}$	α	β
15	5	5.4547	0.050	0.027
15	6	5.4462	0.054	0.025
16	5	5.4383	0.050	0.021
16	6	5.4673	0.042	0.025

An example computation is shown for $n = 15$ and $\alpha = 0.05$ (Fig. 19.7c).

$$\frac{(\overline{X} - 5)\sqrt{15}}{1} = 1.761$$

$$\overline{X} = 5.4547$$

The actual $\beta = 0.027$ is found using $\mu = 6$.

$$t = \frac{(5.4547 - 6)\sqrt{15}}{1} = -2.112$$

The closest n to the desired α and β is $n = 15$.

Problems P19.12–P19.15

19.6 Operating Characteristic Curves

An *operating characteristic* (OC) curve is a graphical presentation of the values for α, β, and n for a particular H_0 and any H_1. The OC curve is a plot of the type II error β versus all values of the parameter in H_1. If the parameter is μ, Fig. 19.8 shows a typical OC curve for a stated α and n. The H_0 value μ_0 has a probability of $1 - \alpha$ of being accepted if $\mu = \mu_0$ is correct. As the correct μ moves away from μ_0, the chances of accepting the false μ_0 diminish. This chance is β.

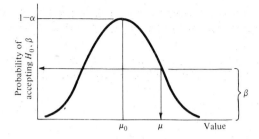

Figure 19.8 A general OC curve for the mean given α and n.

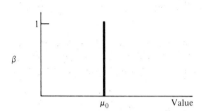

Figure 19.9 An ideal OC curve.

The ideal OC curve (Fig. 19.9) has an abscissa value of 1 for the correct μ_0 and 0 for all other values. Then $\alpha = 0$ and $\beta = 0$ for all $\mu \neq \mu_0$. This is, of course, impossible unless the entire population is tested and the parameter determined without error.

It is possible to use a parameter transformation such as $(\mu - \mu_0)/\sigma$ to plot only the right side of the OC curve. When it is plotted for a set α value and several n values, the OC curve can be used to determine one of α, β, or n given the other two values. A common use is to see what sample size is needed to guarantee specific error probabilities. If a specific β is required for a test, the value must be determined prior to data collection so that the correct n value is used. The β value should be stated along with α in step 2 of the hypothesis testing procedure. Example 19.5 shows how to use an OC curve.

The *power* of a test is determined from the OC curve by plotting $1 - \beta$ on the abscissa. The value $1 - \beta$ is the probability of rejecting H_0 when it is actually false. The power curve is used to compare the decision-making capability of different tests.

Example 19.5 Figure 19.10 is a plot of the OC curve for the two-sided t statistic

$$t = \frac{(\bar{X} - \mu_0)\sqrt{n}}{\sigma}$$

for $\alpha = 0.05$ and several n values. The abscissa scale is

$$d = \frac{|\mu - \mu_0|}{\sigma} \qquad (19.2)$$

where μ is the correct mean value and μ_0 is the hypothesized H_0 value. The test is between H_0: $\mu = 500$ and H_1: $\mu \neq 500$. It is known that $\sigma = 1$.

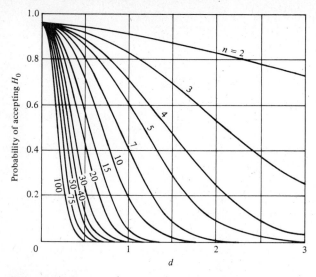

Figure 19.10 The OC curve for the two-sided t statistic and $\alpha = 0.05$. (Reproduced with the permission of the publisher from "Operating Characteristics of the Common Statistical Tests of Significance" by Ferris, C. D., Grubbs, F. E., and Weaver, C. L., *The Annals of Mathematical Statistics*, Vol. 17, No. 2, June 1946.)

(a) What value of n is necessary to achieve $\beta = 0.05$ for $\mu = 501$? For $\mu = 500.5$?

(b) If $n = 5$ and $d = 1.0$, find the value of β if the sample size is doubled. Tripled.

(c) If $d = 1.0$ and $n = 10$, what should happen to β if α is decreased from 0.05 to 0.01?

(d) What is the capability of this test to detect and reject the false value $\mu_0 = 500$ for a sample size of 5 if the true mean is 502?

SOLUTION (a) For $\mu = 501$, $d = 1.0$ from Eq. (19.2). From Fig. 19.10, $n = 15$ will give $\beta = 0.05$ as the probability of accepting the false H_0: $\mu = 500$. For $\mu = 500.5$, $d = 0.5$ and $\beta = 0.05$ at approximately $n = 60$.

(b)

d	n	β
1.0	5	0.62
1.0	10	0.20
1.0	15	0.05

(c) As α is decreased, the t values which define the critical region for a two-sided test move away from 0. If d and n do not change, β must increase as shown in Fig. 19.11, because more of the area under H_1 will be in the acceptance region of H_0.

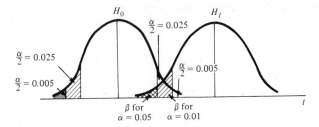

Figure 19.11 Graph of H_0 and H_1 which illustrates that β must increase when α decreases for fixed n and d values on an OC curve.

(d) For $\mu = 502$ and $n = 5$, $d = 2$ and $\beta = 0.10$. The capability to reject a false μ_0 value is given by the power of the test. In this case

$$\text{Power} = 1 - \beta = 1 - 0.10 = 0.90$$

Therefore, 90 out of 100 times this test will correctly reject H_0: $\mu = 500$ if the true mean is H_1: $\mu = 502$.

COMMENT It is not necessary to completely understand how to construct an OC curve to correctly use a hypothesis testing procedure. However, an OC curve can give you, at a glance, the inherent capability of the test you select to accept or reject a hypothesis.

Problems P19.16–P19.21

ADDITIONAL MATERIAL

Hypothesis testing fundamentals (Secs. 19.1 to 19.4): Brownlee [5], pp. 97–105; Gibra [8], Chap. 8; Hoel [13], pp. 106–112; Hogg and Craig [14], Chap. 10; Miller and Freund [18], pp. 194–201; Newton [20], pp. 218–235; Walpole and Myers [25], pp. 238–249.

OC curve fundamentals (Secs. 19.5 and 19.6): Belz [2], pp. 184–193; Bowker and Lieberman [4], pp. 163–178; Kirkpatrick [15], pp. 195–200; Miller and Freund [18], pp. 201–207; Neville and Kennedy [19], pp. 252–260.

PROBLEMS

P19.1 (a) Define the terms *acceptance region* and *critical region*.
(b) Show these regions on a pdf graph to test H_0: $\mu = a$ versus H_1: $\mu > a$, where a is a specific value of the population mean.

P19.2 Write the H_0 and H_1 for the following.
(a) The fraction of a specific attribute present in two samples is equal or the sample 2 fraction exceeds the sample 1 result.
(b) The difference in mean salary at two plants is equal to $150 per month, or the difference is less than $150 per month.
(c) The ratio of two population variances is either equal to or not equal to 1.5. Assume the larger variance would be called V_1.

P19.3 A new product has been thoroughly test-marketed, and sales are expected to follow a normal distribution with $\mu = 2000$ and $\sigma = 350$ units per day. For a sample of 20 days taken in the southern region, the manager wants to test if the mean is larger than the expected 2000 per day.
  (a) Write the expressions for H_0 and H_1.
  (b) Compute the limits of the acceptance region for H_0 using $\alpha = 0.10$ and the standard normal distribution.

P19.4 A total of 15 thermometers is tested for accuracy. In the past the variance in temperature measurement has been 0.7°F. The hypotheses $H_0: \sigma^2 = 0.7°F$ versus $H_1: \sigma^2 \neq 0.7°F$ are to be tested using $1 - \alpha = 0.99$ and the statistic $\chi^2 = (n-1)s^2/\sigma^2$.
  (a) What is the probability of a type I error for this test?
  (b) Determine the values of sample variance s^2 which will require the rejection of H_0. Show these values on the appropriate pdf graph.

P19.5 The following hypotheses are proposed to you by a friend for testing two values of the population mean.

$$H_0: \mu = 25.0 \quad \text{versus} \quad H_1: \mu = 30.0$$

If a sample of size 10 is taken from an $N(25, 15)$ population and the acceptance region is defined as $\bar{X} \leq 28$, (a) graphically illustrate the test similar to Fig. 19.5. Compute the probability of (b) the type I error and (c) the type II error. Use the normal distribution to compute these probabilities.

P19.6 (a) What is the probability of a type II error in Prob. P19.5 if $\bar{X} \leq 28$ defines the acceptance region and the true mean is 29? 28?
  (b) Interpret these probability values for this situation.

P19.7 Marta is a quality assurance manager for the Filthy-Rich Detergent Company. The variance of a particular process is from a normal population and historically has a value of 250 g. A decrease in variance to 150 g is to be tested. Samples of size 17 are taken, and an arbitrarily set value of 205 g is used to define the acceptance region for $H_0: \sigma^2 = 250$ versus $H_1: \sigma^2 = 150$. Marta has asked you to compute the α and β values for this test. Please do so using the statistic $\chi^2 = (n-1)s^2/\sigma^2$.

P19.8 A sample of eight items is taken from an $N(5, 0.3)$ population to test $H_0: \mu = 5$ versus $H_1: \mu \neq 5$ with the actual value of $\mu = 5$ incorrectly rejected 2 out of every 100 times. The t distribution is used, and $s^2 = \sigma^2$.
  (a) Compute the limits of the acceptance region for sample means.
  (b) State the confidence and significance levels of the test.
  (c) Compute the value of β if in reality $\mu = 4$ and the acceptance region in (a) is used to test H_0.

P19.9 Each of the following statements has at least one error. Correct them.
  (a) $H_0: \pi = 0.10$ versus $H_1: \pi \geq 0.10$ is a two-sided test of population variances.
  (b) The acceptance region on the t distribution for a two-sided $\alpha = 0.05$ with $n = 12$ is defined by ± 1.796.
  (c) $H_0: \mu = 25$ versus $H_1: \mu \neq 25$ is a test of two simple hypotheses.

P19.10 State the six steps of the hypothesis testing procedure in your own words. Summarize these steps and place them on a card or sheet of paper for quick reference in the future.

P19.11 A study has been performed to determine the time spent in commuting to work. Given here are the observed times for 450 trips and the expected times if an exponential distribution (Sec. 16.2) is correctly used to explain the probability that a commute time equals a specific value:

Time, minutes	Observed frequency	Expected frequency
0–10	175	208
10–20	125	112
20–30	90	60
30–40	45	32
40–50	15	18
> 50	0	20

When the appropriateness of the exponential distribution is tested using a confidence level of 95 percent, a χ^2 distribution statistic of 47.53 is calculated. For this problem there are 4 degrees of freedom, and the analyst concluded that the exponential is not appropriate. Use the six-step procedure in Sec. 19.4 to show how the analyst came to this conclusion. (Do not worry about the details of the computations—you will learn them later.)

*P19.12 Samples of assembly times are taken from a normal population with a known mean of 18.5 and a standard deviation of 2.4 s. In the last month the assembly time has increased to an estimated 20.0 s with no change in the standard deviation. Use the SND.
    (a) Find the size of a type I error if the critical region is set such that $\beta = 0.305$ for $n = 24$.
    (b) Find the size of a type II error for $\alpha = 0.10$ and $n = 24$.
    (c) Graphically illustrate as in Fig. 19.7 the relationship between the correct H_0 and H_1 and the values of α and β for the two preceding parts.

*P19.13 Dan wants to test the population mean values of 110 for H_0 and 100 for H_1 using a one-tail $\alpha = 0.10$ and $\beta = 0.10$. If the t distribution will be used to perform the test and the parent population is normal with $\sigma = 10$, what sample size is required? The probability of a type I error should be maintained if both α and β cannot be met exactly.

*P19.14 The SND is used to test the hypotheses H_0: $\mu = 25$ versus H_1: $\mu \neq 25$. In a sample of 40 items from an $N(26.5, 9)$ population, H_0 was not rejected with $s^2 = 9$ and $\alpha = 0.05$.
    (a) Compute the probability that this incorrect decision was made.
    (b) By how much would this probability change if $n = 100$ items had been checked?

*P19.15 (a) Rework Prob. P19.14a using $n = 40$, but now assume that the population is $N(25.5, 9)$. Compare the two probabilities.
    (b) Explain why the direction of the change in the probability (that is, increase or decrease) is predictable from the two normal distribution means of 26.5 and 25.5.

P19.16 (a) Use the OC curve in Fig. 19.10 for the two-sided t statistic to find β for the test H_0: $\mu = 25$ versus H_1: $\mu \neq 25$ if a sample of 40 items is taken from an $N(26.5, 9)$ population with $s^2 = 9$ and $\alpha = 0.05$.
    (b) Compare this probability value with that computed in Prob. P19.14a using the SND.

P19.17 Jack Mitchell plans to take a sample from a processing line which manufactures dominoes to ensure that the paint on each dot is applied correctly. The diameter of the paint is supposed to be 6.35 mm, but the line supervisor believes the mean diameter is 6.10 mm. The test of H_0: $\mu = 6.10$ versus H_1: $\mu \neq 6.10$ mm is to be carried out at $\alpha = 0.05$. The diameter is actually from an $N(6.35, 0.30)$ population.
    (a) What sample size is necessary to achieve a type II error probability of 0.40? 0.20? (Use Fig. 19.10.)
    (b) What happens to n for these two probability values if the population is actually $N(6.60, 0.30)$?

P19.18 Explain why the following statement is true: If the true parameter value, say μ, is equal to the null hypothesis value of μ_0, the β value is always equal to $1 - \alpha$, not 1.

P19.19 Determine whether the following are true or false and correct the false statements. In all parts μ is the correct population mean and μ_0 is the value used in H_0.
    (a) If μ_0, α, and n are fixed, an increase in μ will not affect β.
    (b) If μ_0, μ, and n are fixed, an increase in α will decrease β.
    (c) If μ_0, μ, and α are fixed, one way to decrease β is by increasing n.
    (d) If μ and n are fixed, the only way to decrease β is by increasing α.
    (e) The power of a test always increases if α increases for fixed μ_0, μ, and n.

P19.20 (a) For the situation $\mu_0 = 6$ and $\sigma = 0.5$, construct the complete OC curve (both sides) for the two-sided t statistic with $\alpha = 0.05$ and $n = 4$.
    (b) Draw the complete power curve from this OC curve. (You will need a t distribution table with α values up to 0.95 to work this problem correctly.)

P19.21 Construct only the right side of the power curve for the two-sided normal statistic with $\alpha = 0.05$ and $n = 6$. Assume the values $\mu_0 = 20$ and $\sigma = 2$.

CHAPTER
TWENTY

STATISTICAL INFERENCES FOR MEANS

The most commonly used tests for means are discussed in this chapter. You will learn how to test hypotheses involving one sample and two samples. The OC curves for these tests are presented.

CRITERIA

To correctly perform a hypothesis test for means, you must be able to:

1. State the three questions that must be answered to select the appropriate means test.

You must be able to perform all the following tests for means using the hypothesis testing procedure, given the appropriate data:

2. One sample test with the standard deviation known or estimated from a large sample.
3. One sample test with the unknown standard deviation estimated from a small sample.
4. Two independent samples test with the standard deviations known or reliably estimated.
5. Two independent samples test with the standard deviations unknown and assumed equal or unequal.
6. Two nonindependent (paired) samples test.

STUDY GUIDE

20.1 Types of Tests for Means

One of the most common uses of hypothesis testing is to determine if the means of samples are equal to, greater than, or less than one another or some specific value. All these tests are carried out using a form of the z statistic or the t statistic.

$$z = \frac{(\bar{X} - \mu)\sqrt{n}}{\sigma} \qquad (20.1)$$

$$t = \frac{(\bar{X} - \mu)\sqrt{n}}{s} \qquad (20.2)$$

In order to use the correct test, answer the three following questions. The type of statistic and appropriate section are given in Table 20.1. The t statistic is used only when σ is not known and the sample is small.

1. *Are there one or two samples of data?* If there are two samples, are the items from one source? If so, the samples are not independent, they are paired.
2. *Is the standard deviation known or must it be estimated?*
3. *Is the sample size large?* Consider any $n \geq 30$ as a large sample.

Table 20.1 Classification of data for the test for means

Description of sample(s)	Population σ known or unknown	n large or small	Type of test statistic	Section
One	Known	Large or small	z	20.2
	Unknown	Large	z	20.2
	Unknown	Small	t	20.3
Two, independent	Known	Large or small	z	20.4
	Unknown	Large	z	20.4
	Unknown	Small	t	20.5
Two, not independent (paired)	Unknown	Large	z	20.6
	Unknown	Small	t	20.6

20.2 Means Test for One Sample with Standard Deviation Known

When a specific value of the population mean μ_0 is known or desired, an estimated mean μ may be tested using the null hypothesis H_0: $\mu = \mu_0$. The alternative hypothesis H_1 may be $\mu \neq \mu_0$ (two-sided), $\mu > \mu_0$, or $\mu < \mu_0$ (one-sided). Either the standard deviation is known from historical data, or it can be

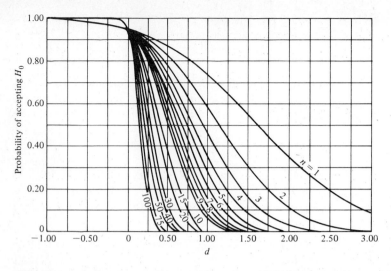

Figure 20.1 Operating characteristic curves for a *one-sided normal* test with $\alpha = 0.05$. (Reproduced with permission from *Engineering Statistics*, 2d ed., by Bowker, A. H., and Lieberman, G. J., Prentice-Hall, Inc., 1972.)

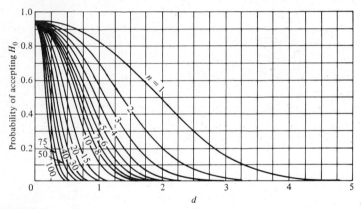

Figure 20.2 OC curves for a *two-sided normal* test with $\alpha = 0.05$. (Reproduced with permission from "Operating Characteristics of the Common Statistical Tests of Significance" by Ferris, C. D., Grubbs, F. E., and Weaver, C. L., *The Annals of Mathematical Statistics*, Vol. 17, No. 2, June 1946.)

estimated by $\hat{\sigma}$ from a large sample ($n \geq 30$) and used in this test, in which case $\hat{\sigma} = \sigma$. The test statistic is the SND variable for sample means.

$$z = \frac{(\mu - \mu_0)\sqrt{n}}{\sigma} \qquad (20.3)$$

where μ is estimated by the sample average $\overline{X}$ and μ_0 is the stated mean value in H_0. Use the hypothesis testing steps of Sec. 19.4 for this and all tests in this chapter.

The probability β of a type II error is given by the OC curve. Figures 20.1 and 20.2 give these curves for one- and two-sided tests, respectively, with

$\alpha = 0.05$. Curves for $\alpha = 0.01$ are available in Hines and Montgomery (see Additional Material). The abscissa scale is different depending on the form of H_1.

Two-sided, $H_1: \mu \neq \mu_0$ $\quad d = \dfrac{|\mu - \mu_0|}{\sigma}$ \hfill (20.4)

One-sided, $H_1: \mu > \mu_0$ $\quad d = \dfrac{\mu - \mu_0}{\sigma}$ \hfill (20.5)

One-sided, $H_1: \mu < \mu_0$ $\quad d = \dfrac{\mu_0 - \mu}{\sigma}$ \hfill (20.6)

The OC curve will give you the β value for the sample size n. If β is deemed too large, a larger sample must be taken and the test redone. If β is selected prior to data collection, the OC curve indicates the required n value. Example 20.1 uses the OC curve for this test.

Example 20.1 An assembly process has a standard time of 0.52 min. A trainee assembled 75 items with an average time of 0.49 min and a computed standard deviation of 0.15 min.
(a) What test is appropriate to determine if the average times are equal?
(b) Perform the tests for the two cases below.

$$H_0: \mu = \mu_0 \quad \text{versus} \quad H_1: \mu \neq \mu_0$$
$$H_0: \mu = \mu_0 \quad \text{versus} \quad H_1: \mu < \mu_0$$

(c) Determine the required n value to ensure that $\beta = 0.20$ for the one-sided test. What is β for the one-sided test above?

SOLUTION The data for this problem is

$$n = 75 \quad \mu_0 = 0.52 \quad \mu = \overline{X} = 0.49 \quad \sigma = \hat{\sigma} = 0.15$$

(a) Answers to the three questions in the previous section are: (1) one sample; (2) σ is estimated by $\hat{\sigma}$; and (3) large sample. The normal test statistic in Eq. (20.3) is appropriate, because σ is reliably estimated from the large sample data.
(b) Use the hypothesis testing steps of Sec. 19.4.

1. The two-sided test is $H_0: \mu = 0.52$ versus $H_1: \mu \neq 0.52$.
2. Use $\alpha = 0.05$ for the type I error probability.
3. The required sample statistics are $\hat{\sigma} = 0.15$ and $\overline{X} = \mu = 0.49$ min.
4. From Eq. (20.3)

$$z = \frac{(0.49 - 0.52)\sqrt{75}}{0.15} = -1.73$$

5. Table 11.2 or Table B-3 is used to determine the acceptance and

critical regions for a two-sided $\alpha = 0.05$. The acceptance region for H_0 is $-1.96 \leq z \leq 1.96$, so the critical region includes all other z values.

6. Since $z = -1.73$ is in the acceptance region, do not reject H_0. Therefore, the sample average of 0.49 is statistically equal to the standard of 0.52 min. The trainee did not beat the standard.

For the next test, steps 1, 5, and 6 change.

1. The one-sided test is $H_0: \mu = 0.52$ versus $H_1: \mu < 0.52$.
5. From Table 11.2 the one-sided, $\alpha = 0.05$ acceptance region is $z \geq -1.645$, and the critical region is $z < -1.645$.
6. The $z = -1.73$ is in the critical region, so reject H_0. Now, the sample average is statistically less than 0.52. The trainee did beat the standard.

Since the trainee could beat or exceed the standard, the two-sided test is more appropriate in this case.

(c) Use Eq. (20.6) to compute $d = 0.2$. Figure 20.1 indicates that $n > 100$ is required for $\alpha = 0.05$ and $\beta = 0.20$. The present $n = 75$ is too small. The probability β for the one-sided test above is approximately 0.30 (Fig. 20.1).

COMMENT For the one-sided test of $H_1: \mu < 0.52$ there is a $5 = 100\alpha$ percent chance that we reject 0.52 when it is true (Fig. 20.3). Using the power of the test, which is the probability of rejecting a false H_0, there is a $1 - \beta = 0.70$ probability of rejecting 0.52 when the true mean is 0.49 min.

Problems P20.1–P20.5

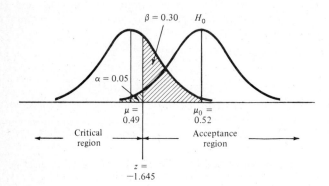

Figure 20.3 The α and β values for the one-sided test of $H_0: \mu = 0.52$ versus $H_1: \mu < 0.52$, Example 20.1.

20.3 Means Test for One Sample with Standard Deviation Unknown and a Small Sample

If σ is unknown, as is the usual case, and a small sample ($n < 30$) has been taken, the unbiased $s = \hat{\sigma}$ is the best estimate of σ, as in Sec. 20.2. But the t distribution is now used to test the hypothesis. The test statistic for one sample is

$$t = \frac{(\mu - \mu_0)\sqrt{n}}{s} \tag{20.7}$$

where μ is estimated by the sample average $\bar{X}$ and μ_0 is the stated mean value in H_0. The t distribution with $\nu = n - 1$ degrees of freedom is used to establish the acceptance region. There are no other changes to the tests performed in Sec. 20.2.

The OC curves for one- and two-sided tests with $\alpha = 0.05$ (Figs. 20.4 and 20.5) are used to determine β or the required n. If a large sample size is needed to ensure a set β value, you can use the SND test of Sec. 20.2.

Example 20.2 The times to failure for five batteries were observed as 32, 41, 42, 49, and 53 h. The manufacturer's warranty is for 50 operating hours. Does it appear that this sample is from a population with a mean of 50? Allow a type I error to be present 10 percent of the time.

SOLUTION Since this is a one-sample, σ unknown, small sample test, Eq. (20.7) is correct for the test statistic.

1. The mean battery life can be on either side of 50 h. Use the two-sided test H_0: $\mu = 50$ versus H_1: $\mu \neq 50$ h.
2. $\alpha = 0.10$; β is not limited.

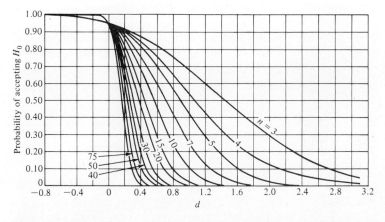

Figure 20.4 OC curves for a *one-sided* t test with $\alpha = 0.05$. (Reproduced with permission from *Engineering Statistics*, 2d ed., Bowker, A. H., and Lieberman, G. J., Prentice-Hall, Inc., 1972.)

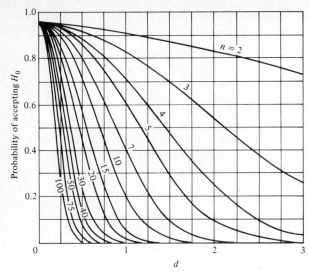

Figure 20.5 OC curves for a two-sided t test with $\alpha = 0.05$. (Reproduced with permission from "Operating Characteristics of the Common Statistical Tests of Significance," by Ferris, C. D., Grubbs, F. E., and Weaver, C. L., *The Annals of Mathematical Statistics*, Vol. 17, No. 2, June 1946.).

3. To perform the t test, we need $\overline{X}$, s, and ν from the sample.

$$\overline{X} = 43.4 \text{ h}$$
$$s = 8.08 \text{ h}$$
$$\nu = n - 1 = 4$$

4. From Eq. (20.7) with $\overline{X}$ used as an estimate of μ,

$$t = \frac{(43.4 - 50.0)\sqrt{5}}{8.08} = -1.83$$

5. For a two-sided test with $\alpha = 0.10$ and $\nu = 4$, the t distribution acceptance region (Table B-5) is $-2.132 \leq t \leq 2.132$.
6. The test statistic value -1.83 is in the acceptance region. Do not reject the null hypothesis that these batteries have a 50-h expected life.

COMMENT If $\sigma = 8.08$ were known and the normal test used, the acceptance region would be $-1.645 \leq z \leq 1.645$. In this case H_0 would be rejected in favor of H_1: $\mu \neq 50$ h. The less that is known about the situation, the more likely it is that H_0 will be accepted because the acceptance region band will be wider.

Problems P20.6–P20.9

20.4 Means Test for Two Independent Samples with Standard Deviation Known

If the means of two, *independently drawn* random samples are to be tested for equality or inequality and the standard deviations of both populations are

20.4 MEANS TEST FOR TWO SAMPLES WITH STANDARD DEVIATION KNOWN

known, the SND is used as a test statistic. The two-sided test for samples 1 and 2 is

$$H_0: \mu_1 = \mu_2 \quad \text{versus} \quad H_1: \mu_1 \neq \mu_2$$

The complete form of the test statistic is

$$z = \frac{(\overline{X}_1 - \overline{X}_2) - (\mu_1 - \mu_2)}{\sigma_d} \tag{20.8}$$

where σ_d is the *standard deviation of the difference* between two variables. According to rule 5 of Sec. 18.7, the variance of the difference equals the sum of the individual, independent variances. Since we are dealing with sample means in Eq. (20.8), σ_d is the square root of the sum of the respective variances of the means σ_i^2/n_i for $i = 1, 2$.

$$\sigma_d = \left(\frac{\text{sample 1 variance}}{\text{size of sample 1}} + \frac{\text{sample 2 variance}}{\text{size of sample 2}}\right)^{1/2}$$

$$= \left(\frac{\sigma_1^2}{n_1} + \frac{\sigma_2^2}{n_2}\right)^{1/2} \tag{20.9}$$

Under H_0 we have $\mu_1 - \mu_2 = 0$, so Eq. (20.8) reduces to the test statistic

$$z = \frac{\overline{X}_1 - \overline{X}_2}{\sigma_d} \tag{20.10}$$

where the $\overline{X}$ values are used as the best estimators of the population means.

If the σ_i values are unknown but n_1 and n_2 exceed 30, the samples are large enough to reliably estimate σ_d using sample s_i values. Equation (20.10) is still used as the test statistic with $s_i = \hat{\sigma}_i = \sigma_i$ in Eq. (20.9).

The OC curves using $\alpha = 0.05$ for these one-sided and two-sided normal tests are given in Figs. 20.1 and 20.2, respectively. The abscissa scale is

$$d = \frac{\mu_1 - \mu_2}{\sqrt{\sigma_1^2 + \sigma_2^2}} \tag{20.11}$$

where $\overline{X}_i$ is used to estimate μ_i. The n value is assumed to be the same for both samples; that is, $n_1 = n_2 = n$. If $n_1 \neq n_2$, select them so that the equivalent n satisfies, as closely as possible, the relation

$$n = \frac{\sigma_1^2 + \sigma_2^2}{\sigma_1^2/n_1 + \sigma_2^2/n_2} \tag{20.12}$$

Example 20.3 Two quality control engineers want to compare the average strength of a plastic made by similar processes at two plants. The processes are well established, and the σ values are known. The data is:

Process	n	$\bar{X}$ (lb/in²)	σ (lb/in²)
1	9	39	3
2	16	35	5

(a) Are the means equal?
(b) What are the required sample sizes to obtain $\alpha = 0.05$ and $\beta = 0.15$ to test the null hypothesis H_0: $\mu_1 - \mu_2 = 5$ lb/in² versus H_1: $\mu_1 - \mu_2 = 10$ lb/in². Perform this test using the same $\bar{X}$ values above and these required n values.

SOLUTION (a) The hypothesis testing steps are used. There are two independent samples, and the σ values are known, so Eq. (20.10) is the test statistic.

1. Two-sided test between the hypotheses

$$H_0: 39 = 35 \text{ lb/in}^2 \quad \text{versus} \quad H_1: 39 \neq 35 \text{ lb/in}^2$$

2. Let $\alpha = 0.05$, and β is unconfined.
3. Sample statistics are listed above. The value of σ_d by Eq. (20.9) is

$$\sigma_d = \left(\frac{3^2}{9} + \frac{5^2}{16}\right)^{1/2} = (2.5625)^{1/2} = 1.60$$

4. The test statistic value is

$$z = \frac{39 - 35}{1.60} = 2.50$$

5. The SND Table 11.2 gives the two-tail acceptance region for $\alpha = 0.05$ as $|z| \leq 1.96$.
6. Since $z = 2.50$ is in the critical region, reject H_0. Therefore, the process averages are not equal.

(b) The one-sided H_0: $\mu_1 - \mu_2 = 5$ (or $\mu_1 > \mu_2$ by 5 lb/in²) is a test that the difference in sample means is statistically equal to 5. See Fig. 20.6. Figure 20.1 is the appropriate OC curve. From Eq. (20.11), $d = 5/\sqrt{34} = 0.86$, and the required sample sizes are $n_1 = n_2 = 9$. To test H_0: $\mu_1 - \mu_2 = 5$ versus H_1: $\mu_1 - \mu_2 = 10$ lb/in², the σ_d value is

$$\sigma_d = \left(\frac{9}{9} + \frac{25}{9}\right)^{1/2} = (3.7778)^{1/2} = 1.94$$

and the test statistic is Eq. (20.8).

$$z = \frac{(39 - 35) - 5}{1.94} = -0.52$$

20.5 MEANS TEST FOR TWO SAMPLES WITH STANDARD DEVIATION UNKNOWN

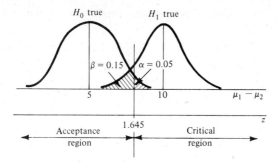

Figure 20.6 A one-sided test for the equality of two sample means, Example 20.3b.

The one-tail acceptance region for $\alpha = 0.05$ is $z \leq +1.645$. Accept the null hypothesis that the difference is 5 lb/in².

Problems P20.10–P20.14

20.5 Means Test for Two Independent Samples with Standard Deviation Unknown and Small Samples

If the population standard deviations σ_1 and σ_2 are not known, it is necessary to assume that they are either equal or unequal and estimate them from the samples. It is possible to determine if they are equal by using the F test of the next chapter prior to performing this test for means.

If *equality* is assumed and the samples are small (n_1 and $n_2 < 30$), the t distribution is used to test H_0: $\mu_1 = \mu_2$ against an alternative hypothesis. The t test statistic and degrees of freedom are

$$t = \frac{\bar{X}_1 - \bar{X}_2}{s_d} \qquad (20.13)$$

$$\nu = n_1 + n_2 - 2 \qquad (20.14)$$

The value of s_d is the standard deviation of the difference of means. The variance s_d^2 is computed as

$$s_d^2 = \frac{\text{combined sum of squares of residuals}}{\text{total degrees of freedom}} \cdot \frac{n_1 + n_2}{n_1 n_2}$$

$$= \frac{\Sigma(X_{i1} - \bar{X}_1)^2 + \Sigma(X_{i2} - \bar{X}_2)^2}{(n_1 - 1) + (n_2 - 1)} \cdot \frac{n_1 + n_2}{n_1 n_2} \qquad (20.15)$$

where X_{ij} ($j = 1, 2$) are the observed values for sample j. It is correct to substitute the unbiased sample variances, Eq. (3.4), in the numerator of Eq. (20.15), because

$$\Sigma(X_{i1} - \bar{X}_1)^2 + \Sigma(X_{i2} - \bar{X}_2)^2 = s_1^2(n_1 - 1) + s_2^2(n_2 - 1) \qquad (20.16)$$

Table 20.2 OC curve relations for two sample means tests with unknown, equal σ values

Alternative hypothesis H_1	Abscissa of OC curve	OC curve sample size	Required sample size
$H_1: \mu_1 \neq \mu_2$ (two-sided)	$d = \dfrac{|\mu_1 - \mu_2|}{2\sigma}$	$n' = 2n - 1$	$n = \dfrac{n' + 1}{2}$
$H_1: \mu_1 > \mu_2$ (one-sided)	$d = \dfrac{\mu_1 - \mu_2}{2\sigma}$	$n' = 2n - 1$	$n = \dfrac{n' + 1}{2}$
$H_1: \mu_2 > \mu_1$ (one-sided)	$d = \dfrac{\mu_2 - \mu_1}{2\sigma}$	$n' = 2n - 1$	$n = \dfrac{n' + 1}{2}$

The OC curves are those for one-sided and two-sided t tests (Figs. 20.4 and 20.5). The abscissa scale d and equivalent n value are shown in Table 20.2. The actual sample size n is used to compute n' for the OC curve assuming $n_1 = n_2 = n$. If $n_1 \neq n_2$, use Eq. (20.12) to compute an equivalent n value.

If population variances are *unequal*, the test must use an approximate t distribution. If there is no good reason to assume that $\sigma_1^2 = \sigma_2^2$, this test should be used. The test statistic and equivalent degrees of freedom are

$$t = \frac{\overline{X}_1 - \overline{X}_2}{\left(s_1^2/n_1 + s_2^2/n_2\right)^{1/2}} \qquad (20.17)$$

$$\nu = \frac{\left(s_1^2/n_1 + s_2^2/n_2\right)^2}{\left(s_1^2/n_1\right)^2/(n_1 + 1) + \left(s_2^2/n_2\right)^2/(n_2 + 1)} - 2 \qquad (20.18)$$

The t test OC curves are not reliable for this test. Example 20.4 illustrates the two tests above.

Example 20.4 Two sample means are to be tested for equality. In a previous test the means were $\overline{X}_1 = 50$ and $\overline{X}_2 = 61$.
(a) If the standard deviations are equal at $s = 3.7$, what sample sizes are needed to obtain $\alpha = 0.05$ and $\beta = 0.10$?
(b) The samples are taken, and the results below are obtained. Assume equal σ values and test for equal means.
(c) Rework part (b) but remove the assumption of equal σ values.

$$\overline{X}_1 = 54 \qquad \overline{X}_2 = 59$$
$$s_1 = 2.9 \qquad s_2 = 4.1$$

SOLUTION (a) The two-sided t test has the OC curve in Fig. 20.5. In Table

20.5 MEANS TEST FOR TWO SAMPLES WITH STANDARD DEVIATION UNKNOWN

20.2, we use the $\overline{X}_i$ values to estimate μ_i and s for σ. The abscissa scale is $d = |50 - 61|/2(3.7) = 1.49$. The equivalent sample size on the OC curve is $n' = 7$, so the required n for each sample is

$$n = \frac{7+1}{2} = 4$$

(b) The hypothesis testing steps are used.

1. H_0: $54 = 59$ versus H_1: $54 \neq 59$ and equal σ values.
2. $\alpha = 0.05$ and $\beta = 0.10$.
3. Equation (20.13) is the test statistic for equal σ values. Compute s_d and ν. Use Eqs. (20.16) and (20.15) for s_d.

$$s_d^2 = \left(\frac{2.9^2(3) + 4.1^2(3)}{3+3}\right)\left(\frac{4+4}{16}\right)$$
$$= (12.61)(0.5) = 6.31$$
$$s_d = 2.51$$
$$\nu = n_1 + n_2 - 2 = 6$$

4. The test statistic value is

$$t = \frac{54 - 59}{2.51} = -1.99$$

5. The acceptance region for a two-tail, $\alpha = 0.05$ t distribution with $\nu = 6$ is $|t| \leq 2.447$.
6. Do not reject H_0 since the computed t is in the acceptance region.

(c) If the σ values are not equal, the test statistic is Eq. (20.17) with ν from Eq. (20.18).

$$t = \frac{54 - 59}{[(2.9)^2/4 + (4.1)^2/4]^{1/2}} = -1.99$$
$$\nu = 7.001$$

For $\nu = 7$ the two-tail acceptance region is now $|t| \leq 2.365$. The H_0 of sample mean equality is still not rejected.

COMMENT In (c) it is possible to interpolate in the t table for a noninteger ν value. However, rounding to the nearest integer is common, unless the computed t value is on the borderline between acceptance and rejection.

Problems P20.15–P20.19

20.6 Means Test for Paired Samples

If the equality of means between two samples is tested, any difference that is present may be due to some factors other than the one of particular interest. For example, suppose the same product is manufactured at two plants and the mean electrical conductivity is tested for equality. If the means are statistically different, one reason might be the difference in the process. However, it is likely that some other aspects of the process—machinery, skill level, or raw materials—may also affect the difference. To test the difference in means resulting from certain treatments, such as the two processes, items from the same source are given the two treatments and the results are measured. The data is then from *paired* samples. The actual pairing is usually accomplished by splitting one item into two pieces, or, in the case of people, using twins as the test subjects.

The test described here is for paired data only—it should *not* be used for testing means of two independent samples. Conversely, the independent sample tests of Secs. 20.4 and 20.5 should not be used for testing paired-data hypotheses.

The paired-data t test is used to test the null hypothesis H_0: $\delta = \delta_0$, where δ is the *mean difference* between the samples. The value δ_0 is the hypothesized value of δ, and usually $\delta_0 = 0$; that is, there is no difference between samples that is caused by the treatments. The mean of the differences is $\bar{d}$, which is used to estimate δ. To obtain $\bar{d}$, compute the differences d_i ($i = 1, 2, \ldots, n$) from the data X_{i1}, and X_{i2}.

$$d_i = X_{i1} - X_{i2}$$

Then
$$\bar{d} = \frac{\Sigma d_i}{n} \tag{20.19}$$

The test statistic is like that of Eq. (20.7), except that the difference δ is used.

$$t = \frac{(\delta - \delta_0)\sqrt{n}}{s_d} \tag{20.20}$$

where $\bar{d} = \delta$. The standard deviation of the difference s_d is unknown, so it is computed by the unbiased formula for standard deviation:

$$s_d = \left(\frac{\Sigma d_i^2}{n-1} - \frac{n}{n-1} \bar{d}^2 \right)^{1/2} \tag{20.21}$$

The hypothesis testing procedure is used for this test.

The degrees of freedom for the t distribution are
$$v = n - 1 \tag{20.22}$$

because there are only n different sample items. This means that the precision of the paired t test is less than the two-sample test where $v = n_1 + n_2 - 2$. Therefore, the OC curve (Figs. 20.4 and 20.5) will show a larger β value for paired data. Note that the abscissa scale on the OC curve for a two-sided paired t test is

$$d = \frac{|\delta - \delta_0|}{s_d}$$

If the sample size is large, the same test statistic, Eq. (20.20), is used, except it is termed z and the SND is used to define the acceptance region.

Example 20.5 An experimental ballistic missile has a part in the guidance system that must be heat-treated. In order to determine whether two methods are equivalent, a metallurgist divided 12 parts and heat-treated each one at a temperature of 60°C or 79°C. The Rockwell hardness values are given in Table 20.3. Is it correct to state that the temperature has no effect on the Rockwell hardness number?

SOLUTION The answer to question 1 in Sec. 20.1 is two samples from one source. The paired t test is applicable here, and the hypothesis testing procedure is used.

1. The two-sided test for differences uses $\delta_0 = 0$ and takes the form

$$H_0: \delta = 0 \quad \text{versus} \quad H_1: \delta \neq 0$$

2. Let $\alpha = 0.05$.
3. The required sample statistics are n, $\bar{d}$, and s_d. From Table 20.3 and Eqs. (20.19) and (20.21)

$$n = 12 \quad \bar{d} = 2.17 \quad s_d = 3.30$$

4. From Eq. (20.20) with $\delta_0 = 0$

$$t = \frac{2.17\sqrt{12}}{3.30} = 2.28$$

Table 20.3 Computations for the paired t test on Rockwell hardness numbers, Example 20.5

Hardness number		Difference	
Sample 1 (60°C)	Sample 2 (79°C)	d_i	d_i^2
81	76	5	25
79	79	0	0
82	80	2	4
79	75	4	16
83	77	6	36
80	74	6	36
83	76	7	49
81	80	1	1
80	82	−2	4
77	79	−2	4
74	75	−1	1
81	81	0	0
		26	176

5. For a two-tail $\alpha = 0.05$ with $\nu = 12 - 1 = 11$, the acceptance region on the t distribution is $|t| \leq 2.201$.
6. Since 2.28 is not in the acceptance region, H_0 is rejected. The statement that temperature has no effect on the hardness number is incorrect because this test shows that $\delta \neq 0$.

COMMENT If the two-sample t test for small samples (Sec. 20.5) had been incorrectly used, the hypothesis H_0: $\mu_1 = \mu_2$ would not have been rejected. The pertinent statistics for Eqs. (20.13) through (20.16) are:

$$\overline{X}_1 = 80.00 \qquad s_1^2 = 6.54$$
$$\overline{X}_2 = 77.83 \qquad s_2^2 = 7.06$$
$$s_d = 1.06 \qquad t = 2.04$$

The acceptance region for $\nu = 12 + 12 - 2 = 22$ is $|t| \leq 2.074$. Not rejecting H_0 is a wrong decision because the data is paired and requires a test specifically designed for it.

Problems P20.20–P20.25

ADDITIONAL MATERIAL

Bowker and Lieberman [4], pp. 183–207, 225–246; Gibra [8], pp. 217–233, 250–261, 265–269; Hines and Montgomery [12], pp. 248–270; Kirkpatrick [15], pp. 201–217; Miller and Freund [18], pp. 211–221; Neville and Kennedy [19], Chap. 13; Volk [24], pp. 109–133.

PROBLEMS

The following situations are used by the indicated problems in this chapter.

A: The number of livestock processed per hour by a meat packing company has averaged 18.7 with a known σ of 1.50. A new method has been instituted and the rates per hour for today were 18, 24, 17, 22, 20, 18, 21, 23, 17.

Problems P20.1, P20.2, P20.6

B: The number of kilograms of pecans harvested per hour averaged 45 last season. A new harvesting method being used this year has averaged 50 kg/h with a standard deviation of 17 kg/h.

Problems P20.3, P20.4, P20.8

C: A city has decided to develop a computer graphics capability for its public works department. This capability includes the entry of the xy coordinates of utility line locations. This computer entry, called digitizing, was performed by two companies bidding for the digitizing contract. The accuracy of the utility line locations was recorded for each company using samples of size 8 for company A and 12 for company B. The accuracy data in meters is presented here:

Company A		Company B		
3.5	3.2	4.2	3.4	1.9
2.4	2.7	3.9	2.9	2.0
1.8	2.4	2.8	4.0	1.6
3.2	1.9	2.5	3.1	3.3
$\bar{X} = 2.64$ m		$\bar{X} = 2.97$ m		
$s = 0.63$ m		$s = 0.85$ m		

Problems P20.10, P20.16, P20.17

D: Cheddar cheese may be aged by two different processes. A new type of cheese is aged by taking a sample of 12 blocks, dividing each one and aging them by the two methods. The time in days to reach a specific degree of sharpness is:

Block number	Process 1	Process 2	Block number	Process 1	Process 2
1	60	73	7	70	69
2	72	59	8	68	58
3	59	62	9	61	81
4	64	70	10	73	64
5	80	66	11	59	68
6	64	58	12	65	65

Problems P20.20, P20.21

Note: Before you work any problem which involves the testing of means, answer the questions in Sec. 20.1.

P20.1 (a) For situation A determine which test statistic should be used to test if the mean rate per hour has changed from the standard for the old method.
 (b) Use this test to determine if the average has increased using $\alpha = 0.05$.
 (c) Use the sample results to place a two-sided 95 percent confidence interval on the population mean.

P20.2 The SND is used to test the hypotheses $H_0: \mu = 18.7$ versus $H_1: \mu \neq 18.7$ with $\alpha = 0.05$ for situation A.
 (a) What is the probability of not rejecting H_0 if the true mean is the observed sample mean?
 (b) By how much must the sample size change to obtain $\beta = \alpha = 0.05$ for this test?

P20.3 If the sample size in situation B is 60, is it correct to say that the new mean is different with a 98 percent level of confidence?

P20.4 What is the β value for the test performed in Prob. P20.3 using situation B?

P20.5 An aircraft manufacturer uses a maximum design life of 3500 flying hours for a particular component. If the sample standard deviation for 35 test components is 425 h and the average life is 3300 h, answer the following for $\alpha = 0.05$.

(a) Is the average component life less than the design life?

(b) What is the power of the test in (a) if the true mean life is 3300 h?

(c) What is the probability of concluding that the correct life is 3500 h in part (a) if in actuality the life is 3400 h? 3250 h?

(d) Based on the sample results, what is the one-sided, upper confidence interval for the mean life?

P20.6 (a) Determine if the sample mean given in situation A is greater than the population mean of 18.7 animals processed per hour. Assume that $\alpha = 0.05$ and that the historical $\sigma = 1.5$ is no longer known to be correct.

(b) Is this decision different from the one made in Prob. P20.1b?

(c) What is the value of β for the test in (a), and how does it compare with the β if the population σ had been known to equal the sample s value?

P20.7 A sample of 12 strength values was taken, and $\bar{X} = 3515$ lb/in^2 and $s = 395$ lb/in^2 computed. Because of misplaced records it is not known whether the shipment is from company D, which has a design strength of 3850 lb/in^2, or company M, which has a design strength of 3400 lb/in^2.

(a) Use a 10 percent significance level to determine which company the shipment is from.

(b) Answer the question in (a) by using the $CI_{0.90}$ on the population mean.

P20.8 For situation B if the sample size is 15 h and $s = 17$, answer the following using $\alpha = 0.05$.

(a) Has the average harvest rate increased from last season?

(b) What is the β value for this test?

(c) If the sample $s = 17$ remains constant, what observed harvest rate this year will result in $\beta = 0.20$? If this harvest rate is, in fact, observed in the sample of 15 h, will the decision reached in part (a) be different?

P20.9 If you want to determine whether a sample mean μ is equal to a specific value μ_0, either the t or the z statistic is used. Detail all the conditions under which each is used.

P20.10 For the accuracy data in situation C, assume that the data is known to be from normal populations with variances $\sigma_A^2 = 0.5$ and $\sigma_B^2 = 0.6$ m^2. Provided this variation difference does not concern the city management, determine if the accuracy data is from populations with the same mean. Use a 98 percent confidence to make this conclusion.

P20.11 Production data has been collected at two offices for the last 45 workdays by Jane Jenkins, the manager of the branch offices for Pay-Too-Much, a tax return preparation firm. The results are:

Statistic	Northside office	Southside office
Average returns per day	75	68
Standard deviation of daily rate	10.3	12.4

(a) Are the mean production rates equal or does the Northside office complete more returns? Use a 5 percent significance level to make this conclusion.

(b) What is the probability of accepting an H_0 of equal population means if, in fact, the two offices have a production rate difference of 5 returns per day? Let $\alpha = 0.05$.

P20.12 A machined dimension is known to be normally distributed with $\sigma = 0.08$ mm. An operator on the dayshift and one on the nightshift use the same machine to produce this dimension. The industrial engineer in the department wants to test for differences in population means. Now, $\alpha = 0.05$ and $\beta = 0.10$ are required to test H_0: $\mu_1 = \mu_2$ versus H_1: $\mu_1 - \mu_2 = 0.085$, where μ_i is the average dimension for operator i ($i = 1, 2$).

(a) Determine the required sample sizes if H_1 is actually correct.

(b) Perform the test assuming the required samples are taken to obtain $\bar{X}_1 = 12.40$ mm and $\bar{X}_2 = 12.36$ mm.

P20.13 Rework Prob. P20.12 with the test being $H_0: \mu_1 - \mu_2 = 0.020$ versus $H_1: \mu_1 - \mu_2 \neq 0.020$ if in actuality $\mu_1 - \mu_2 = 0.085$. Is this conclusion about population means the same as in Prob. P20.12?

P20.14 Mileage samples are taken to determine if population means are equal.

(a) Use the results below to answer the questions in Sec. 20.1 and test if population means are equal with $\alpha = 0.05$.

(b) Determine the sample sizes required to obtain $\beta = 0.20$ and $\alpha = 0.05$ if the true difference is 2 km/h. 4 km/h.

Sample 1	Sample 2
$\bar{X}_1 = 35$ km/h	$\bar{X}_2 = 36.5$ km/h
$s_1^2 = 10$	$s_2^2 = 10$
$n_1 = 40$	$n_2 = 55$

P20.15 Test the hypothesis that population means are equal for the sample mean and variance data of Prob. P20.14 assuming that the variances are equal and that $n_1 = 8$ and $n_2 = 15$. Use $\alpha = 0.02$.

P20.16 For situation C determine if company A is more accurate than company B. Do not assume that population variances are equal, and use a confidence level of 95 percent.

P20.17 (a) Would the conclusion in Prob. P20.16 be any different had the population variances been assumed equal?

(b) What is the type II error probability for this problem (equal variances) if the true difference in accuracy is 0.5 m and the best estimate of population standard deviation is 0.77 m?

(c) If sample sizes are assumed equal, what size is required to obtain $\alpha = 0.05$ and $\beta = 0.20$ in part (b)?

P20.18 The thickness of a paint coating had a standard of 1.2 mm set some time ago. Victor feels that an old machine applies a thinner coat than a newly acquired machine and that both apply coats thicker than the standard. Nothing is known about the population of paint thickness, so the data was collected at random times. Use $\alpha = 0.05$ to help Victor make statistical conclusions about (a) any differences in old and new mean thickness and (b) the mean thickness of both machines compared to the standard of 1.2 mm.

Old machine		New machine		
1.4	1.7	2.0	2.4	1.8
1.0	1.1	1.6	1.0	1.7
1.8	1.4	1.5	1.2	
2.2		2.2	1.4	

P20.19 The number of inoperable cars was recorded for 5 days at two government carpool lots of approximately equal size.

(a) If population variances are assumed to be equal, determine if the mean number of inoperable cars is the same for the lots. Let $\alpha = 0.05$.

Lot A: 5 12 13 19 21
Lot B: 7   7 14 15 23

(b) Determine the value of β for this test if the true difference is four cars and the best estimate of σ is the s for the 10 recorded values.

P20.20 Test the hypothesis that two different cheese aging processes are equivalent in that the mean time to reach a specific degree of sharpness is the same for both. Use the data in situation D, and let $1 - \alpha = 0.80$.

P20.21 (a) Assume the same test as in situation D had been performed on a total of 24 separate blocks of cheese. Use the t test for equal variances to determine if there is a difference in mean aging time to reach a specific degree of sharpness. Let $1 - \alpha = 0.80$.
(b) Is this conclusion different from that of Prob. P20.20?

P20.22 The learning time to perform a certain task was recorded for 40 pairs of twin dogs. One member of a twin set was taught by trainer A, and the other by trainer B. The following difference data was computed.

$$\sum_{i=1}^{40} d_i = -28.4 \text{ h} \qquad \sum_{i=1}^{40} d_i^2 = 132.4 \text{ h}^2$$

where $d_i = X_{iA} - X_{iB}$ and X_{ij} ($j = A$ or B) is the time to train a dog by trainer A or B. Is it correct to state with 95 percent confidence that trainer A is better than B? Use the hypothesis testing procedure to perform this test.

P20.23 The percentage of incoming product that is defective can be determined by one of three methods: chemical, destructive testing, or observation during isolation. To determine the method that is best, five samples of 100 each were divided into thirds and the percentage of defectives recorded for each method. The manufacturer, who uses the chemical method for testing, states that the percentage of defectives for this shipment is 8 percent. The observed percentages are:

	Percent defective		
Sample	Chemical	Destructive	Isolation
1	10	9	6
2	6	8	4
3	13	8	8
4	5	9	9
5	3	10	5

Determine the following for $\alpha = 0.05$.
(a) Is the average percentage defective by the chemical method different from the manufacturer's stated rate of 8 percent?
(b) Is there a difference between the percentage defective rates observed by the chemical and isolation methods?
(c) Quality assurance management has decided to use the destructive testing method because the standard deviation of sample percentage of defectives is smallest. What conclusions can you make about the mean percentage of defectives of this versus the other methods?

P20.24 Summarize the conditions under which the mean test for paired data should be used.

P20.25 George wants to test for the equality of population means for a dimension that is chemically etched by two different processes. Martha states that the paired-data test of means must be used, while George states that it is impossible to observe the same dimension produced by two processes on the same part, so the two-independent-sample test of means should be used. Both Martha and George know that the population variances must be estimated from sample data. Discuss the situation to your satisfaction and determine the conditions under which each of the tests should be used.

CHAPTER TWENTY-ONE

STATISTICAL INFERENCES FOR VARIANCES

The statistical tests used to evaluate variances or standard deviations are explained in this chapter. The variances rather than standard deviations are tested because the test statistics are based on the χ^2 and F distributions, which involve σ^2. The OC curves for these tests are presented.

CRITERIA

To correctly perform a statistical test for standard deviations or variances, you must be able to:

1. State the questions that must be answered to select the appropriate variance test.
2. Use the hypothesis testing procedure to compare a sample variance value to a specific variance value, given the sample size and variance values.
3. Use the hypothesis testing procedure to compare two sample variances, given the sample sizes and variance values.
*4. State what happens to the probability of a type II error when population variances become unequal and there is a constant difference between means.

STUDY GUIDE

21.1 Types of Tests for Variances

In all two-sample tests for means (Chap. 20) some assumption about the equality or inequality of σ^2 values was necessary. The tests of this chapter are used to determine the relation between the variances of two populations.

All the tests for variances use either the χ^2 or the F statistic.

$$\chi^2 = \frac{(n-1)s^2}{\sigma^2} \quad (21.1)$$

$$F = \frac{s_1^2}{s_2^2} \quad (21.2)$$

It is assumed that samples are taken from an $N(\mu, \sigma^2)$ population where μ is not necessarily known. The following questions are used to determine the appropriate test statistic.

1. *Are there one or two samples of data?* If one sample, see Sec. 21.2; if two samples, see Sec. 21.3.
2. *Does the data come from a normal population?* If yes, you can proceed; but if not, an $n \geq 30$ is needed to reliably estimate the variances.

21.2 Variance Test for One Sample

If a sample variance should equal a specific value σ_0^2, the null hypothesis is H_0: $\sigma^2 = \sigma_0^2$ and any one- or two-sided H_1 is testable. The σ_0^2 may be a standard design value, historically observed, or a desired value.

The test statistic is the χ^2 variable with $\nu = n - 1$ degrees of freedom:

$$\chi^2 = \frac{(n-1)s^2}{\sigma_0^2} \quad (21.3)$$

The unbiased sample variance s^2 is computed by Eq. (3.9). The hypothesis testing procedure of Chap. 19 is used to perform this one-sample test.

The OC curves for this test use the ratio of the actual σ value to σ_0 for the abscissa scale. Since σ is usually estimated by s, the abscissa value is actually

$$\lambda = \frac{s}{\sigma_0} \quad (21.4)$$

The OC curves, which are different for each form of H_1, are shown in Figs. 21.1 to 21.3 for $\alpha = 0.05$.

Two-sided:	$H_1: \sigma^2 \neq \sigma_0^2$	(Fig. 21.1)
One-sided lower:	$H_1: \sigma^2 < \sigma_0^2$	(Fig. 21.2)
One-sided upper:	$H_1: \sigma^2 > \sigma_0^2$	(Fig. 21.3)

Example 21.1 The flow rate through a 7.5-cm nozzle has a variance of 0.20 ft^3/s. A new design was tested 30 separate times. If the sample variance was 0.28 ft^3/s, perform (*a*) a one-sided test and (*b*) a two-sided test with $\alpha = 0.05$. (*c*) Determine the probability of a type II error if the true variance is 0.28 ft^3/s.

21.2 VARIANCE TEST FOR ONE SAMPLE

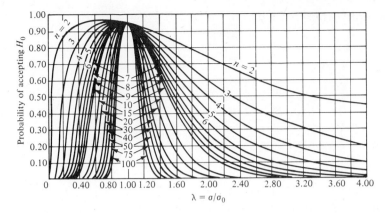

Figure 21.1 OC curves for a *two-sided chi-square* test with $\alpha = 0.05$. (Reproduced with permission from *Engineering Statistics*, 2d ed., by Bowker, A. H., and Lieberman, G. J., Prentice-Hall, Inc., 1972.)

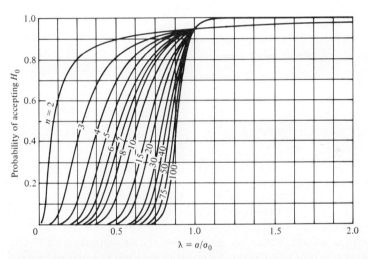

Figure 21.2 OC curves for a *one-sided, lower chi-square* test (H_1: $\sigma^2 < \sigma_0^2$) with $\alpha = 0.05$. (Reproduced with permission from "Operating Characteristics of the Common Statistical Tests of Significance" by Ferris, C. D., Grubbs, F. E., and Weaver, C. L., *The Annals of Mathematical Statistics*, Vol. 17, No. 2, June 1946.)

SOLUTION The questions in Sec. 21.1 are answered first. There is one sample, and $n = 30$ ensures that $s^2 = 0.28$ ft^3/s is a good variance estimate.
(*a*) Use the testing procedure of Sec. 19.4.

1. The one-sided, upper test hypotheses are

$$H_0: \sigma^2 = 0.20 \quad \text{versus} \quad H_1: \sigma^2 > 0.20$$

2. $\alpha = 0.05$ and β is not specified.

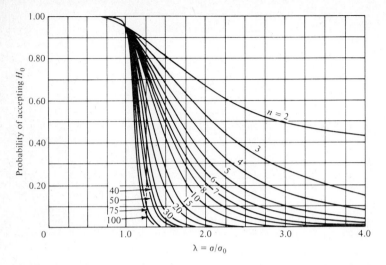

Figure 21.3 OC curves for a *one-sided, upper chi-square* test (H_1: $\sigma^2 > \sigma_0^2$) with $\alpha = 0.05$. (Reproduced with permission from "Operating Characteristics of the Common Statistical Tests of Significance" by Ferris, C. D., Grubbs, F. E., and Weaver, C. L., *The Annals of Mathematical Statistics*, Vol. 17, No. 2, June 1946.)

3. The required statistics and parameters are

$$s^2 = 0.28 \qquad \sigma_0^2 = 0.20 \qquad \nu = 29$$

4. Equation (21.3) is the test statistic:

$$\chi^2 = \frac{29(0.28)}{0.20} = 40.60$$

5. The acceptance region for $\nu = 29$ degrees of freedom includes all $\chi^2(29)$ values not in the upper tail with $\alpha = 0.05$. From Table B-4 this is $\chi^2 \leq 42.56$.
6. Since $\chi^2 = 40.60$ is in the acceptance region, we cannot reject H_0; so we conclude that the nozzles have equal flow-rate variances. Figure 21.4a summarizes this test.

(b) The two-sided test is

$$H_0: \sigma^2 = 0.20 \qquad \text{versus} \qquad H_1: \sigma^2 \neq 0.20$$

The acceptance region (step 5), which now places $\alpha/2 = 0.025$ in each tail, is $16.05 \leq \chi^2 \leq 45.72$. The test statistic 40.60 is again in this range, so H_0 is accepted. See Fig. 21.4b.

(c) From Eq. (21.4), $\lambda = \sqrt{0.28/0.20} = 1.2$. For the one-sided, upper test Fig. 21.3 shows that $\beta = 0.60$ while the two-sided test has $\beta = 0.70$ (Fig. 21.1).

21.3 VARIANCE TEST FOR TWO SAMPLES

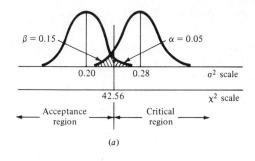

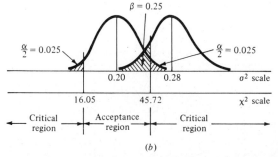

Figure 21.4 Graphical representation of (a) one-sided and (b) two-sided χ^2 tests for variance, Example 21.1.

COMMENT Since it was not known for sure that the sample came from a normal population, the s^2 value was used and the normality assumption made. If study shows the population of flow rates for the new nozzle to be nonnormal, these test results are possibly incorrect.

Problems P21.1–P21.8

21.3 Variance Test for Two Samples

To test for the equality of variances from two normal populations, the F statistic

$$F = \frac{s_1^2/\sigma_1^2}{s_2^2/\sigma_2^2}$$

is used. Under the null hypothesis H_0: $\sigma_1^2 = \sigma_2^2$ the test statistic reduces to

$$F = \frac{s_1^2}{s_2^2} \tag{21.5}$$

which has the distribution $F(\nu_1, \nu_2)$, where

$$\nu_1 = n_1 - 1 \quad \text{and} \quad \nu_2 = n_2 - 1 \tag{21.6}$$

If you plan to test for the equality of two sample means with equal variances, the two-sided H_1: $\sigma_1^2 \neq \sigma_2^2$ must be rejected. The acceptance region has $\alpha/2$ of the area in each tail of the $F(\nu_1, \nu_2)$ distribution. To determine the left limit of the acceptance region use Eq. (15.2). This test is illustrated in Example 21.2.

For a one-sided test, define s_1^2 as the *larger* sample variance, and the alternative hypothesis is always $H_1: \sigma_1^2 > \sigma_2^2$. The test statistic $F \geq 1$ is then

$$F = \frac{\text{larger variance}}{\text{smaller variance}} = \frac{s_1^2}{s_2^2} \tag{21.7}$$

The OC curves use the ratio of standard deviations for the abscissa. This may be expressed as

$$\lambda = \frac{\sigma_1}{\sigma_2} \tag{21.8}$$

The two-sided and one-sided OC curves for $\alpha = 0.05$ are given in Figs. 21.5 and 21.6, respectively. These curves assume that $n_1 = n_2 = n$.

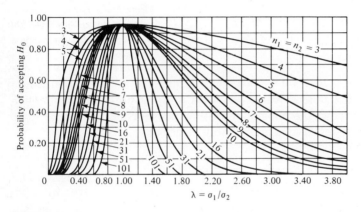

Figure 21.5 OC curves for a *two-sided F* test with $\alpha = 0.05$. (Reproduced with permission from *Engineering Statistics*, 2d ed., by Bowker, A. H., and Lieberman, G. J., Prentice-Hall, Inc., 1972.)

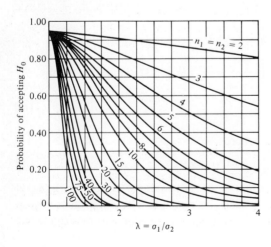

Figure 21.6 OC curves for a *one-sided F* test with $\alpha = 0.05$. (Reproduced with permission from "Operating Characteristics of the Common Statistical Tests of Significance" by Ferris, C. D., Grubbs, F. E., and Weaver, C. L., *The Annals of Mathematical Statistics*, Vol. 17, No. 2, June 1946.)

Example 21.2 Two new methods of measuring chemical reaction temperature are tested. Samples of size 15 and 20 are taken, and the time to determine the correct temperature is recorded. This time is expected to be normally distributed. The variances of the times are 0.15 s² for $n = 15$ and 0.20 s² for $n = 20$. Test for variance equality using (a) two-sided and (b) one-sided tests.

SOLUTION Since there are two samples from normal distributions, the F statistic is used. Define the larger variance as $s_1^2 = 0.20$; then $s_2^2 = 0.15$.

(a) 1. H_0: $\sigma_1^2 = \sigma_2^2$ versus H_1: $\sigma_1^2 \neq \sigma_2^2$.
2. Select $\alpha = 0.10$ with β not specified.
3. The degrees of freedom are $\nu_1 = 19$ and $\nu_2 = 14$.
4. The test statistic is Eq. (21.5):

$$F = \frac{0.20}{0.15} = 1.33$$

5. The acceptance region for $F(19, 14)$ with $\alpha/2 = 0.05$ in each tail is $0.44 \leq F \leq 2.40$, where the lower limit is found by Eq. (15.2):

$$\frac{1}{F(14, 19)} = \frac{1}{2.26} = 0.44$$

6. Since $F = 1.33$ is in the acceptance region, we conclude the variances are equal.

(b) The one-sided test is

$$H_0: \sigma_1^2 = \sigma_2^2 \quad \text{versus} \quad H_1: \sigma_1^2 > \sigma_2^2$$

The acceptance region for $\alpha = 0.10$ is $1.00 \leq F \leq 1.97$, which includes 1.33. Again H_0 is accepted (not rejected).

COMMENT Remember that the OC curves for this test assume that the sample sizes are equal. For example, if $\alpha = 0.05$ and $\beta = 0.10$ are desired in this problem, the value $\lambda = \sqrt{1.33} = 1.15$ shows that n_1 and n_2 are equal and much greater than 100 for the one-sided or two-sided test (Figs. 21.5 and 21.6).

Problems P21.9–P21.14

*21.4 Relation between the Tests for Means and Variances

It is necessary to know if two variances are equal prior to testing for the equality of means because different statistics are used. Figure 21.7a shows two populations with the means 10 units apart. If the σ^2 values are estimated at 5 and assumed equal, the OC curve for a two-sided means test has $\alpha = 0.05$ and

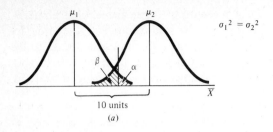

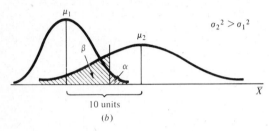

Figure 21.7 Graph showing the increase in β when one population variance exceeds another.

$\beta = 0.20$ for $n = 10$ [Fig. 20.5 with $d = 10/2(5) = 1$]. This implies that the difference in means of 10 units is quite easily detected.

Now, assume that the second population variance is larger than the first; that is, $H_1: \sigma_2^2 > \sigma_1^2$ is accepted in the F test. Figure 21.7b shows that for the same $\alpha = 0.05$ and a 10-unit difference, the type II error probability β is much larger because the second distribution has a larger overlap onto the first. Thus it is much easier to accept the false $H_0: \mu_1 = \mu_2$ (type II error) and not detect the actual difference between means when variances are unequal. This difficulty is overcome to a degree by taking larger samples to get better σ^2 estimates and using the adjusted t statistic, Eq. (20.17), for unknown and unequal variances.

Performing the F test for variances prior to the t test for means is convenient because the F test makes no assumptions about the population means but the t test does require assumptions about the variances.

Problem P21.15

ADDITIONAL MATERIAL

Bowker and Lieberman [4], pp. 207–217, 254–265; Gibra [8], pp. 233–237, 261–265; Hines and Montgomery [12], pp. 270–275; Kirkpatrick [15], pp. 218–231; Miller and Freund [18], pp. 233–238; Neville and Kennedy [19], Chap. 14.

PROBLEMS

P21.1 The life of car batteries, which is normally distributed, was recorded and a sample variance of 60 h^2 computed for 16 batteries. If the historical value of 45 h^2 was expected, use a 5 percent significance level to determine (a) answers to the questions in Sec. 21.1, (b) whether the variance has increased, and (c) the value of β for this test if the actual variance is 60 h^2.

P21.2 The noise level in a manufacturing plant was recorded 10 times. The decibel values were 72, 89, 94, 76, 88, 74, 95, 86, 89, and 96. If the standard is $\sigma = 6$ db and samples are from a normal population, use $\alpha = 0.05$ to answer the following:
(a) Does the observed standard deviation equal the standard?
(b) What is the value of β for the test above if the actual value is $\sigma = 7$ db?
(c) What sample size is required to obtain $\beta = 0.10$ for a two-sided test of variance if the true σ value is 5.5 db?
(d) What are the limits of the two-sided confidence interval on standard deviation using the sample data above?

P21.3 The reliability of an electronic component was measured 15 times and the statistics $\bar{X} = 0.94$ and $s = 0.03$ computed. The government contract under which the component is made requires that the reliability have the population parameters $\mu = 0.96$ and $\sigma = 0.05$. Use $\alpha = 0.05$ to test the sample mean and standard deviation against these standards using (a) the two-sided tests and (b) the appropriate one-sided tests. Assume that the reliability is normally distributed.

P21.4 Test the hypothesis that $\sigma = 0.05$ percent for the percentage defective of an incoming inspection if the sample variance is 0.0018 (percent)^2 for a sample of 31 items. Use $\alpha = 0.05$.

P21.5 A bowler has an average of 235 points per game. During the season she has had a standard deviation of 26. The game scores are normally distributed, and her scores for an eight-game tournament are:

246	258	283	241
212	289	231	192

(a) Has the σ value increased? Let $\alpha = 0.05$.
(b) What is the probability of concluding that $\sigma_0 = 26$ is correct if, in fact, it has decreased to 20 points and a one-sided lower H_0 is used?
(c) Determine the required sample size to obtain a type I error probability of 0.05 and a type II error probability of 0.20 for the situation in (b).

P21.6 Use the hypothesis testing procedure and a significance level of 10 percent in Prob. P13.7 to determine if $\sigma^2 = 0.25$ (percent)^2 or $\sigma^2 > 0.25$ (percent)^2. Use the normal approximation to the χ^2 distribution.

P21.7 The design variance for a mechanical part is 0.0025 mm^2. The last 150 parts were measured, and the sample variance was 0.0032 mm^2. Determine if the sample and design variances are equal with a 98 percent level of confidence.

P21.8 For Prob. P21.7 a 95 percent confidence is required and a χ^2 test is utilized to test H_0: $\sigma^2 = \sigma_0^2$ versus H_1: $\sigma^2 \neq \sigma_0^2$.
(a) Determine if the sample size is sufficiently large to result in $\beta \leq 0.20$ provided the actual value is $\sigma^2 = 0.0032$ mm^2.
(b) Estimate the power of the test if the actual variance value is 0.0020 mm^2.

P21.9 This sample data was obtained when two samples were taken from normal populations:

Sample size	Average	Standard deviation
20	6.8°C	0.39°C
12	7.3°C	0.29°C

(a) Test the hypothesis of equal versus unequal variances for $\alpha = 0.20$.
(b) Are the sample sizes large enough to result in a β value of 0.30 or less and an α value of 0.05 provided that the sample standard deviation values are good estimates of the population σ values?

P21.10 Determine if the correct test for means was used in Prob. P20.16 by using the F test for variances and the hypothesis testing procedure.

P21.11 The mean thicknesses of paint coats were tested for equality in Prob. P20.18 with the assumption that population variances are unknown and unequal. Determine if the variances are actually unequal using a significance level of 0.10.

P21.12 Test the assumption of equal population variances that was made in Prob. P20.19 using $\alpha = 0.05$, a one-sided test, and the assumption that the number of inoperable cars is normally distributed.

P21.13 A landscape architect wants to determine how much people are willing to pay to have a regular-size urban lot well landscaped. To do this, he has collected data from two Southwestern cities.

City A	City B
$n_A = 25$	$n_B = 25$
$\bar{X}_A = \$1575$	$\bar{X}_B = \$1725$
$s_A = \$250$	$s_B = \$150$

If landscape costs are assumed to be normally distributed, do the following.
    (a) Test if the variance of landscaping charges is greater in city A than in city B with $\alpha = 0.01$.
    (b) Find the value of β for the test above when $\alpha = 0.05$.
    (c) Perform the correct test for equality of means for this data using $\alpha = 0.05$.

P21.14 Random samples of sizes 15 and 6 are taken from machines I and II, respectively, by a quality assurance technician. The product's dimension is normally distributed. The technician has concluded that since the sample from machine II is smaller, the variance will probably be larger, so she believes that the sample statistics $s_I^2 = 0.29$ cm^2 and $s_{II}^2 = 0.40$ cm^2 bear out this tendency.
    (a) Use a significance level of 5 percent to determine if machine II has a larger population variance.
    (b) If you took samples of size 15 each from the product of machines I and II and used the sample variances above as correct values of the normal population variances, is the probability of accepting H_0: $\sigma_I^2 = \sigma_{II}^2$ less than or greater than 0.20?

*P21.15 A chemical engineer has done a thorough study of vacuum pressures in two vessels and determined that there is a difference of approximately 1 cm of mercury in the mean vacuum levels. A consultant has performed a test of means, assuming equal variances, on the results of independent samples of size 21 taken from each vessel and computed $s_1^2 = 16$ and $s_2^2 = 8$ cm^2. The actual normal population values for the two vessels are:

Vessel 1	Vessel 2
$\mu_1 = 10$ cm	$\mu_2 = 11$ cm
$\sigma_1^2 = 16$ cm^2	$\sigma_2^2 = 8$ cm^2

At a level of $\alpha = 0.05$, the value $F = 16/8 = 2.0$ is used to accept the equality-of-variance hypothesis H_0: $\sigma_1^2 = \sigma_2^2$. Now the test of means for H_0: $\mu_1 = \mu_2$ versus H_1: $\mu_1 \neq \mu_2$ can be carried out.
    (a) If the actual difference between population means is 1 cm, as determined by the chemical engineer, and if $\alpha = 0.05$, what is the probability of accepting the hypothesis H_0: $\mu_1 = \mu_2$? Let $\hat{\sigma} = s_d$.
    (b) If the variance of vessel 1 has, in fact, increased to $\sigma_1^2 = 24$ cm^2 just prior to the time the consultant performed the test of means, first determine if the hypothesis H_0: $\sigma_1^2 = \sigma_2^2$ is accepted with $\alpha = 0.05$ and with the sample variance $s_1^2 = 24$ cm^2 observed. Second, if the consultant uses the test for means in which equal variances are assumed and a 1-cm difference in means is still present, determine the probability that the hypothesis H_0: $\mu_1 = \mu_2$ is accepted.
    (c) Summarize the results from the two preceding parts in words and by a drawing similar to Fig. 21.7.

CHAPTER TWENTY-TWO

STATISTICAL INFERENCES FOR PROPORTIONS

The statistical tests used to evaluate the proportion of a binomial distribution are discussed in this chapter. The tests explained assume that the samples are large enough that the normal approximation to the binomial may be used.

CRITERIA

To complete this chapter, you must be able to:

1. State the condition under which the tests in this chapter can be accurately used.
2. Use the hypothesis testing procedure to compare a sample proportion value to a specific value, given the sample size and proportion values.
3. Use the hypothesis testing procedure to compare two sample proportion values, given the sample sizes and either the known population proportions or the data to estimate the proportion value.

STUDY GUIDE

22.1 Types of Tests for Proportions

The normal approximation to the binomial (Sec. 11.8) is usually accurate if $np > 5$ and $nq > 5$. This general guide also applies to the testing of population proportions for one or two samples using the normal statistic

$$z = \frac{\text{value} - \text{mean}}{\text{standard deviation}}$$

If the samples are too small to use the normal approximation, the binomial distribution must be used (see the text by Kirkpatrick).

If the normal approximation condition is not violated, the only question to ask before testing a proportion is, *Are there one or two samples of data?* If there is one, see Sec. 22.2; and if there are two, see Sec. 22.3.

22.2 Proportion Test for One Sample

If a sample proportion is supposed to be statistically equal to a specific value p_0, the null hypothesis is H_0: $\pi = p_0$. Any one- or two-sided alternative hypothesis is testable.

The normal approximation test statistic is

$$z = \frac{p - p_0}{\sigma_p} \qquad (22.1)$$

where p is computed from the sample and

$$\sigma_p = \sqrt{\frac{p_0(1 - p_0)}{n}} \qquad (22.2)$$

since under the null hypothesis the true proportion value is p_0. Therefore, each time a new π value is hypothesized for a population, the standard deviation changes. This test is illustrated below using the hypothesis testing procedure.

It is easy to extend the statistic of Eq. (22.1) so that the number, rather than the proportion, is tested. Since the mean and standard deviation for the binomial distribution are np and $\sqrt{np(1 - p)}$, respectively, the test statistic is

$$z = \frac{np - np_0}{\sqrt{np_0(1 - p_0)}} \qquad (22.3)$$

This statistic is applied in the Solved Problem.

Example 22.1 Historically the fraction defective of a product manufactured by the A.B. Sea Company has been 0.02. Five batches sent to a contractor were sampled ($n = 500$) and found to average 0.01 defective. On this basis the contractor plans to sample in the future, expecting to observe this reduced value of fraction defective.
(a) Use $\alpha = 0.05$ to statistically determine if the quality has improved.
(b) Determine the proportion value at which the decision in (a) will change.

SOLUTION The normal approximation test can be used because $np = 500(0.02) = 10$, which exceeds 5. The statistic of Eq. (22.1) is correct because the five batch proportions are averaged and compared to a standard of 0.02.
(a) The hypothesis testing steps are used.

1. This is a one-sided test of

$$H_0: \pi = 0.02 \quad \text{versus} \quad H_1: \pi < 0.02$$

2. $\alpha = 0.05$ is specified.
3. The observed proportion value is $p = 0.01$, and from Eq. (22.2) with $p_0 = 0.02$

$$\sigma_p = \sqrt{\frac{0.02(0.98)}{500}} = 0.0063$$

4. The test statistic value is

$$z = \frac{0.01 - 0.02}{0.0063} = -1.59$$

5. From Table 11.2, the acceptance region is $z \geq -1.645$ for $\alpha = 0.05$.
6. Since $z = -1.59$ is in the acceptance region, H_0 is not rejected. Therefore, the proportion value is not statistically less than 0.02.

(b) When z is less than -1.645, the hypothesis H_1: $\pi < 0.02$ is accepted. We can use Eq. (22.1) to solve for p with $p_0 = 0.02$ to determine a value at which H_1 is accepted.

$$z = \frac{p - 0.02}{0.0063} = -1.645$$

$$p = -1.645(0.0063) + 0.02 = 0.0096$$

Only when the fraction is $p < 0.0096$ can it be stated that the quality has improved.

Example 22.3

Problems P22.1–P22.8

22.3 Proportion Test for Two Samples

To determine if two proportion values are from the same population, the null hypothesis is H_0: $\pi_1 = \pi_2$. The complete test statistic is

$$z = \frac{p_1 - p_2 - (\pi_1 - \pi_2)}{\sigma_{p_1-p_2}} \qquad (22.4)$$

where $\sigma_{p_1-p_2}$ is the *standard deviation of the differences* between the two proportions. Again this is a normal approximation to the binomial and is accurate only if $np_i > 5$ and $nq_i > 5$ for $i = 1, 2$. Under the null hypothesis $\pi_1 - \pi_2 = 0$, the test statistic reduces to

$$z = \frac{p_1 - p_2}{\sigma_{p_1-p_2}} \qquad (22.5)$$

The procedure to determine the value of $\sigma_{p_1-p_2}$ is different if π_1 and π_2 are known rather than estimated from the sample data. *If the values are known* from experience, or by some other means, rule 5 of Sec. 18.7 is applied to compute the variance of the difference of two independent variables. The variances are added to obtain

$$\sigma_{p_1-p_2} = \left[\frac{\pi_1(1-\pi_1)}{n_1} + \frac{\pi_2(1-\pi_2)}{n_2}\right]^{1/2} \qquad (22.6)$$

Each component in Eq. (22.6) is the variance of the fraction defective distribution, as discussed in Sec. 8.8.

If the values of π_1 and π_2 are not known, one estimate of π is obtained by combining the two samples and computing p. Then the estimate of $\sigma_{p_1-p_2}$ is $s_{p_1-p_2}$, where $q = 1 - p$.

$$s_{p_1-p_2} = \left(\frac{pq}{n_1} + \frac{pq}{n_2}\right)^{1/2}$$

$$s_{p_1-p_2} = \left(pq\frac{n_1 + n_2}{n_1 n_2}\right)^{1/2} \quad (22.7)$$

This combining of samples 1 and 2 is possible because H_0 hypothesizes that $\pi_1 = \pi_2$.

Example 22.2 Eight-meter sections of copper wire produced at two different plants have a maximum resistance specification of 1 ohm (Ω). Historically the defective rates at each plant have been statistically equal with 10 percent at plant A and 11 percent at plant B not meeting this specification. Recent samples of 150 and 100 wires at plants A and B, respectively, each had 12 wires that did not meet the specification.
(a) Are the defective rates still statistically equal?
(b) If the historical defective rates are neglected, is the decision in part (a) different?

SOLUTION (a) The two-sample statistic in Eq. (22.5) is correct for this test. The observed defective rates are $p_1 = 12/150 = 0.08$ and $p_2 = 12/100 = 0.12$.

1. H_0: $\pi_1 = \pi_2$ versus H_1: $\pi_1 \neq \pi_2$.
2. Let $\alpha = 0.05$.
3. The historic values of $\pi_1 = 0.10$ and $\pi_2 = 0.11$ are *known* and used in Eq. (22.6) to compute $\sigma_{p_1-p_2}$:

$$\sigma_{p_1-p_2} = \left[\frac{(0.1)(0.9)}{150} + \frac{(0.11)(0.89)}{100}\right]^{1/2} = (0.0016)^{1/2} = 0.04$$

4. From Eq. (22.5) the test statistic is

$$z = \frac{0.08 - 0.12}{0.040} = -1.0$$

5. The acceptance region for $\alpha = 0.05$ and the SND is $|z| \leq 1.96$.
6. Since -1.0 is in the acceptance region, do not reject H_0. Based on these samples, the plants still have statistically equal defective rates.

(b) If the historic defective rates are neglected, Eq. (22.7) is used to estimate $\sigma_{p_1-p_2}$. The value p is computed from the combined sample data.

$$p = \frac{12 + 12}{150 + 100} = 0.096$$

Then $s_{p_1-p_2} = \left[(0.096)(0.904)\left(\frac{150 + 100}{150(100)}\right)\right]^{1/2} = (0.0014)^{1/2} = 0.038$

The test statistic value is now

$$z = \frac{0.08 - 0.12}{0.038} = -1.05$$

The null hypothesis is still accepted.

COMMENT Even if the observed p values are substituted for π_1 and π_2 in Eq. (22.6), $\sigma_{p_1-p_2} = 0.0393$ and the statistic is $z = -1.018$, so H_0 is still accepted. Usually the true values of π are not known, so the two samples must be combined to obtain one p value, making the assumption that H_0: $\pi_1 = \pi_2$ is true.

Problems P22.9–P22.15

SOLVED PROBLEM

Example 22.3 For two years a mechanic has kept statistics on the number of cars he repaired that had a chipped windshield. The long-run average has been 22 percent. In the last 100 cars, 28 have had chips. Has the percentage increased based on this sample?

SOLUTION We can test the hypothesis H_0: $\pi = 0.22$ versus H_1: $\pi > 0.22$ using the fraction defective statistic, Eq. (22.1), or the binomial statistic, Eq. (22.3). For the latter, we have:

1. H_0: $\pi = 0.22$ versus H_1: $\pi > 0.22$.
2. Let $\alpha = 0.05$.
3. The observed $p = 28/100 = 0.28$ and $p_0 = 0.22$.
4. From Eq. (22.3) the test statistic is

$$z = \frac{100(0.28) - 100(0.22)}{\sqrt{100(0.22)(0.78)}} = \frac{28 - 22}{\sqrt{17.16}} = 1.45$$

5. For a one-tail, upper test the acceptance region is $z \leq 1.645$.
6. Since $1.45 < 1.645$, do not reject H_0; so the percentage has not increased.

COMMENT If Eq. (22.3) is used to solve for $100p$ with $n = 100$ and $p_0 = 0.22$, a minimum of 29 chips are needed to reject the null hypothesis.

$$1.645 = \frac{100p - 100(0.22)}{\sqrt{17.16}}$$

$$100p = 28.8$$

Section 22.2

ADDITIONAL MATERIAL

Gibra [8], pp. 241–243, 269–273; Hines and Montgomery [12], pp. 283, 284; Hoel [13], pp. 134–137; Kirkpatrick [15], pp. 232–238; Walpole and Myers [25], pp. 260–265.

PROBLEMS

P22.1 A new medicine for an intestinal disorder has been tested on 1500 patients, of which 585 were cured. The treatment commonly prescribed has a success rate of 41 percent. Use a significance level of 10 percent to determine if the cure rate of the new medicine is (*a*) different from and (*b*) less than that of the commonly used treatment.

P22.2 The company which supplies Frigid County with tire chains has stated that at most 3 percent of the chains shipped will have a flaw in them. Of the 200 chains just received and checked, 8 had at least one poorly welded link.
   (*a*) Does this shipment exceed the agreed-upon percentage defective? Let $\alpha = 0.05$.
   (*b*) Establish the upper 95 percent confidence interval on the population percentage defective and determine how many defective chains are required to change the decision made in (*a*).

P22.3 A warehouse has historically shipped 15 percent of a particular product to the Sun Company. Recent changes in order size and frequency indicate that this percentage has changed, but the direction of change is not certain. John took a sample of size 150 and found that 20.5 percent of the product went to Sun; so he concluded that the percentage increased. Carol sampled 325 items and found that 11 percent went to Sun; so she feels the percentage has decreased. Use a significance level of 5 percent to determine who made the correct conclusion. If both analyses are incorrect, make the correct conclusion.

P22.4 The fraction of people who take a shopping cart upon entering the grocery store has averaged 0.65 in the past. Because of increased business resulting from more sales, it seems this fraction has increased substantially. In a sample of 225 customers, 155 took carts. If $\alpha = 0.02$, test H_0: $\pi = 0.65$ versus H_1: $\pi > 0.65$ by the normal distribution approximation using the statistic for (*a*) the fraction of customers and (*b*) the number of customers taking shopping carts.

P22.5 Rework Prob. P22.4 if the sample size was 2250 and 1550 customers took carts.

P22.6 In the past, 1 out of every 25 bottles of Burpo, the soft drink with spunk, have been underfilled. According to the production manager, a newly installed bottling machine has rectified the problem so that now only 1 out of every 50 are underfilled. Out of the last 5000 bottles, 175 contained less than the correct number of grams. Do the following using a confidence level of 95 percent.
   (*a*) Determine if the underfill rate is equal to or different from 1 in 25 bottles.
   (*b*) Place a confidence interval on the number of underfills per 50 bottles for the new machine using the results of the sample.

(c) Determine if the sample of 5000 bottles was large enough to estimate the true fraction of underfilled bottles within 0.4 percent of the true value.

P22.7 Bicycle races are held each Sunday during the summer months. Usually about 40 percent of the contestants drop out of the race before it is over. In the race today there were 52 contestants and 38 finished. Use a one-sided hypothesis and the test statistic for the number of dropouts to conclude if this race had the same number of dropouts as previous races. Use a 10 percent significance level.

P22.8 For the bicycle race described in Prob. P22.7 determine the lower and upper boundary values for the number of racers who must complete the race to decide that it was representative of past races. Use a 90 percent confidence level.

P22.9 A closely controlled experiment was designed to investigate the response time of electronic guidance gear. Type A responded correctly on 28 of 50 tests, and type B was correct 73 out of 90 times. This is the first time that tests had been performed on these types of guidance mechanisms. Did the two types respond correctly with the same capability at the 95 percent confidence level?

P22.10 The owner of two motels wants to determine if the monthly occupancy rates are the same at two locations. The historic rates and last month's rate are given based on samples of 30 days:

	Location	
Rates	1	2
Historic	85%	92%
Last month	91%	80%

Perform the test for the owner using $\alpha = 0.02$ and the normal approximation to the binomial.

P22.11 In Prob. P22.10 the owner feels that the historic occupancy rates are no longer correct because a new freeway has been opened by location 1 and the amusement park by location 2 burned last week. In discussion it was learned that last month's rates were computed from this data:

	Location	
	1	2
Rooms available	100	80
Days per month	30	30
Total rooms occupied during the month	2730	1920

Now perform the test for equal occupancy rates with the sample sizes being all rooms available for the month. Use a 2 percent significance level and determine if the decision is the same as in Prob. P22.10.

P22.12 The percentage of real meat in Cats' Meow canned food is advertised as 25 percent while Purr-fection claims 33 percent. Using these as good estimators, a nutrition analyst tested 150 cans of Cats' Meow to find an average of 18 percent meat and 90 cans of Purr-fection to find 27 percent meat content on the average.

(a) Is it correct to state with 95 percent confidence that of the two, Purr-fection has a higher meat content?

(b) Disregard the manufacturer's claims and test for a higher meat content in Purr-fection using the test results as estimators of actual meat content. Is the conclusion the same as in (a)?

P22.13 An ME tested the auxiliary pumps on 14 air compressors at plant A and found 4 which had cracked diaphragms, while a similar check of 28 pumps at plant B resulted in 12 cracked diaphragms. Assume that the normal approximation to the binomial is acceptable in an analysis of the fraction of pumps that is inoperable. Use a 10 percent significance to determine if the maintenance of auxiliary pump diaphragms is (a) good enough at each plant to meet the standard that at most 25 percent of the pumps should be inoperable at any one time and (b) the same for both plants, regardless of their ability to meet the maintenance standard.

P22.14 The number of minutes that a train blocks an intersection was electronically measured for two different railway companies for 100 days.

Block time, minutes	Frequency	
	Company A	Company B
0–1	30	8
1–2	40	14
2–3	52	10
3–4	24	20
4–5	30	30
5–6	16	12
6–7	4	6

Use a 5 percent level of significance and the normal approximation to the binomial to determine if the following are true or false.

(a) At least 60 percent of the time company A blocks an intersection less than 4 min.

(b) The percentage of time that a train blocks an intersection for 5 min or more is the same for both companies.

(c) In the past company B has violated the standard of 6-min maximum block time 10 percent of the time. This data indicates that the company has improved in that it violates this standard less often.

(d) The mean block time is the same for both companies.

P22.15 Expand the test statistic in Eq. (22.5) so that the number, not the proportion, is tested. Be sure to develop two standard deviation expressions: one for known and one for unknown population values. Assume that the sample sizes n_1 and n_2 are not equal.

CHAPTER
TWENTY-THREE

STATISTICAL INFERENCES FOR GOODNESS OF FIT

This chapter will acquaint you with two tests used to evaluate the goodness of fit of observed data to a hypothesized statistical distribution. You will also learn to use one of these tests to determine if observed data is statistically independent for two given factors.

CRITERIA

To perform a goodness-of-fit test correctly, you must be able to:

1. State the questions that must be answered to select the appropriate goodness-of-fit test.
2. Use the hypothesis testing procedure to perform a χ^2 goodness-of-fit test for a *discrete* distribution, given the observed data and the hypothesized distribution.
3. Do the same as in 2, except combine expected values so that the minimum frequency restriction is not violated.
4. Do the same as in 2 and 3, except for a *continuous* distribution.
5. Use the hypothesis testing procedure to perform a χ^2 test for two *independent factors*, given the factors and observed values categorized by factor.
*6. Derive the χ^2 goodness-of-fit statistic for two categories by using the normal approximation to the binomial distribution.
7. Use the hypothesis testing procedure to perform a Kolmogorov-Smirnov goodness-of-fit test for a *continuous* distribution, given the observed data and the hypothesized distribution and its parameter values.

STUDY GUIDE

23.1 Types of Goodness-of-Fit Tests

Thus far all our statistical inferences have involved population parameters like means, variances, and proportions. Now we make inferences about the entire population distribution. A sample is taken, and we want to test a null hypothesis of the general form

H_0: sample is from a specified distribution

Any distribution can be tested, regardless of whether it is well known (binomial, Poisson, normal, etc.) or developed by the statistician. The alternative hypothesis is always of the form

H_1: sample is not from a specified distribution

A test of H_0 versus H_1 is called a *goodness-of-fit test*.

Two tests are used to evaluate goodness of fit:

1. The χ^2 test, which is based on an approximate χ^2 statistic.
2. The Kolmogorov-Smirnov (K-S) test. This is called a nonparametric test because it uses a test statistic that makes no assumptions about distribution.

The χ^2 test is best for testing discrete distributions, and the K-S test is best on continuous distributions for which the parameter values are specified independently of the sample data. The χ^2 test is used to quantitatively evaluate the fit to observed data as was subjectively done in Secs. 8.6 (binomial) and 9.6 (Poisson). To determine which test to use, answer the following questions.

1. *Is the distribution to be tested discrete or continuous?*
2. *Are the parameter values for the distribution specified?* If they are not specified, they must be estimated from the sample.

The appropriate test and sections are given in Table 23.1.

It is important to realize that a goodness-of-fit test cannot be used to select the best distribution. After you choose the distribution, the test uses the observed data to statistically determine if the choice was good.

Table 23.1 Appropriate tests for goodness of fit

Discrete or continuous distribution	Parameters specified or estimated	Test to use	Section(s)
Discrete	Specified	χ^2	23.2, 23.3
Discrete	Estimated	χ^2	23.2, 23.3
Continuous	Specified	K-S	23.7
Continuous	Estimated	χ^2	23.4

Problems P23.1, P23.2

23.2 χ^2 Goodness-of-Fit Test for Discrete Distributions

The χ^2 test uses a test statistic that has an approximate χ^2 distribution. More detail about the statistic is given in Sec. 23.6. It is a measure of the relative difference between the observed and expected frequency values for each variable value.

$$\chi^2 = \sum_{i=1}^{k} \frac{(O_i - E_i)^2}{E_i} \qquad (23.1)$$

where $k =$ number of different values of variable
$O_i =$ observed frequency value
$E_i =$ expected or theoretical frequency value

The E_i values are the product of sample size and the probability from the hypothesized distribution.

$$E_i = nP_i \qquad (23.2)$$

Since the χ^2 pdf is used to test the fit, the degree-of-freedom parameter must be determined. It is

$$\nu = k - r - 1 \qquad (23.3)$$

where r is the number of parameters of the hypothesized distribution that is estimated from the sample data.

The hypothesis testing steps of Sec. 19.4 are used to test goodness of fit. The E_i values and the χ^2 value are computed in step 4. This test is always a one-sided test, so the acceptance region for H_0 is all χ^2 values below some χ_0^2 value with $1 - \alpha$ of the area to its left; that is,

$$P(\chi^2 \leq \chi_0^2) = 1 - \alpha$$

Example 23.1 illustrates this test for a situation in which the parameters are specified.

Example 23.1 The personnel director at your company has asked you to verify the statement that absenteeism is twice as bad on Mondays as the rest of the week. You are given personnel records for 3 months covering 890 days of lost work.

Day	Monday	Tuesday	Wednesday	Thursday	Friday
Days lost	304	176	139	141	130

SOLUTION Answers to the questions in Sec. 23.1 indicate that the data is discrete and the parameters are known, so Eq. (23.1) is the correct test statistic. We perform all computations using the hypothesis testing procedure.

1. The hypothesized distribution states that daily absenteeism occurs in the ratios $2:1:1:1:1$, which have a sum of 6. If $i = 1, 2, \ldots, 5$ identifies the day of the week, the distribution for the variable A, which is the fraction of absences occurring on a given day of the week, is

$$f(a) = \begin{cases} \frac{2}{6} & i = 1 \\ \frac{1}{6} & i = 2, 3, 4, 5 \end{cases}$$

The test is between the hypotheses

H_0: A has the distribution $f(a)$
H_1: A does not have the distribution $f(a)$

2. Let $\alpha = 0.05$.
3. No parameters need be estimated.
4. The expected values for each i value are computed using Eq. (23.2) with $n = 890$ and the P_i values from $f(a)$. For example, with $i = 1$, $E_1 = 890(\frac{2}{6}) = 296.7$ days. Table 23.2 shows the observed and expected frequency values. The individual χ^2 values are added to obtain $\chi^2 = 8.54$ for this sample.
5. The degrees of freedom from Eq. (23.3) is $\nu = 5 - 1 = 4$, because there are 5 days and no parameters are estimated from the sample. The acceptance region for $\chi^2(4)$ and $\alpha = 0.05$ is $\chi^2 \leq 9.49$.
6. Since $8.54 < 9.49$, H_0 cannot be rejected. It appears that the absenteeism rate is twice as high on Mondays.

COMMENT As shown in Table 23.2, the E_i values should not be rounded off. If the E_i values are small, rounding tends to change the χ^2 value up or down. If there are E_i values less than 5, the correction discussed in Sec. 23.3 should be used.

Problems P23.3–P23.7

Table 23.2 Computations for the χ^2 goodness-of-fit test for Example 23.1

Day i	Absences Observed, O_i	Expected, E_i	$(O_i - E_i)^2$	χ^2
1	304	296.7	53.29	0.18
2	176	148.3	767.29	5.17
3	139	148.3	86.49	0.58
4	141	148.3	53.29	0.36
5	130	148.3	334.89	2.25
	890	889.9		8.54

23.3 Minimum Frequency Correction to the χ^2 Statistic

All the expected frequency values E_i used in the χ^2 goodness-of-fit test must have a minimum frequency of 5 for the χ^2 approximation to be accurate. If necessary, you should combine the E_i and O_i values of adjacent values to guarantee that all $E_i > 5$. This addition should be done so as to not unrealistically combine the variable values into an implausible arrangement.

Example 23.2 tests a Poisson pdf fit. The correction above is used, and the Poisson parameter is estimated from the sample.

Example 23.2 In Example 9.6 a chemical engineer determined the percentage of sulfur in tires. For 100 days the number of days which violated the 4 percent per tire limit is:

Violations per day	0	1	2	3	4	5	6
Number of days	33	44	10	5	5	2	1

A Poisson pdf was used for computations, and $\hat{\lambda} = \overline{X} = 1.15$ was estimated from this sample. Use the χ^2 goodness-of-fit test to determine if the Poisson was a good choice.

SOLUTION The variable is X, the number of violations per day. Use the hypothesis testing procedure.

1. H_0: X is Poisson-distributed versus H_1: X is not Poisson-distributed.
2. Use $\alpha = 0.05$.
3. The parameter $\hat{\lambda} = 1.15$ has been estimated.
4. Table 23.3 shows all computations for the χ^2 statistic. The E_i values are the product of $n = 100$ and the Poisson $P(x; 1.15)$ values from Table B-2. For $i = 5, 6,$ and 7, $E_i < 5$, so they are combined until $E_4 = 11.0$ is

Table 23.3 Computation for the χ^2 goodness-of-fit test, Example 23.2

i	x	$P(x; 1.15)$	E_i	E_i'	O_i'	χ^2
1	0	0.317	31.7	31.7	33	0.05
2	1	0.364	36.4	36.4	44	1.59
3	2	0.209	20.9	20.9	10	5.68
4	3	0.080	8.0			
5	4	0.023	2.3	11.0	13	0.36
6	5	0.006	0.6			
7	6	0.001	0.1			
						7.68

obtained. The new expected values E_i' and the corresponding O_i' values are used in Eq. (23.1) to obtain $\chi^2 = 7.68$.
5. After combining E_i values, there are $k = 4$ different values. Since $r = 1$ parameter was estimated, Eq. (23.3) gives $\nu = 4 - 1 - 1 = 2$ degrees of freedom. For $\alpha = 0.05$ the acceptance region for $\chi^2(2)$ is $\chi^2 \leq 5.99$.
6. Reject H_0 because $7.68 > 5.99$. The violations per day do not follow a Poisson pdf with $\lambda = 1.15$.

COMMENT Table 23.3 shows that the largest component of $\chi^2 = 7.68$ is from $x = 2$. The observed value of 10 is too far from $E_i' = 20.9$, so H_0 cannot be accepted. If the value $\lambda = 1.15$ is well established and not estimated, then $\nu = 3$ and the acceptance region is $\chi^2 \leq 7.81$, thus allowing H_0 to be accepted.

Do not forget to reduce k when expected values are combined to obtain $E_i > 5$. If you do not, the wrong approximating $\chi^2(\nu)$ distribution will be used.

If there are only two possible values or categories ($k = 2$), the degree of freedom is $\nu = 1$. The normal approximation to the binomial, Eq. (11.6), should be used to test H_0: p_1 of the data is in category 1 and p_2 is in category 2. This is a correct application of the χ^2 approximation as shown in Sec. 23.6.

Problems P23.8–P23.17

23.4 χ^2 Goodness-of-Fit Test for Continuous Distributions

The procedure for testing any continuous distribution is identical to that presented above, except that the data must be grouped into k cells before the E_i values are computed for the hypothesized distribution. There should be at least five cells, and the probability P_i that the variable is in cell i is used to compute E_i with Eq. (23.2). This test is often used to determine if data is from a normal distribution once the expected values are computed by the steps in Sec. 11.7. Example 23.3 illustrates the procedure.

Example 23.3 In Example 11.7 (Sec. 11.7) a normal distribution was fit to flow data in liters per second. The observed and computed expected frequencies are summarized in Table 23.4 from Table 11.3. Test the hypotheses

H_0: flow has a normal distribution
H_1: flow does not have a normal distribution

Table 23.4 Computations for the χ^2 goodness-of-fit test, Example 23.3

Cell i	O_i	E_i	E_i'	O_i'	$(O_i' - E_i')^2$	χ^2
1	5	5.88	5.88	5	0.77	0.13
2	21	21.10	21.10	21	0.01	0.00
3	35	31.08	31.08	35	15.37	0.49
4	15	17.74				
5	3	3.89	21.94	19	8.64	0.39
6	1	0.31				
						1.01

SOLUTION The hypothesis testing steps are used.

1. H_0 and H_1 are given.
2. Let $\alpha = 0.05$.
3. The parameters were estimated as $\hat{\mu} = 8.224$ and $\hat{\sigma} = 0.293$ L/s from the data in Example 11.7.
4. $\chi^2 = 1.01$ is found by adding the components in Table 23.4. Be sure to notice that the minimum frequency restriction of 5 is violated if the last three E_i values are not combined. The observed values are also combined into O_i' values.
5. After combining cells and estimating μ and σ, there is $\nu = 4 - 2 - 1 = 1$ degree of freedom remaining. The acceptance region for $\alpha = 0.05$ and $\chi^2(1)$ is $\chi^2 \leq 3.84$.
6. Since $1.01 < 3.84$, accept H_0 that the data is distributed $N(8.224, 0.293^2)$. This is the same as the subjectively determined conclusion of Example 11.7.

The χ^2 goodness-of-fit test can be used to test any distribution once the parameters are specified or estimated. It is less accurate for continuous distributions, because it is necessary to group the data into fictitious, discrete cells and compare them, whereas discrete data has an inherent grouping by variable value.

Problems P23.18–P23.23

23.5 χ^2 Test for Independence of Factors

One sample may be taken and classified according to two factors, each factor having two or more levels. A χ^2 test may help determine if the observed values are independent of the factors. For example, Table 23.5 gives the observed values for weekly defects by three shifts and two production lines. This is commonly called a *contingency table*. This test will enable you to statistically determine if defects occur independently of shift and line.

Table 23.5 Defects classified by shift and production line

Line	Shift		
	1	2	3
1	10	12	13
2	14	9	12

The general layout of Table 23.6 is used to explain this χ^2 test. The null hypothesis is

$$H_0: p_{ij} = p_i p_j \quad \begin{matrix} i = a, b, \ldots, m \\ j = 1, 2, \ldots, n \end{matrix}$$

The usual steps of the hypothesis testing procedure are used to test H_0. If H_0 is not accepted, the only conclusion is that the factors are not independent, but how and to what degree they are dependent is not determinable by this test. The χ^2 statistic in Eq. (23.1) is used, but written with double subscripts.

$$\chi^2 = \sum_i \sum_j \frac{(O_{ij} - E_{ij})^2}{E_{ij}} \qquad (23.4)$$

The degrees of freedom are

$$\nu = (m-1)(n-1) \qquad (23.5)$$

The E_{ij} values are estimated from the sample data using the following steps and the Table 23.6 format. In all cases $i = a, b, \ldots, m$ rows and $j = 1, 2, \ldots, n$ columns.

Table 23.6 General layout to test for independent factors

	Factor 1 levels, j				Row total	Row proportion
	1	2	$\cdots$	n		
Factor 2 levels, i	a				n_a	p_a
	b				n_b	p_b
	$\cdot$	O_{ij} values			$\cdot$	$\cdot$
	$\cdot$	(E_{ij} values)			$\cdot$	$\cdot$
	$\cdot$				$\cdot$	$\cdot$
	m				n_m	p_m
Column total		n_1 n_2 $\cdots$ n_n			N	1
Column proportion		p_1 p_2 $\cdots$ p_n			1	

1. Compute the row totals n_i, column totals n_j, and the total observations N.
2. Compute the row proportions $p_i = n_i/N$ and column proportions $p_j = n_j/N$.
3. Compute the expected value E_{ij} for each cell.

$$E_{ij} = Np_ip_j \qquad (23.6)$$

This is possible since under the H_0 of independence $p_{ij} = p_ip_j$.

4. Enter E_{ij} in parentheses under the O_{ij} values.

These steps are used in the hypothesis testing procedure as step 3, which is the estimation of parameters.

Example 23.4 Use the data of Table 23.5 to determine if there is any relationship between the shift and the production line.

SOLUTION Apply the hypothesis testing steps of Sec. 19.4.

1. H_0: $p_{ij} = p_ip_j$ for lines $i = 1, 2$ and shifts $j = 1, 2, 3$.
2. Use $\alpha = 0.05$.
3. Table 23.7 gives the results of the four steps above to compute E_{ij}. An example for E_{13} in step 3 using Eq. (23.6) is

$$E_{13} = Np_1p_3 = 70(0.50)(0.36) = 12.6$$

4. The χ^2 test statistic is computed from Eq. (23.4).

$$\chi^2 = \frac{(10 - 11.9)^2}{11.9} + \frac{(12 - 10.5)^2}{10.5} + \cdots + \frac{(12 - 12.6)^2}{12.6}$$
$$= 1.14$$

5. Equation (23.5) results in $\nu = (2 - 1)(3 - 1) = 2$. The $\alpha = 0.05$ acceptance region for $\chi^2(2)$ is $\chi^2 \leq 5.99$.

Table 23.7 Observed and expected values, Example 23.4

Line	Shift			Row total	Proportion p_i
	1	2	3		
1	10 (11.9)	12 (10.5)	13 (12.6)	35	0.50
2	14 (11.9)	9 (10.5)	12 (12.6)	35	0.50
Column total	24	21	25	70	1.00
Proportion p_j	0.34	0.30	0.36	1.00	

6. H_0 is well within the acceptance region, so the defects are independent of shift and production line.

Problems P23.24–P23.27

*23.6 Derivation of the χ^2 Goodness-of-Fit Statistic

This section summarizes the derivation of the χ^2 goodness-of-fit statistic, Eq. (23.1). Start with a sample of n observed values that are placed into one of two categories with probabilities p_1 and $p_2 = 1 - p_1$. If O_1 is the number in category 1, then $O_1 \sim b(o_1; n, p_1)$, and $O_2 = n - O_1$ has $O_2 \sim b(o_2; n, p_2)$. If the normal approximation to the binomial is used for $np_1 > 5$ and $np_2 > 5$, the variable

$$\frac{O_1 - np_1}{\sqrt{np_1 p_2}} \tag{23.7}$$

has an $N(0, 1)$ distribution. If Eq. (23.7) is squared, according to Sec. 13.6 the following relation is correct.

$$\frac{(O_1 - np_1)^2}{np_1 p_2} \sim \chi^2(1) \tag{23.8}$$

Notice that $np_1 = E_1$, the expected frequency for category 1. The use of $np_1 = E_1 > 5$ and $np_2 = E_2 > 5$ is the same minimum frequency restriction used in Sec. 23.3 to guarantee a close χ^2 approximation.

If we use the normal approximation for both categories and the facts that $O_2 = n - O_1$, $E_2 = n - E_1$, and $p_2 = 1 - p_1$, we can write

$$\frac{(O_1 - np_1)^2}{np_1} + \frac{(O_2 - np_2)^2}{np_2} = \frac{(O_1 - np_1)^2}{np_1} + \frac{[n - O_1 - n(1 - p_1)]^2}{n(1 - p_1)}$$

$$= \frac{(O_1 - np_1)^2}{np_1 p_2}$$

This result is identical to Eq. (23.8), so it has a $\chi^2(1)$ distribution. Therefore, the normal distribution is equivalent to the χ^2 goodness-of-fit test for $k = 2$ categories, as mentioned in Sec. 23.3, provided $E_i > 5$ ($i = 1, 2$).

If there are k mutually exclusive categories having proportions $p_1, p_2, \ldots, p_k$, the multinomial distribution, Eq. (18.11), is correct for $O_1, O_2, \ldots, O_k$. It can be shown that the normal approximations for each O_i can be squared and added. The result is

$$\sum_{i=1}^{k} \frac{(O_i - np_i)^2}{np_i}$$

which has an approximate $\chi^2(k - 1)$ distribution. If E_i replaces np_i, this is the χ^2 statistic used to test goodness of fit.

Problems P23.28, P23.29

23.7 K-S Goodness-of-Fit Test

The Kolmogorov-Smirnov (K-S) test should be used instead of the χ^2 test to determine if a sample is from a specified *continuous* distribution. The test is exact for any sample size n, because it does not use an approximate distribution to test the null hypothesis

H_0: data is from a specified distribution with stated parameter values

The test is not reliable if the parameters must be estimated from the sample.

The cumulative distribution function (cdf) of the observed sample and the hypothesized distribution must be determined to carry out a K-S test. The test statistic d is the *maximum absolute difference* between the two cdf's over all observed values. The range on d is $0 \leq d \leq 1$, and the formula is

$$d = \max_x |S(x) - F(x)| \qquad (23.9)$$

where x = each observed value
$S(x)$ = observed cdf at x
$F(x)$ = hypothesized cdf at x

Figure 23.1 is a sketch of $S(x)$ and $F(x)$. At each x value the difference is computed, and the K-S test statistic value is the maximum difference d at x'.

The variable D is the maximum difference in Eq. (23.9). Values of the pdf of D are given in appendix Table B-7 for significance levels of 0.10, 0.05, and 0.01. The K-S test is one-sided, so the area above a certain D value is α. If the computed d does not exceed the tabulated D, accept the null hypothesis.

The hypothesis testing steps are used, except that step 3 involves no statistic or parameter computation. Therefore the values of $S(x)$ and $F(x)$ are determined in this step using the following procedure.

1. Order the sample data from smallest to largest. The ordered values are

$$x_1 \leq x_2 \leq \cdots \leq x_n.$$

2. Compute the sample cdf at each x value using

$$S(x) = \frac{i}{n} \qquad (23.10)$$

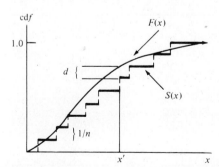

Figure 23.1 Sample cdf $S(x)$ and hypothesized cdf $F(x)$, used in a K-S goodness-of-fit test.

where $i = 1, 2, \ldots, n$ and $x_i \leq x \leq x_{i+1}$. $S(x)$ will jump $1/n$ at each x value (Fig. 23.1).
3. Use the hypothesized cdf and parameters to determine $F(x)$ at each ordered x value. $F(x)$ is drawn as a smooth, approximating, continuous cdf.

Example 23.5 A state vehicle inspection station has been designed so that inspection time follows a uniform distribution with limits of 10 and 15 min. A sample of 10 duration times during low and peak traffic conditions was taken. Use the K-S test with $\alpha = 0.05$ to determine if the sample is from this uniform distribution. The times are:

| 11.3 | 10.4 | 9.8 | 12.6 | 14.8 |
| 13.0 | 14.3 | 13.3 | 11.5 | 13.6 |

SOLUTION Use the hypothesis testing procedure and the steps to determine $S(x)$ and $F(x)$:

1. H_0: sample is from a $U(10, 15)$ distribution versus H_1: sample is not from a $U(10, 15)$ distribution.
2. Let $\alpha = 0.05$.
3. The steps above are used to obtain $S(x)$ and $F(x)$. All column references are to Table 23.8.
    (1.) Column 1 includes the ordered data.
    (2.) The sample cdf $S(x)$ is computed from Eq. (23.10) with $n = 10$ and $i = 1, 2, \ldots, 10$. $S(x)$ is plotted in Fig. 23.2.
    (3.) Equation (16.2) is the cdf for the uniform distribution. Column 3 and

Table 23.8 Computations for the K-S goodness-of-fit test, Example 23.5

Observed times x (1)	$S(x)$ (2)	$F(x)$ (3)	$\lvert S(x) - F(x) \rvert$ (4) = $\lvert (2) - (3) \rvert$
9.8	0.10	0.00	0.10
10.4	0.20	0.08	0.12
11.3	0.30	0.26	0.04
11.5	0.40	0.30	0.10
12.6	0.50	0.52	0.02
13.0	0.60	0.60	0.00
13.3	0.70	0.66	0.04
13.6	0.80	0.72	0.08
14.3	0.90	0.86	0.04
14.8	1.00	0.96	0.04

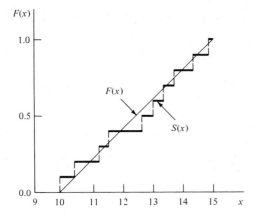

Figure 23.2 Observed and expected cumulative distribution functions, Example 23.5.

Fig. 23.2 include these results. For example, at $x = 12.6$

$$F(12.6) = \frac{12.6 - 10}{15 - 10} = 0.52$$

The time $x = 9.8$ min is less than the lower design limit of 10, so $F(9.8) = 0.00$.
4. Column 4 uses Eq. (23.9) to compute the maximum difference $d = 0.12$ at $x = 10.4$.
5. Table B-7 gives an acceptance region $d \leq 0.41$ for $n = 10$ and $\alpha = 0.05$.
6. Since $0.12 < 0.41$, accept the null hypothesis. The station is able to inspect cars with a time uniformly distributed between 10 and 15 min.

Problems P23.30–P23.34

ADDITIONAL MATERIAL

χ^2 *goodness-of-fit tests* (Secs. 23.1 to 23.6): Belz [2], pp. 201–217; Bowker and Lieberman [4], pp. 458–465; Gibra [8], Chap. 15; Hoel [13], pp. 226–240; Kirkpatrick [15], pp. 239–243; Neville and Kennedy [19], Chap. 12; Newton [20], pp. 278–297; Siegel [21], pp. 42–47, 104–111, 175–179; Volk [24], Chap. 5.

K-S tests (Sec. 23.7): Benjamin and Cornell [3], pp. 466–475; Bowker and Lieberman [4], pp. 454–458; Hoel [13], pp. 324–329; Kirkpatrick [15], pp. 244–246; Siegel [21], pp. 47–52, 127–136.

PROBLEMS

P23.1 Determine which goodness-of-fit test is the most appropriate for these situations.
(a) The Poisson distribution is to be tested as the best distribution to explain newly collected data. The mean occurrence rate to be tested is known from past experience.
(b) The time to failure T can take on the values $t > 0$, and the mean time to failure for the exponential failure model is specified as 550 h.
(c) The normal pdf has been fit to data using μ and σ values estimated from the sample data.

P23.2 Why is it necessary to select the statistical distribution before applying the χ^2 or K-S goodness-of-fit tests?

P23.3 In Example 8.7 (Sec. 8.6) an engineer used the outcomes on four dice to simulate the number of idle lathes.
(a) Use the observed and expected frequency values in Table 8.2 to determine if the binomial $b(x; 4, 0.5)$ is an appropriate distribution at $\alpha = 0.05$. Do not make any corrections for minimum frequency when you work this problem.
(b) Estimate the α value at which the decision made in (a) will change.

P23.4 The response time T of a circuit activator varies between 0.01 and 0.06 s in discrete 0.01-s units. The observed frequencies in 100 tests are given. Use $\alpha = 0.10$ to determine if the observed data follows a discrete uniform distribution of the form

$$f(t) = \begin{cases} 1/k & t = k \text{ different values} \\ 0 & \text{elsewhere} \end{cases}$$

Response time t	0.01	0.02	0.03	0.04	0.05	0.06
Frequency	18	18	15	14	19	16

P23.5 Example 6.7 (Sec. 6.4) describes the number X of problems per 1000 drives for the Hang-It-Up Stapler using the pdf

$$f(x) = \frac{5.255}{x^3} \qquad x = 2, 3, 4, 5, 6$$

A new improved version of the stapler has been sent to you for testing. Because you feel the improvements have not corrected some of the crucial problems, you have collected the following data from 210 people who each drove 1000 staples.

Number of problems per 1000 drives, x	Number of people observing x problems
2	145
3	40
4	12
5	5
6	8
	210

Determine if $f(x)$ can be used as the pdf of X for the new stapler. Use a 90 percent confidence level in making this conclusion.

P23.6 A new hay baling machine requires a wire with a high breaking strength. In a previous sample of 50 types of wire, 30 failed to meet the specifications. Baling machine sales personnel have collected field data from 500 agriculturists on the number of completed bales before a breakage is observed. If X is the bale number on which the first breakage takes place, the observed results may be summarized as follows:

x	1	2	3	4	5	6
Frequency	287	120	40	30	15	8

Determine if the geometric distribution is a good choice for this data if $\alpha = 0.01$.

P23.7 The president of a wholesale distribution company feels that half of all stock items have a turnover ratio in excess of 4 times per year. For the 1200 items in stock, the turnover ratios have been calculated by inventory management personnel. A total of 550 had a ratio in excess of 4, and the remainder turned less than 4 times per year. Use the χ^2 goodness-of-fit test and a 95 percent confidence level to determine if the president is statistically correct.

P23.8 Rework Example 23.2 using a specified parameter value of one violation per day and a significance level of (a) 5 percent and (b) 2.5 percent.

P23.9 The accident data described in Example 9.3 (Sec. 9.3) is summarized here:

Accidents per month	0	1	2	3	4	5
Frequency	27	12	8	2	1	0

(a) Is it necessary to use the minimum frequency correction if the Poisson distribution is tested for goodness of fit for this data?
(b) Does a Poisson distribution explain the number of accidents per month for $\alpha = 0.10$? For $\alpha = 0.05$?

P23.10 Determine if the observed frequencies in Prob. P8.24 follow a binomial pdf using $\alpha = 0.05$.

P23.11 The number of flaws per 100 m² of cloth is to be checked by a quality control chart which assumes an underlying Poisson population. For the last 75 samples the number of flaws have been recorded:

Number of flaws	0	1	2	3	4	5	≥ 6
Frequency	20	30	15	7	2	1	0

If $1 - \alpha = 0.95$, is the Poisson population assumption a reasonable one?

P23.12 Judy is an electrical engineer in component testing. As a result of experimentation, she hypothesizes that when a particular circuit is overloaded, one of the events below occurs with the indicated frequency.

Event	Frequency
Component is degraded	3 in 10 times
Component fails	4 in 10 times
No effect	5 in 20 times
Another component fails	1 in 20 times

A technician has overloaded 100 of the circuits and recorded 30 degradations, 32 failures, 37 no effects, and 1 failure of another component. Do the test results substantiate Judy's predictions at the 99 percent confidence level?

P23.13 Show that the Poisson distribution offers an excellent explanation of the horse-kick data presented in Prob. P9.17.

P23.14 The number of taxis waiting at a cab stand is thought to follow a uniform distribution. The number present was recorded at 50 random times.

Number of taxis	0	1	2	3	4	5	6	7	8
Frequency	4	6	8	5	7	6	6	5	3

(a) Estimate the two parameters of the uniform distribution.
(b) Use the appropriate goodness-of-fit test to determine if the uniform distribution is a good choice at the 5 percent significance level.

P23.15 Car pools are being promoted by the Department of Transportation in the city of Hueco. A traffic engineer observed X, the number of passengers in 80 cars on the highway. Now the engineer wants to fit a discrete distribution to the results.
(a) Select what appears to be an appropriate distribution for the data below.
(b) Estimate the parameters.
(c) Determine if your choice was a good one ($\alpha = 0.05$).

x	0	1	2	3	4	5
Frequency	30	21	11	10	6	2

P23.16 For the data in Prob. P9.6 test the hypothesis that the sample is from a Poisson distribution. Let $\alpha = 0.10$.

P23.17 Determine if the data below follows a binomial distribution for $n = 4$ with a mean of 2 and a standard deviation of 1. Use a 99 percent confidence level and work the problem (a) without the minimum frequency correction and (b) with this correction to determine if the conclusion is different.

Variable value	Observed frequency
0	2
1	12
2	32
3	16
4	2

P23.18 Fit a normal distribution to the data of Prob. P11.5 and perform a goodness-of-fit test on it using a 10 percent significance level.

P23.19 Does the weight distribution of French fries in Prob. P11.26 follow a normal distribution? Let $\alpha = 0.02$.

P23.20 The computations in Example 23.3 using the χ^2 goodness-of-fit test showed that the flow data has a normal distribution with $\mu = 8.224$ and $\sigma = 0.293$ L/s. The manager of the department wants to know if she can generalize and state that the flow distribution is normal with $\mu = 8$ and $\sigma = 0.3$. What is your recommendation to her if a 5 percent significance level is used again?

P23.21 The inventory management consultant in Example 11.9 states that he is 90 percent confident that the printed wiring card demand data follows a normal distribution. Is this statement correct?

P23.22 The time between arrivals at a ticket counter was recorded for 200 customers:

Time, min	1	2	3	4	5	6	7
Frequency	60	30	40	30	20	10	10

Use the χ^2 goodness-of-fit test to determine if the time between arrivals follows an exponential distribution. Let $\alpha = 0.05$.

P23.23 The time that it takes bread to bake in the ovens of the Dough-is-Dough Bakery varies with the water content of the dough, but is always between 15 and 25 min. The times for 20 different bakes are:

21	18	15	19	22
24	23	22	18	17
20	18	15	21	20
20	24	18	19	19

Collect the data into five cells (15 to 17, 17 to 19, and so on), and use the χ^2 goodness-of-fit test to determine at the 90 percent confidence level if the bake times follow the $U(15, 25)$ distribution. Do not use the minimum frequency correction.

P23.24 The average number of back-ordered units is recorded for an item purchased from four vendors under a contract and an as-needed basis. Is there a relationship between the vendor and method of ordering at $\alpha = 0.10$? At $\alpha = 0.05$?

	Vendor			
Orders	1	2	3	4
Contract	12	17	6	11
As needed	25	18	10	5

P23.25 Bananas may be shipped from overseas by one of four methods using any one of three different packaging techniques. In an experiment involving both factors the number of bruised bananas per 100 inspected was recorded. At a 90 percent confidence level determine if the factors of shipping method and packaging technique are independent.

| | Packaging technique | | |
Method of shipment	Box	Cello bag	Bundle
Plane/truck	8	12	15
Plane/train	10	15	18
Ship/truck	6	12	20
Ship/plane	18	10	11

P23.26 A computing science researcher has given a timed programming assignment to university students, first-year university graduates, and industry-trained programmers. One-third of the people were asked to code as they do normally, one-third using structured programming, and one-third using a new technique called no-fault programming. Use the data for number of total coding errors per group to determine at the 10 percent significance level if the factors of programmer background and programming method are independent.

| | Programmer background | | |
Programming method	Student	Graduate	Industry-trained
Normal	32	21	26
Structured	25	18	35
No-fault	19	24	32

P23.27 In an experiment 750 metal specimens were given a breaking strength test. Of the 400 specimens that were heat-treated, 300 passed the test; and of the 350 not heat-treated, 50 did not pass the test. If $\alpha = 0.05$, test the hypothesis that heat treating has no effect on breaking strength.

*P23.28 Summarize the derivation of the χ^2 goodness-of-fit statistic given by Eq. (23.1).

*P23.29 (a) Write the χ^2 goodness-of-fit test statistic for independent factors when three factors are present. Let $k = 1, 2, \ldots, r$ be the subscript for the third factor, which is called the layer. In addition, write expressions for (b) the degrees of freedom ν and (c) the expected value E_{ijk} in terms of the observed frequency totals n_i, n_j, and n_k and the total sample size N.

P23.30 Rework Prob. P23.23 using the K-S goodness-of-fit test and the hypothesis testing procedure to determine if the data is from a $U(15, 25)$ distribution. Let $1 - \alpha = 0.90$.

P23.31 Ten computer processing times were observed. Determine if they come from the distribution

$$f(t) = \frac{10 + t}{2250} \quad 10 \leq t \leq 60 \text{ s}$$

as described in Example 6.10 (Sec. 6.7). Use $\alpha = 0.10$. The times are:

| 10.5 | 20.4 | 25.4 | 35.9 | 38.8 |
| 42.0 | 50.8 | 54.1 | 58.7 | 59.2 |

P23.32 The number of years that eight small businesses were open prior to declaring bankruptcy are recorded. Determine if they follow an exponential distribution with a specified parameter value of (a) $\lambda = 0.3$ or (b) $\lambda = 0.1$. Use $\alpha = 0.10$ for this analysis. The times are 1.7, 4.5, 6.0, 10.0, 12.5, 15.8, 22.3, and 34.2 years.

P23.33 Previous data collected by the marketing department indicates that daily sales are normally distributed with $\mu = 150$ and $\sigma = 25$ units per day. The volumes at 15 branch outlets are listed. Use a 5 percent significance level to determine if the sales data is from the specified normal distribution.

| 172 | 189 | 122 | 198 | 194 | 180 | 125 | 170 |
| 168 | 131 | 170 | 169 | 122 | 138 | 182 | |

P23.34 Why is the K-S test better than the χ^2 goodness-of-fit test for continuous-variable data when the parameters are known, or suspected, to have specific preselected values?

CHAPTER TWENTY-FOUR

GRAPHICAL PROCEDURE FOR FITTING A DISTRIBUTION

This chapter presents the procedure to determine if a particular distribution fits sample data using graphical techniques. These methods are subjective compared to the statistical procedures of the last chapter; however, they do allow us to rapidly determine if a distribution is appropriate and to quickly estimate its parameters.

CRITERIA

To correctly use the graphical techniques discussed in this chapter, you must be able to:

1. State the five steps of the *probability plotting procedure*, and tabulate observed data as it is used in this procedure.

You must be able to use the probability plotting procedure to determine if any of the following distributions are appropriate for a given set of data, and estimate the parameter(s) if the distribution is appropriate:

2. Normal distribution.
3. Exponential distribution.
4. Uniform distribution.
5. Poisson distribution.

STUDY GUIDE

24.1 Probability Plotting Procedure

In the last chapter the χ^2 and K-S tests were used to quantitatively evaluate the null hypothesis

H_0: sample is from a specified distribution

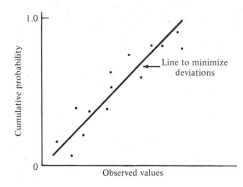

Figure 24.1 Probability plot of the observed values versus cumulative probability with an "eye-balled" straight line.

Graph paper designed for the hypothesized distribution can also be used to evaluate H_0 in a subjective, yet accurate, fashion. As shown in Fig. 24.1, each observed value is plotted versus the cumulative probability, and a straight line is "eye-balled" through the points such that deviations are minimized. If the fit is judged good, the distribution is appropriate based on the sample and the parameters may be estimated using the special graph paper.

Probability paper is commercially available or easily constructed for a number of distributions we have discussed thus far. Some are normal, χ^2, gamma, exponential, log normal, Weibull, beta, uniform, and Poisson. The use of several of these is described in this chapter.

A *probability plotting procedure* is applicable to the probability paper for many of these distributions. The steps are:

1. *Hypothesize the distribution.* Select the distribution which is to be tested for the sample data.
2. *Ordering data.* Order the observed data from smallest to largest. Call the ordered values $x_1, x_2, \ldots, x_n$.
3. *Tabulation.* Determine the *cumulative percentage* points for the sample of size n and $i = 1, 2, \ldots, n$ using the relation

$$100 F(x_i) = \frac{(i - 0.5)100}{n} \tag{24.1}$$

Or, determine the *cumulative probability* points using

$$F(x_i) = \frac{i - 0.5}{n} \tag{24.2}$$

Tabulate the x_i values and the cumulative percentage or probability values. Several variations of this cumulative scale are used depending on the distribution and the layout of the paper. These variations are detailed for each distribution covered in this chapter.
4. *Plotting.* Plot the data above using the graph paper for the selected distribution. Construct the best-fitting straight line through these points.
5. *Conclusion.* Conclude, by observation, if the distribution is appropriate.

6. *Parameter estimation.* If the distribution is appropriate, estimate the parameters. The techniques are covered below for each distribution.

If a large sample is collected and the data is grouped into cells, the upper cell boundaries are tabulated in step 3. The straight line in step 4 is usually constructed with major emphasis on the central points, because the data in the tail areas give a poor fit, especially for small samples.

Example 24.1 The number of incorrectly transmitted signals per 1000 digits was recorded for a newly developed space communication system. In a sample of $n = 10$ transmissions there were 13, 8, 7, 12, 25, 10, 9, 31, 14, and 5 incorrect signals per 1000 digits. Tabulate the observed values and the cumulative percentage values to be used in the probability plotting procedure.

SOLUTION The second and third steps of the procedure apply here. Equation (24.1) is used to compute the cumulative percentages for $i = 1, 2, \ldots, 10$. For example, at $i = 2$, the value is

$$100F(x_2) = \frac{(2 - 0.5)100}{10} = 15.0$$

Table 24.1 presents the results for the ordered data.

COMMENT Cumulative percentages are usually not integers. If $n = 12$, for instance, they are 4.2, 12.5, ..., 95.8 percent. However, they are the same for every situation in which the sample size is constant.

Problems P24.1–P24.3

Table 24.1 Cumulative percentages and ordered data values, Example 24.1

i	x_i	$100F(x_i)\%$
1	5	5.0
2	7	15.0
3	8	25.0
4	9	35.0
5	10	45.0
6	12	55.0
7	13	65.0
8	14	75.0
9	25	85.0
10	31	95.0

24.2 Probability Plotting for the Normal Distribution

The probability paper for a normal distribution is by far the most commonly used. The steps in Sec. 24.1 are used to determine if data is from a normal pdf. The observed values are plotted on an arithmetic scale, and the cumulative probability or percentage scale is extended as shown in Fig. 24.2. A straight line is constructed through the 50 percent point and $z = 0$ of the cdf for the standard normal distribution. The values in each tail are extended. For example, Fig. 24.2 traces the translation of the cumulative percentage point for $100P(z \leq -2) = 2.28$ percent. Figure 24.3 shows data plotted on normal probability paper.

If the points plot close to the eye-balled line, the normal is an appropriate distribution and the parameters μ and σ are estimated as follows.

1. *Estimate of μ.* The mean μ is estimated at $100F(x) = 50$ percent on the cumulative percentage scale. Draw a horizontal line at 50 percent to the fitted line, and drop down to the corresponding data value.
2. *Estimate of σ.* One standard deviation from the mean is found at a cumulative percentage of 15.87 percent (see the tabulation in Fig. 24.2). An estimate of σ is determined by subtracting the variable value x that occurs at 15.87 percent

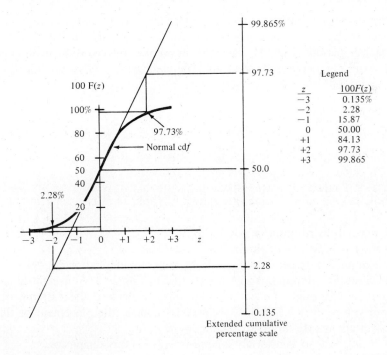

Figure 24.2 Translation from an arithmetic scale to the extended percentage (probability) scale for the normal distribution.

432 GRAPHICAL PROCEDURE FOR FITTING A DISTRIBUTION

(approximately 16 percent) from the estimate of μ.

$$\sigma = \mu - \text{(value with cumulative percentage of 16 percent)}$$
$$= \mu - [100F(x) = 16 \text{ percent}] \quad (24.3)$$

The scatter of points in the tails will commonly be larger than in the central portion of the graph. Even if the normal is an appropriate distribution, the tail points may not fit the straight line because only a small portion of the total probability is present here. The acceptance or rejection of the normal fit should be largely based on the fit of the middle 50 to 80 percent of the data values. The use of normal probability paper is illustrated here.

Example 24.2 Use the probability plotting procedure to determine if the normal distribution is appropriate for the situations below. Estimate the parameters if the fit is good.
(a) Fifteen samples are taken from a processing line for cast aluminum. The sample dimensions in centimeters (cm) are:

10.5	9.6	11.2	8.7	10.6
11.3	10.9	10.5	11.0	10.1
10.3	11.0	9.4	9.9	10.4

(b) A mechanical engineer with a supermarket chain asked the fresh produce managers at 20 stores to record the shelf life of lettuce for a 1-month period. The ME hopes to develop a refrigeration method to substantially increase the shelf life. The managers reported the following average shelf life values in days:

3.00	3.25	3.50	3.50	3.75
4.00	4.25	4.50	5.00	5.25
5.25	5.75	6.00	6.50	6.75
7.00	8.00	8.10	8.50	9.40

SOLUTION (a) All computations are shown in Table 24.2. The steps are:

1. The normal distribution is hypothesized.
2. The data is ordered in column 2 of Table 24.2.
3. The cumulative percentages for $n = 15$ are computed from Eq. (24.1).
4. The values in columns 2 and 3 are plotted in Fig. 24.3. A straight line is constructed through the points, somewhat neglecting the extreme observed values $x_1 = 8.7$ and $x_{15} = 11.3$. Of course, the placement of this line will change slightly with each analyst.
5. The fit is good, so the normal is appropriate for this dimension.
6. At $100F(x) = 50$ percent, the abscissa value is 10.4 cm, which is the

24.2 PROBABILITY PLOTTING FOR THE NORMAL DISTRIBUTION

Table 24.2 Computations for a normal probability plot, Example 24.2a

i (1)	x_i (2)	$100F(x_i)\%$ (3)	i (1)	x_i (2)	$100F(x_i)\%$ (3)
1	8.7	3.3	9	10.5	56.7
2	9.4	10.0	10	10.6	63.3
3	9.6	16.7	11	10.9	70.0
4	9.9	23.3	12	11.0	76.7
5	10.1	30.0	13	11.0	83.3
6	10.3	36.7	14	11.2	90.0
7	10.4	43.3	15	11.3	96.7
8	10.5	50.0			

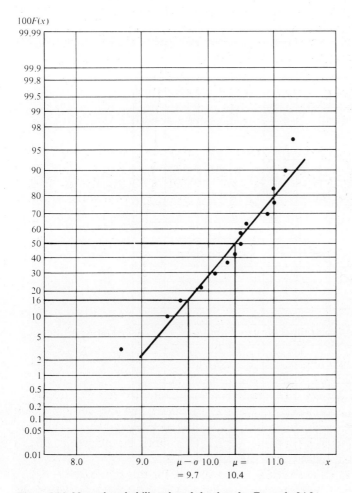

Figure 24.3 Normal probability plot of the data for Example 24.2a.

Table 24.3 Computations for a normal probability plot, Example 24.2b

i	x_i, days	$100F(x_i)\%$	i	x_i, days	$100F(x_i)\%$
1	3.00	2.5	11	5.25	52.5
2	3.25	7.5	12	5.75	57.5
3	3.50	12.5	13	6.00	62.5
4	3.50	17.5	14	6.50	67.5
5	3.75	22.5	15	6.75	72.5
6	4.00	27.5	16	7.00	77.5
7	4.25	32.5	17	8.00	82.5
8	4.50	37.5	18	8.10	87.5
9	5.00	42.5	19	8.50	92.5
10	5.25	47.5	20	9.40	97.5

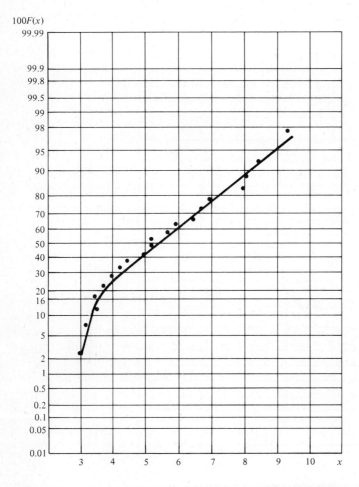

Figure 24.4 Normal probability plot of shelf life data for Example 24.2b.

estimate of μ. At $100F(x) = 16$ percent, the abscissa value is 9.7 cm, so $\sigma = 10.4 - 9.7 = 0.7$ cm. Based on this data, we conclude that an $N(10.4, 0.7^2)$ distribution is appropriate.

(b) The computations for the $n = 20$ shelf life values are summarized in Table 24.3. The plot on normal probability paper (Fig. 24.4) shows a rapidly falling curve in the lower tail. Therefore, the ME should conclude that the normal is not a good distribution for this shelf life data.

COMMENT Some experience in probability plotting will show you the general shape of plots for other distributions on normal paper. The plot in Fig. 24.4 is characteristic of exponential-type data, because the lower tail drops off rapidly. This example is continued in the next section.

Problems P24.4–P24.6

24.3 Probability Plotting for the Exponential Distribution

The steps of Sec. 24.1 are used to plot and test exponential-type data on semilogarithmic paper. Observed X_i values are plotted on the arithmetic scale against the cumulative percentage on the log scale.

Equation (16.4) gives the exponential cdf as

$$F(x) = 1 - e^{-\lambda x}$$

The subscript on x has been dropped for clarity. If we obtain the complement of the cdf, $1 - F(x)$, and take the natural logarithm, we have a relation that is linear in x:

$$\ln[1 - F(x)] = -\lambda x$$

A common log or natural log scale can be used to test for an exponential fit, which will appear as shown in Fig. 24.5. Rather than computing the value of $100[1 - F(x)]$ in step 3, the log scale can be labeled in reverse order and $100F(x)$ versus x plotted directly. This reverse scale is shown in Fig. 24.5 and used in Example 24.3. There are other ways to plot exponential data, but this method is simple and uses the steps of Sec. 24.1.

Once the exponential is determined to fit the data, the parameter λ may be estimated at the x value which has $100F(x) = 63.2$ percent. This is possible because, as shown in Sec. 16.4, the area under the exponential pdf to the left of $1/\lambda$ is 0.632:

$$\int_0^{1/\lambda} \lambda e^{-\lambda x}\, dx = 0.632$$

The estimate obtained from the probability plot is the expected value $1/\lambda$, so the

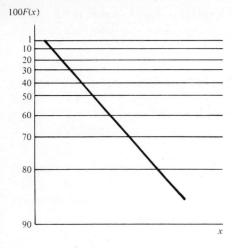

Figure 24.5 General shape of exponential data plotted on semilog paper.

parameter is

$$\lambda = \frac{1}{\text{expected value}} = \frac{1}{\bar{X}} \quad (24.4)$$

Example 24.3 Use the probability plotting procedure on the data in Example 24.2b to determine if it follows an exponential distribution. If so, estimate the parameter λ.

SOLUTION The steps for the exponential plot are:

1. The exponential distribution is hypothesized.
2 and 3. The data is ordered, and the $100F(x)$ values are computed in Table 24.3.
4. Figure 24.6 is a plot of the values in Table 24.3 with a straight-line fit to them. Note that the reverse $100F(x)$ scale is plotted from the top down for each x value.
5. The fit is good except for the last values; therefore, accept the exponential.
6. At $100F(x) = 63.2$, $1/\lambda = 5.9$ days. Therefore, the parameter estimate is

$$\lambda = \frac{1}{5.9 \text{ days}} = 0.17 \text{ per day}$$

The expected shelf life for lettuce is 5.9 days.

COMMENT The value λ can also be estimated from the data once the plot shows the exponential to be appropriate. From Sec. 16.3 the best estimate of

24.4 PROBABILITY PLOTTING FOR THE UNIFORM DISTRIBUTION

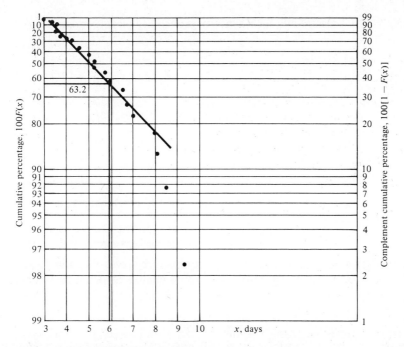

Figure 24.6 Probability plot on exponential paper, Example 24.3.

λ is

$$\hat{\lambda} = \frac{1}{\bar{X}} = \frac{1}{5.56} = 0.18 \text{ per day}$$

which is very close to the estimate from the plot.

It is possible to obtain one form of probability paper which is usable for the exponential, χ^2, and gamma distributions. See the text by Hahn and Shapiro for a complete discussion.

Problems P24.7–P24.12

24.4 Probability Plotting for the Uniform Distribution

If the uniform distribution for a continuous variable is thought to be appropriate, the steps of the probability plotting procedure are used, except in step 3 it is better to compute the cumulative percentage as

$$100F(x_i) = \frac{100i}{n+1} \tag{24.5}$$

The plot of $100F(x_i)$ versus x_i is made on regular, arithmetic graph paper. Again, a good fit to a straight line means that the uniform is appropriate.

438 GRAPHICAL PROCEDURE FOR FITTING A DISTRIBUTION

Table 24.4 Computations for a uniform probability plot, Example 24.4

i	x_i	$100F(x_i)$	i	x_i	$100F(x_i)$
1	985	6.25	9	1101	56.25
2	995	12.50	10	1103	62.50
3	1004	18.75	11	1120	68.75
4	1010	25.00	12	1131	75.00
5	1024	31.25	13	1157	81.25
6	1065	37.50	14	1200	87.50
7	1084	43.75	15	1210	93.75
8	1091	50.00			

The parameters α and β of the uniform pdf, Eq. (16.1), are X_{min} and X_{max}, respectively.

Example 24.4 A large manufacturing company believes that the daily production from a processing line has an equal chance of being between 1000 and 1200. A total of 15 days production is recorded. Test the statement above for correctness.

1120	1084	995	1010	1210
1004	1157	1091	1103	1065
1200	1131	1024	985	1101

SOLUTION The steps of Sec. 24.1 are summarized here. The uniform is hypothesized, and Table 24.4 summarizes the production data and cumulative percentages from Eq. (24.5). Figure 24.7 shows the probability plot and eye-balled straight line. The uniform is an appropriate model. The estimates from the plot are $\alpha = X_{min} = 985$ and $\beta = X_{max} = 1210$, which are close to the 1000 and 1200 figures stated.

Problems P24.13–P24.18

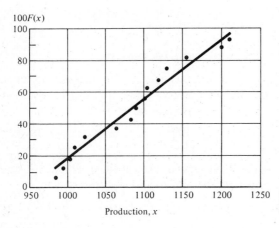

Figure 24.7 Probability plot for a uniform distribution, Example 24.4.

24.5 Probability Plotting for the Poisson Distribution

If the Poisson distribution or the binomial with a small p value ($p \leq 0.10$) is to be tested, special Poisson paper is used. This allows us to graphically test fits to discrete distributions. However, since the variable range is $x = 0, 1, 2, \ldots$ and because the observed frequency for each x value usually exceeds 1, we rewrite the probability plotting steps (especially steps 3 and 4) as follows:

1. The Poisson is hypothesized.
2. Order the observed data with frequencies. Call these x_i and f_i, respectively.
3. Tabulate the cumulative probability for the ordered data.

$$P(X \leq x_i) = \frac{\text{sum of frequencies} - 0.5}{\text{sample size}}$$

or

$$F(x_i) = \frac{\left(\sum_{i=1}^{x_i} f_i\right) - 0.5}{n} \qquad (24.6)$$

4. Plot $F(x_i)$ using the right-side scale on the appropriate curve $x = 0, 1, 2, \ldots$ of the Poisson paper shown in Fig. 24.8. This paper is simply a graphical

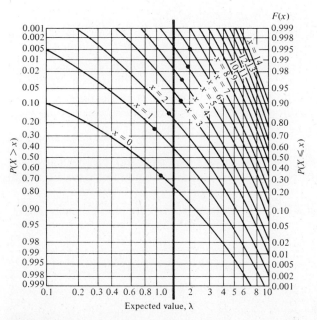

Figure 24.8 Probability plot of sulfur content data on Poisson paper, Example 24.5a. (Adapted from *Applied Statistics for Engineers* by W. Volk, Copyright by McGraw-Hill, 1969. Used with permission of McGraw-Hill Book Co.)

440 GRAPHICAL PROCEDURE FOR FITTING A DISTRIBUTION

equivalent of the cumulative Poisson probability values given in Table B-2. Construct an eye-balled straight, *vertical* line through the points.
5. Conclude that the Poisson is appropriate if the points are close to the vertical line. You may neglect the last one or two x values.
6. If the Poisson is appropriate, an estimate of the parameter λ is given by the intersection of the vertical line and the abscissa.

Example 24.5 illustrates these steps.

Example 24.5 Determine if the Poisson is an appropriate distribution for (*a*) the sulfur content data for tires tested by the χ^2 goodness-of-fit test in Example 23.2 (Sec. 23.3) and (*b*) the number of incorrectly installed rivets in 70 airplane wings. The data is:

Mistakes	0	1	2	3	4	5	6
Number of wings	10	20	15	10	5	7	3

SOLUTION (*a*) The steps above are used.

1 and 2. The Poisson is hypothesized for the percentage data ordered in Table 24.5, as taken from Example 23.2.
3. The cumulative probability $F(x_i)$ using Eq. (24.6) is tabulated in Table 24.5.
4. A plot of x_i versus $F(x_i)$ in Fig. 24.8 is widely dispersed around any vertical line. Note that $F(x_i)$ values are plotted *on* the correct x curve, neglecting the abscissa scale lines at this time.
5. The Poisson is not an appropriate distribution, which is the same conclusion reached by the χ^2 goodness-of-fit test.

(*b*) The same steps are followed here. Table 24.6 and Fig. 24.9 give the results. The cumulative probability values are computed from Eq. (24.6)

Table 24.5 Computations for a Poisson probability plot, Example 24.5a

i	x_i	f_i	Σf_i	$F(x_i)$
1	0	33	33	0.325
2	1	44	77	0.765
3	2	10	87	0.865
4	3	5	92	0.915
5	4	5	97	0.965
6	5	2	99	0.985
7	6	1	100	0.995

24.5 PROBABILITY PLOTTING FOR THE POISSON DISTRIBUTION

Table 24.6 Computations for a Poisson probability plot, Example 24.5b

i	x_i	f_i	Σf_i	$F(x_i)$
1	0	10	10	0.136
2	1	20	30	0.421
3	2	15	45	0.636
4	3	10	55	0.779
5	4	5	60	0.850
6	5	7	67	0.950
7	6	3	70	0.993

as

$$F(x_i) = \frac{(\Sigma f_i) - 0.5}{70}$$

It seems that the Poisson is a good distribution for incorrectly installed rivets, since the points cluster around a vertical line. An estimate of λ (step 6) is $\lambda = 2.2$ rivets per wing. This is an excellent estimate, because the best estimator from the data is $\hat{\lambda} = \overline{X} = 2.18$.

Problems P24.19–P24.22

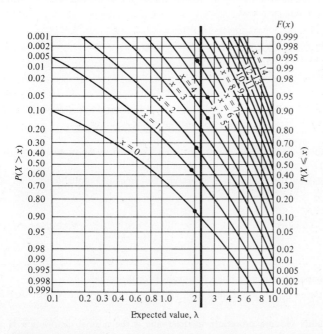

Figure 24.9 Probability plot of incorrectly installed rivets on Poisson paper, Example 24.5b. (Adapted from *Applied Statistics for Engineers* by W. Volk, Copyright by McGraw-Hill, 1969. Used with permission of McGraw-Hill Book Co.)

ADDITIONAL MATERIAL

Belz [2], pp. 109–114; Benjamin and Cornell [3], pp. 447–459; Bowker and Lieberman [4], pp. 453–455; Hahn and Shapiro [10], pp. 260–294; Miller and Freund [18], pp. 143–146; Neville and Kennedy [19], pp. 79, 102–105.

PROBLEMS

The following data sets are used by the indicated problems in this chapter.

A: The time that it took to bake bread was recorded for 20 different bakes. The ordered times in min are:

15	15	17	18	18
18	18	19	19	19
20	20	20	21	21
22	22	23	24	25

Problems P24.4, P24.7, P24.13

B: The diameter of a particular part has an aimed-at value of 15.0 mm. The quality control chart for the mean that is to be used assumes an underlying normal population, so the following 15 readings were taken for distribution testing purposes:

12.2	12.2	12.3	12.5	13.1
13.8	14.6	15.3	15.8	16.8
17.0	17.1	18.2	18.9	19.4

Problems P24.5, P24.8, P24.14

P24.1 Determine the cumulative percentage points that would be used in the probability plotting procedure using Eq. (24.1) for samples of size (a) $n = 8$ and (b) $n = 16$.

P24.2 There are three commonly used relations to determine the cumulative probability values used in plotting procedures:

(a) $F(x_i) = \dfrac{i - 0.5}{n}$

(b) $F(x_i) = \dfrac{i}{n}$

(c) $F(x_i) = \dfrac{i}{n+1}$

Use these three to compute and compare the $F(x_i)$ values for the sample size $n = 10$.

P24.3 Summarize the six steps of the probability plotting procedure on an index card for later reference.

P24.4 (a) Use the probability plotting procedure to determine if the normal distribution is appropriate for the bake times in data set A.

(b) Estimate the parameters for the normal from the probability plot and by the unbiased estimator formulas.

P24.5 Does it appear that the diameter values in data set B are from a normal population with a mean of about 15 mm?

P24.6 Ms. Beluv has developed a rating scale for the efficiency of financial information flow in a company. She feels that the ratings will follow a normal distribution. In a sample of eight separate offices, the ratings computed were 20, 31, 44, 49, 61, 73, 82, and 90.

(a) Use the graphical method to test the hypothesis that the ratings are normally distributed.

(b) What are the best graphically determined estimates of population mean and standard deviation for a normal distribution?

P24.7 Does it appear that the exponential model is appropriate for data set A? Why or why not?

P24.8 Use the plotting procedure to determine if the diameters in data set B follow an exponential distribution.

P24.9 The time in months that 12 employees worked at the All-But-Broke National Bank before receiving a promotion are 4, 5, 7, 9, 10, 11, 12, 12, 15, 22, 30, 35.

(a) Is the time until promotion better explained by the normal or exponential distribution?

(b) Estimate the parameter(s) for the distribution selected in (a).

P24.10 The data shown is from an exponential distribution.

(a) Use the graphical method to estimate the parameter.

(b) Plot these values on normal probability paper and construct the best eye-balled curve through them. Note the shape of the curve of exponential data on normal paper.

10	11	12.2	13.9	15.5
17	19	21	24	26
29	32	35	39	44
49	56	65	78	110

P24.11 Can the exponential distribution be used to explain the efficiency ratings in Prob. P24.6?

P24.12 Use the probability plotting procedure to determine if the data in Prob. P23.32 is better fit by an exponential distribution with a parameter value of $\lambda = 0.3$ or $\lambda = 0.1$.

P24.13 Is the uniform distribution model appropriate for data set A?

P24.14 Fit a uniform distribution to data set B using the probability plotting procedure. Does the data follow a uniform distribution?

P24.15 The data shown is from a uniform distribution.

(a) Plot the data to show it is from the uniform model.

(b) Plot the data on normal probability paper. Note how uniform data looks when plotted on normal paper.

1.8	3.0	4.1	5.4	6.4
7.6	8.8	10.0	11.2	12.2
13.5	14.7	15.8	17.0	18.2

P24.16 Determine if the data in Prob. P23.32 is better explained by the uniform or exponential distribution. Be sure to use the correct formulas when computing the cumulative percentage for these two distributions.

P24.17 (a) Plot the 50 data points in Prob. P23.14 to determine if the uniform distribution is appropriate. Compare this conclusion with the one reached using the goodness-of-fit test in Prob. P23.14b.

(b) Is this a good application of the probability plotting procedure? Why or why not?

P24.18 The service times in min at a full-service gas island were recorded. Is the uniform or exponential distribution more appropriate? Use graphical methods to make this determination.

4.7	6.4	4.2	0.8	7.9	1.4	7.7
9.4	6.7	4.1	8.4	0.6	1.6	3.4
6.7	2.8	0.3	4.0	7.8	2.2	1.6

444 GRAPHICAL PROCEDURE FOR FITTING A DISTRIBUTION

Note: The next four problems use Poisson probability paper. If you do not have access to this paper, you can use the graph in Fig. 24.10.

P24.19 (*a*) Use the graphical procedure to fit a Poisson distribution to the order data in Prob. P9.6.

(*b*) Estimate the Poisson parameter from the probability plot and by the best-estimator equation. Are the two estimates close to each other?

P24.20 Samples of size 20 are taken each hour from a circuit board assembly line. The number of defectives X in a sample has historically followed a $b(x; 20, 0.05)$ distribution. The results of two samples are given: the first from line A and the second from line B. Use the probability plotting procedure and the Poisson approximation to the binomial to determine if both lines are producing defectives according to the $b(x; 20, 0.05)$ distribution.

Number of defectives x	Line A	Line B
0	8	8
1	5	7
2	4	3
3	2	2
4	1	0

P24.21 Use the graphical method to solve Prob. P23.11.

P24.22 (*a*) Use the graphical method to show that the Poisson distribution offers an excellent explanation of the horse-kick data in Prob. P9.17.

(*b*) Estimate the Poisson parameter from the probability plot and compare it with the best-estimator result in Prob. P9.17.

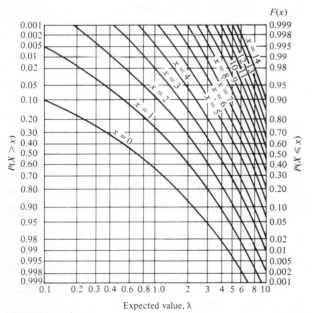

Figure 24.10 Poisson probability paper. (Adapted from W. Volk, *Applied Statistics for Engineers*. Copyright by McGraw-Hill, 1969. Used with permission of McGraw-Hill Book Co.)

CHAPTER
TWENTY-FIVE

STATISTICAL INFERENCE USING NONPARAMETRIC TESTS

Several of the commonly used nonparametric tests are explained in this chapter. Some of these tests are equivalent to those of Chap. 20 which make statistical inferences about means. However, they are all distribution-free; that is, no assumptions are made about the parameters of the population from which the samples are taken.

This chapter serves only as an introduction to the field of nonparametric statistics. A case study is included as the Solved Problem.

CRITERIA

To complete this chapter, you must be able to:

1. State the difference between a parametric and a nonparametric test, and state the questions which must be answered to correctly use the tests in this chapter.

In addition, you must be able to use the hypothesis testing procedure to perform the following tests, given the sample data:

2. *Runs test* for sample *randomness*.
3. One-sample *sign test* of a mean against a specific value, given the values.
4. Two-sample sign test for the equality of means.
5. Two-sample *rank sum test* (Mann-Whitney U test) for the equality of means of *independent* samples.
6. Two-sample *signed rank test* (Wilcoxon test) for the equality of means of *paired* samples.

STUDY GUIDE

25.1 Introduction to Nonparametric Statistics

Virtually all the tests in this section have assumed that samples are taken from a *normal* population. We have, therefore, always been concerned with parameter values, such as μ and σ, to make statistical inferences about the population. Tests which are able to accept or reject a null hypothesis without knowledge of the underlying population and parameters are called *nonparametric*, or *distribution-free*, tests. These tests simply assume that the data is taken from a continuous variable. In addition, some null hypotheses require that the distribution be symmetric about some reference point, for example, the median, mean, or 0. Since the assumptions necessary are few, a penalty must be paid in statistical power. A larger sample size is needed to have a nonparametric test with the same efficiency, that is, the same operating-characteristic (OC) curve (Sec. 19.6) as the corresponding parametric test. However, nonparametric tests are excellent for making conclusions on small to medium-size samples when the underlying population is not known or is known to be nonnormal. Actually, the two goodness-of-fit tests covered in Chap. 23, the χ^2 and the K-S tests, are nonparametric because the statistics do not require any population parameter values.

The tests in this chapter allow you to determine if a sample is actually random and to test the equality of means under various situations. Other less commonly used nonparametric tests are covered in the Additional Material texts. To use the correct test, answer the following questions. A summary of the means tests and appropriate section is given in Table 25.1.

1. *Is the randomness of a sample to be tested?* The runs test of Sec. 25.2 is used for this purpose.
2. *Are there one or two samples of data?*
3. *When testing the means of two samples, is the data naturally paired between samples?*
4. *When testing the means of two samples, do you consider the magnitude of each data value important?* If you do, a rank sum test is required. However, if only relative placement around a reference (mean, 0, or some specified value) is

Table 25.1 Classification of nonparametric tests for means

Description of sample(s)	Is the magnitude of the data important?	Appropriate test	Section
One	—	Sign	25.3
Two, independent	No	Sign	25.4
	Yes	Mann-Whitney U	25.5
Two, paired	No	Sign	25.4
	Yes	Wilcoxon	25.6

important, the direction can be indicated by a plus or minus sign and the sign test is sufficient.

Problem P25.1

25.2 One-Sample Runs Test of Randomness

Usually a sample that is taken from a population should be random. The *runs test* evaluates the null hypothesis

H_0: the order of the sample data is random

The alternative hypothesis is simply the negation of H_0. There is no comparable parametric test to evaluate this null hypothesis.

The *order* in which the data is collected must be retained so that the runs may be developed. A *run* is defined as a sequence of the same symbols. Two symbols are defined, and each sequence must contain a symbol at least once. In the following 14 plus and minus signs, there are a total of 7 runs.

$$\underbrace{+++}_{1}\ \underbrace{-}_{2}\ \underbrace{++}_{3}\ \underbrace{--}_{4}\ \underbrace{+}_{5}\ \underbrace{----}_{6}\ \underbrace{+}_{7}$$

The length of each run is unimportant because the frequency is not tested here as in the goodness-of-fit tests: this test is only for randomness. If there are too few or too many runs, the sample is judged nonrandom.

The one-sample runs test can be performed on different types of data.

1. The order of actual outcomes, where the two symbols might be plus and minus, or a and r for accept and reject.
2. The order of plus and minus signs, where the signs represent the size of the outcomes compared to some reference value.

The hypothesis testing procedure of Sec. 19.4 is used to test for randomness. In step 3, computation of sample statistics, the following are determined.

n_1 = total number of one type of symbol
n_2 = total number of the other symbol
r = total number of runs

The only question to ask prior to performing the test is, *Is the sample size small or large?* The test differs considerably for small and large samples. We will use the guideline that a *small sample* has n_1 and n_2 less than or equal to 15. The value of r is the test statistic. For the two-sided test and $\alpha = 0.05$, Table B-8 gives the lower r_L and upper r_U values of the distribution $f(r)$ with $\alpha/2 = 0.025$ in each tail. That is, as shown in Fig. 25.1,

$$P(r \leq r_L) + P(r \geq r_U) = \sum_{r=2}^{r_L} f(r) + \sum_{r=r_U}^{n} f(r) \doteq 0.05$$

Since $f(r)$ is discrete, the α level may not be exactly 0.05. If r is between these limits, accept H_0; if not, reject H_0 and the sample is not random. A one-sided

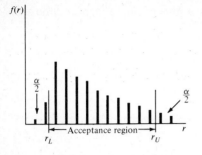

Figure 25.1 Distribution of the number of runs r showing the acceptance region for the test of randomness.

test of too few or too many runs with $\alpha = 0.025$ also uses Table B-8. A small-sample runs test is shown in Example 25.1.

Example 25.1 Carol, a technician, is asked to analyze the results of 22 items made in a preproduction run. Each item has been measured and compared to engineering specifications. The order of acceptances a and rejections r is

aarrrarraaaaarrarraara

Help Carol determine if this is a random sample.

SOLUTION Use the hypothesis testing steps.

1. H_0: the order of results is random; H_1: the order is not random.
2. Let $\alpha = 0.05$ for this two-sided test.
3. The statistics needed for the runs test are

$$n_1 = 12 \; a\text{'s} \qquad n_2 = 10 \; r\text{'s} \qquad r = 11 \text{ runs}$$

The sample is small since n_1 and $n_2 \leq 15$.
4. The test statistic is $r = 11$ runs.
5. The acceptance region for $n_1 = 12$ and $n_2 = 10$ is $7 \leq r \leq 17$ from Table B-8.
6. Since $7 \leq 11 \leq 17$, do not reject the null hypothesis that the preproduction run is random.

COMMENT Either of the one-sided hypotheses that there are too few or too many runs could be substituted for H_1. Then with $\alpha = 0.025$ the same critical values as above would be used. In both cases H_0 would still be accepted because r is in the acceptance region.

If n_1 or n_2 exceeds 15, the sample is considered *large*, in which case a normal approximation to $f(r)$ is used to test H_0 versus H_1. The standard normal distribution (SND) is used:

$$z = \frac{\text{number of runs} - \text{expected number of runs}}{\text{standard deviation of number of runs}}$$

$$= \frac{r - \mu_r}{\sigma_r} \tag{25.1}$$

where the sample size is $n = n_1 + n_2$ and

$$\mu_r = \frac{2n_1 n_2}{n} + 1 \tag{25.2}$$

$$\sigma_r = \left[\frac{2n_1 n_2 (2n_1 n_2 - n)}{n^2 (n - 1)} \right]^{1/2} \tag{25.3}$$

The SND Table B-3 is used to obtain the acceptance and critical regions for one-sided and two-sided tests. This large-sample approximation is applied in the case study.

Problems P25.2–P25.7

25.3 One-Sample Sign Test for the Mean

A specific sample mean value μ_0 can be tested against an observed value $\overline{X}$ using the z or t statistics of Secs. 20.2 and 20.3. An equivalent nonparametric test to evaluate the null hypothesis H_0: $\mu = \mu_0$ is the *one-sample sign test*, which requires only the assumptions that the distribution be symmetric and the variable be continuous.

The sign test is so named because it uses the number of plus and minus signs to determine if H_0 is true. Since only the number of signs is needed, the order in which the sample values were observed is not important, as it was in the runs test. The signs are determined as follows for each sample value X_i ($i = 1, 2, \ldots, n$).

Plus sign: sample value exceeds μ_0; $X_i > \mu_0$

Minus sign: μ_0 exceeds the sample value; $X_i < \mu_0$

No sign: if $X_i = \mu_0$, discard X_i and reduce n by 1

If $\mu = \mu_0$, the probability of a plus or minus sign is $\frac{1}{2}$ each, because the distribution is symmetric. Therefore, this test is equivalent to testing H_0: $p = \frac{1}{2}$ for the binomial parameter p.

The hypothesis testing procedure (Sec. 19.4) is used for the sign test. The alternative hypothesis may be one-sided or two-sided. The test statistic is always the variable S, where

$$s = \text{number of plus signs} \tag{25.4}$$

Because the test is based on the binomial distribution, which is discrete, the values of s which define the acceptance and critical regions will not place exactly α of the area in the critical region. Figure 25.2 shows the area for $s \geq 7$ and $s \geq 8$ for the binomial $b(s; 8, 0.5)$ taken from Table B-1. (Even though S is discrete, inequality symbols are used instead of itemizing each s value. You should, therefore, realize that $s \geq 7$ means $s = 7$ or 8 because $n = 8$.) For a two-sided sign test, because of pdf symmetry, critical regions and associated α

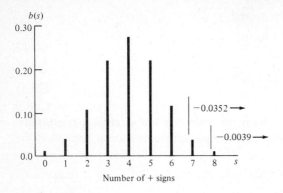

Figure 25.2 The binomial pdf for $n = 8$ and $p = 0.5$ used to establish the acceptance and critical regions for the sign test.

values might be

$$P(1 \geq s \geq 7) = 2(0.0352) = 0.0704$$
$$P(0 \geq s \geq 8) = 2(0.0039) = 0.0078$$

If $\alpha = 0.05$ is the significance level, the second region must be used since the first region will exceed this α level.

It is common to test the one-sided alternative hypothesis $H_1: \mu > \mu_0$ or $H_1: \mu < \mu_0$. In these cases, only the upper or lower tail of the binomial is used for the critical region. If $H_1: \mu > \mu_0$ is tested for $n = 8$, the critical region is $s \geq 7$ and $\alpha = 0.0352$ is the effective significance level. The sign test for one sample is illustrated in Example 25.2.

If the sample size exceeds 20 ($n > 20$), the binomial Table B-1 cannot be used. However, the normal approximation to the binomial (Sec. 11.8) may be applied. The mean and standard deviation of S, the number of plus signs, follows the binomial properties with $p = \frac{1}{2}$.

$$\mu_s = np = \tfrac{1}{2}n \qquad (25.5)$$

$$\sigma_s = \sqrt{npq} = \tfrac{1}{2}\sqrt{n} \qquad (25.6)$$

The test statistic is the standard normal variable Z with a *continuity correction* of 0.5 similar to that of Sec. 11.8. The test statistic is different depending on the size of s.

$$s < \mu_s = \tfrac{1}{2}n: \qquad z = \frac{s + 0.5 - \mu_s}{\sigma_s} \qquad (25.7a)$$

$$s > \mu_s = \tfrac{1}{2}n: \qquad z = \frac{s - 0.5 - \mu_s}{\sigma_s} \qquad (25.7b)$$

No correction is used if $s = \mu_s$. The SND Table B-3 is used to determine the acceptance region.

Example 25.2 A manufacturing company wants to determine the amount of personal time for assembly workers when production standards are set on new processes. The production manager believes that 5 percent of an 8-h day is sufficient. In the last 12 time studies in the plant, the personal time allowances, in percent, were:

| 6.2 | 7.9 | 4.6 | 5.2 | 5.7 | 8.1 |
| 7.4 | 5.0 | 6.8 | 5.0 | 6.1 | 10.2 |

If it cannot be assumed that personal time is set according to a normal distribution, determine if the manager's estimate of 5 percent is (a) acceptable or (b) too low.

SOLUTION First, we determine that the answer to question 2 in Sec. 25.1 is one sample; so, the one-sample sign test is appropriate. The plus and minus signs for this sample are found by comparing each value with $\mu_0 = 5.0$ percent. There are nine values with $X_i > 5.0$ percent, one with $X_i < 5.0$ percent, and two with $X_i = 5.0$ percent. The result is, after reducing the sample size by 2, $n = 10$ and $s = 9$ plus signs.
(a) The hypothesis testing procedure is applied.

1. The two-sided test is

$$H_0: \mu = 5.0\% \quad \text{versus} \quad H_1: \mu \neq 5.0\%$$

2. Select $\alpha = 0.05$.
3. and 4. The sample statistics are $n = 10$ and $s = 9$. The latter is also the test statistic value, according to Eq. (25.4).
5. From Table B-1 with $n = 10$, $p = 0.50$, and $\alpha/2 = 0.025$

$$P(s \leq 1) + P(s \geq 9) = 2(0.0107) = 0.0214$$

defines the integer values of s which will not exceed $\alpha = 0.05$. Therefore, the acceptance region is $2 \leq s \leq 8$.
6. The value of 9 plus signs is not in the acceptance region, so H_0 is rejected. The manager is incorrect. Observation shows that the manager is probably too low with a 5 percent estimate.

(b) To test $H_0: \mu = 5.0$ percent versus $H_1: \mu > 5.0$ percent, the same test as above is performed, except that the critical region is in the upper tail of the $b(s; 10, 0.5)$ distribution. From Table B-1

$$P(s \geq 8) = 0.0547$$

and

$$P(s \geq 9) = 0.0107$$

To not exceed $\alpha = 0.05$, the critical region is $s \geq 9$, so the acceptance region is $s \leq 8$ plus signs. With $s = 9$, reject H_0 and conclude that $\mu > 5.0$ percent.

COMMENT The difference between each sample value and the μ_0 value is not taken into account by the sign test. This magnitude is considered in more powerful nonparametric tests covered later in this chapter.

Remember that this test is equivalent to testing for a binomial parameter value $p = \frac{1}{2}$. Therefore, the one-tail hypothesis $H_1: \mu > \mu_0$ is the same as $H_1: p > \frac{1}{2}$.

An example of the sign test for a large sample is included in the case study.

Problems P25.8–P25.15

25.4 Two-Sample Sign Test for Means

If there are two independent samples and the data values can be matched to one another or if they are actually paired from two nonindependent samples (as in Sec. 20.6), the hypothesis $H_0: \mu_1 = \mu_2$ can be tested by the *two-sample sign test*. Of course, if the two independent samples have $n_1 \neq n_2$, the larger sample must be reduced in size so that $n_1 = n_2 = n$. Once the pairing of values is done, the sign test proceeds in the same way as in Sec. 25.3. The plus and minus signs are determined as follows for the X_{i1} and X_{i2} ($i = 1, 2, \ldots, n$) values in the samples.

Plus sign:	sample 1 greater than sample 2; $X_{i1} > X_{i2}$
Minus sign:	sample 2 greater than sample 1; $X_{i1} < X_{i2}$
No sign:	two values are equal; $X_{i1} = X_{i2}$ are discarded and n reduced by 1

The test statistic is still the variable S as defined in Sec. 25.3:

$$s = \text{number of plus signs} \tag{25.8}$$

Example 25.3 The customer affairs office of a large department store has made a study of the effect of bad employment news on the productivity of 15 salespersons. The daily sales were recorded to the nearest $100 for one week before and one week after the news that overemployment in the recent past would mean the termination of several salespersons. The averages before and after daily sales are given in Table 25.2. Using only the fact that sales went up or down, determine if the announcement affected sales volumes.

SOLUTION If no assumptions are made about sampling from normal distributions of sales volume, the means test of Chap. 20 cannot be used. However, we can always use a nonparametric test. Answers to the four questions of Sec. 25.1 are: (1) randomness need not be checked; (2) two samples; (3) the data is naturally paired by employee; and (4) only sign, not magnitude, of volume change is important. Therefore, the two-sample sign

Table 25.2 Daily sales before and after a personnel cut announcement, Example 25.3

	Sales volume, $/day		Sign of
Employee i	Before, X_{i1}	After, X_{i2}	$X_{i1} - X_{i2}$
1	2,500	2,800	−
2	5,700	5,700	0
3	9,200	9,100	+
4	3,800	4,600	−
5	2,800	1,900	+
6	10,100	9,300	+
7	3,500	3,500	0
8	1,700	2,500	−
9	8,700	8,300	+
10	4,800	5,700	−
11	6,000	7,100	−
12	8,900	9,400	−
13	1,500	1,200	+
14	7,100	6,800	+
15	2,600	2,600	0

test is used. The hypothesis testing steps are:

1. The two-sided test is
$$H_0: \mu_1 = \mu_2 \quad \text{versus} \quad H_1: \mu_1 \neq \mu_2$$
2. Let $\alpha = 0.05$.
3. Table 25.2 shows the plus and minus signs for each employee. Since there are three ties, the sample statistics are $n = 12$ and $s = 6$ plus signs.
4. The test statistic value is $s = 6$.
5. For a two-sided test and $\alpha = 0.05$, Table B-1 is used to determine the acceptance region for $b(s; 12, 0.5)$ as $3 \leq s \leq 9$ plus signs.
6. Do not reject the hypothesis of equal means. The announcement did not have an effect on sales volume per day.

COMMENT If samples have $n > 20$, the normal approximation presented in Eq. (25.7) is used.

Problems P25.16–P25.21

25.5 Two-Sample Rank Sum Test for Means (Mann-Whitney U Test)

The sign tests of the previous sections do not consider the size of the difference between data values, because only plus and minus signs are used to make conclusions. *Rank sum tests*, which are designed to consider the magnitude of the data, are more powerful than sign tests.

The Mann-Whitney U test is used to determine if the means of two independently taken samples are equal. The samples need not be the same size. This is the nonparametric equivalent of the two-sample t test of Sec. 20.5. The test devised here uses the normal approximation for n_1 and $n_2 \geq 9$. If n_1 or n_2 is very small, the exact values of U must be used. See the text by Siegel.

To perform the Mann-Whitney U test, the null hypothesis H_0: $\mu_1 = \mu_2$ is formulated. The following procedure is used in step 3 of the hypothesis testing steps to determine the ranks and the sample statistic U. Example 25.4 explains the tabulation for these steps.

1. List all data values in increasing order for both samples combined, but retain sample identity.
2. Assign ranks 1, 2, ..., $n_1 + n_2$ to the data.
3. Tied data values are each given the *average* rank of those they were initially assigned. For example, if the sixth and seventh values are equal, their rank is the average:

$$\text{Rank} = \frac{6 + 7}{2} = \frac{13}{2} = 6.5$$

4. Sum the final ranks for each sample individually. Compute the sample statistic U, which is the number of times that a data value of one sample precedes a value in the second sample. Rather than counting, U may be computed as

$$u = n_1 n_2 + \frac{n_1(n_1 + 1)}{2} - R_1 \qquad (25.9)$$

where R_1 = sum of ranks of one sample (either sample is acceptable when computing R_1)
n_1 = size of sample for which R_1 is determined

The normal approximation for U has the properties

$$\mu_u = \frac{n_1 n_2}{2}$$

$$\sigma_u = \left[\frac{n_1 n_2 (n_1 + n_2 + 1)}{12} \right]^{1/2}$$

The standard normal distribution (SND) is then used to compute the test statistic

$$z = \frac{u - \mu_u}{\sigma_u} \qquad (25.10)$$

Note that z is the test statistic and u is a sample statistic.

If the null hypothesis $H_0: \mu_1 = \mu_2$ is rejected, we can conclude that the samples do not come from the same population. This test is illustrated in Example 25.4.

Example 25.4 A chemist plans to purchase laboratory supplies from company A or B, depending on which has the better quality. The percentage of defectives of contracted items purchased in the past is shown below for A and B. Use the Mann-Whitney U test to determine if quality levels are equal.

Percentage for A:	7.0	3.5	9.6	8.1	6.2
	5.1	10.4	4.0	2.0	
Percentage for B:	5.5	3.2	4.2	11.0	9.7
	6.9	3.5	4.8	5.5	8.4
	10.1	5.5	12.3		

SOLUTION The Mann-Whitney U test is appropriate here because there are two independent samples and magnitudes are important (see Table 25.1). The hypothesis testing procedure to perform a two-sided test of

$$H_0: \mu_A = \mu_B \quad \text{versus} \quad H_1: \mu_A \neq \mu_B$$

at the $\alpha = 0.05$ level is summarized here. The sample statistic U (step 3) is computed using the four steps above:

1. Table 25.3 shows the combined 22 sample values in increasing order.
2. and 3. Ranks from 1 to 22 are assigned, but they are separately maintained for each sample. Ties occur for 3.5 and 5.5 percent. The two 3.5 percent values are assigned the rank $(3 + 4)/2 = 3.5$, while the three 5.5 percent values are given a rank of $(9 + 10 + 11)/3 = 10$.
4. The sum of the ranks for company A is $R_1 = 95.5$. Then $n_1 = 9$ and $n_2 = 13$. From Eq. (25.9),

$$u = 9(13) + \frac{9(10)}{2} - 95.5 = 66.5$$

The test statistic is now computed by Eq. (25.10) after determining μ_u and σ_u.

$$\mu_u = \frac{9(13)}{2} = 58.5$$

$$\sigma_u = \left[\frac{9(13)(23)}{12}\right]^{1/2} = (224.25)^{1/2} = 14.97$$

$$z = \frac{66.5 - 58.5}{14.97} = 0.53$$

Table 25.3 Assignment of ranks to manufacturer quality levels for the Mann-Whitney U test, Example 25.4

No.	Percentage defective Company A	Percentage defective Company B	Company A ranks	Company B ranks
1	2.0		1	
2		3.2		2
3		3.5		3.5
4	3.5		3.5	
5	4.0		5	
6		4.2		6
7		4.8		7
8	5.1		8	
9		5.5		10
10		5.5		10
11		5.5		10
12	6.2		12	
13		6.9		13
14	7.0		14	
15	8.1		15	
16		8.4		16
17	9.6		17	
18		9.7		18
19		10.1		19
20	10.4		20	
21		11.0		21
22		12.3		22
			95.5	157.5

The acceptance region for the SND with $\alpha = 0.05$ is $|z| \leq 1.96$, so the null hypothesis is accepted. The two companies have equal quality levels.

If the company B rank sum is designated $R_1 = 157.5$, then $U = 50.5$, $n_1 = 13$, and $z = -0.53$. The same conclusion is reached.

Example 25.6 (Case Study)

Problems P25.22–P25.28

25.6 Two-Sample Signed Rank Test for Means (Wilcoxon Test)

If the data is naturally paired and the magnitude of the difference between values is important, the Wilcoxon test of signed ranks should be used instead of the two-sample sign test of Sec. 25.4. The Wilcoxon test, which assumes a

25.6 TWO-SAMPLE SIGNED RANK TEST FOR MEANS (WILCOXON TEST)

continuous variable but does not require that the pdf be symmetric, is a nonparametric equivalent of the paired t test of Sec. 20.6. The null hypothesis is H_0: $\mu_1 = \mu_2$, and the one-sided or two-sided alternative can be used.

The conclusion of the test is based on the sum of the ranks of the differences between paired values with the algebraic sign considered. The variable is

T = smaller sum of absolute value of signed ranks of differences

(The symbol T will be used here for the variable and its value to avoid confusion with the t distribution.) If H_0 is true, this sum should be equal for positively and negatively signed differences. If the sum for one sign is too small, H_0 is rejected. Therefore, the critical region for the null hypothesis is always in the lower tail of the pdf for T (Fig. 25.3). The test is two-sided if H_1: $\mu_1 \neq \mu_2$ is the alternative hypothesis, and is one-sided if H_1: $\mu_1 < \mu_2$ or H_1: $\mu_1 > \mu_2$ is used. However, the one-sided tests will still have the critical region in the lower tail of the pdf, because H_1 simply predicts that the smaller sum of signed ranks will occur for a particular sign.

In the hypothesis testing procedure (step 3) the sample statistics must be computed. The following steps, which are illustrated in Example 25.5, are used in the procedure.

1. Subtract the paired data values for the two samples with $i = 1, 2, \ldots, n$ to obtain the differences

$$d_i = X_{i1} - X_{i2}$$

   Discard all $d_i = 0$ values and reduce n by this number.
2. Arrange the absolute values, that is, $|d_i|$, in increasing order.
3. Assign the ranks $1, 2, \ldots, n$ to the ordered differences. Any ties are given the *average* of the assigned ranks. (This is the same procedure used in the Mann-Whitney U test.)
4. Place the same sign on the ranks as occurs on each corresponding d_i value. Compute the two sums

$$D_1 = \Sigma d_i \quad \text{for all } d_i > 0$$

$$D_2 = \Sigma d_i \quad \text{for all } d_i < 0$$

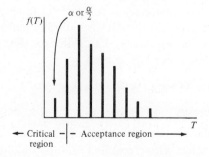

Figure 25.3 The Wilcoxon test has the critical region in the lower tail of the pdf.

To compute the test statistic, it is necessary to answer the question, *Is the sample small* ($n \leq 25$) *or large*? If n is *small*, the test statistic T is the smaller of D_1 and D_2.

$$T = \min [D_1, D_2] \qquad (25.11)$$

The values of T which define the critical and acceptance regions are given in Table B-9 for $n \leq 25$ for two-tail significance levels of 0.05, 0.02, and 0.01. If T *exceeds* the tabulated value T_0, H_0 is accepted. This test result is summarized as follows:

$$T \leq T_0 \text{ means reject } H_0$$
$$T > T_0 \text{ means accept } H_0$$

For a one-sided test, the significance levels tabulated are one-half the two-sided levels above. These levels are also shown in Table B-9.

If the sample is *large*, the SND approximation is used and the test statistic is the standard normal variable.

$$z = \frac{T - \mu_T}{\sigma_T} \qquad (25.12)$$

The mean and standard deviation of T are:

$$\mu_T = \frac{n(n+1)}{4}$$

$$\sigma_T = \left[\frac{n(n+1)(2n+1)}{24} \right]^{1/2}$$

The value of T is computed from Eq. (25.11). The acceptance region is given by the SND Table B-3 with α (two-sided) or $\alpha/2$ (one-sided test) of the area in the lower tail.

Example 25.5 Use the sales volume data of Example 25.3 (Table 25.2) to determine if the daily average sales increased after the personnel reduction announcement. Consider the magnitude of the sales increases and decreases in the computations.

SOLUTION The questions in Sec. 25.1 are answered as follows: (1) randomness need not be checked; (2) two samples; (3) the data is naturally paired; and (4) magnitude of volume change is important. The Wilcoxon test is appropriate here. The hypothesis testing procedure is used.

1. This is a *one-sided* test of

$$H_0: \mu_1 = \mu_2 \qquad \text{versus} \qquad H_1: \mu_1 < \mu_2$$

The volumes after the announcement (sample 2) should increase, so $d_i < 0$ is predicted to occur more often. Therefore, there should be more

ranks with negative signs, and the smaller sum in Eq. (25.11) should occur for the $d_i > 0$ values.
2. Let $\alpha = 0.025$ for this one-sided test.
3. Use the steps above to compute the sample statistics D_1 and D_2. See Table 25.4.
   1. - The differences d_i are given in Table 25.4. Three 0 values occur, so $n = 12$.
   2. - The absolute values $|d_i|$ are ordered for the 12 d_i values.
   3. - Ranks are assigned, and all repeated values are given the average rank.
   4. - Ranks are separated into plus and minus columns (Table 25.4) and summed to obtain $D_1 = 30.5$ and $D_2 = 47.5$.
4. The sample size $n = 12$ is small, so the test statistic is Eq. (25.11).

$$T = \min[30.5, 47.5] = 30.5$$

The positive ranks have a smaller sum, as predicted by the alternative hypothesis.

5. The acceptance region from Table B-9 for a one-tail $\alpha = 0.025$ and $n = 12$ is $T > 14$.
6. Since $30.5 > 14$, H_0 is not rejected. The announcement did not increase sales volume, as was also concluded using the two-sample sign test.

Problems P25.29–P25.34

Table 25.4 Computations for the Wilcoxon test of means, Example 25.5

| Employee i | Sales volume, $/day | | | $\|d_i\|$ ordered | Ranks | |
	Before, X_{i1}	After, X_{i2}	Difference d_i		Plus	Minus
1	2,500	2,800	−300	100	1	
2	5,700	5,700	0	300		3
3	9,200	9,100	+100	300	3	
4	3,800	4,600	−800	300	3	
5	2,800	1,900	+900	400	5	
6	10,100	9,300	+800	500		6
7	3,500	3,500	0	800		8
8	1,700	2,500	−800	800		8
9	8,700	8,300	+400	800	8	
10	4,800	5,700	−900	900		10.5
11	6,000	7,100	−1100	900	10.5	
12	8,900	9,400	−500	1,100		12
13	1,500	1,200	+300		30.5	47.5
14	7,100	6,800	+300			
15	2,600	2,600	0			

460 STATISTICAL INFERENCE USING NONPARAMETRIC TESTS

SOLVED PROBLEM (Case Study)

Example 25.6 Henry Roberts is a business consultant working on the design and implementation of a new drug distribution system in a triservice military hospital. The total time to deliver and administer an emergency injection of a drug to a patient is collected for 30 patients using the present distribution system. The times are given in minutes. The standard time was set at 1.5 min when this method was initiated.

1.1	0.8	2.1	0.4	0.9	1.3
1.2	2.1	3.0	1.0	0.6	2.5
0.5	0.9	1.8	2.1	1.0	0.8
3.1	1.0	3.0	0.9	0.2	2.1
1.4	2.0	1.5	1.4	0.8	1.0

Henry does not expect the times to be normally distributed because the present procedure is not standardized on all floors of the hospital. Use nonparametric statistics to answer the following questions for Henry.

(a) Is the sample random about the mean? The order of collection is retained across the rows.
(b) Is the average time equal to the standard time?
(c) A new distribution method is implemented and 30 trials are observed after training is completed. Are the sample means equal if only sign differences are considered?

0.8	0.7	1.5	2.0	1.7	0.8
0.6	1.5	0.9	1.9	3.1	0.2
1.0	1.9	0.7	1.1	2.0	1.1
0.6	0.4	0.8	0.6	0.4	1.9
2.0	0.9	0.6	1.0	1.5	0.6

(d) Test for equality of means, only consider the size of the differences in times.
(e) Use the appropriate test from Chap. 20 to test for equality of means.

SOLUTION First, you should answer the questions in Sec. 25.1 and refer to Table 25.1 for the appropriate test. We give only the results from Table 25.1. The significance level is $\alpha = 0.05$ for all parts, and the results of the hypothesis testing procedure are summarized. The samples are large so the normal approximation tests are used.

(a) Question 1 on randomness is answered yes for a large sample. In Sec. 25.2, Eq. (25.1) is the test statistic. The average time for the old method is $\bar{X} = 1.42$ min. When plus and minus signs are determined using $\bar{X}$ as a reference value, we have

$n_1 = 19$ minus signs $\qquad n_2 = 11$ plus signs $\qquad r = 17$ runs

Therefore $\mu_r = 14.93$, $\sigma_r = 2.49$, and

$$z = \frac{17 - 14.93}{2.49} = +0.83$$

This value is in the acceptance region $|z| \leq 1.96$: do not reject H_0, so the sample is random.

Section 25.2

(b) For one sample (question 2) the sign test is used with

$$H_0: \mu = 1.5 \quad \text{versus} \quad H_1: \mu \neq 1.5 \text{ min}$$

From Eq. (25.4) and the signs prepared in (a), $s = 11$ plus signs. The test statistic is Eq. (25.7a) because $s = 11 < \mu_s = 15$. With $\sigma_s = 2.74$,

$$z = \frac{11.5 - 15}{2.74} = -1.28$$

Again, with the acceptance region of $|z| \leq 1.96$, H_0 is accepted. The old method of distribution seems to still comply with the standard of 1.5 min.

Section 25.3

(c) For two independent samples the sign test is appropriate to test H_0: $\mu_1 = \mu_2$, where the subscripts 1 and 2 refer to the old and new methods, respectively. The plus and minus signs below are determined using the criteria in Sec. 25.4, that is, plus sign if the old method took longer than the new method. The data values are matched across rows to determine the signs.

```
+ + + − − +
+ + + − − +
− − + + − −
+ + + + − +
− + + + − +
```

There are $s = 19$ plus signs in this sequence. Since the samples are large, the mean and standard deviation of s, the number of plus signs, by Eqs. (25.5) and (25.6), are $\mu_s = 15$ and $\sigma_s = 2.74$. The test statistic is Eq. (25.7b):

$$z = \frac{18.5 - 15}{2.74} = +1.28$$

The acceptance region is again $|z| \leq 1.96$, so H_0 is not rejected. The two methods have statistically equal distribution times.

Section 25.4

(d) To test for mean equality of two samples accounting for the magnitude of differences requires that the Mann-Whitney U test be used for H_0: $\mu_1 = \mu_2$. Table 25.5 gives the results of all computations outlined in the steps in Sec. 25.5. The sample values are ordered and ranked, and the

Table 25.5 Comparison of two methods of drug distribution, Example 25.6

Ordered data		Ranks		Ordered data		Ranks	
Old	New	Old	New	Old	New	Old	New
0.2		1.5			1.0		28.5
	0.2		1.5	1.1		33	
	0.4		4		1.1		33
	0.4		4		1.1		33
0.4		4		1.2		35	
0.5		6		1.3		36	
0.6		9.5		1.4		37.5	
	0.6		9.5	1.4		37.5	
	0.6		9.5	1.5		40.5	
	0.6		9.5		1.5		40.5
	0.6		9.5		1.5		40.5
	0.6		9.5		1.5		40.5
	0.7		13.5		1.7		43
	0.7		13.5	1.8		44	
0.8		17.5			1.9		46
0.8		17.5			1.9		46
0.8		17.5			1.9		46
	0.8		17.5	2.0		49.5	
	0.8		17.5		2.0		49.5
	0.8		17.5		2.0		49.5
0.9		23			2.0		49.5
0.9		23		2.1		53.5	
0.9		23		2.1		53.5	
	0.9		23	2.1		53.5	
	0.9		23	2.1		53.5	
1.0		28.5		2.5		56	
1.0		28.5		3.0		57.5	
1.0		28.5		3.0		57.5	
1.0		28.5		3.1		59.5	
	1.0		28.5		3.1		59.5
						1014.0	816.0

ranks summed. If the new method is called sample 1, $R_1 = 816$ and the sample statistic is

$$u = 30(30) + \frac{30(31)}{2} - 816 = 549$$

The normal approximation gives

$$\mu_u = 450 \qquad \sigma_u = 67.64 \qquad z = 1.46$$

Again H_0 is accepted since 1.46 is in the acceptance region $|z| \leq 1.96$.

Therefore, all tests have shown that the old standard time of 1.5 min is still good and the new method offers no significant improvement.

Section 25.5

(e) The questions in Sec. 20.1 and Table 20.1 indicate that the (parametric) normal test statistic for large samples in Sec. 20.4 may be used to test H_0: $\mu_1 = \mu_2$. The sample means and variances are needed to compute z.

Old method: $\quad\quad\quad \mu_1 = 1.42$ min $\quad$ and $\quad \sigma_1^2 = 0.628$ min^2

New method: $\quad\quad\quad \mu_2 = 1.16$ min $\quad$ and $\quad \sigma_2^2 = 0.442$ min^2

The standard deviation of the difference, by Eq. (20.9), is

$$\sigma_d = \left(\frac{0.628}{30} + \frac{0.442}{30}\right)^{1/2} = 0.189$$

The test statistic is given by Eq. (20.10)

$$z = \frac{1.42 - 1.16}{0.189} = 1.38$$

which is in the two-tail acceptance region $|z| \leq 1.96$. Again, the new method is no better or worse than the old one.

Section 20.4

COMMENT If the data were paired in this problem, the two methods would have been used to distribute each emergency injection. This is probably an unrealistic assumption, but it would allow use of the Wilcoxon test for H_0: $\mu_1 = \mu_2$. If the steps of Sec. 25.6 are followed, the sample statistic T is

$$T = \min\left[D_1 = 287, D_2 = 178\right] = 178$$

and the test statistic from Eq. (25.12) is

$$z = \frac{178 - 232.5}{48.62} = -1.12$$

The means are still judged to be equal.

Of course, the consultant on this hospital job would not use all these tests to compare mean values; however, this does give you an idea of how many nonparametric ways there are to test the same hypothesis evaluated by parametric methods.

ADDITIONAL MATERIAL

Bowker and Lieberman [4], pp. 246–254; Brownlee [5], Chap. 7; Gibra [8], Chap. 11; Lipson and Sheth [17], Chap. 8; Miller and Freund [18], Chap. 10; Siegel [21], all chapters; Volk [24], Chap. 11; Wilks [26], Chap. 14.

PROBLEMS

The following situations are used by the indicated problems in this chapter.

A: Ms. Confidence is a salesperson with a computer software company. She estimates that she is on the telephone an average of 60 minutes per workday. A timing mechanism on her phone has recorded the connect times for the last 24 days. The order of observation is retained across the rows.

38	65	80	76	115	52
72	28	97	52	41	58
102	41	39	50	78	89
32	110	92	67	30	70

Problems P25.3, P25.12, P25.21

B: Mr. Confidence owns several tree nurseries in town. He puts up temporary structures in new subdivision areas to take advantage of landscaping needs when the new residents move in. The number of large trees sold each week at the north and west locations are:

North:	10	23	15	20	38	45
	28	12	29	10	18	41
West:	15	21	18	35	49	57
	45	22	25	30	23	60
	57	30	18	21	32	44

Problems P25.16, P25.23

C: Two different chemical machining processes were used on samples of 10 similar alloys. The times in seconds to obtain the required measurement are:

Alloy number	Contact time, s	
	Chemical 1	Chemical 2
1	78	95
2	105	69
3	60	52
4	85	97
5	108	119
6	34	52
7	68	40
8	76	82
9	89	88
10	52	60

Problems P25.19, P25.25, P25.29

P25.1 What is the basic difference between a parametric and a nonparametric statistical test?

P25.2 The number of persons per car pool given in Table 2.1 (Sec. 2.1) is in order of collection by row. Determine if the sample is random about the mean (3.14 persons per car) using (a) a two-sided test with $\alpha = 0.05$ and (b) a one-sided test that there are too many runs for $\alpha = 0.025$.

P25.3 Is the data presented in situation A random if (a) the mean and (b) the median is used as the reference point? Let $\alpha = 0.05$ for this test.

P25.4 The flow rate in a pipe must be closely controlled. The prescribed value is 580 L/s. This value was used as a reference to determine the plus and minus signs below. Do they represent a random sample at the $\alpha = 0.05$ level?

$$+\ -\ -\ +\ +\ +\ -\ +\ -\ -\ +\ +\ -\ -\ -\ +\ -\ +\ -\ -$$

P25.5 Assume that the order of collection is preserved across rows for the bake time data in Prob. P23.23. Determine if the sample is random around each of the following reference points: (a) the median, which is 19.5, (b) the mode, and (c) the value of 22 min. Let $\alpha = 0.05$ for all these tests. Use plus and minus signs to perform these tests and eliminate any data values which equal the particular reference point.

P25.6 The Mini Business Administration considers a small business a success if it meets certain criteria in sales, profit, and other factors. For the last 25 businesses which this administration granted a loan, the following success (s) and failure (f) ratings were determined.

ssffffffssssssffsssfffss

Are there too few runs to consider this a random sample if the significance level is 2.5 percent?

P25.7 Automobiles traveling on a road come to a traffic light and must turn right (r) or left (l). A technician observed the following sequence of turns for 40 cars.

lllrrlrrrrllrlrrrrrl
lllrlrllrrrlllllrlrr

Is this a random sample if $\alpha = 0.05$?

P25.8 Why is the binomial distribution with $p = 0.50$ always used to determine the acceptance region for the one-sample sign test?

P25.9 The business manager for the Hi Neighbor Campgrounds has collected the occupancy rates on 12 consecutive Mondays: 0.42, 0.28, 0.61, 0.31, 0.24, 0.69, 0.18, 0.35, 0.43, 0.54, 0.60, 0.57. The Monday rates last year averaged 0.30 and did not follow a normal distribution.

(a) Answer the questions in Sec. 25.1 to determine which test is correct to determine if the average occupancy rate for Monday has changed.

(b) Perform this test using a 4 percent significance level.

(c) Select and perform the appropriate test if the Monday occupancy rates are known to be normally distributed.

P25.10 Flip a coin 15 times and use a 5 percent significance level to test the hypotheses (a) that the sample is random and (b) that the probability of a head is $\frac{1}{2}$. Assign a plus sign to a head and a minus sign to a tail.

P25.11 Rework Prob. P25.10 for a sample of 30 flips.

P25.12 Is the average daily connect time for situation A equal to 60 min if the normal distribution assumption (a) is not made and (b) is made? Let $\alpha = 0.05$ for both tests.

P25.13 The cost of a car tuneup (in dollars) was collected from eight different garages: 45, 50, 52, 60, 48, 39, 60, and 54. If the expected cost was $48, use a 90 percent confidence level to determine if the average cost will exceed this value assuming that any one of the garages is randomly selected to perform the tuneup. Perform this test using (a) the appropriate nonparametric test and (b) the appropriate parametric test. (c) Since the nonparametric test is based on a discrete distribution, what is the actual confidence level used to perform this test?

P25.14 A civil engineering consulting firm needs the average commute time for drivers in a specific city. Person A believes it is 15 min and person B wants to use 20 min. A technician collects data for 450 trips as presented in Prob. P19.11. Unfortunately the times are not known to be normally distributed, and the data is grouped into cells. Use the most appropriate nonparametric test with $\alpha = 0.05$ to determine if either person's estimate is correct. When performing these tests, assume that the commute times are uniformly distributed within the cells used in Prob. P19.11 and that no times equal exactly 15 or 20 min.

P25.15 Test the hypotheses H_0: $\mu = 4.5$ versus H_1: $\mu \neq 4.5$ years for Prob. P23.32 using a nonparametric test with $\alpha = 0.20$. With $\alpha = 0.05$.

P25.16 Use the two-sample sign test for the sales data in situation B to determine if the means are equal at the 5 percent significance level. Eliminate the last six values for the west location.

P25.17 If the number of inoperable cars in Prob. P20.19 is not normally distributed, determine if the averages for the two lots are equal. Let $\alpha = 0.10$ and do not take different magnitudes into account.

P25.18 The Red Hot Chili Company has just harvested its peppers for chili powder. In an experiment 25 peppers were divided and dried by two different methods for 5 days. The percentage of water remaining is given for each method. Is the chemical method more effective in removing water if a 15 percent significance level is used to test the means and the normal assumption cannot be made? Do not consider the percentage of moisture difference between methods when performing this test.

Pepper number	Method (%)		Pepper number	Method (%)	
	Natural	Chemical		Natural	Chemical
1	4.5	3.2	16	8.4	10.8
2	6.2	9.8	17	10.0	10.0
3	4.8	5.2	18	3.5	4.7
4	6.0	6.0	19	5.6	2.8
5	7.1	3.5	20	3.8	7.2
6	7.9	5.2	21	9.7	8.5
7	10.2	6.8	22	10.3	2.6
8	8.9	7.6	23	6.4	8.0
9	9.8	4.3	24	5.4	9.6
10	7.5	7.6	25	7.2	4.9
11	8.9	2.9			
12	6.9	10.4			
13	7.2	8.0			
14	16.1	9.3			
15	10.8	12.4			

P25.19 Assume that the alloy samples in situation C are not paired, and use a 12 percent significance level to test the hypothesis that means are equal. Do not consider the size of the difference between the contact times for the two chemicals.

P25.20 Use the data in situation D of Chap. 20 to test (*a*) if the mean of process 1 is equal to 62 days and 59 days and (*b*) if the means of process 1 and process 2 are equal. Use nonparametric tests and an 85 percent confidence level. Assume that the magnitude of the differences is of no consequence when testing for the equality of process means.

P25.21 At the same time that the connect times were recorded for Ms. Confidence in situation A, the connect times for another saleperson were recorded.

17	27	56	31	46	52
56	34	57	102	19	50
38	57	34	42	34	52
52	41	21	67	119	33

Use only the fact that one number is larger or smaller than another, a confidence level of 95 percent and do not assume sampling from a normal distribution when working this problem.
  (a) Determine if the average of the connect times above is statistically equal to 60 min.
  (b) Determine if the mean connect time here is smaller than that in situation A.

P25.22 What is the basic difference between the two-sample sign test and the Mann-Whitney U test?

P25.23 Are the sales volumes of large trees at the two nursery locations in situation B the same or does the north location sell more than the west location? Let $\alpha = 0.02$ for this hypothesis test and consider the size of the differences in sales between the two locations.

P25.24 If $\alpha = 0.05$ and the magnitude of numbers is important, determine if the percentage of water remaining in the chilis of Prob. P25.18 is the same for both drying methods. Neglect the fact that the data is paired when you perform this test. (See Prob. P25.32 for further analysis.)

P25.25 (a) Use the appropriate rank sum test to determine if the average contact times in situation C are equal. Assume that the alloy samples are unpaired and that $\alpha = 0.12$.
  (b) Is the conclusion the same as the one in Prob. P25.19?

P25.26 A total of 189 passengers may be carried on aircraft which fly between two cities. A survey of several flights by two airline companies produced the following number of unfilled seats per flight.

High-Road Airlines:	12	29	18	35	12
	16	42	8	21	37
Low-Road Airlines:	34	6	19	31	14
	17	29	12	18	42
	10	7	18		

The number of unfilled seats per flight is not guaranteed to be normally distributed. Use a 5 percent significance level to determine if the samples came from populations with the same mean under the following conditions:
  (a) The differences in the number of unfilled seats is of no concern.
  (b) The difference in the number of unfilled seats is definitely of concern.
  (c) The normality assumption is made and variances are assumed equal.

P25.27 Use a nonparametric test and a 2 percent significance level to determine if students at Johnson public school read faster than students at Crockett public school. The time in minutes it took to read a specific passage is:

Johnson:	15.1	10.3	12.4	18.1	10.7
	12.1	13.0	16.9	11.5	8.9
Crockett:	16.1	10.3	9.5	17.6	16.9
	13.0	13.8	14.4	12.4	9.6

The differences in reading rates are important to the teachers who have requested this analysis.

P25.28 In the Mann-Whitney U test statistic, Eq. (25.9), it is possible to select the sum of either sample's ranks as R, and obtain the same conclusion in the hypothesis test. Why is this choice possible?

P25.29 If the 10 alloy samples in situation C are divided and each half treated with a chemical, and if the time differences are important, use a nonparametric test and $\alpha = 0.05$ to determine if the two chemicals are equal in their machining ability.

P25.30 Julie is a business analyst for a large retail store. She wants to determine if the present line of plastic flowers should be carried or discontinued. Weekly sales data one week before and two weeks after a sales promotion was collected for 10 different items.

Item	Number	Weekly sales, $	
		Before	After
Roses	1	30	45
Tulips	2	20	15
Marigolds	3	35	17
Sweetpeas	4	18	42
Mums	5	40	40
Ferns	6	60	120
Plants	7	70	85
Trees	8	192	175
Vines	9	35	68
Berries	10	15	12

(a) Did sales improve after the promotional campaign? Use a 1.0 percent level of significance and consider the magnitude of the differences between the two samples.

(b) Rework the problem without considering the magnitude of the differences. Is the result the same as in (a)?

P25.31 The amount per year invested in the commodity market by specific speculators was sampled under the conditions of "bullish" and "bearish" markets. The differences are given in the table. Use them to determine at the 5 percent significance level if there is a difference in the invested amounts. You cannot assume that commodity market investments are normally distributed, and the magnitude of investment differences is definitely important.

Speculator	Difference	Speculator	Difference	Speculator	Difference
1	$ 250	11	$ − 600	21	$ 800
2	− 1000	12	− 1550	22	1800
3	800	13	950	23	− 1500
4	750	14	350	24	− 2500
5	− 100	15	425	25	2400
6	− 850	16	− 725	26	250
7	1500	17	− 850	27	− 400
8	− 550	18	− 1750	28	300
9	900	19	− 950	29	850
10	1800	20	− 150	30	− 1100

P25.32 (a) Apply the two-sided Wilcoxon test to the data in Prob. P25.18 with $\alpha = 0.05$.

(b) The Mann-Whitney U test was used on the same data in Prob. P25.24, but the pairing of data was purposely neglected. Are the conclusions of these two tests the same?

P25.33 Two rapid weighing techniques—electrical and mechanical—are being tested on bottles of ketchup which are suppose to contain 1000 g. Thirty identifiable bottles were mixed in with 970 correctly filled bottles and exposed to both techniques. The weight readouts are given. If the size of the differences in weights is important, use a 95 percent confidence level to determine if the two techniques are equally accurate.

Bottle number	Weight (g)		Bottle number	Weight (g)	
	Electrical	Mechanical		Electrical	Mechanical
1	992	996	16	996	991
2	982	980	17	980	988
3	1004	999	18	1001	1010
4	995	995	19	1010	1000
5	1001	1000	20	989	989
6	990	996	21	991	1004
7	1008	1007	22	995	996
8	989	996	23	1000	989
9	1000	1003	24	980	980
10	1004	999	25	999	1008
11	989	991	26	1000	1001
12	997	1003	27	985	985
13	986	999	28	992	998
14	1003	991	29	1002	996
15	984	987	30	1004	1000

P25.34 The number of defective items per production run was recorded for 12 runs using two different inspectors.

Production run	Inspector	
	1	2
1	51	39
2	85	73
3	46	21
4	33	55
5	55	92
6	34	62
7	88	80
8	39	44
9	50	32
10	26	45
11	29	46
12	19	25

The items are not paired when they are inspected. Answer the following using $\alpha = 0.05$.

(a) Are the two samples random about their respective means?
(b) Are the sample means equal if only sign differences are considered?
(ci) Are the sample means equal if the magnitude of the differences in number of defective items found is important?
(d) If the items had been individually tracked to obtain paired data and magnitudes are important, are the means equal?

CHAPTER
TWENTY-SIX

STATISTICAL INFERENCES USING BAYESIAN ESTIMATES

This chapter introduces the method of Bayesian statistics for estimating population parameters. Loss and risk functions as well as prior and posterior distributions are discussed and used in the development of Bayesian estimates. This chapter is only introductory to the subject of Bayesian statistics.

CRITERIA

To have a basic understanding of the technique of Bayesian estimation, you should be able to:

1. State the difference between the Bayesian method and classical method of parameter estimation.
2. State the equations for and compute values for the *loss function*, *risk function*, and *mean risk*, given all possible parameter values and decisions, the prior probability distribution, and the conditional probabilities of making a decision for each parameter value.
3. State the relation between *prior* and *posterior* distributions for normal, beta, and gamma priors, and compute the posterior distribution parameters, given the prior parameter values and the appropriate sample statistics.
4. State the formula for and compute the Bayesian estimate for the population parameters—normal μ, binomial π, and Poisson λ—given a quadratic loss function, the prior distribution parameters, and the appropriate sample statistics.

STUDY GUIDE

26.1 Classical and Bayesian Methods

In previous chapters we estimated population parameters using some sample statistics such as $\overline{X}$ for $\hat{\mu}$ and s for $\hat{\sigma}$. These point estimates, which are best estimators (Sec. 6.9), assume that the parameter is a *fixed* population value. Even if an interval estimate is made, the true population value is assumed to be a fixed value somewhere in this interval. This entire procedure is called the *classical* approach to estimation.

If the parameter is viewed as a random variable, there is a recognition that a pdf can be constructed for the parameter value. In the *Bayesian* approach to estimation, the parameter pdf takes a particular form, which is then used in conjunction with some sample statistic to give a better (Bayesian) estimate of the parameter value. The pdf formulated for the parameter is based on prior experience with the parameter or a subjective feel for its shape. Accordingly, the name *prior distribution* is given to this pdf.

The basic difference, therefore, between the two methods is stated as follows: In the classical approach the parameter is viewed as a fixed value to be estimated by a sample statistic, whereas in the Bayesian approach any information that is known about the parameter is used to improve the sample statistic estimate.

Bayesian methods have been developed for point estimates, interval estimates, and hypothesis testing. If no good information or subjective feel for the prior distribution is available, classical methods must be used. In fact, the logical reason that Bayesian methods are not used more commonly is because of the inability of the analyst to easily formulate a prior distribution in such a way that Bayesian theory can better estimate the parameter.

Example 26.1 Bill wants to estimate the mean thickness μ of mirror glass produced by a new piece of equipment using a sample of 50 mirrors. In a phone conversation with Joan at another plant, Bill learned that she had the same equipment and had trouble estimating the mean. In fact, the 25 estimates she made had a mean of 0.52 cm and a variance of 0.02 cm^2. Explain the classical and Bayesian approaches that Bill might use in estimating μ.

SOLUTION The classical approach entails the computation of $\overline{X} = \hat{\mu}$ from the sample of 50 mirrors. The Bayesian approach requires that a prior distribution be assumed for μ such as a normal with mean 0.52 and variance 0.02, that is $\mu \sim N(0.52, 0.02)$. Then the technique of Bayesian estimates for means as discussed below can be used to estimate μ.

You must keep in mind that the $N(0.52, 0.02)$ pdf has been assumed for μ, not the glass thickness itself. Of course, the thickness may also be normally distributed.

Problem P26.1

26.2 Loss and Risk Functions

In order to understand how Bayesian estimates are made, the concepts of loss and risk functions need to be introduced. In the classical method, the decision can be made to estimate a population parameter θ with a sample statistic d computed from a random sample of the variable X; that is, d is a function of the observed values $x_1, x_2, \ldots, x_n$, or $d = d(x_1, x_2, \ldots, x_n)$. For example, if $\overline{X}$ estimates the parameter μ, the decision is $d = \overline{X} = \hat{\mu}$. A *loss function* results in a numerical value of the penalty incurred when the (wrong) decision d is accepted as the parameter estimate and the actual value is θ. Therefore, a loss function $L(d, \theta)$ is a function of the statistical decision d and the correct value θ. The two common mathematical loss functions described here use c as a weighting factor, or what might be called a loss coefficient.

1. *Quadratic.* This is used if large errors are considered very serious. It is also called the *error-squared loss function*.

- $$L(d, \theta) = c(\theta - d)^2 \qquad (26.1)$$

2. *Absolute error.* If large errors are not so serious, this form is used.

- $$L(d, \theta) = c|\theta - d| \qquad (26.2)$$

Since d and θ are random variables, $L(d, \theta)$ is a random variable. The expected value of the loss function is called the *risk function* $R(d, \theta)$.

- $$R(d, \theta) = E[L(d, \theta)] \qquad (26.3)$$

There is a risk value for each value $\theta = \theta_j$ where $j = 1, 2, \ldots, l$; therefore, $R(d, \theta_j)$ may be written as the loss function for each decision $d = d_i$ ($i = 1, 2, \ldots, k$) times the conditional probability $P(d_i/\theta_j)$ that the decision is d_i if the parameter value is actually θ_j. See Sec. 5.6 for a review of conditional probability. The risk $R(d, \theta_j)$ is

- $$R(d, \theta_j) = \sum_{i=1}^{k} L(d_i, \theta_j) P(d_i/\theta_j) \qquad (26.4)$$

We hope to develop some statistical decision d that will minimize the risk $R(d, \theta_j)$ for all θ_j values. Unfortunately, in most cases there is no d that can minimize risk for all θ_j. Our approach is to place a probability distribution on θ and try to minimize the *mean risk* $r(d, \theta)$, which is the expected value of the risk function.

$$r(d, \theta) = E[R(d, \theta)]$$

$$r(d, \theta) = \sum_{j=1}^{l} R(d, \theta_j) P(\theta_j) \quad (26.5)$$

The $P(\theta_j)$ are the probabilities that the parameter equals θ_j. It is, therefore, a *prior probability* from the prior distribution, meaning it is determined before a sample is taken.

Example 26.2 illustrates the computations and graphs of the loss and risk functions in inventory control.

Example 26.2 A toy store manager wants to know how many deluxe swing sets to stock for the summer. There is space available for no more than five sets, and only one order can be placed for the entire summer. Past data indicates that the summer demand follows the Poisson pdf with a mean of 1.8 sets, but has never exceeded 5. The cost per set is $70, and the sales price is $110. If the stocking decision d is less than the actual demand θ, the profit of $40 is lost; but if too many sets are stocked, the profit decreases to $20 and an extra $10 advertisement and inventory cost is incurred. An absolute error loss function is used by the manager.

The manager realizes that the stocking decision d is conditional on the actual demand θ, so the probability $P(d/\theta)$ will be assumed to follow a Poisson pdf for each $\theta = 0, 1, 2, \ldots, 5$. Do the following.
(a) State the loss function.
(b) Compute and graph the loss function for the different stocking decisions d.
(c) Compute the risk function for each demand value.
(d) Determine the mean risk.

SOLUTION In summary, we presently know the following.

d: stocking decision with six possible values $d_i = 0, 1, \ldots, 5$

θ: actual demand parameter with possible values $\theta_j = 0, 1, \ldots, 5$ (A Poisson pdf with mean 1.8 is the prior distribution used to compute $P(\theta)$ for each θ value.)

$L(d, \theta)$: absolute error as in Eq. (26.2), with different loss coefficients for over- and understocking

$P(d/\theta)$: conditional probability for each stocking decision d (A Poisson pdf is assumed for each θ value.)

(a) If understocking takes place $(d < \theta)$, the profit is lost; and if overstocking occurs $(d > \theta)$, a total of $30 is lost. The loss function from Eq.

(26.2) is

$$L(d, \theta) = \begin{cases} \$40|\theta - d| & d < \theta \\ 0 & d = \theta \\ \$30|\theta - d| & d > \theta \end{cases} \quad (26.6)$$

(b) The values of the loss function $L(d, \theta)$ are computed from Eq. (26.6) and presented in Table 26.1. The 0 values on the diagonal occur when the stocking decision equals the actual demand. These V-shaped, linear loss functions are plotted in Fig. 26.1 for $d = 0, 1, 3,$ and 5.

(c) To compute the risk function for each θ value θ_j by Eq. (26.4), the conditional probabilities $P(d/\theta)$ are determined from the Poisson Table B-2. The probability rows in Table 26.2 give these values. For instance,

$$P(d = 0/\theta = 3) = \text{Poisson with } \theta = 3 \text{ for } d = 0$$

$$= \frac{e^{-3}3^0}{0!} = 0.05$$

$$P(d = 2/\theta = 4) = \frac{e^{-4}4^2}{2!} = 0.15$$

All Poisson probability left over for $d > 5$ is added to $P(d = 5/\theta)$, thus increasing the $d = 5$ column values to ensure that the sum for each row is 1. This is necessary because a maximum of only five sets can be in inventory.

The components of $R(d, \theta)$ are computed from Eq. (26.4) using the $L(d, \theta)$ values in Table 26.1. Summing across the columns gives $R(d, \theta)$

Table 26.1 Absolute error loss function values, Example 26.2

Actual demand, θ	$L(d, \theta)$					
	Stocking decision d					
	0	1	2	3	4	5
0	$ 0	$ 30	$ 60	$90	$120	$150
1	40	0	30	60	90	120
2	80	40	0	30	60	90
3	120	80	40	0	30	60
4	160	120	80	40	0	30
5	200	160	120	80	40	0

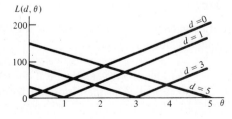

Figure 26.1 Graphs of the absolute error loss function, Example 26.2.

26.2 LOSS AND RISK FUNCTIONS

Table 26.2 Computation of the risk function values, Example 26.2

Actual demand θ	Probability or risk	Stocking decision d						Risk function $R(d, \theta)$
		0	1	2	3	4	≥ 5	
0	$P(d/0)$	1	0	0	0	0	0	
	$R(d, 0)$	0	0	0	0	0	0	$0
1	$P(d/1)$	0.37	0.37	0.18	0.06	0.02	0	
	$R(d, 1)$	14.80	0	5.40	3.60	1.80	0	25.60
2	$P(d/2)$	0.14	0.27	0.27	0.18	0.09	0.05	
	$R(d, 2)$	11.20	10.80	0	5.40	5.40	4.50	37.30
3	$P(d/3)$	0.05	0.15	0.22	0.22	0.17	0.19	
	$R(d, 3)$	6.00	12.00	8.80	0	5.10	11.40	43.30
4	$P(d/4)$	0.02	0.07	0.15	0.20	0.20	0.36	
	$R(d, 4)$	3.20	8.40	12.00	8.00	0	10.80	42.40
5	$P(d/5)$	0.01	0.03	0.08	0.14	0.18	0.56	
	$R(d, 5)$	2.00	4.80	9.60	11.20	7.20	0	34.80

for each $\theta = 0, 1, \ldots, 5$. A search of Table 26.2 indicates that the minimum risk of 0 occurs for each θ value at a different d, thus substantiating the earlier statement that usually no one decision will minimize risk for all parameter values.

(d) The Poisson prior probabilities used to compute mean risk are given in Table 26.3. They are the $P(\theta; 1.8)$ values from Table B-2 rounded to two decimals. From Eq. (26.5) and Table 26.3 the average risk taken by the toy store manager for all stocking decisions d and possible demands θ is

$$r(d, \theta) = \$28.69$$

Problems P26.2–P26.7

Table 26.3 Computation of the mean risk, Example 26.2

θ	$P(\theta)$	$R(d, \theta)$	$R(d, \theta)P(\theta)$
0	0.17	$0	$0
1	0.30	25.60	7.68
2	0.27	37.30	10.07
3	0.16	43.30	6.93
4	0.07	42.40	2.97
5	0.03	34.80	1.04
			$28.69

26.3 Prior and Posterior Distribution Relations

If the logic of Sec. 5.7 on Bayes' formula is used, and if x is an observed sample statistic taken from a population with a parameter value θ, the probability of observing x given θ may be written using the general conditional probability forms in Eqs. (5.2) and (5.4).

$$P(x/\theta) = \frac{P(x \cap \theta)}{P(\theta)} = \frac{P(\theta/x)P(x)}{P(\theta)} \qquad (26.7)$$

The statistic x is the same one used in making the decision d of Sec. 26.2. The extreme-right expression in Eq. (26.7) is the usual Bayes' formula form. If x and θ are values from continuous random variables, it is possible to write an expression similar to Eq. (26.7) for distributions using the conditional distribution terminology presented in Sec. 18.4.

$$f(x/\theta) = \frac{f(\theta/x)f(x)}{f(\theta)} \qquad (26.8)$$

where $f(x/\theta)$ = conditional pdf of X for fixed θ
$f(x)$ = marginal pdf of X
$f(\theta/x)$ = conditional pdf of θ for fixed x (This is a *posterior* distribution.)
$f(\theta)$ = specified pdf for the parameter θ (This is a *prior* distribution.)

You learned about the prior distribution $f(\theta)$ in Sec. 26.1; now, the posterior distribution $f(\theta/x)$ is discussed. First, solve Eq. (26.8) for the posterior pdf:

$$f(\theta/x) = \frac{f(\theta)f(x/\theta)}{f(x)} \qquad (26.9)$$

The posterior pdf is important because it is the pdf of the parameter θ for a given statistic x.

Statistical theory can be used to show that *certain prior distributions naturally lead to certain posterior distributions*. The three prior-posterior relations explained below are called *natural conjugates* because the prior and posterior are in the same family, the only difference being parameter values. A subscript 0 is used for prior pdf parameters, and a 1 is for the posterior pdf. These relations will be used later to make Bayesian estimates for population parameters.

1. *Normal prior–normal posterior relation.* If the population mean μ from a normal distribution has a normal prior $f(\mu)$, the posterior pdf $f(\mu/\overline{X})$ is also normal and the sample statistic is $\overline{X}$ (Fig. 26.2a). A summary is:

    Parameter estimated: normal mean μ
    Parameter range: $-\infty \leq \mu \leq \infty$
    Prior pdf: $\mu \sim N(\mu_0, \sigma_0^2)$
    Posterior pdf: $f(\mu/\overline{X})$ is normal with parameters μ_1 and σ_1^2

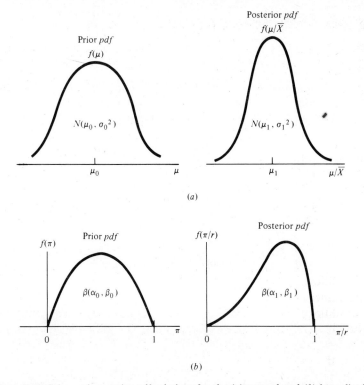

Figure 26.2 Prior and posterior pdf relations for the (*a*) normal and (*b*) beta distributions.

The posterior pdf parameters are a weighted function of the prior parameters and the statistics $\overline{X}$ and $s^2 = \hat{\sigma}^2$.

$$\mu_1 = \frac{n\overline{X}\sigma_0^2 + \mu_0\hat{\sigma}^2}{n\sigma_0^2 + \hat{\sigma}^2} \qquad (26.10)$$

$$\sigma_1^2 = \frac{\hat{\sigma}^2\sigma_0^2}{n\sigma_0^2 + \hat{\sigma}^2} \qquad (26.11)$$

2. *Beta prior–beta posterior relation.* If the population proportion π (or p) for a *Bernoulli* (or binomial) distribution has the beta prior $f(\pi)$ (see Table 16.1), the posterior pdf $f(\pi/r)$ is also beta, where r successes are observed in a sample of size n (Fig. 26.2*b*).

Parameter estimated:	Bernoulli proportion π
Parameter range:	$0 \leq \pi \leq 1$
Prior pdf:	$f(\pi)$ is beta (β) with parameters α_0 and β_0
Posterior pdf:	$f(\pi/r)$ is beta with parameters α_1 and β_1

The posterior pdf parameters are

$$\alpha_1 = \alpha_0 + r \qquad (26.12)$$
$$\beta_1 = \beta_0 + n - r \qquad (26.13)$$

3. *Gamma prior–gamma posterior relation.* If the rate λ in the *Poisson* distribution has a gamma prior $f(\lambda)$ as in Sec. 16.6, the posterior pdf $f(\lambda/x)$ is also gamma, where there are x occurrences of the event in a sample of t time units.

Parameter estimated:	Poisson rate λ
Parameter range:	$\lambda > 0$
Prior pdf:	λ is gamma (γ) with parameters n_0 and ϵ_0 (ϵ_0 is a substitute for λ in Sec. 16.6)
Posterior pdf:	$f(\lambda/x)$ is gamma with parameters n_1 and ϵ_1

The posterior pdf parameters are

$$n_1 = n_0 + x \qquad (26.14)$$
$$\epsilon_1 = \epsilon_0 + t \qquad (26.15)$$

The normal prior–posterior relation is explained in Example 26.3.

Example 26.3 The resistance in copper rods has a normal distribution with some mean μ, which is to be estimated. From previous experiments the analyst suspects that the mean itself is normal, that is, $\mu \sim N(10{,}000, 500)$. A sample of 25 readings has $\overline{X} = 8500\ \Omega$ and $s = 52\ \Omega$. What are the updated mean and variance values for the distribution of μ?

SOLUTION The normal prior–posterior relation can be used to compute the posterior pdf parameters μ_1 and σ_1^2. The prior parameters $\mu_0 = 10{,}000$ and $\sigma_0^2 = 500\ \Omega^2$ with the sample results $\overline{X} = 8500$ and $s^2 = \hat{\sigma}^2 = 2704\ \Omega^2$ are used in Eqs. (26.10) and (26.11). The posterior pdf is normal; that is, $\mu/\overline{X} \sim N(8{,}767, 89)$.

$$\mu_1 = \frac{25(8500)(500) + 10{,}000(2704)}{25(500) + 2704} = 8767$$

$$\sigma_1^2 = \frac{2704(500)}{25(500) + 2704} = 89$$

COMMENT The distribution of μ has changed quite a lot as a result of the sample. The mean decreased from 10,000 to 8767, and the variance has greatly decreased from 500 to 89, thus making the pdf $f(\mu/\bar{X})$ less dispersed than $f(\mu)$, as shown in Fig. 26.2a.

Problems P26.8–P26.11

26.4 Bayesian Estimates for Quadratic Loss Functions

You have probably wondered why we learned about loss and risk functions and prior-posterior relations! It is simply to make the following statement (without proof) about a Bayesian estimate.

> If a parameter θ has the prior pdf $f(\theta)$ and a quadratic loss function $L(d, \theta) = c(\theta - d)^2$ is used, the Bayesian estimate which minimizes the mean risk $r(d, \theta)$ is the expected value $E[\theta/d]$ of the posterior distribution $f(\theta/d)$.

Therefore, if a quadratic loss function is appropriate and one of the prior pdf's from Sec. 26.3 are acceptable, the Bayes estimate of the population parameter is the *expected value of the posterior pdf*. The Bayes estimate for each prior pdf studied is given in Table 26.4. If a different prior pdf is assumed, it is necessary to determine the posterior pdf and its expected value. This is often difficult, but the procedure is covered in texts on Bayesian statistics.

> **Example 26.4** A congressional election candidate has contracted with a polling company to estimate the fraction π of eligible voters supporting her views on international relations. A beta prior with $\alpha_0 = 5.0$ and $\beta_0 = 1.5$ (Fig. 26.3) is placed on π by the pollster because a preponderance of voters

Table 26.4 Bayesian estimates for specific prior distributions and a quadratic loss function

Parameter	Distributions		Sample statistic	Posterior expected value formula	Bayes estimate of the parameter
	Prior	Posterior			
Normal μ	Normal $N(\mu_0, \sigma_0^2)$	Normal $N(\mu_1, \sigma_1^2)$	$\bar{X}$ from n observations	μ_1	$E[\mu/\bar{X}] = \dfrac{n\bar{X}\sigma_0^2 + \mu_0\hat{\sigma}^2}{n\sigma_0^2 + \hat{\sigma}^2}$
Binomial $\pi(p)$	Beta $\beta(\alpha_0, \beta_0)$	Beta $\beta(\alpha_1, \beta_1)$	r successes in n trials	$\dfrac{\alpha_1}{\alpha_1 + \beta_1}$ (Table 16.1)	$E[\pi/r] = \dfrac{\alpha_0 + r}{\alpha_0 + \beta_0 + n}$
Poisson λ	Gamma $\gamma(n_0, \epsilon_0)$	Gamma $\gamma(n_1, \epsilon_1)$	x occurrences in t units	$\dfrac{n_1}{\epsilon_1}$ (Sec. 16.6)	$E[\lambda/x] = \dfrac{n_0 + x}{\epsilon_0 + t}$

480 STATISTICAL INFERENCES USING BAYESIAN ESTIMATES

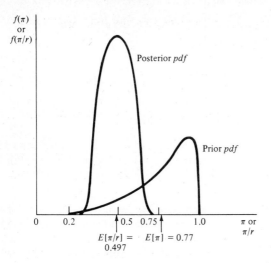

Figure 26.3 Beta prior and posterior distributions, Example 26.4.

usually support this candidate's views. In a sample of 1000 voters, 495 people voiced support of the opinions of the candidate.
(a) What is the Bayes estimate of the fraction of supporters?
(b) Compare the posterior and prior pdf and discuss the differences made by the sample results.

SOLUTION (a) The Bayes estimate of π is the expected value of the beta posterior distribution $f(\pi/r)$. The statistic $r = 495$ supporters in a total of $n = 1000$ voters is used in the Table 26.4 formula for $E[\pi/r]$:

$$E[\pi/r] = \frac{5 + 495}{5 + 1.5 + 1000} = \frac{500}{1006.5} = 0.497$$

The Bayes estimate is that 49.7 percent of the voters support the candidate's view.
(b) The Bayes estimate 0.497 is much different from the prior pdf expected value, which is

$$E[\pi] = \frac{\alpha_0}{\alpha_0 + \beta_0} = \frac{5.0}{6.5} = 0.77$$

The posterior beta parameters, computed from Eqs. (26.12) and (26.13), are $\alpha_1 = 500$ and $\beta_1 = 506.5$. The general shape of this pdf (Fig. 26.3) is symmetric with the expected value (which is the Bayesian estimate of π) at 0.497 and a variance of 0.00025 (Table 16.1) compared to the prior variance of 0.024. In conclusion, the sample has drastically changed the mean and variance of the distribution of π.

COMMENT Regular unbiased estimates could have been used here with good results because the proportion mean and variance (Sec. 8.6) are

$$\hat{\pi} = \frac{r}{n} = p = 0.495$$

$$\hat{\sigma}^2 = \frac{pq}{n} = \frac{0.495(0.505)}{1000} = 0.00025$$

These are virtually identical to the posterior pdf parameters of 0.497 and 0.00025.

Problems P26.12–P26.17

ADDITIONAL MATERIAL

Benjamin and Cornell [3], Chap. 5; Bowker and Lieberman [4], pp. 134–162, 308, 309; Hines and Montgomery [12], pp. 451–460; Hoel [13], pp. 348–356; Hogg and Craig [14], pp. 250–253, 278–283; LaValle [16], Chaps. 15–19; Miller and Freund [18], pp. 188–191, 246–249; Wilks [26], Chap. 16; Winkler [27], Chaps. 3 to 7.

PROBLEMS

P26.1 What is the purpose of developing the prior distribution when the Bayesian approach is utilized?

P26.2 A couple is considering investing in a manufacturing and sales company. A total of $15,000 and the sale of 5000 units are needed for the company's breakeven. If the couple invests the entire $15,000, their profit will be $3 per unit for all sales levels below 5000 units and $2 per unit for all levels at and above 5000 units. The $1 per unit difference is added expense for extra units. The couple has the option of investing only $7500 and realizing one-half of the profits for all sales levels. However, the couple must agree to bear the entire $1 per unit extra cost if sales are above 5000. Write and graph two loss functions $L(d, \theta)$, where θ is the actual sales volume and there are two possible decisions: (a) d_1, which is to invest the entire $15,000, and (b) d_2, which is to invest $7500.

P26.3 A building may actually take $\theta = 0, 1, 2, 3$ or 4 weeks to construct, with 0 meaning the job is canceled. The builder must contract with some employees for the number of weeks he thinks the job will take; that is, the decision values are $d = 0, 1, 2, 3$, or 4. The absolute error loss function has been formulated as

$$L(d, \theta) = \begin{cases} 1|d - \theta| & d \geq \theta \\ 2|d - \theta| & d < \theta \end{cases}$$

where the loss coefficient represents the number of weeks of wages lost. The prior distribution on the actual completion time is $b(\theta; 4, 0.3)$; and if the contractor knew the value of θ, he would tend to overestimate construction time according to a discrete uniform distribution with $a = \theta$ and $b = 4$.

(a) Interpret the loss function in words.

(b) Determine the values of the loss function and conditional probabilities $P(d/\theta)$ for all possible (d, θ) pairs.

(c) Compute the value of the risk function for all possible construction times.

(d) If the weekly wages for the employees involved are $5000, find the mean risk in dollars.

P26.4 Rework Example 26.2 using a quadratic loss function.

P26.5 A suction pump company which makes 10,000 pumps per year has detailed the following choices for next year's production level: remain constant, increase to 20,000, or drop to 5000 units. Actual demand next year is estimated to be 10,000 with a probability of 0.5; 20,000 with a probability of 0.2; and 5000 with a probability of 0.3. The loss function for making the wrong decisions has been estimated by management as follows:

Production level d	Demand level θ		
	10,000	20,000	5,000
10,000	$ 0	$10,000	$15,000
20,000	5,000	0	25,000
5,000	10,000	20,000	0

Management is expected to make the same production level decision regardless of the actual demand. The chances for these decisions are:

Production level decision	Probability
10,000	0.20
20,000	0.75
5,000	0.05

Compute (a) the risk function and (b) the mean risk.

P26.6 A management consultant has been hired by the firm in Prob. P26.5 to determine if the data and mean risk are correct. The consultant found that the expected losses were wrong, so she redefined the loss functions as follows:

$$L(d, \theta) = \begin{cases} 0.50|\theta - d| & d < \theta \\ 0 & d = \theta \\ 0.75|\theta - d| & d > \theta \end{cases}$$

(a) Compute the new loss function values for all (d, θ) pairs.

(b) Determine the new value of the mean risk using the same probabilities stated in Prob. P26.5. Compare the mean risk values.

P26.7 There are two possible investment decisions $d_1 = 10$ and $d_2 = 20$.

(a) Graph the quadratic loss functions for $\theta = 5, 10,$ and 12:

$$L_1(d_1, \theta) = \begin{cases} 400(d_1 - \theta)^2 & \theta < 10 \\ 0 & \theta \geq 10 \end{cases}$$

$$L_2(d_2, \theta) = \begin{cases} 0 & \theta < 10 \\ 100(d_2 - \theta)^2 & \theta \geq 10 \end{cases}$$

(b) Compute the risk function if the decision d_1 is made 80 percent of the time that $\theta \leq 10$ and 40 percent of the time that $\theta > 10$.

(c) Compute the mean risk if the θ values are observed with the same frequency as a variable X, which follows a binomial distribution with parameters $n = 2$ and $p = 0.5$. The correspondence is $x = 0$ and $\theta = 5$, etc.

P26.8 The mean of a population is assumed to follow a normal distribution with $\mu_0 = 25$ and $\sigma_0 = 4$ cm. A sample of 100 measurements resulted in $\bar{X} = 27.5$ and $s^2 = 36.0$ cm^2. If the mean of the population is to be updated using this new sample information, determine the distribution and its parameters.

P26.9 If a sample is taken, the number of families that will purchase a natural Christmas tree follows a binomial distribution. The binomial parameter π, which is the probability that any randomly selected family will buy a natural tree, has historically followed a symmetric beta distribution with parameters $\alpha_0 = \beta_0 = 2$. A survey this year showed that 325 of 500 families plan to buy a natural tree.

  (a) Determine the updated values of the parameters using the results of the survey.
  (b) Compute the mean of the posterior distribution and determine if it is skewed right or left.

P26.10 The A. Q. Rat Precision Company has supplied a product which averages 3.75 percent defective. This defective rate varies according to a $\beta(3, 77)$ distribution. A new company is now used as a supplier, and they have promised the same percentage defective. In a sample of 50 items, 10 were defective. Determine (a) the parameters and (b) the properties of the posterior distribution of the fraction defective if the sample results are used for a parameter update of the previous suppliers' defective rate.

P26.11 The number of automobile wrecks per hour on a city freeway follows a Poisson distribution with a mean of 3 throughout the day. However, when examined on a hour-by-hour basis, the number still follows a Poisson, but the parameter varies according to a gamma distribution with parameters $n_0 = 3$ and $\epsilon_0 = 1$. In the last 48 h, 125 accidents occurred. What are the parameter values of the posterior distribution of the Poisson parameter given this new accident information?

P26.12 If a quadratic loss function is used, what is the Bayesian estimate of the population mean for Prob. P26.8?

P26.13 Determine the Bayesian estimate of (a) π for Prob. P26.9 and (b) the fraction defective for Prob. P26.10.

P26.14 Compute the probability of observing at least 3 accidents per hour for Prob. P26.11 using (a) the Poisson distribution prior to the 48-h sample and (b) the Poisson distribution updated by the Bayesian estimate and the sample results.

P26.15 Given that the variable $X \sim N(\mu, \sigma^2)$ and that $\mu \sim N(5, 0.16)$, use the sample statistics $n = 10$, $\bar{X} = 8.5$, and $s^2 = 0.2$ to determine the Bayesian estimate of μ. Assume that a quadratic loss function is appropriate.

*P26.16 The average time between arrivals to a truck unloading station, which is 0.25 h, follows an exponential distribution: The parameter λ of the distribution for the number of arrivals per hour follows a $\gamma(2, 0.5)$ distribution. If in the last 8 h 40 trucks arrived at the station, use Bayes' procedures to update (a) the value of λ and (b) the estimate of the average time between arrivals.

*P26.17 This problem will lead you through the derivation of the Bayesian estimate of a parameter θ for a quadratic loss function $L(x, \theta) = (x - \theta)^2$ where x is a single observation of the variable X. Assume that all the following are continuous distributions on the real number line: $f(\theta), f(x), f(x/\theta), f(\theta/x)$, and $f(x, \theta)$.

  (a) Start with the loss function and show that the mean risk may be written

$$r(x, \theta) = \int \int (x - \theta)^2 f(x, \theta) \, dx \, d\theta$$

  (b) Rewrite $f(x, \theta)$ in $r(x, \theta)$ and show that it may be expressed as

$$r(x, \theta) = \int f(x) \left[\int (\theta - x)^2 f(\theta/x) \, d\theta \right] dx$$

What is the expression in brackets?

  (c) Use the definition of an efficient estimator (Sec. 6.9 and Prob. P6.36) and the Bayesian estimate statement in Sec. 26.4 to write the logic used to determine that $E[\theta/x]$ is Bayes' estimate of θ

LEVEL FOUR
STATISTICAL ANALYSIS TECHNIQUES

The four chapters in this level are introductions to analysis techniques commonly used to evaluate the results of experimentation. Regression and correlation are studied first. These tools investigate the mathematical relation between variables once sample observations have been taken.

Quality control has many aspects. The control of a manufacturing process via statistical distributions and the acceptance or rejection of lots of items are also discussed in this level. Finally, a brief introduction to the procedures of the analysis-of-variance (ANOVA) technique is presented. The use of ANOVA on the results of a regression analysis is discussed.

Because these techniques utilize much of the information in the prior levels of this text, it is recommended that, as a minimum, the reader be familiar with the material in the chapters indicated in the table before reading the chapters of this level.

Level Four chapter	Required chapters												
	4	6	8	9	11	12	13	14	15	17	18	19	27
27		X			X			X		X		X	
28		X			X			X			X	X	X
29	X		X	X	X	X						X	
30					X		X		X			X	X

CHAPTER TWENTY-SEVEN

CURVE FITTING BY LEAST-SQUARES REGRESSION

The procedure used to fit a line to a set of data points using simple linear regression is explained in this chapter. You will learn how to test hypotheses and develop confidence intervals for this straight line. The topics of multiple linear regression and stepwise regression are also introduced.

CRITERIA

To correctly use the techniques of least-squares regression, you must be able to:

1. State the definition and assumptions of the least-squares regression technique.
*2. Derive the least-squares estimators for linear regression with the origin located at the points (0, 0) and $(\bar{X}, \bar{Y})$.
3. Compute the simple linear regression equation for a *straight line* using the least-squares estimators, given the observed data for two variables.
4. Use the hypothesis testing procedure to perform tests on the intercept and slope of a regression line, given the equation of the fitted line and the hypothesized values.
5. Compute and graph the confidence interval for the following four values, given the observed data and the equation of the regression line:
   a. *Mean* of the observed values for each value of the independent variable X
   b. *Single* estimated value for a specific X value
   c. *Intercept* of the regression line
   d. *Slope* of the regression line

6. State the general model for *multiple linear regression* and the inputs for computer solution, and compute the variance around the regression surface, given the observed data and the regression equation.
*7. State a definition of *stepwise multiple linear regression*, and state the steps in the procedure used to stepwise select independent variables for inclusion in a regression model.
*8. Compute the *curvilinear* regression equation for any of the following models, given the observed data for two variables: exponential, power, hyperbolic, and *k*-degree polynomial.
*9. State the equation used to partition the *total sum of squares* for any set of data, and compute the individual sum of squares and the *mean squares*, given the data and regression equation.

STUDY GUIDE

27.1 Introduction to Least-Squares Regression

The construction of an eye-balled line through points plotted on a graph is one way to fit a line to data. However, if the line should be the best possible and if an equation of the line is needed, regression is the correct approach.

> *Regression* is a statistical technique to determine the equation of the line or curve which minimizes the *deviations* between the observed data and the regression equation values.

Regression is, therefore, based on the *least-squares* principal to minimize the error (residual) sum of squares between the observed values Y_i and the values $\hat{Y}_i$, which are estimated by the regression equation: In symbols, for $i = 1, 2, \ldots, n$, if $\epsilon_i = Y_i - \hat{Y}_i$, this is

- Minimize $$S = \Sigma \epsilon_i^2 = \Sigma (Y_i - \hat{Y}_i)^2 \tag{27.1}$$

The square of the ϵ_i values is used to eliminate the plus and minus signs on the deviations.

Regression, or curve fitting as it is also called, uses pairs of data (X_i, Y_i) and Eq. (27.1) to determine an equation which explains the mathematical relation between the two variables. X is commonly the *independent* variable because it has specific preselected and fixed (nonrandom) values. Y is the *dependent*, or response, variable which is random for each X value. Figure 27.1 is a graph of the residuals ϵ_i around a fitted line for several (X_i, Y_i) pairs.

The least-squares technique makes certain assumptions:

1. There is no error in the X_i values, because they are set prior to the experiment. The X variable is, therefore, *not random*.
2. A normal distribution is observable at each X_i if many Y_i values are recorded.
3. The normal distributions are *independent* of one another.

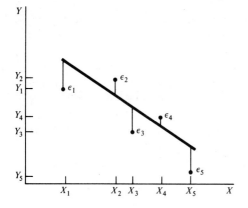

Figure 27.1 The deviations or residuals around a straight-line fit to (X_i, Y_i) data pairs.

4. The variance σ^2 of the normal at each X_i value is the same.
5. The least-squares fit is through the mean μ_i of the normal distribution at each X_i.

Figure 27.2 presents these assumptions with a straight-line fit through the means of the $N(\mu_i, \sigma^2)$ distributions.

Because the least-squares regression line or curve has a mathematical relation developed using only specified X_i values, extrapolation beyond this range is risky. There is no assurance that the normal distribution for the Y values will continue to follow the assumptions above. Extrapolation is commonly done by practitioners, but it *must* be done realizing that no statistical confidence can be placed on the forecasted results.

Problem P27.1

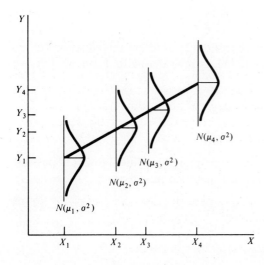

Figure 27.2 Straight-line fit through the means of independent, normally distributed variables.

*27.2 Derivation of the Linear Regression Estimators

If a straight-line relation is assumed to be the best between X and Y, the mathematical model for $i = 1, 2, \ldots, n$ is

$$Y_i = \alpha + \beta X_i + \epsilon_i$$

where the ϵ_i values are the random errors about the line. The best estimators of α and β are $\hat{\alpha} = a$ and $\hat{\beta} = b$ from the equation

$$\hat{Y}_i = a + bX_i \tag{27.2}$$

where $\hat{Y}_i$ is the estimate from the regression equation. Equations (27.1) and (27.2) may be used to determine a and b using the least-squares method.

Minimize
$$S = \Sigma(Y_i - \hat{Y}_i)^2$$
$$= \Sigma[Y_i - (a + bX_i)]^2$$

The subscript i is dropped for simplicity, and the partial derivatives with respect to a and b are set equal to 0. This procedure results in the two *normal equations*, Eqs. (27.3) and (27.4).

$$\frac{\partial S}{\partial a} = -2\Sigma(Y - a - bX) = 0$$

or
$$\Sigma Y = na + b\Sigma X \tag{27.3}$$

Also
$$\frac{\partial S}{\partial b} = -2\Sigma X(Y - a - bX) = 0$$

or
$$\Sigma XY = a\Sigma X + b\Sigma X^2 \tag{27.4}$$

Simultaneous solution for a and b from the normal equations gives

$$a = \frac{\Sigma X^2 \Sigma Y - \Sigma X \Sigma XY}{n\Sigma X^2 - (\Sigma X)^2} \tag{27.5}$$

$$b = \frac{n\Sigma XY - \Sigma X \Sigma Y}{n\Sigma X^2 - (\Sigma X)^2} \tag{27.6}$$

The values of a and b are substituted into Eq. (27.2) to get the equation for $\hat{Y}_i$.

If the first normal equation, Eq. (27.3), is rewritten with $\overline{Y}$ on the left, we have

$$\frac{\Sigma Y}{n} = \frac{na}{n} + b\frac{\Sigma X}{n}$$

$$\overline{Y} = a + b\overline{X}$$

Therefore, the centroid point $(\overline{X}, \overline{Y})$ is on the regression line. Now it is possible to move the origin to $(\overline{X}, \overline{Y})$ and write the assumed model as

$$\hat{Y} - \overline{Y} = a' + b(X - \overline{X}) \tag{27.7}$$

Figure 27.3 shows this translation. The purpose of this movement is to obtain

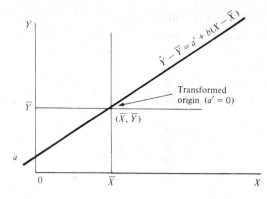

Figure 27.3 The linear regression model with the origin moved to $(\bar{X}, \bar{Y})$.

easier-to-use relations for the constants. If we substitute the fact that

$$\Sigma(X - \bar{X}) = 0 \quad \text{and} \quad \Sigma(Y - \bar{Y}) = 0$$

into Eqs. (27.5) and (27.6), they reduce to

- $a' = 0$

$$b = \frac{\Sigma(X - \bar{X})(Y - \bar{Y})}{\Sigma(X - \bar{X})^2}$$

Substitution into Eq. (27.7) gives the correct form of the line. The technique is illustrated in Sec. 27.3.

Problem P27.2

27.3 Determination of the Regression Line

If simple linear regression is used, a linear relation is assumed between the variables X and Y, and the model for $i = 1, 2, \ldots, n$ is

$$Y_i = \alpha + \beta X_i$$

where α is the y intercept and β is the true slope. If the least-squares estimators $\hat{\alpha} = a$ and $\hat{\beta} = b$ are used, it is possible to move the origin to the point $(\bar{X}, \bar{Y})$, as shown in Fig. 27.3, and derive quite simple relations for the constants. Of course, a calculator that has linear regression preprogrammed eliminates the need to compute a and b manually; however, you do need to know the meaning of the regression equation and the correct way to use it.

Without the origin translation mentioned, the model is

- $$\hat{Y}_i = a + bX_i \qquad (27.8)$$

and with the translation it is

- $$\hat{Y}_i - \bar{Y} = a' + b(X_i - \bar{X}) \qquad (27.9)$$

The equations for a' and b are given with the subscript i omitted:

$$a' = 0$$

$$b = \frac{\Sigma(X - \bar{X})(Y - \bar{Y})}{\Sigma(X - \bar{X})^2} \qquad (27.10)$$

These are least-squares estimators which are derived in the previous (optional) section using Eq. (27.1). Since $a' = 0$, the value of b is substituted into Eq. (27.9), and the original $\hat{Y} = a + bX$ equation is obtained.

After a and b are determined, any value of X can be used in the regression equation to determine the expected value $\hat{Y}$, which is the mean value of a normal distribution placed at the X value. Example 27.1 illustrates the regression technique using the following steps.

1. Compute $\bar{X}$ and $\bar{Y}$.
2. Obtain the values and sums for $X - \bar{X}$, $(X - \bar{X})^2$, and $Y - \bar{Y}$.
3. Calculate b using Eq. (27.10).
4. Substitute b into Eq. (27.9) and rewrite the equation in the form $\hat{Y} = a + bX$.
5. Solve for $\hat{Y}$ using any given specific X values and/or plot the equation.

Example 27.1 An engineering statistics student has a summer job with the forestry service. A new variety of tree was planted 6 years ago, and the trunk diameters were taken after each year of growth:

Year	1	2	3	4	5	6
Diameter (cm)	1.3	2.5	3.7	5.3	6.4	7.2

Neglect all environmental factors.
(a) Determine the regression line equation.
(b) Estimate the average diameter for 3.5-year-old trees.
(c) Compute the error sum of squares for the regression line.

SOLUTION (a) The steps above and the computations in Table 27.1 are used to determine the regression line. The year is the independent variable X, since it is fixed, and the response variable Y is the diameter.

1 and 2. The means of $\bar{X} = 3.5$ years and $\bar{Y} = 4.4$ cm are used in Table 27.1 to obtain the correct differences.
3. From Eq. (27.10)

$$b = \frac{21.40}{17.50} = 1.223$$

27.3 DETERMINATION OF THE REGRESSION LINE

Table 27.1 Computations to determine the value of b for a regression line, Example 27.1

Year X	Diameter Y	$X - \bar{X}$	$Y - \bar{Y}$	$(X - \bar{X})(Y - \bar{Y})$	$(X - \bar{X})^2$	$\hat{Y}$	$(Y - \hat{Y})^2$
1	1.3	−2.5	−3.1	7.75	6.25	1.34	0.0016
2	2.5	−1.5	−1.9	2.85	2.25	2.57	0.0049
3	3.7	−0.5	−0.7	0.35	0.25	3.79	0.0081
4	5.3	0.5	0.9	0.45	0.25	5.01	0.0841
5	6.4	1.5	2.0	3.00	2.25	6.24	0.0256
6	7.2	2.5	2.8	7.00	6.25	7.46	0.0676
21	26.4			21.40	17.50		0.1919

$$\bar{X} = 3.5 \quad \bar{Y} = 4.4$$

4. The equation for the line using $a' = 0$ in Eq. (27.9) is
$$\hat{Y} - 4.4 = 1.223(X - 3.5)$$
$$\hat{Y} = 0.120 + 1.223X \qquad (27.11)$$

The $\hat{Y}$ values are given in Table 27.1 and plotted with the original data in Fig. 27.4. The fit looks very good because the deviations are small.

(b) The average diameter for $X = 3.5$ years is found from Eq. (27.11):
$$\hat{Y} = 0.120 + 1.223(3.5) = 4.4$$

This is the $(\bar{X}, \bar{Y}) = (3.5, 4.4)$ point which always lies on the regression line.

(c) The error sum of squares is found using the sum in Eq. (27.1). From Table 27.1
$$\Sigma \epsilon_i^2 = \Sigma(Y_i - \hat{Y})^2 = 0.1919$$

COMMENT Remember that X is the preselected, nonrandom variable years

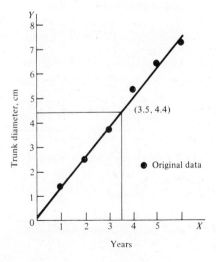

Figure 27.4 Plot of the original data and the regression line $\hat{Y} = 0.120 + 1.223X$, Example 27.1.

and Y is the dependent, random variable. If a $\hat{Y}$ value is computed for X above 6 years, there is an implied assumption that linearity continues past $X = 6$ years, an assumption not verifiable by the collected data.

The centroid point $(\bar{X}, \bar{Y})$ will always be on the line. In fact, the easiest way to plot the regression line is to spot the points $(X, Y) = (0, a)$ and $(\bar{X}, \bar{Y})$ and draw a line between them.

Problems P27.3–P27.6

27.4 Hypothesis Tests of the Intercept and Slope Values

In order to make any type of statistical conclusions about the regression, it is necessary to compute the *variance around the regression line* s_Y^2, which is

$$s_Y^2 = \frac{\text{error sum of squares}}{\text{degrees of freedom}} = \frac{\Sigma \epsilon_i^2}{\nu}$$

$$= \frac{\Sigma(Y_i - \hat{Y}_i)^2}{n - 2} \qquad (27.12)$$

The numerator is Eq. (27.1) and $\nu = n - 2$ because it requires two points to determine a specific line, so 2 degrees of freedom are lost.

It is possible to test the a and b values of the fitted line $\hat{Y} = a + bX$ against any hypothesized value a_0 or b_0. This is useful if some design criteria have been set on a fitted line or another set of data is to be compared with it. The t distribution is used to perform tests for each of these values. If a hypothesized value $b_0 = 0$ is accepted, the regression line has a slope of 0, so the variables X and Y are *independent* and the fitted line is of no value.

The two-sided hypotheses, the t test statistics, and the degrees of freedom are summarized in Table 27.2. In either case, if the computed t exceeds the Table B-5 value for a two-tail α, then H_0 is rejected. The hypothesis testing procedure, which is used for these tests, is illustrated in the next example.

● **Table 27.2 Test statistics for testing the intercept and slope of a regression line**

Hypothesis	Standard deviation of a or b	t test statistic	Degrees of freedom
$H_0: a = a_0$ $H_1: a \neq a_0$	$s_a = \left[s_Y^2 \left(\dfrac{1}{n} + \dfrac{\bar{X}^2}{\Sigma(X_i - \bar{X})^2} \right) \right]^{1/2}$	$t = \dfrac{|a - a_0|}{s_a}$	$\nu = n - 2$
$H_0: b = b_0$ $H_1: b \neq b_0$	$s_b = \left[s_Y^2 \left(\dfrac{1}{\Sigma(X_i - \bar{X})^2} \right) \right]^{1/2}$	$t = \dfrac{|b - b_0|}{s_b}$	$\nu = n - 2$

Example 27.2 In Example 27.1 on trunk diameters the line fit to six data points has the equation $\hat{Y} = 0.120 + 1.223X$ with an error sum of squares of 0.1919. The line for a similar tree was found to be $\hat{Y} = 0.200 + 0.95X$.
(a) Test if the two lines have the same intercept and slope.
(b) Is the value $b = 1.223$ statistically different from 0?

SOLUTION (a) The hypothesis testing steps are detailed for the test of the intercept.

1. The two-sided test is
$$H_0: 0.12 = 0.20 \quad \text{versus} \quad H_1: 0.12 \neq 0.20$$
2. Let $\alpha = 0.05$.
3. The two sample statistics needed are $\nu = 6 - 2 = 4$ and s_a. To obtain s_a, the values of $\bar{X} = 3.5$ and $\Sigma(X - \bar{X})^2 = 17.5$ are obtained from Table 27.1, and the value of s_Y^2 is computed from Eq. (27.12):
$$s_Y^2 = \frac{0.1919}{4} = 0.048$$
$$s_a = \left[0.048\left(\frac{1}{6} + \frac{(3.5)^2}{17.5}\right)\right]^{1/2} = (0.0416)^{1/2}$$
$$= 0.204$$
4. The test statistic value from Table 27.2 is
$$t = \frac{|0.12 - 0.20|}{0.204} = 0.392$$
5. The acceptance region for a two-tail $\alpha = 0.05$ and $\nu = 4$ on the t distribution is $|t| \leq 2.776$.
6. Since $0.392 < 2.776$, H_0 is not rejected.

Using the same procedure, the test for the slope is $H_0: 1.223 = 0.950$ versus $H_1: 1.223 \neq 0.950$. From Tables 27.1 and 27.2 the standard deviation of b is
$$s_b = \left[0.048\left(\frac{1}{17.5}\right)\right]^{1/2} = (0.0027)^{1/2} = 0.052$$

The test statistic value is
$$t = \frac{|1.223 - 0.950|}{0.052} = 5.25$$

The acceptance region is again $|t| \leq 2.776$, and H_0 is rejected for the slope. So, the lines have the same intercept but different slopes.
(b) To test the hypothesis $H_0: b = 0$ versus $H_1: b \neq 0$, the test statistic from Table 27.2 is
$$t = \frac{|1.223 - 0|}{0.052} = 23.52$$

H_0 is definitely rejected, thus the variables of trunk diameter and age are not independent.

COMMENT A test useful in determining if a set of data is best fit using a linear relationship is covered in Chap. 30 on analysis of variance. A test which simultaneously checks values of a and b may be performed using an F distribution (see the Bowker and Lieberman text).

Problems P27.7–P27.10

27.5 Confidence Intervals for Regression Lines

The regression line is fit, as explained in Sec. 27.1, through the mean of the normal distribution at each observed X value. The data, the equation of the line $\hat{Y} = a + bX$, and the variance around the regression line s_Y^2 are useful in setting $100(1 - \alpha)$ percent confidence intervals on at least four different values. The first we will discuss is the interval estimate on $\hat{Y}$, which is the *mean* of the observed values at each preset X value. The limits are found on $\hat{Y}$ for each X value, and the symmetric limits are drawn as shown in Fig. 27.5. The variance for each mean value $\hat{Y}$ is

$$s_{\hat{Y}}^2 = s_Y^2 \left[\frac{1}{n} + \frac{(X - \bar{X})^2}{\Sigma(X - \bar{X})^2} \right] \tag{27.13}$$

The X in the numerator is the value for which the interval is to be set, so a new $s_{\hat{Y}}^2$ is computed for each X value.

The interval estimate, which is set using the t distribution with $\nu = n - 2$ and a two-tail α value, is

$$\hat{Y} \pm t s_{\hat{Y}} \tag{27.14}$$

The interval is always the narrowest at $\bar{X}$. Therefore, to maintain the same confidence as we go away from the $\bar{X}$ value, the interval gets wider. This fact is illustrated in Example 27.3.

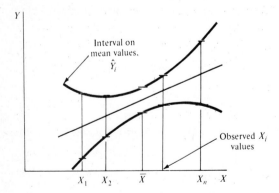

Figure 27.5 Confidence interval on the mean values $\hat{Y}_i$ at each observed X_i value.

27.5 CONFIDENCE INTERVALS FOR REGRESSION LINES

If a new X value X_0 is substituted in the regression equation and the $\hat{Y}$ is used as a *predicted* value for the future, the interval is set on the *single* estimated value $\hat{Y}_0$. This interval, which looks like the one above, is wider simply because it is computed for a single $\hat{Y}$ value, not for a mean. Often values of X not observed when the data was collected are used in this computation. The variance for a single $\hat{Y}_0$ value will be symbolized by s_0^2 and is equal to

$$s_0^2 = s_Y^2 \left[1 + \frac{1}{n} + \frac{(X_0 - \bar{X})^2}{\Sigma(X - \bar{X})^2} \right] \qquad (27.15)$$

Again, the interval is set by the t distribution for $\nu = n - 2$ and a two-tail α:

$$\hat{Y}_0 \pm ts_0 \qquad (27.16)$$

where $\hat{Y}_0 = a + bX_0$ is from the regression equation. It is common, as shown below, to compute these single $\hat{Y}$ interval limits for specific X_0 values rather than constructing the entire fan-shaped curves of Fig. 27.5. Note that s_0^2 in Eq. (27.15) is larger than $s_{\hat{Y}}^2$, Eq. (27.13), by the amount s_Y^2.

Interval estimates may also be established for the intercept a and the slope b using the t distribution with $\nu = n - 2$ degrees of freedom. The formulas and a figure reference are given here:

Parameter	Interval estimate	Figure
Intercept a	$a \pm ts_a$	27.6a
Slope b	$b \pm ts_b$	27.6b

The standard deviations s_a and s_b are computed from the formulas in Table 27.2. Note in Fig. 27.6b that the slope interval limits go through the point $(\bar{X}, \bar{Y})$.

Example 27.3 A research team has tested a new airfoil design at seven speeds at $-17°C$ and observed the temperature at the forward edge (Table 27.3). The regression equation is

$$\hat{Y} = -31.671 + 0.074X \qquad (27.17)$$

(a) Compute and graph the 95 percent confidence interval for the mean values of the temperatures.
(b) Estimate the specific value of temperature at 425 mi/h with 95 percent confidence.

SOLUTION (a) Table 27.3 gives the estimates $\hat{Y}$ using Eq. (27.17) and the error sum of squares $\Sigma\epsilon_i^2 = 14.52$. The variance around the regression line

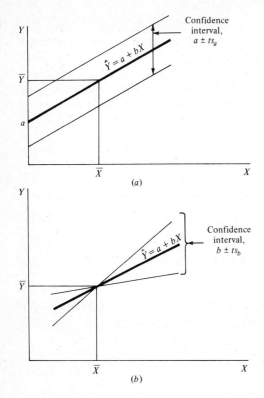

Figure 27.6 Confidence intervals for (a) the intercept and (b) the slope of a regression line.

Table 27.3 Confidence interval computation for the mean estimated $\hat{Y}$ values, Example 27.3

Speed, X mi/h (1)	Temp, $Y°C$ (2)	$\hat{Y}$ (3)	$(Y - \hat{Y})^2 = \epsilon^2$ (4)	$(X - \bar{X})^2$ (5)	Standard deviation $s_{\hat{Y}}$ (6)	Confidence interval $\hat{Y} \pm ts_{\hat{Y}}$ (7)
300	−10.0	−9.5	0.25	22,500	1.16	−12.48, −6.52
350	−3.9	−5.8	3.61	10,000	0.91	−8.14, −3.46
400	−2.5	−2.1	0.16	2,500	0.72	−3.95, −0.25
450	1.7	1.6	0.01	0	0.64	−0.05, 3.25
500	3.9	5.3	1.96	2,500	0.72	3.45, 7.15
550	7.2	9.0	3.24	10,000	0.91	6.66, 11.34
600	15.0	12.7	5.29	22,500	1.16	9.72, 15.68
			14.52	70,000		

by Eq. (27.12) is

$$s_Y^2 = \frac{14.52}{7-2} = 2.90$$

A 95 percent interval estimate must be computed for each X value using Eq. (27.13) to compute the standard deviation $s_{\hat{Y}}$ and Eq. (27.14) for the interval limits (Table 27.3). For example, if $X = 400$ mi/h with $\bar{X} = 450$, the interval is computed as

$$s_{\hat{Y}} = \left[2.90\left(\frac{1}{7} + \frac{(400-450)^2}{70,000}\right)\right]^{1/2} = (0.518)^{1/2}$$

$$= 0.72$$

Limits: $-2.1 \pm 2.571(0.72) = -3.95$ and -0.25

The t value 2.571 is for a two-tail $\alpha = 0.95$, and $\nu = 7 - 2 = 5$. Figure 27.7, which is a graph of the confidence interval, shows that the fan-shaped interval is symmetric about the line and is the narrowest at $\bar{X} = 450$ mi/h.

(b) A 95 percent interval estimate of the *single* $\hat{Y}_0$ value at $X_0 = 425$ is computed from Eqs. (27.15) for s_0^2 and (27.16) for the interval limits. Use $X_0 = 425$, $\hat{Y}_0 = -31.671 + 0.074(425) = -0.22$, and

$$s_0 = \left[2.90\left(1 + \frac{1}{7} + \frac{(425-450)^2}{70,000}\right)\right]^{1/2} = (3.34)^{1/2}$$

$$= 1.83$$

Limits: $-0.22 \pm 2.571(1.83) = -4.92$ and 4.48

This interval is also graphed in Fig. 27.7. As you can see, the interval is

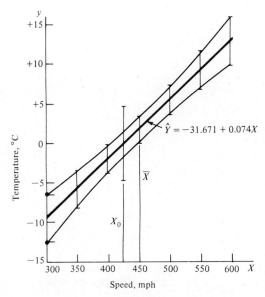

Figure 27.7 Confidence intervals for the mean values on the regression line and the individual estimate at $X = 425$ mi/h, Example 27.3.

much wider for the single $\hat{Y}$ estimate than for the mean estimate at $X_0 = 425$ mi/h.

COMMENT You should now compute the 95 percent intervals for a and b. The answers are:

Limits for intercept: $\quad -39.30 \leq a \leq -24.04$
Limits for slope: $\quad\quad\quad\; 0.057 \leq b \leq 0.091$

Problems P27.11–P27.15

27.6 Multiple Linear Regression

If more than one independent variable is considered in the linear model, multiple linear regression is used. For k independent variables, the model is

$$Y_i = \alpha + \beta_1 X_{i1} + \beta_2 X_{i2} + \cdots + \beta_k X_{ik}$$

The assumptions listed in Sec. 27.1 are correct for Y and each X_i variable in this model. The least-square estimators for the α and β parameters are called *regression coefficients* and are obtained by minimizing the error (residual) sum of squares.

- Minimize $\quad\quad S = \Sigma \epsilon_i^2 = \Sigma (Y_i - \hat{Y}_i)^2 \quad\quad$ (27.18)

where the model for $\hat{Y}$ is (subscript i omitted)

- $\quad\quad\quad \hat{Y} = a + b_1 X_1 + b_2 X_2 + \cdots + b_k X_k$

If a three-dimensional model is used, there are two X variables, X_1 and X_2, and a plane is fit through the points to minimize the error sum of squares (Fig. 27.8). The model

$$\hat{Y} = a + b_1 X_1 + b_2 X_2 \quad\quad (27.19)$$

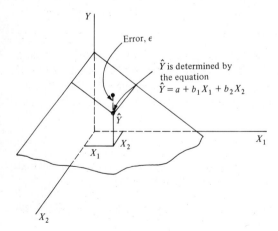

Figure 27.8 Multiple linear regression for two independent variables.

is used in Eq. (27.18) to derive the normal equations by setting the three partial derivatives with respect to each coefficient equal to 0 and simultaneously solving the three normal equations:

$$\Sigma Y = na + b_1 \Sigma X_1 + b_2 \Sigma X_2$$
$$\Sigma X_1 Y = a \Sigma X_1 + b_1 \Sigma X_1^2 + b_2 \Sigma X_1 X_2 \qquad (27.20)$$
$$\Sigma X_2 Y = a \Sigma X_2 + b_1 \Sigma X_1 X_2 + b_2 \Sigma X_2^2$$

Setting up the sums and solving the system in Eq. (27.20) is not too difficult for two X variables, but the availability of computer software (programs) to perform multiple linear regression for any number of variables makes the job much easier. The inputs to a computerized regression package are:

1. Number of observations n
2. Number of variables $k + 1$
3. Which variable is the response variable Y (The variables are called $X_1, X_2, \ldots, X_{k+1}$ in the programs.)
4. Values for each point $(Y, X_1, X_2, \ldots, X_k)$ for all n observations

One important output will be the value of the coefficients $a, b_1, \ldots, b_k$ for the multiple regression model. Another output will be the error sum of squares for the fitted model, that is, the value $\Sigma \epsilon_i^2$ from Eq. (27.18). The same procedure is used here to obtain $\Sigma \epsilon_i^2$ as it was in simple linear regression. The $\hat{Y}$ values are obtained from the fitted equation, and the squared differences $(Y - \hat{Y})^2$ are summed. The variance around the regression surface may be written

$$s_Y^2 = \frac{\Sigma (Y - \hat{Y})^2}{n - k - 1} = \frac{\Sigma \epsilon_i^2}{n - k - 1} \qquad (27.21)$$

where k is the number of independent variables. This variance allows us to perform hypothesis tests on and set confidence intervals for the regression coefficients using a t distribution with $\nu = n - k - 1$ degrees of freedom.

Example 27.4 Judi hypothesizes that the speed Y of a certain reaction depends on three variables: temperature of chemical A (X_1), temperature of chemical B (X_2), and flow rate (X_3). (This data is taken from Neville and Kennedy [1]). A four-dimensional multiple linear regression model has been computer-fit to the data (Table 27.4) to obtain the equation

$$\hat{Y} = 83.08 - 0.32 X_1 + 0.55 X_2 + 0.18 X_3 \qquad (27.22)$$

(a) Explain the meaning of Eq. (27.22).
(b) Compute the variance around the regression surface.

SOLUTION (a) Equation (27.22) is the least-squares fit which minimizes the

Table 27.4 Observed and estimated values for a multiple linear regression fit, Example 27.4

X_1	X_2	X_3	Y	$\hat{Y}$	$(Y - \hat{Y})^2$
11	58	11	126	113.44	157.75
32	21	13	92	86.73	27.77
15	22	28	107	95.42	134.10
26	55	27	120	109.87	102.62
9	41	21	103	106.53	12.46
31	18	20	84	86.66	7.08
12	56	20	113	113.64	0.41
29	40	27	110	100.66	87.24
13	57	30	104	115.67	136.19
10	21	12	83	93.59	112.15
33	40	19	85	97.94	167.44
31	58	29	104	110.28	39.44
			1,231		984.65

error sum of squares $\Sigma\epsilon^2 = \Sigma(Y - \hat{Y})^2$ using the model

$$\hat{Y} = a + b_1 X_1 + b_2 X_2 + b_3 X_3$$

(b) The $\hat{Y}$ values given in Table 27.4 are substituted into Eq. (27.21) to compute the variance for $n = 12$ and $k = 3$:

$$s_Y^2 = \frac{984.65}{12 - 3 - 1} = 123.08$$

The numerator of s_Y^2 is the error sum of squares $\Sigma(Y - \hat{Y})^2$.

Problems P27.16–P27.18

*27.7 Stepwise Multiple Linear Regression

When the independent variables that should be included in the regression model are not known, stepwise multiple linear regression is the most efficient way to determine the variables and regression coefficients. Because of the number of computations, a computer must be used to perform the analysis. We cover only the logic of the method here. By definition,

> *Stepwise* multiple linear regression is a technique that places only the independent variables into the regression equation that remove a significant portion of the variation in Y, the dependent variable. The criterion used is the amount of variability (sum of squares) removed by each variable, as measured by the F test, *independent* of the order in which the variable enters solution.

Initially the candidate variable must have the maximum correlation between itself and Y to be selected. Correlation, a quantitative measure of the relation between variables, is studied in Sec. 28.2. The steps below track the progress of a computerized stepwise program. For convenience, variables are called Y (response), X_1 (first entering variable), X_2 (second entry), etc. The value of the regression coefficients $a, b_1, \ldots$ changes with each new model.

1. Enter into solution the X variable having the highest correlation with Y, because it will explain the most variability in Y. The model is now
$$\hat{Y} = a + b_1 X_1$$
2. Select as X_2 the X variable having the highest correlation with Y after the effect of X_1 is accounted for by the regression model. Use the F test to determine if more variability could have been explained if X_2 had entered *prior* to X_1. If it had, X_1 is removed and the model is
$$\hat{Y} = a + b_2 X_2$$
If more is removed with the order X_1 than X_2, the model is
$$\hat{Y} = a + b_1 X_1 + b_2 X_2$$
3. Select as X_3 the X variable having the largest correlation with Y after X_1 and X_2 are accounted for. Again use the F test to see if some other entry order of X_1 and X_2 removes more variability. The orders checked are X_1 after X_2 and X_3, and X_2 after X_1 and X_3. Remove X_1 and/or X_2 from the model if the test does not justify their presence.
4. Continue this routine until either no more X variables remove a significant amount of variability or no new X variables remain to be considered.

See the Draper and Smith text for complete details and computer examples of stepwise regression.

Problem P27.19

*27.8 Curvilinear Regression for Two Variables

Often data for two variables X and Y is clearly not linear, so the simple linear regression method of Sec. 27.3 is not appropriate. Curvilinear (or nonlinear) regression can be performed like simple linear regression if the model can be linearized. For example, the *exponential* curve model can be linearized by taking the base 10 logarithm. The original and linearized model are (subscript i omitted on Y and X)

$$Y = ab^X$$
$$\log Y = \log a + X \log b \qquad (27.23)$$

Equation (27.23) may be written $Z = a' + b'X$, where $Z = \log Y$, $a' = \log a$, and $b' = \log b$. This is a straight line when X and Z are plotted.

The relations given by Eqs. (27.5) and (27.6) with the origin at (0, 0) are easily transformed to find the intercept a and slope b of the linearized model. Table 27.5 summarizes the models, graphs, and equations for a and b for three models, including the exponential.

One other commonly used nonlinear model is the k-degree *polynomial*, which takes the form

$$Y = a + b_1 X + b_2 X^2 + \cdots + b_k X^k \qquad (27.24)$$

The least-squares method is applied to minimize the error sum of squares and develop $k + 1$ normal equations which are then simultaneously solved for the regression coefficients $a, b_1, \ldots, b_k$. For example, if a second-degree polynomial is used, a parabola of the form

$$Y = a + b_1 X + b_2 X^2$$

is fit to the data, and the normal equations are

$$\Sigma Y = na + b_1 \Sigma X + b_2 \Sigma X^2$$
$$\Sigma XY = a\Sigma X + b_1 \Sigma X^2 + b_2 \Sigma X^3 \qquad (27.25)$$
$$\Sigma X^2 Y = a\Sigma X^2 + b_1 \Sigma X^3 + b_2 \Sigma X^4$$

The pattern in Eq. (27.25) continues for polynomial fits of higher degrees.

Care must be taken in the degree of polynomial fit to data so that a "good" fit is obtained with the lowest degree possible. Figure 27.9 shows the general shape of several polynomial models. It is always possible to exactly fit n data points with a polynomial of degree $n - 1$ or less. However, since the aim of regression is to use the simplest model possible to fit data, a change in polynomial models when new data is collected for the same variables is not a good method.

In all cases the best fit among several linear and nonlinear models is obtained by using the model which results in the *smallest error sum of squares*; that is, the criterion used to select a regression model is to minimize

$$\Sigma \epsilon_i^2 = \Sigma (Y_i - \hat{Y}_i)^2 \qquad (27.26)$$

Example 27.5 In a study of scrap rate and production rate, the following figures were collected. One analyst plans to fit an exponential model while another plans to use a second-degree polynomial. Determine which model offers the better fit. Use production rate as the nonrandom variable.

Production rate, X units/week	1000	2000	3000	3500	4000	4500	5000
Scrap rate, Y % of production	5.2	6.5	6.8	8.1	10.2	10.3	13.0

Table 27.5 Summary of curvilinear regression for the exponential, power, and hyperbolic models

Name	Original and transformed model	Graphs — Original	Graphs — Transformed	Equations for intercept and slope
Exponential	$Y = ab^X$ $\log Y = \log a + X \log b$	(curves: $b > 1$, $b < 1$ vs X)	(lines: $b > 1$, $b < 1$; intercept $\log a$ vs X)	$\log a = \dfrac{\Sigma X^2 \Sigma \log Y - \Sigma X \Sigma X \log Y}{n\Sigma X^2 - (\Sigma X)^2}$ $\log b = \dfrac{n\Sigma X \log Y - \Sigma X \Sigma \log Y}{n\Sigma X^2 - (\Sigma X)^2}$
Power	$Y = aX^b$ $\log Y = \log a + b \log X$	(curves: $b > 1$, $b < 1$ vs X)	(lines: $b > 1$, $b < 1$; intercept $\log a$ vs $\log X$)	$\log a = \dfrac{\Sigma(\log X)^2 \Sigma \log Y - \Sigma \log X \Sigma (\log X \log Y)}{n\Sigma(\log X)^2 - (\Sigma \log X)^2}$ $b = \dfrac{n\Sigma(\log X \log Y) - \Sigma \log X \Sigma \log Y}{n\Sigma(\log X)^2 - (\Sigma \log X)^2}$
Hyperbolic	$Y = a + b/X$ $Y = a + bU$ where $U = 1/X$	(curve vs X)	(line vs U)	$a = \dfrac{\Sigma(1/X)^2 \Sigma Y - \Sigma(1/X) \Sigma(Y/X)}{n\Sigma(1/X)^2 - [\Sigma(1/X)]^2}$ $b = \dfrac{n\Sigma(Y/X) - \Sigma(1/X) \Sigma Y}{n\Sigma(1/X)^2 - [\Sigma(1/X)]^2}$

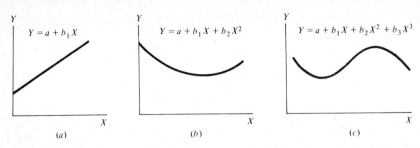

Figure 27.9 Polynomial curves of degree (a) one, a straight line; (b) two, a parabola; and (c) three.

SOLUTION The sums needed for the *exponential* fit are

$$\Sigma X = 23{,}000 \qquad \Sigma \log Y = 6.41$$
$$\Sigma X^2 = 8.75 \times 10^7 \qquad \Sigma X \log Y = 22{,}180.94$$

Substitution into the relations for log a and log b (Table 27.5) give the following solution.

$$\log a = \frac{(8.75 \times 10^7)(6.41) - 23{,}000(22{,}180.94)}{7(8.75 \times 10^7) - (23{,}000)^2} = 0.607$$

$$a = 4.05$$

$$\log b = \frac{7(22{,}180.94) - 23{,}000(6.41)}{7(8.75 \times 10^7) - (23{,}000)^2} = 9.4 \times 10^{-5}$$

$$b = 1.0002$$

The original exponential regression equation $\hat{Y}_e$ is

$$\hat{Y}_e = 4.05(1.0002)^X$$

The sums for the second-degree *polynomial* fit are

$$\Sigma X = 23{,}000 \qquad \Sigma X^2 = 8.75 \times 10^7 \qquad \Sigma X^3 = 3.59 \times 10^{11}$$
$$\Sigma X^4 = 1.54 \times 10^{15} \qquad \Sigma XY = 2.19 \times 10^5 \qquad \Sigma X^2 Y = 8.88 \times 10^8$$
$$\Sigma Y = 60.1$$

The relations in Eq. (27.25) may be solved to obtain

$$a = 6.585 \qquad b_1 = 0.00035 \qquad b_2 = 6.75 \times 10^{-8}$$

Therefore, the equation of the polynomial regression curve $\hat{Y}_p$ is

$$\hat{Y}_p = 6.585 + 0.00035X + (6.75 \times 10^{-8})(X^2)$$

Table 27.6 presents the data, $\hat{Y}_e$ and $\hat{Y}_p$ values, and the terms for the error sum of squares.

Exponential: $\Sigma \epsilon_i^2 = 6.11$

Polynomial: $\Sigma \epsilon_i^2 = 17.50$

Since the exponential fit has the smaller $\Sigma \epsilon_i^2$, it is the better fit.

Table 27.6 Error sum of squares for an exponential and a polynomial ($k = 2$) fit, Example 27.5

Production rate X	Scrap rate Y	Exponential $\hat{Y}_e$	ϵ^2	Polynomial ($k = 2$) $\hat{Y}_p$	ϵ^2
1000	5.2	4.95	0.06	7.00	3.24
2000	6.5	6.04	0.21	7.56	1.12
3000	6.8	7.38	0.34	8.25	2.10
3500	8.1	8.16	0.003	8.64	0.29
4000	10.2	9.01	1.42	9.07	1.28
4500	10.3	9.96	0.12	9.53	0.60
5000	13.0	11.01	3.96	10.02	8.87
			6.113		17.50

COMMENT If more X values were used in the data above, the polynomial (or some other model) could easily be better because the values of the error sum of squares (Table 27.6) are close. Remember that the $\Sigma \epsilon^2$ is the *unexplained*, or residual, sum of squares. This is why it should be small when an appropriate model is selected.

Problems P27.20–P27.28

*27.9 Partitioning of the Sum of Squares

The total sum of squares SS_T for any data may be computed as the sum of squares of the differences between each Y_i and the mean $\bar{Y}$.

$$SS_T = \Sigma(Y_i - \bar{Y})^2 \qquad (27.27)$$

SS_T may be partitioned, in general, as

$$SS_T = \text{sum of squares explained by regression} + \text{sum of squares unexplained by regression}$$

The sum of squares explained by regression, call it SS_R, is the amount removed by performing the regression and is calculated by using the differences between the estimate $\hat{Y}$ and the mean $\bar{Y}$:

$$SS_R = \Sigma(\hat{Y}_i - \bar{Y})^2$$

The sum of squares unexplained by regression, call it SS_E, is the amount remaining after the regression is performed. It is the well-known error sum of squares, which comes from $\hat{Y}_i$ and each observed Y_i value

$$SS_E = \Sigma(Y_i - \hat{Y}_i)^2 = \Sigma \epsilon_i^2$$

Therefore, the total sum of squares may be written as either

$$SS_T = SS_R + SS_E$$

or

$$\Sigma(Y_i - \bar{Y})^2 = \Sigma(\hat{Y}_i - \bar{Y})^2 + \Sigma(Y_i - \hat{Y}_i)^2 \quad (27.28)$$

In practice the SS_R term is usually found by subtraction; that is, $SS_R = SS_T - SS_E$. For simple linear regression SS_R may be rewritten using the model form in Eq. (27.9), that is, $\hat{Y}_i - \bar{Y} = a' + b(X_i - \bar{X})$. Since $a' = 0$,

$$\Sigma(\hat{Y}_i - \bar{Y})^2 = b^2 \Sigma(X_i - \bar{X})^2 \quad (27.29)$$

The partitioning of sum of squares is used in a later chapter on analysis of variance. Another useful value is the *mean squares* (*MS*) which are computed as

$$\text{mean square} = \frac{\text{sum of squares}}{\text{degrees of freedom}} = \frac{SS}{\nu}$$

The ν values and MS terms for each sum of squares in Eq. (27.28) are given as:

Sum-of-squares term	Degrees of freedom	Mean square term
Regression, SS_R	1	$MS_R = SS_R$
Error, SS_E	$n - 2$	$MS_E = \dfrac{SS_E}{n - 2}$
Total, SS_T	$n - 1$	$MS_T = \dfrac{SS_T}{n - 1}$

Note that only the SS_T and the $n - 1$ degrees of freedom terms are the sum of the regression and error components. The mean square of the total MS_T is not usually needed in the analysis of sum of squares removed by regression.

Example 27.6 Use the data of Example 27.3 (Table 27.3) to (*a*) demonstrate that Eq. (27.28) is correct and (*b*) compute the mean square terms for regression, error, and total.

SOLUTION (*a*) Table 27.7 shows all the computations to demonstrate that $SS_T = SS_R + SS_E$ as follows (subscript *i* omitted).

$$\Sigma(Y - \bar{Y})^2 = \Sigma(\hat{Y} - \bar{Y})^2 + \Sigma(Y - \hat{Y})^2$$
$$397.85 = 383.33 + 14.52$$

The regression sum-of-squares term can also be computed from Eq. (27.29) with

Table 27.7 Computation of the total regression and error sum of squares ($\bar{Y} = 1.63$)

X	Y	$\hat{Y}$	SS_T $(Y - \bar{Y})^2$	SS_R $(\hat{Y} - \bar{Y})^2$	SS_E $(Y - \hat{Y})^2$
300	−10.0	−9.5	135.26	123.88	0.25
350	−3.9	−5.8	30.58	55.20	3.61
400	−2.5	−2.1	17.06	13.91	0.16
450	1.7	1.6	0.00	0.00	0.01
500	3.9	5.3	5.15	13.47	1.96
550	7.2	9.0	31.03	54.32	3.24
600	15.0	12.7	178.77	122.54	5.29
	11.4		397.85	383.33	14.52

$b = 0.074$ and $\Sigma(X - \bar{X})^2 = 70,000$ from Table 27.3:

$$SS_R = \Sigma(\hat{Y} - \bar{Y})^2 = (0.074)^2(70,000) = 383.32$$

(b) The mean square terms are computed using $n = 7$ data points:

Term	Sum of squares	Degrees of freedom	Mean square Value	Symbol
Regression	383.33	1	383.33	MS_R
Error	14.52	5	2.90	MS_E
Total	397.85	6	66.31	MS_T

COMMENT The total mean square is not the sum of the regression and error mean squares, as is made clear by the fact that division takes place to obtain the values. Since the mean squares are actually *variances*, the ratio MS_R/MS_E is a calculated F distribution value (Sec. 15.1) which can be used to test whether the amount of variation removed by the regression equation is statistically significant. This is discussed later.

Problems P27.29–P27.32

REFERENCE

1. Neville, A. M., and Kennedy, J. B., *Basic Statistical Methods for Engineers and Scientists*, Intext Book Company, New York, 1970, p. 219.

ADDITIONAL MATERIAL

Simple linear regression (Secs. 27.1 to 27.5): Belz [2], pp. 331–358; Bowker and Lieberman [4], pp. 325–362; Draper and Smith [6], Chaps. 1 to 3; Duncan [7], pp. 744–767; Gibra [8], pp. 408–433; Hines and Montgomery [12], pp. 323–341; Hoel [13], pp. 169–174; Miller and Freund [18], pp. 289–302; Neville and Kennedy [19], Chap. 15; Newton [20], pp. 328–344; Walpole and Myers [25], pp. 280–303.

Multiple linear regression (Secs. 27.6 and 27.7): Draper and Smith [6], Chaps. 4, 6 to 10; Duncan [7], pp. 776–805; Hines and Montgomery [12], pp. 341–353; Hoel [13], pp. 174–177; Miller and Freund [18], pp. 306–312; Neville and Kennedy [19], Chap. 17; Newton [20], pp. 352–369; Walpole and Myers [25], pp. 315–333.

Curvilinear regression (Sec. 27.8): Draper and Smith [6], Chap. 5; Duncan [7], pp. 805–814; Hoel [13], pp. 177–181; Miller and Freund [18], pp. 312–317.

PROBLEMS

The following data sets are used by the indicated problems in this chapter.

A: The kilometer-per-liter (km/L) figures for a new engine are recorded for fixed speeds between 56 and 112 km/h.

Speed (km/h)	Mileage (km/L)
56	14.7
104	13.2
64	14.5
88	13.2
112	12.8
96	13.4
84	13.3
68	14.5
80	13.8
100	13.0
60	14.6
72	14.3

Problems *P27.3, P27.7, P27.11, P27.12, *P27.21, *P27.22, *P27.30*

B: The percentage of moisture remaining in smoked sausage was determined for 10 different duration times using a specific smoke-curing method.

Sausage number	Time (h)	Percentage of moisture
1	115	21
2	125	19
3	185	11
4	200	10
5	75	50
6	80	42
7	150	13
8	175	12
9	90	26
10	100	24

Problems P27.4, P27.8, P27.13, P27.16, *P27.23, *P27.24, *P27.31

C: The production rate data shown is the result of using a new teaching technique developed by a methods analyst. The output in units per hour is measured after each 25 units to determine when the worker has reached the standard of 40 units per hour. The data for eight new employees is given:

Total number of units produced	Units per hour produced
25	16
50	28
75	28
100	29
125	30
150	32
175	40
200	42

Problems P27.5, P27.9, P27.14, P27.18, *P27.25, *P27.32

P27.1 (a) Rewrite the assumptions of the least-squares technique if Y is the fixed variable and X is the response variable. This is called the *regression of X on Y*, whereas the use of a fixed X leads to the regression of Y on X.

(b) Construct drawings similar to Figs. 27.1 and 27.2 for the regression of X on Y. Retain the x axis as horizontal for these new figures.

(c) Would you expect the regression lines of Y on X and X on Y to be coincident? Why or why not?

P27.2 Summarize the assumptions and theoretical development of the least-squares method for fitting a straight line to the observed values of two variables.

P27.3 (a) Use the steps of the regression technique to determine the best-fit straight line for the mileage results in data set A.

(b) If you have a calculator with regression capability check your answer.
(c) Compute the mean estimated mileage figure for a speed of 85 km/h.
(d) For what speed does the regression line estimate a mileage of 14.5 km/L? How close is this to the actual test results?

P27.4 (a) Determine the regression line equation for data set B.
(b) Compute the error sum of squares.
(c) If the sausage were smoked for only 50 h, what is the estimated average water content?
(d) What assumption is made in the computation of part (c)?

P27.5 (a) Use linear regression and data set C to estimate the average units per hour produced after 115 units are made.
(b) What is the error sum-of-squares component for 125 units produced?
(c) Show that the regression line goes through the centroid point.

P27.6 (a) What is the regression line equation for the monthly sales data shown?
(b) Show that the regression line passes through the point $(\bar{X}, \bar{Y})$.

Month of 1980	Sales ($ + 10,000)
1	15.2
2	19.4
3	14.5
4	18.4
5	13.9
6	12.4
7	16.7
8	15.0
9	18.1
10	10.3
11	9.8
12	11.4

P27.7 (a) Use the regression equation for data set A (Prob. P27.3) to compute the variance around the regression line.
(b) Is the slope for this line statistically equal to 0 for $\alpha = 0.05$?
(c) Regression on test results from another engine indicated an intercept value of 18.0. Are this value and the intercept for the data set A regression statistically equal at $\alpha = 0.05$?

P27.8 Test the intercept and slope values in the regression for data set B with the value $b_0 = -0.3$ and $a_0 = 65$ percent, respectively, using a 90 percent level of confidence.

P27.9 Use the results of the linear regression analysis on data set C and a 5 percent significance level (a) to determine the maximum computed value of the slope to not reject the null hypothesis that the number of units produced and the production rate are independent and (b) to test the hypotheses $H_0: a = 12$ versus $H_1: a > 12$ units per hour.

P27.10 For a particular set of 25 data points, the regression equation was $\hat{Y} = 1256.9 + 2.59X$ with an error sum of squares of 8291.4. If $\bar{X} = 7.2$ and $\Sigma(X_i - \bar{X})^2 = 957.5$, test if the following are statistically correct for $\alpha = 0.05$: (a) $b = 4.0$ and (b) $a = 1255$.

P27.11 For data set A compute the 90 percent confidence interval for (a) the mean mileage values for the speeds of 60, 70, 80, 90, and 100 km/h and (b) the individual estimate of mileage at 90 km/h. Compare the two confidence intervals for 90 km/h. Use the value n = 12 when working this problem.

P27.12 Calculate and graph the 90 percent confidence intervals for (a) the intercept and (b) the slope for data set A. The standard deviations were computed in Prob. P27.7.

P27.13 Do the following for data set B using $\alpha = 0.05$.
  (a) Compute the confidence interval on the slope.
  (b) Determine if the moisture value of 45 percent is within the interval estimate for the intercept.
  (c) A sample of 10 sausages were smoke-cured for 100 h, and the following moisture percent values were determined: 18, 22, 28, 16, 31, 22, 26, 19, 20, 21. Is the sample mean of 22.3 percent in the interval estimate for the mean percent moisture at 100 h?
  (d) Are all the sample values in (c) within the interval estimate for a single moisture observation at 100 h?

P27.14 (a) For data set C calculate and graph the complete 90 percent confidence interval for the mean number of units produced each hour.
  (b) Determine and graph the 90 percent confidence interval for the slope of the regression line for data set C.

P27.15 Do the following for the monthly sales data in Prob. P27.6. Use a 10 percent level of significance.
  (a) Compute the error sum of squares.
  (b) Make an interval estimate of the mean sales for the fifth month.
  (c) Determine the interval estimate for the slope of the regression line in Prob. P27.6.
  (d) Describe the necessary assumption(s) if the regression line is used to predict monthly sales for 1981.

P27.16 The weight of each sausage prior to smoke-curing was taken in data set B. In numerical order they were 1.2, 1.0, 1.0, 0.8, 1.5, 1.6, 0.9, 1.2, 1.3, and 0.9 kg.
  (a) Find the linear regression equation which may be used to predict percentage of moisture Y, given weight X_1 and smoke-curing time X_2.
  (b) Estimate the mean moisture content for a 1.0-kg sausage that is smoke-cured for 110 h.

P27.17 A couple who does market analyses hypothesizes that a linear relation exists between the average number of hours Y per day that a household television set is on and two variables: family income X_1 and the total number X_2 of persons residing in the house.
  (a) Use the multiple linear regression equation for the data shown to predict television viewing time for a family of three with a $16,000 per year income.
  (b) Compute the variance around the regression surface.

Income ($/yr + 1000)	Number in household	Average hours per day
28.4	2	3
22.2	2	9
16.0	3	6
30.5	3	4
24.0	4	8
14.5	4	8
12.1	4	12
10.3	5	13
18.9	5	12
26.1	6	10

P27.18 In addition to the productivity values in data set C, the number of months X_2 of experience in similar work has been recorded: 24, 14, 60, 39, 6, 18, 22, 42. (These values, in order, form another column entitled "Experience in months" in data set C.)
  (a) Determine the multiple linear regression equation if average units per hour is the observed variable Y.

(b) Compute the error sum-of-squares component for this regression surface for 39 months' experience and 100 units produced.

*P27.19 In the stepwise multiple linear regression procedure, the order in which variables enter solution is checked after each new variable is placed into the regression model. Why is it necessary to check the entry order?

*P27.20 (a) Fit a hyperbolic model to the data shown which was collected from the business research sections in nine cities. The data shows the percentage of discretionary income that goes for entertainment versus the percentage spent on housing. Consider the housing expenditure as the independent variable.

(b) Would a linear fit be better than the hyperbolic fit?

Percentage of income on housing	Percentage of income on entertainment
16	19
18	18
25	15
20	18
19	16
15	18
13	20
22	13
24	15

*P27.21 Does a polynomial of degree $k = 2$ offer a better fit for data set A than the linear model fit in Prob. P27.3? (*Note:* If you worked Prob. P27.7, you already have the error sum of squares for the linear model.)

*P27.22 Compare a hyperbolic model fit with the linear model fit for data set A. The linear model was fit in Prob. P27.3.

*P27.23 (a) Fit the power model $Y = aX^b$ to data set B.
(b) Is the power model a better fit than the linear model in Prob. P27.4?

*P27.24 (a) Fit the hyperbolic model $Y = a + b/X$ to data set B.
(b) Is the hyperbolic model a better fit than the linear model in Prob. P27.4?

*P27.25 For data set C determine which of the following gives the best fit: (a) linear (see Prob. P27.9), (b) power, or (c) polynomial of degree $k = 2$.

*P27.26 Compare the linear and exponential model fits for the data in Prob. P27.6. (*Note:* The error sum of squares for the linear model was computed in Prob. P27.15a.)

*P27.27 The average pollution index and the population for urban areas were recorded for 13 cities. Use this data to determine if a linear or exponential model best explains the increase in the index with a growing population.

Population (+1000)	Index	Population (+1000)	Index
50.2	50	249.8	72
58.9	51	321.6	80
67.3	52	324.9	76
109.8	52	378.6	95
153.4	58	429.8	115
217.8	68	531.6	130
221.4	70		

*P27.28 (a) Plot the data shown and select which of the following models you think will fit best: linear, exponential, power, or polynomial of degree $k = 2$.

(b) Write a computer program that fits these four models and selects the best fit on the basis of the error sum of squares. Was your selection correct?

X	Y	X	Y	X	Y
50	44	96	49	64	39
26	44	76	47	14	41
78	35	40	51	88	36
10	38	84	42	98	56
80	47	6	42	32	47
58	46	42	46	5	33
18	47	90	45	94	60

*P27.29 Explain why it is possible to utilize Eq. (27.28) to partition the total sum of squares for nonlinear as well as linear fits, but that it is only for the linear model that Eq. (27.29) is correct.

*P27.30 For data set A compute (a) the total sum of squares, (b) the sum of squares explained by the linear regression in Prob. P27.3, and (c) the mean square terms for these sums of squares.

*P27.31 Determine the fraction of the total sum of squares for data set B explained by (a) the linear fit in Prob. P27.4 and (b) the hyperbolic fit in Prob. P27.24. (c) How can these fractions be used to select the better-fitting model?

*P27.32 (a) Compute all the mean square terms for the linear regression on data set C as determined in Prob. P27.5.

(b) Calculate the corresponding mean square terms for data set C using the power model in Prob. P27.25. Compare the two sets of values.

CHAPTER
TWENTY-EIGHT

CORRELATION ANALYSIS

The material in this chapter will help you conduct a correlation study and interpret its results. Use of the linear correlation coefficient is presented, and nonlinear and multiple linear correlation analysis is introduced. The use of correlation analysis in conjunction with regression analysis is discussed.

CRITERIA

To correctly perform a correlation analysis, you must be able to:

1. Define the term *correlation analysis*, construct a *scattergram*, and state the cause-and-effect implication in a correlation analysis.
2. Compute the *sample correlation coefficient*, and state the properties of this coefficient, given the data for two random variables.
3. Use the hypothesis testing procedure to determine if the correlation coefficient is significantly different from 0 or some specific nonzero value, given the sample size, sample correlation coefficient, and specific value.
*4. Define, compute, and interpret the *coefficient of determination* and the *generalized correlation coefficient*, given the data and a linear or nonlinear regression equation.
*5. Graph the relation between two normal distributions for different values of the population correlation coefficient. Define *covariance* and state the equations for correlation and covariance that result when two normal variables are independent.
*6. Define and compute the *multiple* and *partial linear correlation coefficients*, given the data and equation or variation terms from a multiple linear regression analysis for more than two variables. Also, write a *correlation matrix* in the correct form and interpret its values.

STUDY GUIDE

28.1 Introduction to Correlation Analysis

Correlation analysis is the determination of a numerical measure of the relationship between two *random* variables. It may be used to study the relation between such variables as IQ and education, fatigue and production rate, temperature and pressure, etc.

Like regression analysis, correlation studies usually concentrate on the *linear* relationships. The outcome of the correlation study is a good indicator, as you will see, of how well a regression line explains the variation in the response variable Y. Even though in regression Y is random and the independent variable X is fixed, correlation analysis can give useful information for the regression study.

A graph of the observed data for two random variables X and Y is called a *scattergram* (Fig. 28.1). This graph will give an immediate feel for the linear or nonlinear relation between the two variables. Although nonlinear correlation and regression analysis are possible, much of the time a simple linear relation is sufficient to explain a large part of the variation in the data. You should always plot a scattergram first to see if your data is close to being or obviously not linear.

No cause-and-effect relation is implied in a correlation study. If an analyst studies, for example, the grade point average (GPA) of engineering students versus the number of hours the student works, she or he may very well find that the students with lower grades have jobs at which they work more hours than the student with higher grades. This correlation analysis does not imply that the job causes lower grades. There may well be other factors acting to cause this effect. Therefore, correlation cannot show cause-and-effect reasons; it can only

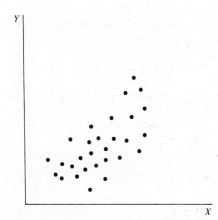

Figure 28.1 General graph of a scattergram used in correlation analysis.

estimate the mathematical relation between the data. It is up to the analyst to determine cause and effect.

Problem P28.1

28.2 The Correlation Coefficient and Its Properties

The relation between two variables is measured by the *sample correlation coefficient r*, also called the Pearson product-moment correlation coefficient. In general, r is defined as

$$r = \left(\frac{\text{explained variation}}{\text{total variation}}\right)^{1/2} \quad (28.1)$$

For *linear* correlation the data pairs (X_i, Y_i) for $i = 1, 2, \ldots, n$ observations are used to compute r (subscript i omitted):

$$r = \frac{n\Sigma XY - \Sigma X \Sigma Y}{\{[n\Sigma X^2 - (\Sigma X)^2][n\Sigma Y^2 - (\Sigma Y)^2]\}^{1/2}} \quad (28.2)$$

Some of the properties of r, with figure references, are

Range:	$-1 \le r \le +1$
$r = \pm 1$:	perfect linear correlation (Fig. 28.2a)
$r = 0$:	no linear relation, that is, uncorrelated (Fig. 28.2b)
$r > 0$:	positive slope between X and Y
r close to $+1$:	strong positive linear trend (Fig. 28.2c)
$r < 0$:	negative slope between X and Y
r close to -1:	strong negative linear trend (Fig. 28.2d)
Dimension:	none, regardless of dimensions of X and Y
Translation:	r remains the same even if variables are multiplied by a constant or have a constant added to them

If $r = \pm 1$ and a regression line were fit to the data, the fit would be perfect and the error sum of squares, Eq. (27.1), would equal 0. If $r = 0$, the variables are *uncorrelated*; that is, there is no discernible linear trend. The variables are *not necessarily independent* just because $r = 0$, but it is correct to state that two independent variables must have $r = 0$ (Sec. 28.5).

If a preprogrammed computer package is used to perform the regression analysis, the r value is usually printed for you. If you have a hand-held calculator with a correlation key or program, the output is the r value computed by Eq. (28.2). Example 28.1 illustrates the computation and interpretation of r.

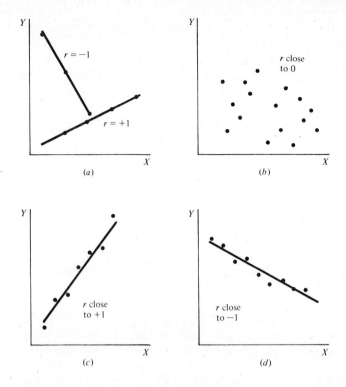

Figure 28.2 Correlation coefficient scattergrams for (a) perfectly linear, (b) linearly uncorrelated, (c) strong positive trend, and (d) strong negative trend.

Example 28.1 An environmental engineer has hypothesized that the pollution content in the Mississippi River can be explained by a linear relation between BOD_5 (biochemical oxygen demand) and distance on the river. Fifteen samples were taken at random locations along a 150-km distance, and the BOD readings were recorded in the coded form 10^{-4} mg/L (Table 28.1). Determine the degree of linear correlation between the two variables.

SOLUTION To determine the degree of correlation, compute r from Eq. (28.2). The necessary summations are presented in Table 28.1.

$$r = \frac{15(498.21) - 961(5.42)}{\{[15(92,143) - (961)^2][15(2.7658) - (5.42)^2]\}^{1/2}}$$

$$= \frac{2264.53}{[458,624(12.1106)]^{1/2}} = 0.961$$

The value $r = 0.961$ means that there is a high linear correlation between BOD reading and distance. Figure 28.3, which is a scattergram of the data

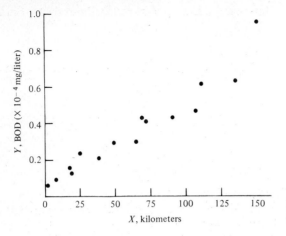

Figure 28.3 Scattergram of BOD content on a river, Example 28.1.

in Table 28.1, shows the close grouping about some eye-balled line which you may draw.

COMMENT There is no implication that BOD readings get larger as the distance on the river is increasing just because the correlation analysis results in a large, positive linear effect. Of course, we know that distance in itself does not cause pollution; it takes people, society, and nature to cause the effects of pollution.

Table 28.1 Computation of r for BOD content in a river, Example 28.1

Kilometers X	BOD Y	XY	X^2	Y^2
2	0.06	0.12	4	0.0036
9	0.10	0.90	81	0.0100
18	0.16	2.88	324	0.0256
20	0.12	2.40	400	0.0144
25	0.24	6.00	625	0.0576
38	0.21	7.98	1,444	0.0441
50	0.29	14.50	2,500	0.0841
65	0.31	20.15	4,225	0.0961
69	0.43	29.67	4,761	0.1849
71	0.41	29.11	5,041	0.1681
92	0.43	39.56	8,464	0.1849
107	0.47	50.29	11,449	0.2209
110	0.61	67.10	12,100	0.3721
135	0.63	85.05	18,225	0.3969
150	0.95	142.50	22,500	0.9025
961	5.42	498.21	92,143	2.7658

At this point actually only two conclusions can be made about data from a computation of r:

1. Positive ($r > 0$) or negative ($r < 0$) trend of data
2. Degree of correlation of the two variables and how closely the data is clustered about some linear relation of the variables

Problems P28.2–P28.5

28.3 Hypothesis Test for the Correlation Coefficient

To determine if the computed r value is large enough to be statistically signficant or equal to a certain value ρ_0, the hypothesis testing procedure (Sec. 19.4) is used to test

$$H_0: \rho = \rho_0 \quad \text{versus} \quad H_1: \rho \neq \rho_0$$

where ρ is the true population value of the correlation coefficient as estimated by the sample value r. This hypothesis test is illustrated below; however, a different test statistic from either the t or standard normal (z) distribution must be used to test H_0 depending on the value of ρ_0. If the specific value ρ_0 is 0 and if H_0 is rejected, the correlation of the data is significant with $100(1 - \alpha)$ percent confidence. To determine the correct statistic, first answer these questions and then use the correct relation from Table 28.2.

1. *Is the sample size small ($n < 30$) or large?* If the sample is small, but a normal distribution can be assumed for the data, use the statistic for large samples. If $\rho_0 \neq 0$ and $n < 30$, take more data or assume normality.
2. *In the null hypothesis $H_0: \rho = \rho_0$, is ρ_0 equal to 0, or some nonzero value?*

The t statistic in Eq. (28.3) has $\nu = n - 2$ degrees of freedom. The tests may be

- **Table 28.2 Test statistics for testing the correlation coefficient**

Sample size	ρ_0 value in H_0	Type of test statistic	Test statistic	Equation number
Small	0	t	$\dfrac{r\sqrt{n-2}}{\sqrt{1-r^2}}$	(28.3)
Large	0	z	$\dfrac{r\sqrt{n-2}}{\sqrt{1-r^2}}$	(28.4)
Large	Nonzero	z	$\dfrac{\sqrt{n-3}}{2} \ln\left[\left(\dfrac{1+r}{1-r}\right)\left(\dfrac{1-\rho_0}{1+\rho_0}\right)\right]$	(28.5)

one-sided, but it is more common to perform a two-sided test for correlation. If the computed t on z values exceed the tabulated values, H_0 is rejected.

Example 28.2 A correlation study between decreasing chemical potency and time exposure to artificial room light resulted in $r = 0.68$ for a sample size of 24.
(a) Is there a significant correlation present?
(b) In similar experiments it has been found that the data is approximately normal, with the average $r = 0.72$. Is the value of 0.68 statistically different?

SOLUTION The hypothesis testing procedure is detailed for part (a) only. Let $\alpha = 0.05$ for both parts.
(a) Before the test is performed, the two questions are answered as follows: (1) small sample; (2) $\rho_0 = 0$. Therefore, the t statistic in Eq. (28.3) is correct.

1 and 2. The two-sided test with $\alpha = 0.05$ is
$$H_0: \rho = 0 \quad \text{versus} \quad H_1: \rho \neq 0$$
3. The sample statistics needed are $n = 24$ and $r = 0.68$.
4. From Eq. (28.3) the test statistic is
$$t = \frac{0.68\sqrt{24 - 2}}{\sqrt{1 - (0.68)^2}} = 4.35$$
5. For the t distribution with $\nu = 24 - 2 = 22$ and a two-tail $\alpha = 0.05$, the acceptance region for H_0 is $|t| \leq 2.074$.
6. Since $4.35 > 2.074$, $H_0: \rho = 0$ is rejected and the correlation between chemical potency and time exposure is statistically significant.

(b) The initial questions are answered: (1) large, because the normal assumption is appropriate; (2) nonzero. The two-sided hypothesis
$$H_0: \rho = 0.72 \quad \text{versus} \quad H_1: \rho \neq 0.72$$
is tested using Eq. (28.5):
$$z = \frac{\sqrt{24 - 3}}{2} \ln \frac{(1 + 0.68)(1 - 0.72)}{(1 - 0.68)(1 + 0.72)} = -0.36$$

From the SND Table 11.2, the acceptance region for a two-tail $\alpha = 0.05$ is $|z| \leq 1.96$. H_0 is not rejected: the new coefficient is statistically the same as before.

COMMENT You can perform a one-tail test on r by using the same procedure and $H_1: \rho > \rho_0$ or $H_1: \rho < \rho_0$ with α in only one tail.

It is possible to determine the minimum sample size n needed to show that a significant correlation is present by solving Eq. (28.4) for n. From the probability statement below, n can be determined using the two-tail z value for α from the SND table.

$$P\left(z \leq \left|\frac{r\sqrt{n-2}}{\sqrt{1-r^2}}\right|\right) = 1 - \alpha$$

or

$$n \geq \frac{z^2(1-r^2)}{r^2} + 2 \qquad (28.6)$$

For example, to show a significant correlation for $r = 0.4$ with $\alpha = 0.05$, the two-tail $z = 1.96$, and by Eq. (28.6) an n of at least 23 (rounded up) is required.

Problems P28.6–P28.10

*28.4 Coefficient of Determination and the Generalized Correlation Coefficient

The general definition of the correlation coefficient r, Eq. (28.1), may be used to compute the *fraction* of total variation in the data that is removed by any linear or nonlinear regression equation. The square of r is called the *coefficient of determination*:

$$r^2 = \frac{\text{explained variation}}{\text{total variation}}$$

For a quick review of the types of variations present in a regression analysis, see optional Sec. 27.9. In sum-of-squares terms from Eq. (27.28), r^2 may be written

$$r^2 = \frac{SS_R}{SS_T} = \frac{\Sigma(\hat{Y} - \overline{Y})^2}{\Sigma(Y - \overline{Y})^2} \qquad (28.7)$$

where $\hat{Y}$ = estimates from regression equation
$\overline{Y}$ = average of observed Y values

The value of r^2, which has a range $0 \leq r^2 \leq 1$, is included in the printout of most computerized regression packages.

You should realize that the square root of Eq. (28.7) and the correlation coefficient in Eq. (28.2) coincide only when *linear* regression is used. Equation (28.2) will give the same numerical result regardless of what curve is actually fit, because only original data pairs (X_i, Y_i) are used, not the estimates $\hat{Y}$ as in Eq. (28.7). Therefore, if any type of nonlinear equation is to be fit to the data, r is correctly computed as the square root of r^2 in Eq. (28.7). Therefore, the relation

$$r = \left[\frac{\Sigma(\hat{Y} - \overline{Y})^2}{\Sigma(Y - \overline{Y})^2} \right]^{1/2} \tag{28.8}$$

is called the *generalized correlation coefficient*.

It is now possible to use the result of Eq. (28.8) as a measure of how well a specific regression equation explains the observed variation. This was not recommended in Sec. 27.8 because of the linear correlation coefficient "trap" just discussed. However, now it should be clear from Eq. (27.28) that a minimum error sum of squares, $\Sigma(Y - \hat{Y})^2$, and a maximum r from Eq. (28.8) are equivalent measures for regression analysis. Example 28.3 discusses the correct and incorrect uses of r and r^2.

Example 28.3 Nonlinear regression has been used to fit a *hyperbolic* equation of the form $\hat{Y} = a + b/X$ where X is degrees Fahrenheit divided by 100 and Y is the dissolving time of a chemical in seconds. The least-squares result, which is plotted in Fig. 28.4, is

$$\hat{Y} = -3.05 + \frac{21.01}{X} \tag{28.9}$$

Use the observed data (Table 28.3) to compute r^2 and r for (*a*) the hyperbolic fit and (*b*) the linear fit. (*c*) Explain why the linear r is incorrect if a nonlinear regression analysis is performed.

SOLUTION (*a*) The coefficient of determination is computed by Eq. (28.7)

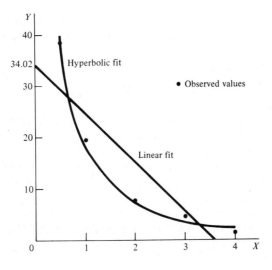

Figure 28.4 Hyperbolic and linear models fit to a set of data, Example 28.3.

28.4 COEFFICIENTS OF DETERMINATION AND GENERALIZED CORRELATION

Table 28.3 Computation for correlation coefficients in Example 28.3 ($\bar{Y} = 14.1$)

X	Y	Total variation $(Y - \bar{Y})^2$	Hyperbolic $\hat{Y}$, Eq. (28.9)	Explained variation $(\hat{Y} - \bar{Y})^2$	XY	X^2	Y^2
0.5	38.3	585.64	38.97	618.52	19.15	0.25	1466.9
1.0	19.5	29.16	17.95	14.82	19.50	1.00	380.3
2.0	7.5	43.56	7.46	44.09	15.00	4.00	56.3
3.0	4.2	98.01	3.95	103.02	12.60	9.00	17.6
4.0	1.0	171.61	2.20	141.61	4.00	16.00	1.0
10.5	70.5	927.98		922.06	70.25	30.25	1922.1

using the results in Table 28.3.

$$r^2 = \frac{922.06}{927.98} = 0.994$$

Then $\quad r = \sqrt{0.994} = 0.997$

This means that 99.4 percent of the total variation (927.98) is explained by the hyperbolic fit.

(b) If a straight line is fit to the data, the observed data is substituted into Eq. (28.2).

$$r = \frac{5(70.25) - 10.5(70.5)}{\{[5(30.25) - 10.5^2][5(1922.1) - 70.5^2]\}^{1/2}} = -0.892$$

Then $\quad r^2 = (-0.892)^2 = +0.796$

Alternatively, a straight line may be fit to the data and r computed from the general formula Eq. (28.8). The linear fit (Fig. 28.4) has the equation

$$\hat{Y} = 34.02 - 9.49X$$

and $\quad r^2 = 0.796 \quad r = 0.892$

Therefore, 79.6 percent of the variation is explained by a straight line, compared to the 99.4 percent by the hyperbolic fit.

(c) If a nonlinear fit is planned, it is wrong to compute the linear $r = -0.892$ as a judge of how well a *curve* will fit the data, because this is only for a *linear* relation. The curve (hyperbolic in this case) must be fit and r computed from Eq. (28.8).

COMMENT The correct results for the hyperbolic and linear fits are summarized as follows.

		Correlation, r		Coefficient of
Type	Equation	Value	Equation	determination r^2
Hyperbolic	$\hat{Y} = -3.05 + 21.01/X$	+0.997	(28.8)	0.994
		+0.892	(28.8)	0.796
Linear	$\hat{Y} = 34.02 - 9.49X$			
		−0.892	(28.2)	0.796

Since Eq. (28.8) defines r in terms of variation rather than original data, r is always positive. This is why $r = +0.892$ for the linear fit even though the slope is negative and $r = -0.892$ from the original data Eq. (28.2). Therefore all r values computed in terms of variation have a range $0 \leq r \leq 1$, whereas the linear r computed from the original data has a range $-1 \leq r \leq 1$.

Problems P28.11–P28.17

*28.5 Development of the Population Correlation Coefficient Using Covariance

The sample correlation coefficient r is an estimate of the population correlation coefficient ρ. If you take many samples from two random-variable populations, compute an r value for each sample, and plot a histogram of r versus the frequency $f(r)$, you approximate the distribution of ρ. The pdf $f(\rho)$ is a symmetric distribution if $\rho = 0$, but it becomes highly skewed for $\rho \neq 0$ (Fig. 28.5). Actually, only one value from this $f(\rho)$ distribution is obtained when data is

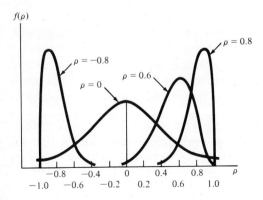

Figure 28.5 General shape of the pdf of ρ for several different values.

28.5 DEVELOPMENT OF THE POPULATION CORRELATION COEFFICIENT

collected and r computed. The statistics presented in Table 28.2 to test different values of ρ are approximations to the distribution $f(\rho)$.

The population correlation coefficient may be introduced into the linear regression relation. If X and Y have normal populations, then

$$E[Y/X] = E[Y] + \rho\sqrt{\frac{V[X]}{V[Y]}}\,(X - E[X]) \qquad (28.10)$$

where $E[Y/X]$ is the conditional expected value of Y given X. Regression assumes a normal distribution at each X value (Fig. 27.2); but if there are two *independent* distributions, the correlation between them is 0 and $\rho = 0$. This may be shown as two normal curves with *unequal* means and variances at 90° to each other (Fig. 28.6a). Then in Eq. (28.10) we have $\rho = 0$ and $E[Y/X] = E[Y]$, as it should be. As ρ moves away from 0, the variables become correlated and the axes have less than 90° between them (Fig. 28.6b). If $\rho = +1$, the two

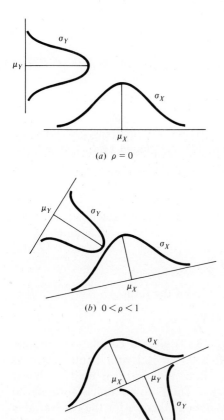

Figure 28.6 Graphs of the correlation relation between two normally distributed variables which are (a) uncorrelated, (b) partially correlated, and (c) perfectly correlated.

distributions fall on the same line, which is the regression line (Fig. 28.6c). Thus, if we have two normal variables, which is the bivariate normal (Sec. 18.5), the formula for r gives an unbiased estimate of ρ. We do not usually know if the variables are normal, but we use r anyway because it is the best known measure of statistical correlation.

The definition of ρ for a bivariate normal distribution is

$$\rho = \frac{\text{covariance of } X \text{ and } Y}{\sqrt{V[X]V[Y]}} \qquad (28.11)$$

where the covariance of X and Y is defined as

$$\text{Cov}[X, Y] = E\{(X - E[X])(Y - E[Y])\} = E[XY] - E[X]E[Y] \qquad (28.12)$$

The covariance is a numerical measure of the *dependence* of two variables which may be used in ρ to show correlation. The reason that $\rho = 0$ implies uncorrelated variables is because the covariance is 0 and Eq. (28.11) is therefore 0. A $\text{Cov}[X, Y] = 0$ in Eq. (28.12) occurs for two independent variables because, as discussed in Sec. 18.6, rule 5, $E[XY] = E[X]E[Y]$ for such variables.

Problems P28.18, P28.19

*28.6 Multiple and Partial Correlation Coefficients

If the relationship between more than two variables is studied, the correlation of all variables is measured by the *sample multiple correlation coefficient R*. The same general definition as for r is used here.

$$R = \left(\frac{\text{explained variation}}{\text{total variation}}\right)^{1/2}$$

$$= \left[\frac{\Sigma(\hat{Y} - \bar{Y})^2}{\Sigma(Y - \bar{Y})^2}\right]^{1/2} \qquad (28.13)$$

For example, if a linear model in four variables of the form

$$\hat{Y} = a + b_1 X_1 + b_2 X_2 + b_3 X_3$$

is fit using multiple linear regression, Eq. (28.13) gives a correlation measure between all four variables. The range of R is $0 \leq R \leq 1$ with 1 interpreted as perfect correlation and 0 as no correlation. Equation (28.13) is used for linear and nonlinear multiple correlation analysis. There is no quick formula to compute R for linear models using original data as there is for the two-variable case.

28.6 MULTIPLE AND PARTIAL CORRELATION COEFFICIENTS

The value R^2 is called the *coefficient of multiple determination* and is usually expressed as a percentage. It gives the fraction of variation accounted for by the regression equation fit to the observed data.

Even when multiple correlation analysis is utilized, it is of interest to know the correlation between any two variables in the total set $(Y, X_1, \ldots, X_m)$ with all others held constant. This correlation measure for the linear case is explained by the *partial correlation coefficients* R_{ij} for variables X_i and X_j. The response variable Y is identified by a subscript 0, so R_{01} is the coefficient for Y and X_1 with $X_2, X_3, \ldots, X_m$ held constant. The linear R_{ij} for i and $j = 0, 1, \ldots, m$ is

$$R_{ij} = \frac{\Sigma(X_i - \overline{X}_i)(X_j - \overline{X}_j)}{\left[\Sigma(X_i - \overline{X}_i)^2 \Sigma(X_j - \overline{X}_j)^2\right]^{1/2}} \quad (28.14)$$

where the sum is over the sample size n for each variable. The range is $-1 \leq R_{ij} \leq 1$.

A multiple linear regression computer program usually prints out the R_{ij} values as a *partial correlation matrix* in the general format

$$\begin{bmatrix} 1 & R_{12} & \cdots & R_{1m} \\ R_{21} & 1 & \cdots & R_{2m} \\ \vdots & \vdots & & \vdots \\ R_{m1} & R_{m2} & \cdots & 1 \end{bmatrix}$$

The values on the main diagonal are $R_{11} = R_{22} = \cdots = R_{mm} = 1$. The matrix is symmetric since $R_{ij} = R_{ji}$. It is assumed that all variables not mentioned in the subscripts are held constant; that is, their effect is not measured. Example 28.4 discusses the interpretation of the multiple and partial correlations and utilizes a partial correlation matrix.

Example 28.4 An industrial engineer is in the process of studying the productivity of the testing laboratory of a hospital. A four-dimensional linear model has been computer-fit using a multiple linear regression package. The result is

$$\hat{Y} = 2.053 + 2.101X_1 - 2.546X_2 + 0.134X_3 \quad (28.15)$$

where Y = number of tests performed per person per day
X_1 = experience of employee in months
X_2 = percentage of time spent in duties other than testing
X_3 = salary in dollars per week

The computer printout contains the following information:

Total variation: 2519.69

Error sum of squares: 1218.59

Correlation matrix:
$$\begin{bmatrix} 1 & 0.53 & -0.62 & 0.31 \\ 0.53 & 1 & 0.94 & 0.90 \\ -0.62 & 0.94 & 1 & 0.91 \\ 0.31 & 0.90 & 0.91 & 1 \end{bmatrix}$$

Help the IE by doing the following:
(a) Compute the multiple correlation coefficient and compare this value with the two-variable coefficients of Y with X_1, X_2, and X_3. Comment on the results.
(b) Determine the amount of variation in Y that is explained by each of the X variables.
(c) Find the value 0.94 in the correlation matrix, write the formula used for its computation, and interpret its meaning.

SOLUTION (a) In order to use Eq. (28.13) to compute R, the explained variation is computed from

explained variation = total variation − unexplained variation
= total variation − error sum of squares
= 2519.69 − 1218.59 = 1301.10

Therefore, $$R = \left(\frac{1301.10}{2519.69}\right)^{1/2} = 0.72$$

Comparison with the two-variable R_{ij} values is easier if the correct subscripts are placed on the correlation matrix.

$$\begin{array}{c|cccc} & 0 & 1 & 2 & 3 \\ \hline 0 & 1 & 0.53 & -0.62 & 0.31 \\ 1 & 0.53 & 1 & 0.94 & 0.90 \\ 2 & -0.62 & 0.94 & 1 & 0.91 \\ 3 & 0.31 & 0.90 & 0.91 & 1 \end{array}$$

The R_{0j} ($j = 1, 2, 3$) values in the first row are the partial correlations to be compared with $R = 0.72$. For R_{02} we must compare R with $|R_{02}|$ since the minus sign shows negative correlation and $0 \leq R \leq 1$ is due to the definition in terms of variation, not original data. $R = 0.72$ exceeds all three $|R_{0j}|$ values, as is expected, since the addition of more relevant X variables usually explains more variation than any one X variable.
(b) The fraction of variation in Y removed by each X variable with the other two held constant is obtained by squaring the R_{0j} ($j = 1, 2, 3$) coefficients in the correlation matrix. The results are summarized as:

Variable X_j	R_{0j}	R_{0j}^2	Interpretation of $100R_{0j}^2$
X_1	0.53	0.28	28 percent of the variation in the number of tests per person per day (Y) is removed by the experience variable
X_2	−0.62	0.38	38 percent of the variation in the number of tests is removed by the percentage-of-time-on-other-work variable
X_3	0.31	0.10	10 percent of the variation in the number of tests is removed by the salary variable

These values may be compared with $R^2 = 0.52$, which means that 52 percent of the total variation in Y is removed by all the X variables.

(c) The value $R_{12} = R_{21} = 0.94$ is the two-variable correlation coefficient of X_1 and X_2 without the response variable Y. The formula used takes its form from Eq. (28.14).

$$R_{12} = \frac{\Sigma(X_1 - \bar{X}_1)(X_2 - \bar{X}_2)}{\left[\Sigma(X_1 - \bar{X}_1)^2 \Sigma(X_2 - \bar{X}_2)^2\right]^{1/2}} = 0.94$$

where the sums are over the entire sample of size n.

The square, $R_{12}^2 = 0.88$, means that 88 percent of the variation between the experience variable X_1 and the percentage-of-time-on-other-work variable X_2 would be explained if X_1 and X_2 were fit with a straight line. Note that this result has nothing to do with the response variable Y; therefore, it is of no value with respect to the four-variable linear regression problem. Similar formulas and interpretations may also be developed for R_{13} and R_{23} and the associated X variables.

Problems P28.20–P28.22

ADDITIONAL MATERIAL

Linear correlation (Secs. 28.1 to 28.3): Draper and Smith [6], pp. 33–35; Gibra [8], pp. 435–443; Hoel [13], pp. 163–169; Kirkpatrick [15], pp. 333–337; Miller and Freund [18], pp. 321–327; Neville and Kennedy [19], Chap. 16; Newton [20], pp. 345–352; Walpole and Myers [25], pp. 303–308.

Multiple and partial correlation (Sec. 28.6): Brownlee [5], pp. 429–431; Draper and Smith [6], pp. 147–150; Duncan [7], pp. 787–793; Neville and Kennedy [19], p. 215; Walpole and Myers [25], pp. 333–338.

Generalized correlation and covariance (Secs. 28.4 and 28.5): Brownlee [5], pp. 77–80; Hoel [13], pp. 149–151; Hogg and Craig [14], pp. 61–65.

PROBLEMS

The following data sets are used by the indicated problems in this chapter. These data sets are the same as those in the previous chapter, so they are in summary form here.

A: Mileage figures in kilometers per liter (km/L):

Speed (km/h)	Mileage (km/L)
56	14.7
104	13.2
64	14.5
88	13.2
112	12.8
96	13.4
84	13.3
68	14.5
80	13.8
100	13.0
60	14.6
72	14.3

Problems P28.3, *P28.11, *P28.16

B: Percentage of moisture in sausage after smoke-curing for a certain time:

Time (h)	Percentage of moisture
115	21
125	19
185	11
200	10
75	50
80	42
150	13
175	12
90	26
100	24

Problems P28.4, P28.7, *P28.12, *P28.20

C: Output rate after producing a certain number of units using a new teaching technique:

Total units produced	Units per hour
25	16
50	28
75	28
100	29
125	30
150	32
175	40
200	42

Problems P28.6, *P28.13, *P28.17

P28.1 Explain the role of correlation analysis in making cause-and-effect conclusions between two random variables.

P28.2 Summarize the properties of the linear correlation coefficient and develop a graphical example to illustrate each point in your summary.

P28.3 Compute the linear correlation coefficient for data set A using Eq. (28.2). Check this value on a calculator if you have one with a correlation routine.

P28.4 (a) Compute the r value for data set B.
(b) Interpret its meaning.

P28.5 Two sets of data were collected and the linear correlation coefficients computed to be $r_1 = 0.24$ and $r_2 = -0.89$. What differences in the data sets can you conclude from these values?

P28.6 If the production results in data set C are taken from a normal population, determine if the linear correlation coefficient is statistically equal to 0.8 at an 80 percent level of confidence.

P28.7 Determine if there is a significant linear correlation between the moisture content and smoke-curing times in data set B. Let $1 - \alpha = 0.95$.

P28.8 The sales data and months of the year given in Prob. P27.6 have an $r = -0.6$. Is this correlation significant if 98 percent confidence is required?

P28.9 Depending on the processing time, different amounts of a particular by-product result. Jerry recorded the amount for 40 different processing times at plant A, while Judy observed the amounts for 18 processing times at plant B, which uses an alternate processing method. The linear correlations between processing time and kilograms of by-product were $r_A = 0.35$ (plant A) and $r_B = 0.42$ (plant B).
(a) Test the hypotheses $H_0: \rho = 0$ versus $H_1: \rho \neq 0$ for both plants using a 5 percent level of significance.
(b) Is the value $r_A = 0.35$ statistically equal to 0.40 at the $\alpha = 0.05$ level?

P28.10 If a 98 percent confidence level is required, what is the minimum sample size to show the significance of a linear correlation of (a) 0.75 and (b) -0.50?

*****P28.11** The linear regression equation for data set A from Prob. P27.3 is

$$\text{Mileage} = 16.77 - 0.0365 \text{ speed}$$

(a) Show that the linear correlation coefficient $r = -0.956$ can be obtained using the generalized correlation coefficient Eq. (28.8).

(b) Compute and explain the meaning of r^2 for data set A.

P28.12 From optional Sec. 27.9 it is possible to write the explained sum of squares SS_R in terms of the total and error sum of squares.

(a) Use this fact to develop a relation for the generalized correlation coefficient equivalent to Eq. (28.8).

(b) Use data set B to show that the value $r = -0.865$ from Prob. P28.4 is also obtained by your new formula. The linear regression equation from Prob. P27.4 is

$$\text{Percentage of moisture} = 56.2 - 0.26 \text{ time}$$

P28.13 The linear regression equation for data set C is

$$\text{Units per hour} = 16.96 + 0.121 \text{ units produced}$$

(a) Find the power model regression equation (Prob. P27.25) for this data.

(b) Use the appropriate formulas to compute the correlation coefficients for the two models.

(c) Determine which is a better-fitting model.

(d) Compute and interpret the coefficients of determination for both models.

P28.14 (a) Fit an exponential model of the form $Y = ab^X$ to the data in Prob. P27.27.

(b) Determine the correlation coefficient.

(c) Use this coefficient to decide if a linear or exponential model is better to explain the relationship between population and the pollution index.

(d) If an exponential model is used, is it correct to use the square of Eq. (28.2) to determine what fraction of total variation is removed by the regression equation? Why or why not?

P28.15 Five very expensive response-time tests have been performed on prototype models of a new underwater pipeline safety mechanism.

Age (months)	Response time (s)
1	0.10
5	0.15
12	0.18
18	0.24
24	0.95

Use the correlation coefficient to determine which of the following models can be used to best predict the response time: (a) linear, (b) exponential, or (c) hyperbolic.

P28.16 Compute and interpret the coefficient of determination for data set A using a polynomial model of degree $k = 2$ (see Prob. P27.21).

P28.17 For data set C the value $r = 0.93$ is obtained from Eq. (28.2).

(a) Is this number useful in determining which of the following regression models removes more of the total sum of squares: polynomial of degree $k = 2$, power, or linear?

(b) Explain why your answer is correct.

P28.18 Explain how it is possible to have two variables, X and Y, which are not independent yet are uncorrelated.

P28.19 If the bivariate distribution for X and Y is

$$f(x, y) = 2 \qquad 0 \le x \le y \le 1$$

find the value of the correlation coefficient ρ using Eqs. (28.11) and (28.12).

*P28.20 The sausage weights prior to smoking were recorded for data set B: 1.2, 1.0, 1.0, 0.8, 1.5, 1.6, 0.9, 1.2, 1.3, and 0.9 kg. In Prob. P27.16 you found the multiple linear regression equation for this data.
  (a) Determine the sample multiple correlation coefficient.
  (b) Compute and interpret the coefficient of multiple determination.
  (c) Compute the linear partial correlation coefficient between weight and percentage of moisture with weight considered the independent variable.

*P28.21 For the television running time data in Prob. P27.17:
  (a) Use the multiple linear regression equation to compute the multiple correlation coefficient.
  (b) Compare this value with the two partial correlation coefficients of viewing time and the two independent variables to see how much more of the total variation is explained when the two independent variables are taken together.

*P28.22 The observed time Y to complete a professional registration test was related to total months or experience in the profession X_1 and preparation and study time in months X_2, using multiple linear regression. The following regression coefficients were computed.

Overall correlation:	0.85
Test time versus experience:	−0.20
Test time versus preparation:	0.71
Experience versus preparation:	−0.12

  (a) Construct the correlation matrix and interpret the values in it.
  (b) Compare the variation removed by the multiple linear regression and the simple linear regressions in which each independent variable is treated separately.

CHAPTER
TWENTY-NINE

QUALITY CONTROL ANALYSIS

This chapter introduces the techniques used to control the quality of a product by the use of statistics. The most commonly used process control charts and product acceptance sampling procedures are discussed and illustrated.

CRITERIA

To complete this chapter, you must be able to:

1. State the definition of *quality control*, and name the primary quality control techniques and the properties evaluated by these techniques.
2. State the three ways in which process mean and dispersion can change and be able to set up $\bar{X}$ and R charts using the specified procedure, given the observed sample values or the sample $\bar{X}$ and R values with a sample size.
3. Use the *process capability* procedure to determine the expected percentage of defective product and the *natural process limits*, given the data necessary to construct $\bar{X}$ and R charts.
4. Compute the control limits for a *fraction defective* chart, given the sample size and number of defectives in each sample. Determine if the process is in control.
5. Compute the control limits for a *number of defects* chart for a constant and varying sample size, given the observed number of defects and the sample sizes. Determine if the number of defects is in control.

6. Compute the values for, graph, and interpret an *operating characteristic (OC) curve*, given the sample size and acceptance number for a single sampling plan. In addition, be able to state the *producer's risk* and *consumer's risk* for a single sampling plan, given the OC curve values or graph and two fraction-defective values.
7. State the primary uses of the three sampling plan standards MIL-STD-105D, Dodge-Romig, and MIL-STD-414. Be able to determine a single sampling plan using MIL-STD-105D, given the lot size, inspection level, and the acceptable quality level for normal inspection.

STUDY GUIDE

29.1 Introduction to Quality Control

The field of quality control uses many different statistical techniques to determine when a product has met a stated quality level and can be delivered to the customer.

> *Quality control* is the application of statistical procedures to determine if a *stable system of chance causes* is present when a process or product is tested against a quality standard.

"A stable system of chance causes" means that the quality characteristic measured (weight, length, flow rate, etc.) has only random variation according to the statistical distribution used to measure quality. If the probability that an *observed* outcome should occur is quite small, the quality (null) hypothesis of this system is rejected. In this case an *assignable cause* is present, and the reason for it is sought. Therefore, quality control is the application of the hypothesis testing procedure each time a sample is taken; however, the use of graphical analysis and tabulated data simplifies the procedure to the extent that a nonstatistician can easily make the required quality inferences.

For our purposes, we will divide quality control into two major categories:

1. Control of a *process*
2. Acceptance (or rejection) of a *product*

In each category there are statistical techniques to control by *variables* (actual measurements) and *attributes* (good or bad). Table 29.1 summarizes the primary quality control techniques discussed in this chapter. The first four properties are evaluated by process control chart techniques, and the last two use acceptance sampling techniques.

Table 29.1 Primary quality control techniques

Process or product property	Random variable controlled	Distribution used	Technique name
Mean	$\bar{X}$	Normal	$\bar{X}$ chart
Dispersion	R or s	Normal	R or s chart
Fraction defective	p	Binomial	p-chart
Number of defects	c	Poisson	c-chart
Fraction defective of product lots	p	Poisson	Acceptance sampling for attributes
Mean and dispersion of product lots	$\bar{X}$ and R or s	Normal	Acceptance sampling for variables

29.2 Control of the Process Mean and Dispersion

Variables such as dimensions and weights which are used to control a production process usually have random errors that follow a normal distribution with mean μ and standard deviation σ. To determine if a stable system of chance causes is present, both μ and σ must be monitored. The mean is controlled by an $\bar{X}$ chart and the dispersion by an R (range), or sometimes an s (standard deviation), chart. We discuss $\bar{X}$ and R charts here. These two are used together to determine if and when one of the following has occurred:

1. The process mean has shifted.
2. The process dispersion has changed.
3. Both the process mean and dispersion have changed.

Figure 29.1 illustrates these possibilities using the general format of a control chart.

Each item produced is supposed to meet stated upper and/or lower specification limits. To develop a control chart, samples of size n are selected at random from the production line at specific times, and the quality characteristic of interest is measured. In order to determine if the process mean is in control, the null hypothesis is formulated:

$$H_0: \mu = \text{some specific value } \bar{\bar{X}}$$

If H_0 is not rejected, the mean is in control; if it is rejected, the alternative H_1: $\mu \neq \bar{\bar{X}}$ is correct and the process is out of control, which indicates the presence of an assignable cause.

To control the process mean, the $\bar{X}$ distribution, not the X distribution of original data, is used. Figure 29.2 shows that the upper and lower $\bar{X}$ control limits, $UCL_{\bar{X}}$ and $LCL_{\bar{X}}$, respectively, are set at $\pm 3\sigma_{\bar{X}}$ from the mean of the

29.2 CONTROL OF THE PROCESS MEAN AND DISPERSION

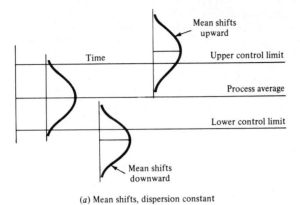

(a) Mean shifts, dispersion constant

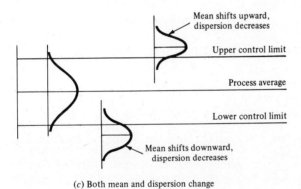

(b) Mean constant, dispersion changes

(c) Both mean and dispersion change

Figure 29.1 The three different ways in which the mean and dispersion of a process variable may change.

sample means $\bar{\bar{X}}$. That is, the control limits are

$$CL_{\bar{X}} = \bar{\bar{X}} \pm 3\sigma_{\bar{X}} \tag{29.1}$$

where

$$\bar{\bar{X}} = \frac{\text{sum of sample } \bar{X} \text{ values}}{k \text{ samples}}$$

$$= \sum_{i=1}^{k} \bar{X}_i / k \tag{29.2}$$

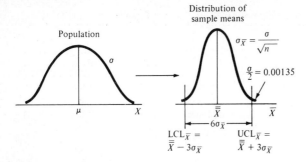

Figure 29.2 Control chart limits for the mean are set $3\sigma_{\bar{X}}$ from the mean of sample means $\bar{\bar{X}}$ of the $\bar{X}$ distribution.

Some multiple of $\sigma_{\bar{X}}$ other than 3 can be used, but 3 is most common. This makes $\alpha = 0.0027$, so the quality null hypothesis H_0: $\mu = \bar{\bar{X}}$ is incorrectly accepted 0.27 percent of the time. Therefore, a type I error occurs, on the average, 27 out of every 10,000 samples. Two other limits used are $\bar{\bar{X}} \pm 3.09\sigma_{\bar{X}}$, which have an $\alpha = 0.001$ and are called *probability limits*, and $\bar{\bar{X}} \pm 1.96\sigma_{\bar{X}}$ limits with $\alpha = 0.05$ which are used as *warning limits*. Only the control or probability limits are used to control quality, whereas the warning limits may be added to indicate when the process is tending toward an out-of-control situation.

The central limit theorem (CLT) is used to compute the limits for the $\bar{X}$ chart using Eq. (12.2), $\sigma_{\bar{X}} = \sigma/\sqrt{n}$, where n is the sample size. Since σ, the population standard deviation, is usually not known, it is common to estimate it with $\hat{\sigma}$ from the sample range values.

$$\hat{\sigma} = \bar{R}/d_2 \qquad (29.3)$$

where $\bar{R}$ = average of the sample range values

$$= \sum_{i=1}^{k} R_i/k$$

d_2 = conversion factor that is dependent on the sample size n

Table 29.2 gives values of d_2 for $n = 2$ to 15. Now the $\bar{X}$ control limits in Eq. (29.1) can be rewritten as follows:

$$CL_{\bar{X}} = \bar{\bar{X}} \pm 3\sigma_{\bar{X}}$$

$$= \bar{\bar{X}} \pm \frac{3\sigma}{\sqrt{n}}$$

$$= \bar{\bar{X}} \pm \frac{3\bar{R}}{d_2\sqrt{n}}$$

$$CL_{\bar{X}} = \bar{\bar{X}} \pm A_2\bar{R} \qquad (29.4)$$

Values of $A_2 = 3/d_2\sqrt{n}$ are given in Table 29.2. Remember these A_2 values are correct only for the 3-standard-deviation limits on $\bar{X}$.

29.2 CONTROL OF THE PROCESS MEAN AND DISPERSION

Table 29.2 Factors used in the preparation of $\overline{X}$, R, and s charts

Sample size, n	d_2	$\overline{X}$ chart A_2	R chart D_4	D_3	s chart B_4	B_3
2	1.128	1.880	3.267	0	3.267	0
3	1.693	1.023	2.574	0	2.568	0
4	2.059	0.729	2.282	0	2.266	0
5	2.326	0.577	2.114	0	2.089	0
6	2.534	0.483	2.004	0	1.970	0.030
7	2.704	0.419	1.924	0.076	1.882	0.118
8	2.847	0.373	1.864	0.136	1.815	0.185
9	2.970	0.337	1.816	0.184	1.761	0.239
10	3.078	0.308	1.777	0.223	1.716	0.284
11	3.173	0.285	1.744	0.256	1.679	0.321
12	3.258	0.266	1.717	0.283	1.646	0.354
13	3.336	0.249	1.693	0.307	1.618	0.382
14	3.407	0.235	1.672	0.328	1.594	0.406
15	3.472	0.223	1.653	0.347	1.572	0.428

Adapted from ASTM STP 15D, *Manual on Presentation of Data and Control Chart Analysis*, Copyright American Society for Testing and Materials, 1916 Race St., Philadelphia, Pa. 19103.

Control limits for the range are 3 standard deviations of the range σ_R away from the mean $\overline{R}$.

$$CL_R = \overline{R} \pm 3\sigma_R$$

The limits are actually computed using the D_4 and D_3 factors in Table 29.2:

- $$UCL_R = D_4 \overline{R} \qquad LCL_R = D_3 \overline{R} \qquad (29.5)$$

The distribution of R is like that of the variance—nonsymmetric—as discussed in Chap. 13. Therefore the limits to control the dispersion measure are not symmetric about $\overline{R}$. In fact, for $n \leq 6$, $D_3 = 0$ so the lower limit is $LCL_R = 0$.

The two charts, $\overline{X}$ and R, are sufficient to control one quality characteristic. Using the two charts, the null hypothesis is more correctly worded as

$$H_0: \text{process is in control}$$

If either chart has an out-of-control point for a particular sample, H_0 is rejected and troubleshooting to find the problem should be started.

The steps below are used to set up trial charts for $\overline{X}$ and R. It is first necessary to ensure that the process is in a stabilized situation so that the statistics collected reflect the actual process characteristics.

1. Take k random samples each of size n from the process. Without past experience start with $n = 5$ and $k = 25$.
2. Compute $\overline{X}$ and R for each sample.

3. Compute $\bar{\bar{X}}$ and $\bar{R}$ for the k samples.
4. Determine the upper and lower control limits for the $\bar{X}$ chart using Eq. (29.4) and for the R chart using Eq. (29.5).
5. Plot the limits for the $\bar{X}$ chart and all sample $\bar{X}$ values.
6. Plot the limits for the R chart and all sample R values.
7. Accept the hypothesis H_0: process is in control if all $\bar{X}$ and R points are between their control limits, and reject H_0 for all sample points outside.
8. Look for the assignable cause for all sample values outside the limits. Remove the out-of-control points and recompute new limits until control is obtained.

Example 29.1 The Second Chance Company wants to closely control the percentage of artificial coloring in a new product. Sixteen samples of 5 each were taken and the sample means and ranges computed (Table 29.3).
(a) Construct the $\bar{X}$ and R charts.
(b) If any points are out of control, remove them and recompute the control limits to determine if the process is now in control.

SOLUTION (a) In this example $n = 5$ and $k = 16$. The results of the first two steps of the procedure are already presented in Table 29.3.

3. The $\bar{\bar{X}}$ and $\bar{R}$ values are computed.

$$\bar{\bar{X}} = \frac{31.3}{16} = 1.9563$$

$$\bar{R} = \frac{10.5}{16} = 0.6563$$

4. The $\bar{X}$ chart limits use $A_2 = 0.577$ for $n = 5$. From Eq. (29.4)

$$CL_{\bar{X}} = 1.9563 \pm 0.577(0.6563)$$

Upper limit: $UCL_{\bar{X}} = 2.33$
Lower limit: $LCL_{\bar{X}} = 1.58$

Table 29.3 Sample $\bar{X}$ and R values for percentage of artificial color, Example 29.1

Sample	$\bar{X}$ (%)	R (%)	Sample	$\bar{X}$ (%)	R (%)
1	2.4	0.3	9	1.8	0.2
2	2.3	0.2	10	1.5	0.4
3	1.8	0.5	11	2.0	1.2
4	1.4	0.4	12	1.7	0.9
5	2.0	0.5	13	1.5	1.0
6	3.6	1.9	14	1.8	0.6
7	1.5	0.6	15	2.0	0.4
8	2.3	0.9	16	1.7	0.5

29.2 CONTROL OF THE PROCESS MEAN AND DISPERSION

The R chart limits from Eq. (29.5) are

$$UCL_R = 2.114(0.6563) = 1.39$$
$$LCL_R = 0(0.6563) = 0$$

5, 6, and 7. Figure 29.3 is a plot of the $\overline{X}$ and R charts. Six $\overline{X}$ values and one R value are out of control, so the percentage of color added needs troubleshooting and correcting.

(b) Assume that troubleshooting (step 8) has uncovered reasons why the points are outside the limits on both charts. We remove the six $\overline{X}$ and R values for samples 1, 4, 6, 7, 10, and 13 from both sets of data. Now, we have

$$\overline{\overline{X}} = \frac{19.4}{10} = 1.94 \qquad \overline{R} = \frac{5.90}{10} = 0.59$$

$$UCL_{\overline{X}} = 1.94 + 0.577(0.59) \qquad UCL_R = 2.114(0.59)$$
$$= 2.28 \qquad\qquad\qquad\quad = 1.25$$

$$LCL_{\overline{X}} = 1.94 - 0.577(0.59) \qquad LCL_R = 0(0.59)$$
$$= 1.60 \qquad\qquad\qquad\quad = 0$$

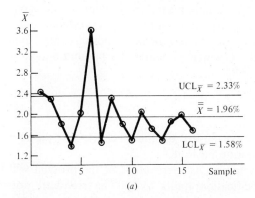

(a)

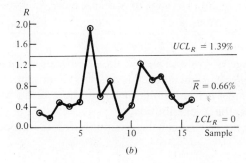

(b)

Figure 29.3 Control charts for (a) $\overline{X}$ and (b) R for percentage of artificial color, Example 29.1

The range is now in control, but the two samples with $\bar{X} = 2.3$ are still above the upper limit, so the average is still out of control.

COMMENT To initially obtain control, the points that are outside the limits on either chart are removed. If too many points must be removed to obtain control, some new samples should be taken after the process is corrected.

If the sample standard deviation s, rather than the range, is used to control process dispersion, an s chart is used. The control limits are

$$UCL_s = B_4 \bar{s}$$
$$LCL_s = B_3 \bar{s}$$

where $\bar{s}$ is the average of the sample s values. The B_4 and B_3 factors are included in Table 29.2. Setup and interpretation of the points on the s chart are identical to those for the R chart.

Problems P29.1–P29.6

29.3 Process Capability and Natural Process Limits from $\bar{X}$ and R charts

Once the mean and dispersion of the sample means are in control, the capabilities of the process itself can be analyzed and used for future production design. Control on the R chart is obtained first by removing all out-of-control points. The revised $\bar{R}$ is used to compute $\hat{\sigma}$, which estimates the population standard deviation σ.

$$\hat{\sigma} = \frac{\bar{R}}{d_2} \quad (29.6)$$

The revised $\bar{R}$ is also used to obtain control on the $\bar{X}$ chart, and the revised $\bar{\bar{X}} = \hat{\mu}$ is an estimate of the population mean μ.

The values $\hat{\sigma}$ and $\hat{\mu}$ help determine what percentage of the product will be within or outside of preset specification limits (SL). The standard normal distribution (SND) value

$$z = \frac{SL - \hat{\mu}}{\hat{\sigma}} \quad (29.7)$$

is used to determine the percentage above and below the specification limits. This is called *process capability analysis*. Remember, this analysis determines the capability of the process itself to meet certain limits; it has nothing to do with the capability of the means, that is, the $\bar{X}$ distribution.

Finally, the *natural process limits* (NPL) can be computed for the process using the values of $\bar{\bar{X}} = \hat{\mu}$ and $\hat{\sigma}$:

$$NPL_X = \hat{\mu} \pm 3\hat{\sigma} \quad (29.8)$$

29.3 PROCESS CAPABILITY AND NATURAL PROCESS LIMITS FROM $\bar{X}$ AND R CHARTS

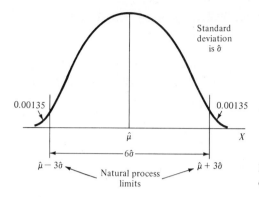

Figure 29.4 Natural process limits for a production process computed using $\hat{\mu} \pm 3\hat{\sigma}$.

The subscript X is a reminder that these limits are for the process itself, not $\bar{X}$ values for sample means. Figure 29.4 illustrates these limits and the fact that there is a six-standard-deviation ($6\hat{\sigma}$) band between these natural limits. If upper and lower specification limits have been firmly established, the limits in Eq. (29.8) should be compared with them. If specification limits have not been established, they should be no tighter than the natural process limits to avoid the production of excessive defective product, as shown in Fig. 29.5.

The above analysis is summarized by the procedure to study process capability:

1. Obtain control of the dispersion via the R chart.
2. Compute $\hat{\sigma}$ by Eq. (29.6) to estimate the population standard deviation.
3. Obtain control of the mean via the $\bar{X}$ chart and the revised $\bar{R}$ value.
4. Use the revised $\bar{\bar{X}}$ as an estimate $\hat{\mu}$ of the population mean.
5. Use the specification limits and the SND value from Eq. (29.7) to estimate the percentage of acceptable and defective product.
6. Compute the natural process limits by Eq. (29.8) as guides for specification limits which will minimize defective product.

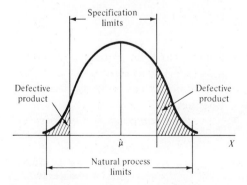

Figure 29.5 Comparison of natural process limits and too-tight specification limits.

This procedure is illustrated in Example 29.2.

Example 29.2 Use the process capability procedure and the results of Example 29.1 to determine the percentage of defective product and the natural process limits. Specification limits on the percentage of artificial color have been previously set at 1.75 ± 0.50 percent.

SOLUTION All six steps are detailed:

1. Control on the R chart resulted in $\bar{R} = 0.59$.
2. With $d_2 = 2.326$ for $n = 5$,

$$\hat{\sigma} = \frac{0.59}{2.326} = 0.254$$

3 and 4. New $\bar{X}$ control limits in Example 29.1 placed the two $\bar{X} = 2.3$ values out of control. Removal of them obtains control for $\bar{X}$ with $\bar{\bar{X}} = \hat{\mu} = 1.85$.

5. The z value from Eq. (29.7) is computed for the lower specification limit $LSL = 1.25$ and the upper specification limit $USL = 2.25$ using $\hat{\mu} = 1.85$ and $\hat{\sigma} = 0.254$ (Fig. 29.6).

Lower: $$z = \frac{1.25 - 1.85}{0.254} = -2.36$$

$$P(z \leq -2.36) = 0.0091$$

Upper: $$z = \frac{2.25 - 1.85}{0.254} = +1.57$$

$$P(z \geq 1.57) = 0.0582$$

A total of $0.91 + 5.82 = 6.73$ percent of the product is expected to have too little or too much artificial color. This is more

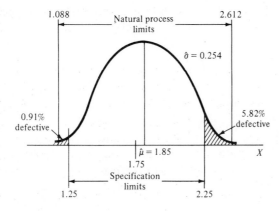

Figure 29.6 Comparison of the specification limits and the natural process limits for Example 29.2.

defective product than is expected, because the specification limits should be set to have virtually all acceptable product.

6. The natural process limits are $3\hat{\sigma}$ on each side of $\hat{\mu}$.

$$\hat{\mu} \pm 3\hat{\sigma} = 1.85 \pm 3(0.254)$$
$$= 1.85 \pm 0.762$$
$$= 1.088 \text{ and } 2.612\%$$

Therefore, if the specification limits were set at the natural process limits (Fig. 29.6) and the process mean were 1.85 percent, rather than 1.75 percent, the expected percentage defective would be 0.27 percent.

Problems P29.7–P29.11

29.4 Control of the Process Fraction Defective

An item is considered *defective* if one or more quality standards are not met. To control the fraction defective or number of defectives made by a production line, an *attribute* chart is used. The *fraction defective*, or *p*-chart is an attribute chart based on the normal approximation to the binomial distribution. Rather than use the binomial directly, the distribution of proportions discussed in Sec. 8.8 is used. The mean $\mu = p$ and standard deviation $\sigma = \sqrt{pq/n}$ are estimated by

$$\hat{\mu} = \bar{p}$$

$$\hat{\sigma} = \sqrt{\frac{\bar{p}\bar{q}}{n}}$$

where

$$\bar{p} = \frac{\text{total number of defectives}}{\text{total items inspected}} \qquad (29.9)$$

$$\bar{q} = 1 - \bar{p}$$

The null hypothesis tested each time a sample is taken is H_0: $\mu =$ some specific value $\bar{p}$. The control limits CL_p for the fraction defective p are 3 standard deviations from $\bar{p}$:

● $$CL_p = \bar{p} \pm 3\sqrt{\frac{\bar{p}\bar{q}}{n}} \qquad (29.10)$$

If the lower control limit LCL_p is computed as less than 0 by Eq. (29.10), it is set equal to 0 since a negative value is not observable. The hypothesis H_0 is not accepted whenever a point is outside these limits. A value below the LCL_p should be checked to see why the improvement in quality occurred. Possibly the reason can be used in the future to reduce the fraction defectives of this or other products.

In order to control the *number of defective* items rather than fraction defective, the binomial distribution is used with the estimates for μ and σ being

$$\hat{\mu} = n\bar{p}$$

$$\hat{\sigma} = \sqrt{n\bar{p}\bar{q}}$$

The control limits are now set for the *np-chart* for the number of defectives.

$$CL_{np} = n\bar{p} \pm 3\sqrt{n\bar{p}\bar{q}} \qquad (29.11)$$

The limits in Eq. (29.11) are interpreted in the same way as the *p*-chart limits.

Example 29.3 Mrs. Jameson, a quality control engineer, wishes to develop a fraction-defective chart for a printed-circuit board used in telephone switching equipment. For the past 20 days samples of 50 boards were taken each day and tested. The number of boards failing the test, which involves several voltage and resistance checks, is recorded in Table 29.4. Compute the *p*-chart limits and determine if the fraction defective is in control.

SOLUTION The $\bar{p}$ value is found by Eq. (29.9):

$$\bar{p} = \frac{\text{total number of defectives}}{(20 \text{ days})(50 \text{ items/day})}$$

$$= \frac{61}{1000} = 0.061$$

Then $\bar{q} = 1 - 0.061 = 0.939$

The *p*- chart limits from Eq. (29.10) are:

$$CL_p = 0.061 \pm 3\sqrt{\frac{(0.061)(0.939)}{50}} = 0.061 \pm 0.102$$

$$= 0.0 \text{ and } 0.163$$

Table 29.4 Number of defective circuit boards in samples of size 50, Example 29.3

Day	Number defective	Fraction defective	Day	Number defective	Fraction defective
1	3	0.06	11	0	0.00
2	4	0.08	12	1	0.02
3	0	0.00	13	2	0.04
4	5	0.10	14	3	0.06
5	3	0.06	15	2	0.04
6	4	0.08	16	4	0.08
7	3	0.06	17	1	0.02
8	4	0.08	18	5	0.10
9	5	0.10	19	4	0.08
10	6	0.12	20	2	0.04
				61	

The $LCL_p = 0.061 - 0.102 = -0.041$ value is set to 0. The p-chart may be drawn in a manner similar to the $\overline{X}$ or R chart. However, simple observation indicates that all fraction-defective values in Table 29.4 are well within the control limits.

> COMMENT If Mrs. Jameson had controlled using an np-chart, the limits would have been computed using Eq. (29.11) and the number of defectives in Table 29.4 would be plotted. You should compute the np-chart limits now.
>
> If the sample size n is not constant from one sample to the next, the p-chart limits in Eq. (29.10) may be (1) computed for each n value, in which case they vary with each sample, or (2) computed using the average value $\bar{n}$. The $\bar{n}$ method is commonly used, but it is necessary to compute the appropriate limit with the correct n value when a point falls close to the $\bar{n}$ limit. This is required to make the final determination whether the point is in or out of control.

Problems P29.12–P29.18

29.5 Control of the Number of Defects per Standard Unit

In practice, the Poisson has been found to be an excellent distribution for controlling the number of defects. A *defect* is a flaw: it takes one or more defects to make a defective product. A defect may be a hole in a sheet of material, a bubble in a glass pane, or an incorrectly inserted screw in a sheet-metal assembly. If the sample size is one *standard unit*, say, 10 m², the defects are controlled by a *c-chart* (number-of-defects chart), whereas control of defects when the sample size varies is accomplished by a *u*-chart. Both are discussed here.

After k standard units have been inspected, the average number of defects per unit $\bar{c}$ is computed using x_i ($i = 1, 2, \ldots, k$) as the defects per unit:

$$\bar{c} = \frac{\text{total number of defects}}{\text{number of standard units}}$$

$$\bar{c} = \frac{\sum_{i=1}^{k} x_i}{k} \tag{29.12}$$

The control limits for the c-chart are

● $$CL_c = \bar{c} \pm 3\sqrt{\bar{c}} \tag{29.13}$$

As before, a lower control limit of 0 is used if Eq. (29.13) results in $LCL_c < 0$. A point outside the limits requires rejection of the null hypothesis

H_0: number of defects is in control

550 QUALITY CONTROL ANALYSIS

If the sample size is other than the standard unit, a number-of-defects-per-unit chart, or *u-chart*, is used. For each sample i determine m_i, the number of standard units. For instance, if the standard unit is 25 items and the sample size is 40, then $m_i = 40/25 = 1.6$. The average number of defects per sample for k samples is

$$\bar{u} = \frac{\Sigma x_i}{\Sigma m_i} \qquad (29.14)$$

where the sum is over the k samples and there are x_i defects in sample i. The limits for the *u*-chart must be computed for each sample.

• $$CL_{u_i} = \bar{u} \pm 3\sqrt{\bar{u}/m_i} \qquad (29.15)$$

Each x_i value is compared with its appropriate limits to determine if the defects are in control.

Example 29.4 A nationwide discount store system has two distribution centers that receive 60-cm, three-speed room fans from Universal Manufacturing. Twenty-five fans are supposed to be inspected from each shipment. The number of defects per sample is recorded for paint chips, broken knobs, bent grills, etc. Warehouse A inspects in standard units of 25 fans, but warehouse B does not. Table 29.5 gives the defects for the last 10 shipments at each warehouse. Prepare the correct number of defect charts for each warehouse.

SOLUTION The *c*-chart is used for warehouse A because all samples are 1 standard unit of 25 fans. From Eq. (29.12) for $k = 10$

$$\bar{c} = \frac{164}{10} = 16.4 \text{ defects per 25 fans}$$

Table 29.5 Number of defects per lot, Example 29.4

	Warehouse A		Warehouse B	
Sample	Number inspected	Number of defects	Number inspected	Number of defects
1	25	12	25	13
2	25	14	40	26
3	25	10	10	3
4	25	18	50	29
5	25	21	70	41
6	25	16	35	24
7	25	10	40	18
8	25	19	25	9
9	25	23	60	26
10	25	21	100	52
		164		241

29.6 ACCEPTANCE SAMPLING PLANS FOR ATTRIBUTE DATA

Table 29.6 Computation of u-chart control limits with $\bar{u} = 13.2$, Example 29.4

Sample i (1)	Number of defects per sample x_i (2)	Number of standard units m_i (3)	Number of defects per standard unit (4) = (2)/(3)	$3\sqrt{\bar{u}/m_i}$ (5)	Control limits $\bar{u} \pm 3\sqrt{\bar{u}/m_i}$	
					LCL_u (6)	UCL_u (7)
1	13	1.0	13.00	10.9	2.3	24.1
2	26	1.6	16.25	8.6	4.6	21.8
3	3	0.4	7.50	17.2	0.0	30.4
4	29	2.0	14.50	7.7	5.5	20.9
5	41	2.8	14.64	6.5	6.7	19.7
6	24	1.4	17.14	9.2	4.0	22.4
7	18	1.6	11.25	8.6	4.6	21.8
8	9	1.0	9.00	10.9	2.3	24.1
9	26	2.4	10.83	7.0	6.2	20.2
10	52	4.0	13.00	5.4	7.8	18.6
	241	18.2				

The control limits are computed by Eq. (29.13).

$$CL_c = 16.4 \pm 3\sqrt{16.4}$$
$$= 4.3 \text{ and } 28.5$$

All values are within the control limits.

For warehouse B the sample size varies, so m_i ($i = 1, 2, \ldots, 10$) is computed first and the u-chart is used to control the defects per sample. Table 29.6 gives all computations for the control limits for each sample. The average defects per sample by Eq. (29.14) is

$$\bar{u} = \frac{241}{18.2} = 13.2$$

For sample $i = 2$, for example,

$$LCL_u = 13.2 - 3\sqrt{13.2/1.6} = 4.6$$

Comparison of the number of defects per standard unit (column 4) and the control limits shows that all samples are in control.

Problems P29.19–P29.22

29.6 Acceptance Sampling Plans for Attribute Data

Many products are manufactured and shipped in lots which are inspected using specific quality characteristics. If all items in the lot are checked, the process is called 100 percent inspection. If less than 100 percent inspection is used to accept or reject an entire lot, *acceptance sampling* procedures for attribute data (each item being good or bad) or variable data are used.

A *single sampling plan* requires the testing of each item in a random sample of n items from a lot of N items. If the number of defective items d is less than or equal to a predetermined value c, called the *acceptance number*, the entire lot is accepted. If d exceeds c, the lot is rejected.

If, after a lot is rejected, it is 100 percent inspected and all bad items are replaced with good items, the term *rectified inspection* applies.

The ability of a sampling plan to discriminate between good and bad lots is best explained by an operating characteristic (OC) curve (Fig. 29.7) which is a plot of different fraction-defective values p on the abscissa and the probability of accepting the lot P_a on the ordinate. This OC curve is much like those used in the hypothesis testing chapters. The steeper the OC curve, the better its discriminating power; an ideal curve, which is vertical at some desirable fraction-defective level p_0 (Fig. 29.7), is obtainable only for error-free inspection.

The P_a values are correctly computed using the hypergeometric pdf $h(d; n, p, N)$, where d is the number of defectives in n items from a p-defective lot of N items (Sec. 7.7). However, if n is small compared to N and if p is small, the *Poisson* pdf $P(d; np)$, where np is the mean number of defectives in n items, gives a good approximation. The following steps are used to compute and graph the OC curve values, given n and c.

1. List several p values which cover the expected range of fractive defectives.
2. Compute the Poisson mean $\lambda = np$ for each p value.
3. Set $c = x'$ and use the Poisson Table B-2 to determine $P_a = P(d \leq x')$. This will give the probability of observing less than or equal to c defective items. By the Poisson this is

$$P_a = \sum_{x=0}^{x'} \frac{e^{-np}(np)^x}{x!}$$

4. Plot p versus P_a.

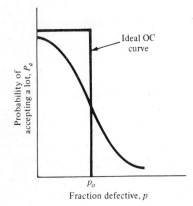

Figure 29.7 General shape of an operating characteristic (OC) curve and an ideal OC curve.

5. Construct a smooth, continuous curve through the points. More p values might be needed to obtain a well-shaped OC curve.

Example 29.5 illustrates these steps.

Example 29.5 The incoming materials inspection department plans to use the sampling plan $n = 100$, $c = 2$ to accept and reject lots of $\frac{1}{6}$-hp motors from the DC Company.
(a) Graph the OC curve for this plan.
(b) How would the shape of the OC curve change if the plan $n = 150$, $c = 2$ were used?

SOLUTION (a) The steps above are used, and the results are given in Table 29.7.

1 and 2. Values of p varying from 0.005 to 0.070 (0.5 to 7 percent) are listed with the Poisson means $\lambda = np = 100p$.
3. From Table B-2 the P_a values for $c = x' = 2$ are recorded for each np value.
4 and 5. The values are plotted in Fig. 29.8 and a smooth curve approximated through them.

Table 29.7 Computations for the OC curve for the sampling plan $n = 100$, $c = 2$, Example 29.5

p	np	P_a	p	np	P_a
0.005	0.5	0.986	0.030	3.0	0.423
0.008	0.8	0.953	0.035	3.5	0.322
0.010	1.0	0.920	0.040	4.0	0.238
0.015	1.5	0.809	0.050	5.0	0.125
0.020	2.0	0.677	0.060	6.0	0.062
0.025	2.5	0.544	0.070	7.0	0.030

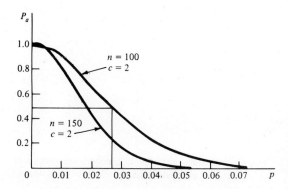

Figure 29.8 OC curve for the sampling plans $n = 100$, $c = 2$ and $n = 150$, $c = 2$, Example 29.5.

This OC curve can also be used to determine the p value necessary to accept a certain percentage of the submitted lots. For example, 50 percent of the lots which have $p = 0.027$ are accepted because $P_a = 0.50$ for this p value.

(b) When the n value is increased, the plan is better able to discriminate good and bad lots, so the OC curve becomes steeper. Figure 29.8 includes the OC curve for $n = 150$, $c = 2$.

If the acceptance number c is decreased and n is held constant, the OC curve moves closer to the origin (Fig. 29.9).

A sampling plan is often designed such that the OC curve goes through or very near two specific points. These points and the effects on the resulting plan are explained here.

1. *Acceptable Quality Level (AQL)*. This is a small p value such that most of the lots submitted with this quality level are accepted. The P_a for the AQL is commonly 0.95. This probability $1 - P_a$, which is 0.05 if $P_a = 0.95$, is called the *producer's risk* α, because it is the chance that a good-quality lot will be rejected. This is the probability of a type I error for the null hypothesis

$$H_0: \text{fraction defective} \leq AQL$$

Figure 29.10 indicates the point $(AQL, 1 - \alpha)$.

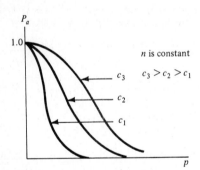

Figure 29.9 OC curves for increasing values of c with n held constant.

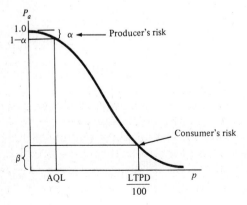

Figure 29.10 OC curve indicating the producer's risk α and consumer's risk β.

2. *Lot Tolerance Percent Defective* (*LTPD*). This is a relatively large p value which has $P_a = \beta$ of being accepted. Figure 29.10 shows the point ($LTPD/100, \beta$). The probability β is usually 0.10 and is called the *consumer's risk* because it is the chance that a poor lot will be accepted. The value β is the type II error probability for the alternative hypothesis

$$H_1: \text{fraction defective} \geq \frac{LTPD}{100}$$

There are many more interesting facts and computations that can be discussed about sampling plan design for attributes and variables. Since these are best explained in the texts on quality control where space is available, you should refer to them for more details.

Example 29.6 A large wholesale outlet uses the plan $n = 100$, $c = 2$ to inspect incoming steel pipe. The vendor-buyer contract requires that lots of 0.8 percent or less defective be accepted and of 5 percent or more defective be rejected. Determine the α and β values for this plan.

SOLUTION The OC curve values for $n = 100$, $c = 2$ in Fig. 29.9 and Table 29.7 are used to determine α and β. The specified *AQL* is 0.008 which has $P_a = 0.953$, so the α value is $1 - 0.953 = 0.047$. The $LTPD/100$ of 0.05 results in a consumer's risk of $\beta = 0.125$. Therefore, about 5 percent of the good lots are rejected and about 12.5 percent of the bad lots will be accepted using the plan $n = 100$, $c = 2$.

COMMENT A sampling plan may be designed by specifying the points ($AQL, 1 - \alpha$) and ($LTPD/100, \beta$). Because the discrete Poisson pdf is used to obtain the OC curve, both points cannot be achieved exactly, so a slight compromise in α and/or β is necessary.

Problems P29.23–P29.29

29.7 Published Sampling Plans

Several, well-accepted publications are available which may be used to select a predesigned sampling plan [1, 2, 3]. Three of these are outlined:

1. *Military Standard MIL-STD-105D*. These are attribute sampling plans that provide an n and c value for a specified *AQL*. Rejected lots are not 100 percent inspected if this standard is used, but three different types of inspection are available—normal, tightened, and reduced. If the quality history has been good, fewer items are inspected under reduced inspection, while poor-quality history indicates the need for a tightened acceptance criteria, so c is decreased. Rules are specified for switching between the three

types of inspection, but the goal is to use reduced inspection whenever possible. MIL-STD-105D, which is the most commonly used attribute sampling plan standard, is illustrated in Example 29.7.

2. *Dodge-Romig Plans.* These attribute plans minimize the average number of items tested if rejected lots are 100 percent inspected (rectified sampling). The user specifies (1) the *average outgoing quality level*, for example, 0.02 or 2 percent, that should result from the inspection process and (2) a range of fraction-defective values expected from the vendor. The average outgoing quality value is the maximum fraction defective that will be observed using the sampling plan.

3. *Military Standard MIL-STD-414.* The sampling plans in this standard are for variables data with the assumption that the quality characteristic is normally distributed in the lot. There are plans for both known and unknown population standard deviation with normal, tightened, and reduced sampling; all plans require that the AQL be selected. The standard normal distribution variable Z is computed to determine if the estimated fraction defective exceeds the specified AQL. Both single and double specification limits can be handled by MIL-STD-414.

In most standards, like MIL-STD-105D, double and multiple sampling plans are given. For a double sampling plan two samples n_1 and n_2, are allowed, and two acceptance numbers listed. For instance,

$$n_1 = 125 \quad c_1 = 1$$
$$n_2 = 125 \quad c_2 = 5$$

If the number of defective items d in n_1 is $d \leq c_1$, accept the lot; and if $d > c_2$, reject the lot. But if $c_1 < d \leq c_2$, take a second sample of size n_2. If the total defectives in $n_1 + n_2$ is $d \leq c_2$, accept the lot; but if $d > c_2$, reject it. Double sampling usually requires that fewer items be inspected on the average than single sampling requires.

Example 29.7 Mr. Juniper receives lots of size 5000 spring steel fasteners at his plant each week. There are two types of defects which cause the fasteners to be scrapped, one critical and one minor. If the AQL values are 0.65 percent for the critical and 4.0 percent for the minor defect, use MIL-STD-105D to determine the n and c values for single sampling plans.

SOLUTION To obtain a plan for MIL-STD-105D, it is necessary to use the following procedure:

1. The inspection level to be used is in Table 29.8. General inspection level II is most commonly used. Level I gives less discriminating and level III gives more discriminating sampling plans.
2. The sample size code letter from Table 29.8 is based on the submitted lot size.

Table 29.8 Sample size code letters for MIL-STD-105D

Lot size	Special inspection levels				General inspection levels		
	S-1	S-2	S-3	S-4	I	II	III
2–8	A	A	A	A	A	A	B
9–15	A	A	A	A	A	B	C
16–25	A	A	B	B	B	C	D
26–50	A	B	B	C	C	D	E
51–90	B	B	C	C	C	E	F
91–150	B	B	C	D	D	F	G
151–280	B	C	D	E	E	G	H
281–500	B	C	D	E	F	H	J
501–1,200	C	C	E	F	G	J	K
1,201–3,200	C	D	E	G	H	K	L
3,201–10,000	C	D	F	G	J	L	M
10,001–35,000	C	D	F	H	K	M	N
35,001–150,000	D	E	G	J	L	N	P
150,001–500,000	D	E	G	J	M	P	Q
500,001 and over	D	E	H	K	N	Q	R

3. The AQL level is selected.
4. The use of normal, tightened, or reduced inspection is made. Initially normal inspection is selected (Table 29.9) and the switching rules published in the standard then guide the user.

For the fasteners submitted in lots of 5000 if level II inspection is used, Table 29.8 gives a code letter L. For normal inspection, the single-sampling plans from Table 29.9 for the critical and minor defects are:

Critical defect	Minor defect
$AQL = 0.65\%$	$AQL = 4.0\%$
$n = 200$	$n = 200$
$c = 3$	$c = 14$

A sample of 200 fasteners is randomly selected; and if the number of critical defects is 3 or less and the number of minor defects is 14 or less, the lot is accepted.

COMMENT Tables 29.8 and 29.9 are only samples from MIL-STD-105D. The actual standard includes plans for single, double, and multiple sampling for normal, tightened, and reduced inspection. Complete instructions and specimen OC curves are given in the standard.

Table 29.9 Master table for normal inspection (single sampling) for MIL-STD-105D

| Sample size code letter | Sample size | \multicolumn{2}{c}{0.010} | 0.015 | 0.025 | 0.040 | 0.065 | 0.10 | 0.15 | 0.25 | 0.40 | 0.65 | 1.0 | 1.5 | 2.5 | 4.0 | 6.5 | 10 | 15 | 25 | 40 |

Acceptable quality levels (AQL)

Code	n	0.010 Ac Re	0.015 Ac Re	0.025 Ac Re	0.040 Ac Re	0.065 Ac Re	0.10 Ac Re	0.15 Ac Re	0.25 Ac Re	0.40 Ac Re	0.65 Ac Re	1.0 Ac Re	1.5 Ac Re	2.5 Ac Re	4.0 Ac Re	6.5 Ac Re	10 Ac Re	15 Ac Re	25 Ac Re	40 Ac Re
A	2	↓	↓	↓	↓	↓	↓	↓	↓	↓	↓	↓	↓	↓	↓	0 1	↑	↑	2 3	3 4
B	3	↓	↓	↓	↓	↓	↓	↓	↓	↓	↓	↓	↓	↓	0 1	↑	↑	1 2 2 3	3 4	5 6
C	5	↓	↓	↓	↓	↓	↓	↓	↓	↓	↓	↓	↓	0 1	↑	↑	1 2	2 3	3 4	7 8
D	8	↓	↓	↓	↓	↓	↓	↓	↓	↓	↓	↓	0 1	↑	↑	1 2	2 3	3 4	5 6	7 8 10 11
E	13	↓	↓	↓	↓	↓	↓	↓	↓	↓	↓	0 1	↑	↑	1 2	2 3	3 4	5 6	7 8	10 11 14 15
F	20	↓	↓	↓	↓	↓	↓	↓	↓	↓	0 1	↑	↑	1 2	2 3	3 4	5 6	7 8	10 11	14 15 21 22
G	32	↓	↓	↓	↓	↓	↓	↓	↓	0 1	↑	↑	1 2	2 3	3 4	5 6	7 8	10 11	14 15	21 22 ↑
H	50	↓	↓	↓	↓	↓	↓	↓	0 1	↑	↑	1 2	2 3	3 4	5 6	7 8	10 11	14 15	21 22	↑
J	80	↓	↓	↓	↓	↓	↓	0 1	↑	↑	1 2	2 3	3 4	5 6	7 8	10 11	14 15	21 22	↑	
K	125	↓	↓	↓	↓	↓	0 1	↑	↑	1 2	2 3	3 4	5 6	7 8	10 11	14 15	21 22	↑		
L	200	↓	↓	↓	↓	0 1	↑	↑	1 2	2 3	3 4	5 6	7 8	10 11	14 15	21 22	↑			
M	315	↓	↓	↓	0 1	↑	↑	1 2	2 3	3 4	5 6	7 8	10 11	14 15	21 22	↑				
N	500	↓	↓	0 1	↑	↑	1 2	2 3	3 4	5 6	7 8	10 11	14 15	21 22	↑					
P	800	↓	0 1	↑	↑	1 2	2 3	3 4	5 6	7 8	10 11	14 15	21 22	↑						
Q	1250	0 1	↑	↑	1 2	2 3	3 4	5 6	7 8	10 11	14 15	21 22	↑							
R	2000	↑	↑	1 2	2 3	3 4	5 6	7 8	10 11	14 15	21 22	↑								

↓ = Use first sampling plan below arrow. If sample size equals, or exceeds, lot or batch size, do 100 percent inspection.
↑ = Use first sampling plan above arrow.
Ac = Acceptance number.
Re = Rejection number.

If tightened inspection is used for the situation above, the sample size is still 200 but the acceptance numbers are reduced to 2 for critical and 12 for minor defects. This reduction in c moves the OC curve toward the origin, as shown in Fig. 29.9.

Problems P29.30–P29.33

REFERENCES

1. *Military Standard 105D—Sampling Procedures and Tables for Inspection by Attributes*, Superintendent of Documents, Government Printing Office, Washington, 1963.
2. Dodge, H. F., and Romig, H. G., *Sampling Inspection Tables—Single and Double Sampling*, Wiley, New York, 1959.
3. *Military Standard 414—Sampling Procedures and Tables for Inspection by Variables for Percent Defective*, Superintendent of Documents, Government Printing Office, Washington, 1957.

ADDITIONAL MATERIAL

Process control (Secs. 29.1 to 29.5): Belz [2], Chap. 20; Bowker and Lieberman [4], Chap. 12; Duncan [7], Chaps. 18–23; Gibra [8], Chap. 16; Grant and Leavenworth [9], Chaps. 1 to 11; Miller and Freund [18], pp. 419–435; Neville and Kennedy [19], Chap. 19.

Acceptance sampling plans (Secs. 29.6 and 29.7): Belz [2], Chap. 9; Bowker and Lieberman [4], Chap. 13; Duncan [7], Chaps. 7 to 10, 16, 17; Gibra [8], pp. 490–509; Grant and Leavenworth [9], Chaps. 12 to 16; Miller and Freund [18], pp. 436–446.

Variables sampling plans: Bowker and Lieberman [4], Chap. 14; Duncan [7], Chaps. 11 to 15; Gibra [8], pp. 509–515; Grant and Leavenworth [9], Chap. 17.

PROBLEMS

P29.1 Assume that an $\bar{X}$ control chart has been set up using $3\sigma_{\bar{X}}$ limits. A point has been plotted above the $UCL_{\bar{X}}$ line. Write out all the steps of the hypothesis testing procedure for the rejection of the quality hypothesis for this particular value.

P29.2 What is the purpose of using warning limits on $\bar{X}$ and R control charts?

P29.3 Thirty samples of size $n = 5$ were taken over a 3-day period on a product manufactured by government-trained machinists. The sums of sample averages and ranges were

$$\sum_{i=1}^{30} \bar{X}_i = 644.9 \text{ cm} \qquad \sum_{i=1}^{30} R_i = 2.1 \text{ cm}$$

Complete the control limits for (*a*) the $\bar{X}$ and (*b*) the R control charts.

P29.4 The specification for filling cans of almonds is 170 ± 3 g. Samples taken each hour generated the weights shown (recorded as deviations from 170 in 0.1-g units). Set up control charts for process mean and dispersion and obtain control for the weight. Assume that troubleshooting efforts are successful in finding assignable courses for out-of-control points.

Sample number	Number of can			
	1	2	3	4
1	22	−18	10	−5
2	16	33	−19	−14
3	−14	41	−18	−4
4	3	−12	22	−7
5	−6	41	3	18
6	7	34	−31	19
7	25	9	14	45
8	−38	24	12	−9
9	11	−31	−11	7
10	−27	−62	−39	−30
11	11	27	48	−19
12	17	23	−56	−15
13	−31	−44	−34	−15
14	30	−37	22	−1
15	−17	−16	29	49

P29.5 The number of volts generated for a specific use must be closely controlled. Samples of size 3 were taken for a period of 12 days and the sample means and ranges computed.
  (a) Use the data shown to obtain control of the output.
  (b) Set up tentative $\bar{X}$ control chart limits for the next period. Would you consider these reliable estimates of $\bar{X}$ control limits for the next 12 samples? Why or why not?

Sample	$\bar{X}$	R
1	604	2.2
2	592	1.8
3	612	1.9
4	601	2.8
5	591	1.9
6	594	2.2
7	610	2.9
8	600	1.7
9	597	2.4
10	608	1.8
11	599	2.1
12	603	3.6

P29.6 Compute (a) the warning limits and (b) the 0.001 probability limits for the 12 $\bar{X}$ values in Prob. P29.5. (c) How many $\bar{X}$ values are outside the probability limits compared to the number outside the $3\sigma_{\bar{X}}$ limits first computed in Prob. P29.5?

P29.7 Control has been obtained on the production cost for a new product with

$$n = 8 \text{ units per sample}$$

$$\bar{\bar{X}} = \$5.85 \text{ per unit}$$

$$\bar{R} = \$0.45$$

The specification limits on unit production costs have been previously set by management as $5.60 and $6.00 per unit.

(a) What are the current $\bar{X}$ and R chart limits?

(b) Use the process capability procedure to determine the percentage of product that will have a production cost within the limits set by management.

(c) By what amount would the width of the present specification limit band have to be changed so that it would equal the natural cost limit band of the process?

P29.8 In Prob. P29.4 $\bar{X}$ and R charts were used to control the weight of almonds in a 170-g container. With the process mean and dispersion in control, use the process capability procedure to determine the following.

(a) All cans filled above the upper limit of 173 g are passed and sold as is, but an estimated 3¢ per can profit is lost. All cans below the lower limit of 167 are sold as seconds at a profit loss of 8¢ per can. In 10,000 cans estimate the average loss in profit due to overfilling or underfilling.

(b) Determine the natural process limits and compare them to the specification limits of 170 ± 3g.

P29.9 If the reading in Prob. P29.5 goes above 602 V, the machinery will possibly be damaged. Use the controlled sample values to do the following.

(a) If you observe the voltage at any time, compute the probability that it will exceed this limit.

(b) Calculate the natural limits of the voltage readings.

P29.10 The percentage of isomers in an insect repellent is supposed to be between 2.4 and 2.6 percent. The chemical mixing process is in statistical control with an average of 2.48 and an estimated process standard deviation of 0.05 percent.

(a) The quality assurance supervisor wants no more than 5 percent of the product to have below 2.4 percent isomers and no more than 2 percent above 2.6 percent isomers. Has this level of accuracy been reached?

(b) Compute the natural process limits and compare them graphically with the specification limits.

P29.11 Explain in detail the difference between control limits and natural process limits.

P29.12 Samples of 500 items have been taken for 20 days from a printshop's output. The number of items that have some type of defect—printing, folding, coloring—are recorded with the order retained by row:

7	3	5	7	4	9	10	11	2	7
3	5	4	12	7	5	2	6	10	8

Use the fraction-defective chart to determine if the printing process is in control.

P29.13 (a) Ms. Thomas took samples of size 100 for 10 days. The fraction-defective values are 0.04, 0.08, 0.10, 0.03, 0.05, 0.07, 0.09, 0.11, 0.01, 0.07. Determine if the fraction defective of the process is in control.

(b) When Ms. Thomas was on vacation for a week, Mr. James took samples each day, but the sizes varied. The sample sizes and number of defects were:

Day	Sample size	Number of defects
Monday	75	6
Tuesday	115	5
Wednesday	75	11
Thursday	50	10
Friday	115	7

Is the process still in control using the average fraction defective computed in (a) but accounting for the variations in sample size?

P29.14 A consultant to the state college system wants to use the fraction-defective chart to monitor the dropout rate of students. The first-year class size and number of dropouts were determined for 12 semesters at one college.

Semester	Enrollment	Dropouts
1	1,500	270
2	1,725	360
3	1,380	275
4	1,790	320
5	1,850	390
6	1,635	160
7	1,800	270
8	1,905	380
9	1,625	260
10	1,550	295
11	1,725	430
12	1,850	375
	20,335	3,785

(a) Compute the p-chart control limits based on the average enrollment value $\bar{n}$ and plot the dropout rates on the chart.

(b) Determine the semesters for which the dropout rate is clearly outside the control limits and compute the exact limits for those semesters with rates close to a control limit.

(c) When a plotted value is just outside a control limit based on the average sample size, and the actual sample size n exceeds the average sample size $\bar{n}$, that is, $n > \bar{n}$, is it necessary to compute the exact control limit to determine that the point is actually out of control? Why or why not?

(d) Answer the questions in (c) if the point plots just inside the control limit and $n > \bar{n}$.

P29.15 A new zipper insertion method has been initiated at the Slinky Pants Company. The scrap rate for samples of 200 pants was recorded for 20 days (listed in order by row):

| 0.12 | 0.16 | 0.13 | 0.14 | 0.12 | 0.10 | 0.07 | 0.06 | 0.06 | 0.05 |
| 0.05 | 0.04 | 0.05 | 0.03 | 0.04 | 0.03 | 0.06 | 0.03 | 0.04 | 0.03 |

(a) Use the p-chart to obtain control by successively removing all points that are out of control during these 20 days.

(b) Use the new process average in (a) to determine if the next 5 days of production are in control when samples of 100 each have the following scrap rates: 0.04, 0.05, 0.03, 0.07, 0.02.

(c) After observing the 25 scrap rate values, explain what you think has happened to the zipper processing line.

P29.16 Use the np-chart to determine if the number of defectives in Prob. P29.12 is in control.

P29.17 For the dropout data in Prob. P29.14 develop the number-of-defectives chart based on the average sample size $\bar{n}$.

P29.18 If a process has an average fraction defective of 4 percent with $n = 200$, use the Poisson approximation to the binomial to estimate the probability that a point will fall above the upper control limit of the p-chart.

P29.19 Men's suits are inspected and the number of defects recorded. Determine if the number of defects for 25 suits (in order by row) is in control.

8	5	8	10	4	6	9	3	9	4
6	3	8	11	2	9	4	6	8	5
12	8	2	1	14					

P29.20 Crates of apples are delivered to a market each morning. The number of damaged fruit in 10 bushel baskets inspected each day is recorded for 12 days:

73	56	92	49	81	73
59	102	71	64	50	31

Is the number of damaged fruit in statistical control? Which days, if any, are out of control?

P29.21 Even though 10 baskets were supposed to be checked each day in Prob. P29.20, the technician was new to the job and inspected a different number on some days. The number inspected and the number of damaged fruit were actually recorded as follows:

Crates inspected	Damaged fruit
10	73
9	56
12	92
15	49
10	81
12	73
8	59
20	102
6	71
10	64
8	50
4	31

Determine if the conclusions about the statistical control of the number of damaged fruit are the same as in Prob. P29.20.

P29.22 Stereo components are inspected using a standard unit of eight components. Use the number of defects and number of standard units inspected to determine if the process is in control.

Day	Standard units	Number of defects
1	2.00	16
2	1.25	10
3	1.50	18
4	0.75	8
5	1.00	10
6	2.00	18
7	0.50	4
8	1.00	13

P29.23 Explain the use of an OC curve in acceptance sampling.

P29.24 Compute the probability of acceptance P_a for the following single sampling plans:
  (a) $n = 50, c = 2, p = 0.03$
  (b) $n = 175, c = 11, p = 0.08$
  (c) $n = 15, c = 0, p = 0.004$

P29.25 (a) Plot the OC curves for $n = 75, c = 1$ and $n = 50, c = 1$ on the same graph.
  (b) Compare the consumer's risk values for 8 percent defective lots.

P29.26 An electrical supply house inspects incoming switches using the sampling plan $n = 80, c = 3$ with an $AQL = 1.5$ percent. Compute the following.
  (a) Probability of accepting a lot which has a percent defective value equal to the AQL
  (b) Producer's risk of this plan for $p = 0.015$
  (c) The percent defective of a lot which has a 50 percent chance of being rejected
  (d) The percent defective for a consumer's risk of 0.10

P29.27 Draw the general shape of an OC curve which goes through the two points (0.012, 0.94) and (0.07, 0.05). State the exact meaning of the P_a values on the OC curve at these two points.

P29.28 The State of Buffaloed purchasing agency wants to purchase safe vehicles, so the brakes are thoroughly checked on new cars using an $n = 50, c = 1$ sampling plan. A new safety standard requires that all purchases be inspected using a plan which meets the following criteria as closely as possible:

$$LTPD = 7\% \quad \text{at } P_a = 0.05$$
$$AQL = 1\% \quad \text{at } P_a = 0.98$$

How does the present plan measure up against these criteria?

P29.29 Demonstrate the statement in Example 20.5b that the OC curve moves closer to the origin for constant n values as c decreases. Use $n = 100$ for your computations.

P29.30 Use MIL-STD-105D to find single-sampling plans for the inspection of bottles of chemicals which are received at the weekly rate of 150 cases of 24 bottles each. Use level II, normal inspection and an AQL of (a) 4 percent and (b) 2.5 percent.

P29.31 In MIL-STD-105D what happens to the acceptance number c when going from (a) normal to tightened and (b) normal to reduced inspection? (c) What effect should each of these have on the shape of the OC curve for a fixed-sample-size code letter?

P29.32 The following facts are stated: MIL-STD-105D; normal inspection; level II; single-sampling plans; and an AQL of 1.5 percent. At military depot A, 500 parts are received each week; and at depot B, 1500 are received.
  (a) What sampling plans should be used?
  (b) Which plan has the smaller producer's risk?

P29.33 If a lot of size 5000 is submitted to the Picky-Picky Company, the AQL used is 0.65 percent and the acceptance sampling plan is $n = 200, c = 3$. However, the same lot sent to the Liberal Company will be subjected to the sampling plan $n = 200, c = 14$ with $AQL = 4.0$ percent. Your company has in the past sent first-line merchandise to Picky-Picky and seconds to Liberal. Which company's sampling plan offers these products the larger probability of acceptance if the percent defectives equal the respective AQL values?

CHAPTER
THIRTY

ANALYSIS OF VARIANCE

This chapter discusses the technique and uses of analysis of variance (ANOVA). The mathematical model that is assumed, the hypothesis to be tested, and the complete ANOVA table are presented for one- and two-factor ANOVA. The ANOVA test to determine if the coefficients in a linear regression equation are significant is also covered.

CRITERIA

To complete this chapter, you must be able to:

1. State a fundamental model and hypothesis used in ANOVA. Also, be able to write the basic equation to partition the total variation of observed data, and write the general form of an ANOVA table.
2. Use the *single-factor, completely randomized* ANOVA procedure to test the hypothesis of no treatment effect, given the observed data for all treatments.
3. Use the *multiple linear regression* ANOVA procedure to test for the significance of a regression equation, given the observed values and the resulting regression equation with coefficients.
4. Use the *single-factor, randomized-blocks* ANOVA procedure to test the hypotheses of no treatment and no block effects, given the observed data for all treatments by blocks.
5. Define the terms *factorial experiment* and *interaction*, and use the *two-factor* ANOVA procedure to test the hypotheses of no treatment effect for each factor and no interaction, given the observed data for all levels of each factor.

STUDY GUIDE

30.1 The Analysis of Variance Rationale

To understand the basic rationale of analysis of variance (ANOVA), consider the following situation. Four machines can be used to manufacture the same aluminum product. Samples of five thicknesses i ($i = 1, 2, 3, 4, 5$) have been taken from each machine j ($j = 1, 2, 3, 4$). You want to know if the thicknesses from all machines are statistically equal or if there is a "machine effect" that makes some thicknesses different from others. We refer to the machines as *treatments*, because each machine is a different way of treating the product. If there are only two machines, the test of means in Chap. 20 can be used, but greater than two are analyzed more efficiently using the analysis of variance technique.

In this introductory section we briefly discuss (1) an elementary ANOVA model, (2) the hypothesis tested, (3) the partitioning of the total variation, and (4) how an ANOVA table is prepared and used to obtain results pertinent to the hypothesis. These are all necessary whenever an ANOVA is performed, regardless of the complexity of the experimental design.

For the four-machine problem above, a *linear model* may be formulated.

$$X_{ij} = \mu + T_j + \epsilon_{ij} \qquad (30.1)$$

where X_{ij} = ith observed value from machine j
μ = overall mean thickness for all machines
T_j = effect from machine j which is in addition to overall mean μ
ϵ_{ij} = random effect for the ith observation from machine j

For instance, if the mean thickness is 5 cm and product $i = 3$ made on machine $j = 2$ has a machine effect of -0.3 cm and a random effect of $+0.1$ cm, then $X_{32} = 5.0 - 0.3 + 0.1 = 4.8$ cm. The algebraic sum of all treatment effects is assumed to be 0 when ANOVA is used. This is written as

$$\sum_{j=1}^{k} T_j = 0$$

where $j = 1, 2, \ldots, k$ are the treatments, each of which comes from a separate population. These populations are the aluminum products made by each machine in this example.

The treatment effect T_j is a random variable, so Eq. (30.1) is sometimes called the random effects model. In ANOVA each T_j is assumed to have an $N(0, \sigma_T^2)$ distribution. Since the variance σ_T^2 is the same for all treatments, the *null hypothesis* tested is that no treatment effect is present. Therefore

$$H_0: T_j = 0 \quad \text{for all treatments } j$$

is tested against the alternative hypothesis

$$H_1: T_j \neq 0 \quad \text{for some treatment } j$$

If one or more machines make products that are statistically different from the others, H_0 is rejected.

The *partitioning of the total variation* is accomplished by separating the total sum of squares SS_T for the observed data. (If you read optional Sec. 27.9 on partitioning SS_T for regression, you will see that much the same logic is used here.) The SS_T is always the sum of squares of the differences between each X_{ij} and the grand mean $\bar{\bar{X}}$.

$$SS_T = \sum_j \sum_i \left(X_{ij} - \bar{\bar{X}} \right)^2 \qquad (30.2)$$

The grand mean $\bar{\bar{X}}$ is

$$\sum_{j=1}^{k} \sum_{i=1}^{n} \frac{X_{ij}}{kn} \qquad (30.3)$$

where there are k treatments and n observations per treatment. There are two sources of variation in the model Eq. (30.1) that comprise SS_T. One of them, the differences caused by the machine effect T_j, is the *between-treatment* sum of squares SS_{Tr}, which may be written

$$SS_{Tr} = \sum_j n \left(\bar{X}_j - \bar{\bar{X}} \right)^2 \qquad (30.4)$$

where n = number of observations per sample ($i = 1, 2, \ldots, n$)
$\bar{X}_j$ = average of observed values for each treatment j

The second is the random effect ϵ_{ij} that occurs *within* each machine (treatment) because of its specific characteristics. This results in the *within-treatment*, or error, sum of squares SS_E.

$$SS_E = \sum_j \sum_i \left(X_{ij} - \bar{X}_j \right)^2 \qquad (30.5)$$

where $\bar{X}_j$ is defined as before. It can be shown that SS_T is the sum of Eqs. (30.4) and (30.5):

$$SS_T = SS_{Tr} + SS_E$$

• $$\sum_j \sum_i \left(X_{ij} - \bar{\bar{X}} \right)^2 = \sum_j n \left(\bar{X}_j - \bar{\bar{X}} \right)^2 + \sum_j \sum_i \left(X_{ij} - \bar{X}_j \right)^2 \qquad (30.6)$$

These summations are used in later sections to obtain ANOVA results; however, they will be rewritten in easier-to-compute forms.

The ANOVA results are obtained by calculating an F distribution value for each source of variation except the total and random error. This requires that the mean square (MS) be computed from each sum of square except the total. The MS values are nothing more than variance estimates, so they are the sum of squares divided by the correct degrees of freedom ν. For MS_{Tr} the between-

● **Table 30.1 General form of an ANOVA table**

Source of variation	Degrees of freedom	Sum of squares	Mean squares	F value
Between treatments (T_j)	$k - 1$	SS_{Tr}	$MS_{Tr} = \dfrac{SS_{Tr}}{k-1}$	$\dfrac{MS_{Tr}}{MS_E}$
Error (ϵ_{ij})	$k(n-1)$	SS_E	$MS_E = \dfrac{SS_E}{k(n-1)}$	
Total	$kn - 1$	$SS_T = SS_{Tr} + SS_E$		

treatment degrees of freedom ν_{Tr} is the number of treatments k minus 1:

$$MS_{Tr} = \frac{SS_{Tr}}{\nu_{Tr}} = \frac{SS_{Tr}}{k-1} \tag{30.7}$$

For the error mean square ν_E is the total degrees of freedom, $nk - 1$, minus ν_{Tr}, or $\nu_E = k(n - 1)$.

Then
$$MS_E = \frac{SS_E}{\nu_E} = \frac{SS_E}{k(n-1)} \tag{30.8}$$

The results are all entered into an ANOVA table similar to Table 30.1. If the calculated F value exceeds the Table B-6 value for a selected α value with the degrees of freedom ν_{Tr} and ν_E, H_0 is not accepted and a treatment effect is present. In the problem presented, this conclusion means that at least one of the machines is producing different-thickness material.

Problem P30.1

30.2 Single-Factor (One-Way) ANOVA—Completely Randomized

The treatments are different levels of a *factor*, which is the thing of interest in an ANOVA. If only one factor, such as machine type, pressure level, temperature level, etc., is present, a single-factor ANOVA is used. This is also called a one-way ANOVA. Further, the design is completely randomized if no restrictions are placed on the selection of the random samples from the k treatment levels. This elementary ANOVA assumes the model in Eq. (30.1) with the results presented in the format of Table 30.1. It is possible to rewrite the sum-of-squares terms in Eqs. (30.2) through (30.5) in easier-to-use forms. The steps below present these rewritten forms and outline the procedure to perform a one-way ANOVA for $i = 1, 2, \ldots, n$ observations and $j = 1, 2, \ldots, k$ treatments.

1. Write the assumed model, the null hypothesis, and the α level.

2. Compute the sum of all observations and call it the correction term T:

$$T = \sum_j \sum_i X_{ij} \qquad (30.9)$$

3. Calculate the total sum of squares:

$$SS_T = \sum_j \sum_i X_{ij}^2 - \frac{T^2}{nk} \qquad (30.10)$$

where T is from Eq. (30.9). The degrees of freedom are $\nu_T = nk - 1$.

4. Calculate the between-treatment sum of squares:

$$SS_{Tr} = \sum_j \frac{X_j^2}{n} - \frac{T^2}{nk} \qquad (30.11)$$

where X_j^2 = square of sum of X_{ij} for each treatment j
$= \left(\sum_i X_{ij} \right)^2$

The degrees of freedom are $\nu_{Tr} = k - 1$.

5. Determine the error sum of squares and degrees of freedom by subtraction:

$$SS_E = SS_T - SS_{Tr} \qquad (30.12)$$
$$\nu_E = \nu_T - \nu_{Tr}$$

6. Place the results into Table 30.1 format and compute the mean square values by Eqs. (30.7) and (30.8).
7. Determine the F value for the treatment effect and compare with the tabulated value for $\nu_1 = k - 1$ and $\nu_2 = k(n - 1)$. Accept or reject the hypothesis of no treatment effect.

Example 30.1 illustrates this procedure.

Example 30.1 A safety engineer is testing four different types of smoke alarm systems. After installing five of each type in a smoke chamber, he introduced smoke to a uniform level, electrically connected the alarms, and observed the reaction time in seconds (Table 30.2). Is there a significant difference in the reaction time of the four types?

Table 30.2 Reaction time in seconds of smoke alarm systems, Example 30.1

Observation	Alarm type			
	1	2	3	4
1	5.2	7.4	3.9	12.3
2	6.3	8.1	6.4	9.4
3	4.9	5.9	7.9	7.8
4	3.2	6.5	9.2	10.8
5	6.8	4.9	4.1	8.5

SOLUTION The factor of interest is alarm type. No restrictions on randomization are present in the smoke chamber, so a single-factor, completely-randomized ANOVA is appropriate.

1. The model is
$$X_{ij} = \mu + T_j + \epsilon_{ij}$$
for $n = 5$ observations (subscript i) and $k = 4$ types. The null hypothesis is the equality of reaction times for all four alarm types:
$$H_0: T_1 = T_2 = T_3 = T_4 = 0$$
Let $\alpha = 0.05$ be the significance level.
2. From Eq. (30.9) the correction term is $T = 139.5$.
3. The sum of X_{ij}^2 is 1078.07, so the SS_T is from Eq. (30.10) and $\nu_T = (5)(4) - 1 = 19$.
$$SS_T = 1078.07 - \frac{(139.5)^2}{5(4)} = 105.06$$
4. The sums of the five reaction times for each type are squared and substituted into Eq. (30.11) for SS_{Tr}. The degrees of freedom are $\nu_{Tr} = 4 - 1 = 3$.

Type j	1	2	3	4
X_j	26.4	32.8	31.5	48.8
X_j^2	696.96	1075.84	992.25	2381.44

$$SS_{Tr} = \frac{1}{5}(696.96 + \cdots + 2381.44) - \frac{(139.5)^2}{(5)(4)}$$
$$= 1029.30 - 973.01 = 56.29$$
5. By subtraction the error sum of squares is
$$SS_E = 105.06 - 56.29 = 48.77$$
The degrees of freedom are $\nu_E = 19 - 3 = 16$.
6. Table 30.3 presents the ANOVA computations including the mean square values.

Table 30.3 ANOVA table for Example 30.1

Source of variation	Degrees of freedom	Sum of squares	Mean squares	F value
Between types	3	56.29	18.76	6.15
Error	16	48.77	3.05	
Total	19	105.06		

7. The value $F = 6.15$ for the treatment effect is compared with the Table B-6 value for $\alpha = 0.05$ with $\nu_1 = \nu_{Tr} = 3$ and $\nu_2 = \nu_E = 16$, which is $F = 3.24$. Since $6.15 > 3.24$, H_0 is rejected; there is a difference in reaction time between smoke alarm types.

COMMENT Since the mean squares are variance estimates (Sec. 30.1), why can they not be added to obtain a total variance value as the sum of squares are added? Here is why! The error mean square $MS_E = 3.05$ is an estimate $\hat{\sigma}_\epsilon^2$ of the variance of the error term ϵ_{ij}; but ϵ_{ij} is assumed to be $N(0, \sigma_\epsilon^2)$, where σ_ϵ^2 is the same for all treatments. Since each treatment effect is assumed to have an $N(0, \sigma_{Tr}^2)$ distribution, the between-type mean square $MS_{Tr} = 18.76$ is an estimate of the entire between-treatment variance; that is, $\sigma_\epsilon^2 + n\sigma_{Tr}^2$. The treatment variance estimate $\hat{\sigma}_{Tr}^2$ is calculated as

$$18.76 = \hat{\sigma}_\epsilon^2 + n\hat{\sigma}_{Tr}^2 = 3.05 + 5\hat{\sigma}_{Tr}^2$$

or

$$\hat{\sigma}_{Tr}^2 = \frac{18.76 - 3.05}{5} = 3.14$$

Now the *total variance estimate* $\hat{\sigma}_{Tot}^2$ for this data is the sum of the error and treatment variances.

$$\hat{\sigma}_{Tot}^2 = \hat{\sigma}_\epsilon^2 + \hat{\sigma}_{Tr}^2$$
$$= 3.05 + 3.14 = 6.19$$

Therefore, you can see that it is not possible to simply add the variance estimate terms given by the mean squares to obtain the total variance. Only the sum-of-squares terms can be added to obtain the total sum of squares.

Problems P30.2–P30.6

30.3 Analysis of Variance for a Linear Regression Equation

The single-factor ANOVA of Sec. 30.2 may be used to test the significance of a multiple (or simple) linear regression model of the general form

● $$Y_i = a + b_1 X_{i1} + b_2 X_{i2} + \cdots + b_k X_{ik}$$

where Y_i $(i = 1, 2, \ldots, n)$ is the observed value analogous to X_{ij} in Sec. 30.2. The b_j values, which are estimates of the β_j coefficients, are found by the procedures discussed in Chap. 27. The significance of the entire regression reduces to testing the hypothesis

$$H_0: b_1 = b_2 = \cdots = b_k = 0$$

Not accepting H_0 by the method of ANOVA means that at least one b_j is not statistically equal to 0, so at least one X_j is not independent of Y and therefore the regression is meaningful. Of course, several other X_j variables may add little or nothing to the explanation of the relationship with Y.

To perform on ANOVA on the regression equation, consider the treatments to be the regression coefficients b_j and use the Table 30.1 format, but replace the between-treatment variation with regression. The steps of Sec. 30.2 are rewritten here with the correct equations. A review of optional Sec. 27.9 is suggested before you go on.

1. Write the regression model, the null hypothesis, and the selected α level.
2. Compute the correction term T which is the sum of all observed values Y:

$$T = \sum_i Y_i$$

3. Calculate the total sum of squares using

$$SS_T = \sum_i Y_i^2 - \frac{T^2}{n} \qquad (30.13)$$

and the degrees of freedom $\nu_T = n - 1$.

4. Determine the error sum of squares by

$$SS_E = \sum_i (Y_i - \hat{Y}_i)^2 \qquad (30.14)$$

and the degrees of freedom $\nu_E = n - k - 1$, where k is the number of b values in H_0.

5. Compute the regression sum of squares SS_R by subtraction:

$$SS_R = SS_T - SS_E \qquad (30.15)$$

Degrees of freedom are $\nu_R = k$. Alternatively, SS_R may be computed using the equation

$$SS_R = \sum_i (\hat{Y}_i - \bar{Y})^2$$

6. Place the results into the Table 30.1 format and compute the mean squares for regression and error.
7. Determine the F value for the regression effect and compare with the Table B-6 value for $\nu_1 = \nu_R$ and $\nu_2 = \nu_E$. Accept H_0 if the computed value does not exceed the table value at the α significance level.

These steps are also used for simple linear regression equations with $k = 1$.

Example 30.2 In Example 27.4 a multiple linear regression was performed on 12 observations of three independent variables. The data and error sum of squares $SS_E = 984.65$ are shown in Table 27.4, and the regression equation is (subscript i omitted)

$$\hat{Y} = 83.08 - 0.32X_1 + 0.55X_2 + 0.18X_3 \qquad (30.16)$$

Use ANOVA to determine if this regression is significant.

SOLUTION The steps above are used with $n = 12$ observations and $k = 3$ treatments or variables.

Table 30.4 ANOVA table for multiple linear regression, Example 30.2

Source of variation	Degrees of freedom	Sum of squares	Mean squares	F value
Regression	3	1204.27	401.42	3.26
Error	8	984.65	123.08	
Total	11	2188.92		

1. The model is Eq. (30.16), and the null hypothesis is that statistically
$$H_0: b_1 = b_2 = b_3 = 0$$
Let $\alpha = 0.05$ be the significance level.
2. From Table 27.4, $T = 1231$.
3. By Eq. (30.13) the total sum of squares is
$$SS_T = 128{,}469 - \frac{(1231)^2}{12} = 2188.92$$
and
$$\nu_T = 12 - 1 = 11$$
4. The error sum of squares is $SS_E = 984.65$ with $\nu_E = 12 - 3 - 1 = 8$ degrees of freedom.
5. By the subtraction in Eq. (30.15)
$$SS_R = 2188.92 - 984.65 = 1204.27$$
with $\nu_R = 3$ degrees of freedom.
6. Table 30.4 presents the ANOVA results.
7. For $\alpha = 0.05$ with $\nu_1 = 3$ and $\nu_2 = 8$, Table B-6 gives $F = 4.07$. Since $3.26 < 4.07$, the null hypothesis that all regression coefficients are 0 is not rejected. This means that Y is independent of all X_j variables; that is, there is no treatment effect. Therefore, the regression Eq. (30.16) does not explain a significant amount of the variation in the data.

Problems P30.7–P30.11

30.4 Single-Factor (Two-Way) ANOVA—Randomized Blocks

Often the design of an experiment calls for tests to be separated into two or more areas (called *blocks* in ANOVA) because of the lack of room in one block. Some examples are:

Observed data	Treatments	Blocks
Alarm reaction time	Smoke alarms	Smoke chambers
Tire mileage	Tire brands	Cars
Defective product	Processing temperature	Processing lines

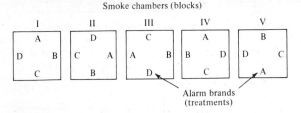

Figure 30.1 Four treatments (alarms) randomly placed within each block (chamber).

Because there are experiments performed in each block, there is possibly a block effect that must be included and analyzed in the ANOVA. As an example, suppose four smoke alarm brands are tested in five different smoke chambers rather than all in one chamber. The layout of alarms (Fig. 30.1) is randomized within each block (chamber), and one of each type of alarm is installed in each chamber.

In the ANOVA a single factor with treatment levels T_j ($j = 1, 2, \ldots, k$) is still of prime interest, but the assumed model now includes the block effect B_i ($i = 1, 2, \ldots, n$):

$$X_{ij} = \mu + T_j + B_i + \epsilon_{ij} \tag{30.17}$$

For the model in Eq. (30.17) all blocks contain k observations, one from each treatment. If not all treatments can be included in each block, an incomplete randomized-block analysis is performed (see the text by Hicks).

The null hypothesis of no treatment effect is, as before,

$$H_0: T_1 = T_2 = \cdots = T_k = 0$$

The hypothesis of no block effect, that is,

$$H_0: B_1 = B_2 = \cdots = B_n = 0$$

may also be tested. Because of T_j and B_i effects, the ANOVA is called *two-way* analysis, and the sum of squares for blocks SS_{Bl} is obtained by reducing only the error sum of squares SS_E, not any others. The steps given in Sec. 30.2 for one-way analysis are corrected with the following three alterations. First, rename step 4 to Step 4a and insert step 4b.

4b. Compute the between-block sum of squares

$$SS_{Bl} = \sum_i \frac{X_i^2}{k} - \frac{T^2}{nk} \tag{30.18}$$

where X_i^2 = squares of sum of X_{ij} for each block i

$$= \left(\sum_j X_{ij}\right)^2$$

The degrees of freedom are $\nu_{Bl} = n - 1$.

30.4 SINGLE-FACTOR (TWO-WAY) ANOVA—RANDOMIZED BLOCKS

Second, step 5 is rewritten to reduce the error sum of squares to accommodate SS_{Bl}:

5. Determine SS_E and ν_E by subtraction:

$$SS_E = SS_T - SS_{Tr} - SS_{Bl} \qquad (30.19)$$
$$\nu_E = \nu_T - \nu_{Tr} - \nu_{Bl} = (k-1)(n-1)$$

Finally, the degrees of freedom in step 7 for the F value are now $\nu_1 = k - 1$ and $\nu_2 = (k-1)(n-1)$.

The format of Table 30.1 is expanded to include the between-blocks variation as shown in Example 30.3.

Example 30.3 A two-way ANOVA on the smoke alarm reaction time data presented in Table 30.5 is to be performed. This is the same data as in Table 30.2 except that five chambers are used as pictured in Fig. 30.1. Perform the analysis for both treatment and block effects.

SOLUTION The results of the steps in Sec. 30.2 as updated above are summarized. If the computation was performed in Example 30.1, the result is repeated here.

1. The model is given by Eq. (30.17), and the treatment and block hypotheses are, respectively,

$$H_0: T_1 = T_2 = T_3 = T_4 = 0$$
$$H_0: B_I = B_{II} = B_{III} = B_{IV} = B_V = 0$$

Let $\alpha = 0.05$ for both tests.

2, 3, and 4a. From Example 30.1

$$T = 139.5 \qquad SS_T = 105.06 \qquad SS_{Tr} = 56.29$$

Table 30.5 Smoke alarm reaction times using five smoke chambers, Example 30.3

Block i (chamber)	Alarm type, j				Block total X_i
	1	2	3	4	
I	5.2	7.4	3.9	12.3	28.8
II	6.3	8.1	6.4	9.4	30.2
III	4.9	5.9	7.9	7.8	26.5
IV	3.2	6.5	9.2	10.8	29.7
V	6.8	4.9	4.1	8.5	24.3

4b. The block sum of squares from Eq. (30.18) uses the block totals given in Table 30.5:

$$SS_{Bl} = \frac{1}{4}(28.8^2 + \cdots + 24.3^2) - \frac{(139.5)^2}{5(4)}$$

$$= 979.08 - 973.01 = 6.07$$

The degrees of freedom are $\nu_{Bl} = 5 - 1 = 4$.

5. By the subtraction in Eq. (30.19)

$$SS_E = 105.06 - 56.29 - 6.07 = 42.70$$

The degrees of freedom are $\nu_E = 19 - 3 - 4 = 12$.

6. Table 30.6 presents the ANOVA results. All sum-of-squares terms are the same as in Table 30.3, except that SS_{Bl} has been removed from SS_E.

7. The treatment effect F value of 5.27 exceeds the Table B-6 value of 3.49 for $\nu_1 = 3$ and $\nu_2 = 12$. A treatment effect is present. This conclusion is predictable because the one-way ANOVA gave the same result and MS_E has decreased in this two-way analysis.

The null hypothesis of no block effect is accepted because 0.43 is less than 3.26, which is the Table B-6 value for $\nu_1 = 4$ and $\nu_2 = 12$ with $\alpha = 0.05$. In fact, there is no difference between smoke chambers simply because the F value is less than 1.

COMMENT In this problem and Example 30.1 the hypothesis of no treatment effect was rejected. It is quite possible that we could conclude no treatment effect in a one-way analysis and find that it is present in a two-way analysis simply because the block effect removes sums of squares and degrees of freedom from the error variation. Then the treatment F value MS_{Tr}/MS_E increases, thus increasing the possibility that the treatment H_0 is rejected in the two-way analysis. However, remember that the tabulated F value also increases as ν_2 is reduced from $k(n - 1)$ to $(k - 1)(n - 1)$.

Problems P30.12–P30.15

Table 30.6 ANOVA table for Example 30.3

Source of variation	Degrees of freedom	Sum of squares	Mean squares	F value
Between types (treatments)	3	56.29	18.76	5.27
Between chambers (blocks)	4	6.07	1.52	0.43
Error	12	42.70	3.56	
Total	19	105.06		

30.5 ANOVA for Multifactor Experiments

If two or more factors are involved in the ANOVA, it is called a factorial analysis.

> A *factorial experiment* is one in which all treatment levels of each factor are combined with all levels of all other factors with repeated observations for each combination.

Each factor combination with n observations is called a *cell*. A 2×3 factorial layout with three observations per cell is presented in Fig. 30.2. Factor A has two treatment levels and factor B has three levels, thus the name 2×3.

The assumed model is

$$X_{ijk} = \mu + A_i + B_j + AB_{ij} + \epsilon_{ijk} \qquad (30.20)$$

where AB_{ij} is the *interaction* between factors A and B. An interaction is present if the factors are dependent, that is, if the observed values for different levels of one factor are altered by the presence of another factor. Figure 30.3 shows the graphs of two factors each at two levels. If the observed values fall on lines which are approximately parallel, no interaction is present; but if the lines are not parallel or cross, interaction is present. The ANOVA presented here tests for these interactions.

The ANOVA for two factors is developed below. The hypotheses tested are factor A effect, factor B effect, and interaction AB effect. Table 30.7 presents the general format for the ANOVA results. The between-treatment sum of squares SS_{Tr} used in the single-factor analyses has been separated into the factor and interaction sums of squares SS_A, SS_B, and SS_{AB}. That is, SS_{Tr} is the sum of these three terms. The degrees of freedom $\nu_{Tr} = ab - 1$ is also the sum of the degrees-of-freedom values for these sources of variation.

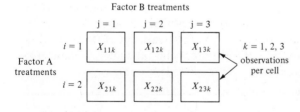

Figure 30.2 Layout for a two-factor experiment with three observations per cell. This is a 2×3 factorial experiment.

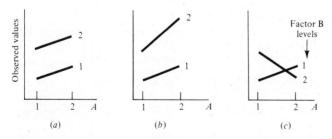

Figure 30.3 Factors A and B with (a) no interaction, (b) interaction with magnitude difference, and (c) interaction with direction change.

● **Table 30.7 Format for a two-factor ANOVA**

Source of variation	Degrees of freedom	Sum of squares	Mean squares	F value
Factor A	$a - 1$	SS_A	$MS_A = \dfrac{SS_A}{a - 1}$	$\dfrac{MS_A}{MS_E}$
Factor B	$b - 1$	SS_B	$MS_B = \dfrac{SS_B}{b - 1}$	$\dfrac{MS_B}{MS_E}$
AB interaction	$(a - 1)(b - 1)$	SS_{AB}	$MS_{AB} = \dfrac{SS_{AB}}{(a - 1)(b - 1)}$	$\dfrac{MS_{AB}}{MS_E}$
Error	$ab(n - 1)$	SS_E	$MS_E = \dfrac{SS_E}{ab(n - 1)}$	
Total	$abn - 1$	SS_T		

The procedure below is used to perform a two-factor ANOVA with sums taken over $i = 1, 2, \ldots, a$ (factor A); $j = 1, 2, \ldots, b$ (factor B); and $k = 1, 2, \ldots, n$ (observations per cell). As in previous procedures, the absence of the subscript i, j, or k on an X^2 means the sum must be taken over it before squaring.

1. State the model, hypothesis, and α level.
2. Compute the correction term:

$$T = \sum_i \sum_j \sum_k X_{ijk} \qquad (30.21)$$

3. Calculate the total sum of squares:

$$SS_T = \sum_i \sum_j \sum_k X_{ijk}^2 - \frac{T^2}{nab} \qquad (30.22)$$

4. Determine the factor A sum of squares:

$$SS_A = \sum_i \frac{X_i^2}{nb} - \frac{T^2}{nab} \qquad (30.23)$$

5. Determine the factor B sum of squares:

$$SS_B = \sum_j \frac{X_j^2}{na} - \frac{T^2}{nab} \qquad (30.24)$$

6. Calculate the between-treatment sum of squares and use it to obtain the AB interaction sum of squares by subtraction:

$$SS_{\text{Tr}} = \sum_i \sum_j \frac{X_{ij}^2}{n} - \frac{T^2}{nab} \qquad (30.25)$$

$$SS_{AB} = SS_{\text{Tr}} - SS_A - SS_B \qquad (30.26)$$

7. Compute the error sum of squares by subtraction.

$$SS_E = SS_T - SS_{\text{Tr}} \qquad (30.27)$$

8. Place the results in the Table 30.7 format, compute the degrees of freedom and mean square values.
9. Calculate the F values for factors A and B and the interaction AB. Reject or do not reject each null hypothesis.

Example 30.4 illustrates a two-factor ANOVA.

Example 30.4 Life tests have been performed on spacecraft solar power generators. Factor A, hours used per week, has two levels, and factor B, collector size, has three levels. Three observations per combination (cell) have been made to estimate the hours of useful life. Table 30.8 presents these values coded in thousands of hours.
(a) Use the two-factor ANOVA procedure to test for factor and interaction effects.
(b) Plot the cell totals to graphically verify the interaction decision made above.

SOLUTION (a) Table 30.9 gives all the necessary sums and squares to perform the ANOVA. The following subscript legend is good for reference.

Subscript	Variable	Upper Limit
i	Factor A	$a = 2$
j	Factor B	$b = 3$
k	Cell values	$n = 3$

1. Equation (30.20) is the model, and the three null hypotheses tested at $\alpha = 0.05$ are

$$H_{01}: \text{all } A_i = 0 \quad H_{02}: \text{all } B_j = 0 \quad H_{03}: \text{all } AB_{ij} = 0$$

2. Using Eq. (30.21), $T = 127.3$.

Table 30.8 Estimated life (in 1000 h) of power generators with two factors varied

Factor A	Factor B		
	1	2	3
1	3.4	7.6	6.1
	5.1	12.4	2.1
	4.2	8.9	7.5
2	2.9	5.6	8.1
	4.8	11.2	12.0
	10.0	9.5	5.9

Table 30.9 Sums necessary to perform a two-factor ANOVA, Example 30.4

Factor A (A_i)	Sums	Factor B (B_j) 1	2	3	X_i	X_i^2
1	X_{1jk}	3.4 5.1 4.2	7.6 12.0 8.9	6.1 2.1 7.5		
	X_{1j}	12.7	28.5	15.7	56.9	3237.61
	X_{1j}^2	161.29	812.25	246.49		
2	X_{2jk}	11.2 12.4 10.0	5.6 2.9 9.5	8.1 4.8 5.9		
	X_{2j}	33.6	18.0	18.8	70.4	4956.16
	X_{2j}^2	1128.96	324.0	353.44		
	X_j	46.3	46.5	34.5	127.3	8193.77 = ΣX_i^2
	X_j^2	2143.69	2162.25	1190.25		5496.19 = ΣX_j^2

3. If all X_{ijk} values in Table 30.9 are squared, Eq. (30.22) is used to obtain

$$SS_T = 1066.73 - \frac{(127.3)^2}{18} = 166.44$$

4. The factor A sum of squares from Eq. (30.23) is

$$SS_A = \frac{8193.77}{9} - \frac{(127.3)^2}{18} = 10.13$$

5. The factor B sum of squares is

$$SS_B = \frac{5496.19}{6} - \frac{(127.3)^2}{18} = 15.74$$

6. The between-treatment value is used only to obtain the interaction sum of squares.

$$SS_{Tr} = \frac{1}{3}(161.29 + 812.25 + \cdots + 353.44) - \frac{(127.3)^2}{18}$$
$$= 1008.81 - 900.29 = 108.52$$
$$SS_{AB} = 108.52 - 10.13 - 15.74 = 82.65$$

7. The error sum of squares by Eq. (30.27) is

$$SS_E = 166.44 - 108.52 = 57.92$$

8. The ANOVA results are given in Table 30.10.

Table 30.10 Two-factor ANOVA for Example 30.4

Source of variation	Degrees of freedom	Sum of squares	Mean squares	F value	Result
A	1	10.13	10.13	2.10	Accept H_{01}
B	2	15.74	7.87	1.63	Accept H_{02}
AB	2	82.65	41.33	8.56	Reject H_{03}
Error	12	57.92	4.83		
Total	17	166.44			

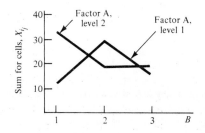

Figure 30.4 Interaction graphs of the sum of observed values per cell, Example 30.4.

9. The hypotheses of no factor A and factor B effect are accepted because the F values do not exceed the Table B.-6 values for $\alpha = 0.05$. Therefore, neither of the two usage levels in factor A significantly changes the generator's life, nor does collector size (factor B). However, there is a significant AB interaction term, as expected because collector size and required usage are necessarily related.

(b) Figure 30.4 is a plot of the cell totals X_{ij} from Table 30.9 at each level of factors A and B. Since the lines cross, an interaction term is expected to be present.

It is possible to analyze the effects of any number of factors and interactions by ANOVA. More advanced models are discussed in several of the bibliography texts. Computerized analysis by ANOVA is well developed and should be used whenever the problem becomes too large for simple manual solution.

Problems P30.16–P30.20

ADDITIONAL MATERIAL

Elementary ANOVA (Secs. 30.1, 30.2, 30.4): Bowker and Lieberman [4], pp. 377–405; Gibra [8], Chap. 13; Hicks [11], Chaps. 3, 4; Hines and Montgomery [12], Chap. 11; Kirkpatrick [15] pp. 353–366; Miller and Freund [18], Chap. 12; Siegel [21], pp. 184–194; Snedecor and Cochran [22], Chaps.

10, 11; Steel and Torrie [23], Chap. 7, pp. 132–145; Volk [24], pp. 149–196.

Regression ANOVA (Sec. 30.3): Draper and Smith [6], pp. 53–56, Chap. 9; Hicks [11], pp. 35–41, 52–54; Hines and Montgomery [12], pp. 330–334, 347–353; Snedecor and Cochran [22], Chap. 13; Steel and Torrie [23], pp. 287–303.

Advanced ANOVA (Sec. 30.5 and more): Bowker and Lieberman [4], pp. 405–436; Hicks [11], Chaps. 5 to 17; Hines and Montgomery [12], Chap. 13; Hogg and Craig [14], pp. 320–326; Miller and Freund [18], Chap. 13; Snedecor and Cochran [22], Chaps. 12, 14, 16; Steel and Torrie [23], pp. 146–159, Chaps. 11 to 13, 15; Volk [24], pp. 196–258.

PROBLEMS

P30.1 Write out from memory (*a*) the basic linear model hypothesized for analysis of variance and (*b*) the complete general form of an ANOVA table which includes total, between-treatment, and within-treatment variation.

P30.2 A nationwide advertising campaign was conducted by a firm to boost sales of a shampoo. Twenty-five cities in five different regions were randomly selected and the campaigns conducted in a completely randomized order. The percentage increase in sales by region is given:

		Region		
I	II	III	IV	V
2.4%	4.8%	7.2%	3.8%	2.0%
3.8	1.6	0.2	9.2	1.9
2.9	0.2	5.3	1.4	5.4
4.6	3.9	2.9	0.1	7.2
3.5	5.1	1.6	1.5	3.7

Use the one-way analysis-of-variance procedure to determine if the average percentage increase in sales was the same in all regions using a 5 percent level of significance.

P30.3 Use the summation terms in Prob. P30.2 to show that:
   (*a*) The total sum of squares by Eq. (30.10) is the same as by Eq. (30.2).
   (*b*) The between-treatment sum of squares by Eq. (30.11) is the same as by Eq. (30.4).
   (*c*) The error sum of squares resulting from the subtraction in Eq. (30.12) is the same as by Eq. (30.5).

P30.4 A fertilizer, which is supposed to contain 20 percent nitrogen, was tested after being stored for 1 year. In the experiment samples from four different manufacturers were tested using specimens from eight bags for each manufacturer. The results for the percentage of nitrogen were:

				Manufacturer			
1		2		3		4	
21	12	14	20	25	24	14	26
25	15	16	19	22	20	18	25
19	25	19	15	19	18	22	22
22	20	22	18	26	15	12	20

If $\alpha = 0.05$ and if the samples were selected at random from one warehouse, test the hypothesis that average potency retention of nitrogen does not vary with manufacturer.

P30.5 In Prob. P20.19 the average number of inoperable cars in two government car pool lots was found to be equal with $\alpha = 0.10$ ($t = 0.195$).

(a) Use one-way analysis of variance to test for the equality of means for the data in Prob. P20.19.

(b) In optional Sec. 15.5 the relation between the F and t distributions $F(1, \nu) = [t(\nu)]^2$ for a given ν was stated. Show that this relation is correct for the analysis-of-variance results above.

P30.6 An electrical engineer determined the magnetic induction in webers per square meter (Wb/m^2) around similar magnets from three different sources. The readings were obtained for six specimens from each source. Is the average strength of the field the same for all sources if the experiment was completely randomized? Let $\alpha = 0.02$.

	Source		
Number	1	2	3
1	1.5	1.9	1.1
2	1.4	1.2	2.0
3	1.0	2.0	1.5
4	2.6	1.5	1.7
5	1.0	1.2	1.9
6	2.5	1.8	2.1

P30.7 Use a confidence level of 95 percent to test the significance of the linear regression in Prob. P27.3.

P30.8 Test the significance of the linear regression determined in Prob. P27.4. Let $\alpha = 0.05$.

P30.9 The simple linear regression for the data given is $\hat{Y} = -0.43 + 2.11X$.

X	Y
4	9
2	4
6	12
3	5

(a) Use the analysis-of-variance procedure for linear regression to test the significance of this regression. Be sure to write out the null hypothesis. Let $\alpha = 0.05$.

(b) The hypothesis tested above can also be evaluated using the procedure in Sec. 27.4. Perform this test and show that the square of the t statistic is equal to the F statistic in (a).

P30.10 Test the significance of the multiple linear regression equation determined in Prob. P27.16 at the 5 percent level of significance

P30.11 Is the regression equation in Prob. P27.17 significant if $\alpha = 0.10$?

P30.12 The temperature in degrees Celsius at which a particular chemical reaction takes place was recorded for five different catalysts. The refining company had the test repeated at each of its three refineries to be sure the reaction took place at all of them. Use a 10 percent significance level to determine if (a) the average temperature is the same for all catalysts and (b) the location causes a refinery effect to be present.

	Catalyst type				
Refinery	1	2	3	4	5
U.S.	66°C	58°C	70°C	64°C	68°C
Canada	71	74	75	69	69
Mexico	54	60	62	59	67

P30.13 In Prob. P30.2 the average increase in sales resulting from advertising campaigns in five different regions was analyzed by a one-way ANOVA. Because of the geographic location of the 25 cities studied, five separate divisions of the advertising firm were involved. Identify the percentage increases given in Prob. P30.2 by division by adding the column $i = 1, 2, 3, 4, 5$ down the left side of the data.

(a) Perform a two-way ANOVA with $\alpha = 0.05$ to evaluate the region and division effects.

(b) Is the treatment effect conclusion here the same as in Prob. P30.2?

P30.14 The magnetic inductions in Prob. P30.6 were taken 6 at a time using magnets from all three sources; however, they were taken at six different locations in the plant.

(a) If the numbers indicate the correct location of the tests, use analysis of variance to determine if the average strength of the field is the same for all sources. Let $\alpha = 0.02$.

(b) Using the six different locations as the blocks in this randomized-block design, construct a drawing showing one possible layout of the three sources randomized at each location.

P30.15 A total of 16 people took a speed reading course using four different techniques. The length of the in-class time was varied from 2 weeks to 5 weeks. A randomized-block design was used for the experiment, which was performed primarily to study the effect of the different techniques on the average percentage increase in reading speed.

(a) Use a two-way analysis of variance to determine if the average percentage increase is the same for all techniques at the 5 percent significance level.

(b) Is there a block effect present?

	Technique's name			
Weeks	Reading dynamo (1)	Reading upgrade (2)	Fast-n-dirty (3)	Rapid read (4)
2	75%	60%	82%	90%
3	90	70	95	70
4	90	100	110	95
5	105	80	80	115

(c) Of the 16 people involved in this test there were 4 each of male teenagers (MT), female teenagers (FT), female adults (FA), and male adults (MA). Draw a possible layout for the experiment using a randomized-block design.

P30.16 State the meaning of the terms (a) *factorial experiment* and (b) *factor interaction*. (c) Write all the sources of variation for a three-factor analysis of variance with the factors being A, B, and C.

P30.17 The time for equally skilled stenographers to type a particular technical letter was observed using the factors of typewriter brand and noise environment. There were three brands (factor A), four different noise levels (factor B), and three observations per cell. The times in minutes are recorded:

	Factor A		
Factor B	1	2	3
1	7.8 8.3 7.2	10.9 8.3 7.4	7.5 7.2 9.1
2	8.4 9.2 7.5	7.4 11.8 9.4	8.2 10.1 9.4
3	6.5 7.4 9.3	13.1 8.2 7.5	6.1 8.5 10.8
4	10.4 8.9 7.3	10.2 8.9 9.4	8.1 7.2 11.8

(a) Plot the average typing time observed in each cell and use this graph to judge if there is an interaction effect.

(b) Use a two-factor ANOVA to test for factor and interaction effects at the 5 percent level of significance.

P30.18 This problem is designed to show you the relation between sum-of-square terms in a one-way and a two-way ANOVA. The reaction time in seconds when two chemicals are mixed is observed at three different temperatures. The experiment is repeated six times for each treatment (temperature):

	Temperature		
Observation	65°C	70°C	75°C
1	10	8	7
2	13	7	9
3	15	11	8
4	12	10	9
5	10	8	10
6	18	12	12

Use the one-way ANOVA to (a) compute the total sum of squares SS_{T1} and (b) the treatment sum of squares SS_{Tr1}.

Now assume that the same reaction times are observed, but two factors are present: factor A is temperature at three levels, factor B is humidity at two levels, and there are three observations per cell.

Humidity	Temperature (factor A)		
(factor B)	65°C	70°C	75°C
Low	10 13 15	8 7 11	7 9 8
High	12 10 18	10 8 12	9 10 12

Do the following for the two-way ANOVA:
(c) Compute the total sum of squares SS_{T2}. Compare with SS_{T1}.
(d) Compute the factor sum of squares SS_A and SS_B.
(e) Compute the interaction sum of squares using the formula

$$SS_{AB} = \sum_i \sum_j \frac{X_{ij}^2}{n} - \sum_i \frac{X_i^2}{nb} - \sum_j \frac{X_j^2}{na} + \frac{T^2}{nab}$$

(f) Compute the treatment sum of squares SS_{Tr2}.
(g) Show that the following relation is correct

$$SS_{Tr2} = SS_A + SS_B + SS_{AB}$$

(h) Compare the treatment sum of squares SS_{Tr1} and SS_{Tr2}. Do they have the same numerical value? Explain why using the data tables above.

P30.19 The hothouse germination rate per 100 pine tree seeds was observed under the factors of insect treatment and depth of planting. The levels are:

Factor A: Untreated, treated chemically, treated organically

Factor B: 0.6, 1.3, 1.9, and 2.2 cm

(a) Perform a two-way analysis of variance on this data at the 2 percent level of significance.
(b) Plot the cell totals and comment on the interaction effect conclusion made above.

Factor A	Factor B			
	1	2	3	4
1	72	50	74	65
	71	68	85	75
	68	72	81	80
	70	52	88	70
2	95	71	81	75
	64	82	75	82
	74	85	89	56
	80	70	92	75
3	78	90	84	89
	82	65	90	62
	90	71	75	70
	65	82	82	94

P30.20 The refining company in Prob. P30.12 has decided that the refinery location should be treated as a factor, not merely as a block effect.
(a) What implication does this have on the analysis-of-variance procedure you would use to analyze the effects of the two factors, catalyst type and refinery location?
(b) The experiment was repeated and the following results obtained:

Refinery location	Catalyst type				
	1	2	3	4	5
U.S.	66°C	58°C	70°C	64°C	68°C
	60	74	72	62	69
Canada	71	74	75	69	69
	75	65	72	71	67
Mexico	54	60	62	59	67
	56	72	61	63	61

Perform a two-way analysis of variance to determine the effect of catalyst type, refinery location, and interaction of catalyst and refinery. Let $\alpha = 0.05$.

(c) Plot cell totals and comment on the conclusion about the interaction effect made in (b).

EPILOG:
WHAT IS NEXT IN STATISTICS?

You have learned a lot about statistics in this text, but this is only the beginning. There are many applications and theoretical developments not even introduced here. Depending on your likes, dislikes, and future need for statistics, you can continue to study statistical applications and theory. This concluding chapter gives you the names of some of these areas. For each application there is, of course, background statistical theory, but you should now possess the knowledge to understand virtually any of these applications. The statistical theory areas mentioned are based on the material in this book. Most of the theory, you will discover, is used to develop techniques covered in this text or some application discussed below.

All the material is easily found in texts on statistical applications and mathematical statistics. None of these are beyond your comprehension if you have a good understanding of the subjects covered in this book.

Further Statistics Applications

There are a lot of applications of statistics beyond what you have learned thus far in the areas of:

Quality control using process control and acceptance sampling techniques
Analysis of variance for many different experimental designs which have real-world complications not accounted for in simple ANOVA methods
Regression and correlation techniques for nonlinear and stepwise methods, many of which use already developed computer software (programs)
Nonparametric testing procedures usable when the necessary underlying distributions cannot be assumed
Extensive applications of the normal and other distributions in the areas of mechanical design and safety factor determination; production and inventory control; work methods analysis; chemical analysis; noise, air, and

water pollution; actuarial and insurance risk analysis; business and tax analysis; etc.

A few areas not discussed here directly, but which may interest you, are:

Reliability analysis, which investigates the probability that a component or item will operate successfully for at least a stated period of time
Queuing analysis, which is the study of waiting lines at all types of service counters
Forecasting, econometric, and time-series analysis used to account, as best possible, for past trends, seasonal effects, and random fluctuations in data when predicting the future
Analysis of covariance, which is an extension of ANOVA to include variables that cannot be controlled in the experimental design
Simulation techniques and the method of Monte Carlo sampling from probability distributions which represent observed phenomena

Further Statistical Theory

You may want to learn more about the development of areas such as:

Probability theory
Distribution theory to include new distributions, multivariate distributions, and additional parameters which may be included in ones already discussed
Generating functions for distributions to include moment generating and characteristic functions
Parameter estimators derived by methods such as maximum likelihood, the method of moments, and Bayesian statistics
Design of best tests to be used in hypothesis testing procedures
Transforms of one or more variables to determine the distribution of the function of variables

Some advanced areas of statistical theory not discussed here are named. Most of these are covered in texts on mathematical statistics.

Stochastic processes
Stochastic convergence for limiting distributions
Renewal theory
Order statistics and their distributions
Markov theory and Markov chains
Queuing theory used to study waiting lines
Multivariate theory behind ANOVA and other techniques

Many more topics could be listed in each area, but this gives you an idea of the tremendous breadth of subjects in which statistics plays a very important part.

APPENDIX
A

FINAL ANSWERS TO SELECTED PROBLEMS

Chapter 1

P1.1 (*a*) Probabilistic; (*b*) deterministic; (*c*) probabilistic; (*d*) probabilistic; (*e*) deterministic; (*f*) probabilistic. (Sec. 1.1)

P1.5 One in eleven. (Sec. 1.2)

P1.6 No, should be 50 heads and 50 tails. (Sec. 1.2)

P1.9 (*a*) For weeks 1 through 12 the probabilities are 0.80, 0.72, 0.79, 0.70, 0.77, 0.77, 0.73, 0.77, 0.73, 0.74, 0.76 and 0.75; (*b*) 0.75. (Sec. 1.3)

P1.10 (*a*) 0.231, frequency; (*b*) 0.167, classical; (*c*) 0.21, frequency. (Sec. 1.3)

P1.13 (*a*) Sample; (*b*) no. (Sec. 1.5)

P1.16 (*a*) True; (*b*) false; (*c*) false; (*d*) true; (*e*) false. (Sec. 1.6)

P1.19 (*a*) Descriptive probability; (*b*) computational probability; (*c*) descriptive statistics; (*d*) descriptive statistics; (*e*) inferential statistics. (Sec. 1.7)

Chapter 2

P2.2 Frequency values for this ungrouped data are (Sec. 2.2)

Bacteria ($\times 1/100$)	Frequency	Bacteria ($\times 1/100$)	Frequency
24	3	30	5
25	5	31	4
26	0	32	4
27	3	33	0
28	2	34	2
29	2		

P2.4 (*a*) $c = 78.5$ Hz. If 80 Hz is used, boundaries may be 127.5–207.5, (Sec. 2.3)
207.5–287.5, ... , 847.5–927.5; (*b*) $c = 52.3$ Hz. If 50 Hz is used, 16 cells
may be defined as 127.5 – 1777.5, 177.5 – 227.5, ... , 877.5 – 927.5.

P2.5 For $c = 80$ Hz, the boundaries are 128–208, 208–288, ... , 848– (Sec. 2.3)
928.

P2.6 (*a*) $c = 0.12°$ (Sec. 2.3)

Cell	Boundaries	Midpoint
1	1.405–1.525	1.465
2	1.525–1.645	1.585
3	1.645–1.765	1.705
4	1.765–1.885	1.825
5	1.885–2.005	1.945
6	2.005–2.125	2.065

(*b*) Move all values up or down 0.005°.

P2.9 (*a*) (Sec. 2.4)

Histogram cells	Polygon midpoint	Frequency
1–4	2.5	2
4–7	5.5	8
7–10	8.5	15
10–13	11.5	23
13–16	14.5	30
16–19	17.5	22
19–22	20.5	35
22–25	23.5	12
25–28	26.5	3

(*b*) 25.3%.

P2.11 (*a*) Graph the frequencies in data set *D*; (Sec. 2.4)
(*b*)

Rating, %	Frequency
30–50	10
50–70	42
70–90	11
90–110	2
110–130	0
130–150	1

Characteristic shape of the histogram is unchanged.

P2.12 (a)   (Sec. 2.4)

Cells, kpl	Midpoints, kpl	Frequency
5.5– 6.0	5.75	2
6.0– 6.5	6.25	7
6.5– 7.0	6.75	7
7.0– 7.5	7.25	6
7.5– 8.0	7.75	8
8.0– 8.5	8.25	11
8.5– 9.0	8.75	12
9.0– 9.5	9.25	12
9.5–10.0	9.75	8
10.0–10.5	10.25	7

P2.14 (a) Discrete and grouped;   (Sec. 2.5)
(b)

Days	Cumulative frequency	Days	Cumulative frequency
3	5	10	43
4	9	13	55
5	19	18	62
7	34	20	72

(c) 26.4 percent.

P2.15 (a) 125; (b) 48; (c) 16; (d) 47.   (Sec. 2.5)

P2.16 (a) 540; (b) 20; (c) 95; (d) 0.067; (e) 0.0125 g.   (Sec. 2.5)

P2.18 (a)   (Sec. 2.5)

kpl	Cumulative frequency	kpl	Cumulative frequency
5.5–6.0	2	8.0– 8.5	41
6.0–6.5	9	8.5– 9.0	53
6.5–7.0	16	9.0– 9.5	65
7.0–7.5	22	9.5–10.0	73
7.5–8.0	30	10.0–10.5	80

(b) 9.75 kpl.

FINAL ANSWERS TO SELECTED PROBLEMS **593**

P2.19 (a) (Sec. 2.6)

Days	Frequency	Relative frequency	Cumulative frequency	Cum. rel. frequency
3	5	0.07	5	0.07
4	4	0.06	9	0.13
5	10	0.14	19	0.27
7	15	0.21	34	0.48
10	9	0.12	43	0.60
13	12	0.16	55	0.76
18	7	0.10	62	0.86
20	10	0.14	72	1.00
	72	1.00		

P2.21 (a) (Sec. 2.6)

Upper cell boundary	Cum. rel. frequency	Upper cell boundary	Cum. rel. frequency
4	0.01	19	0.67
7	0.07	22	0.90
10	0.17	25	0.98
13	0.32	28	1.00
16	0.52		

(b) 0.48.

P2.22 (a) 0.9; (b) 0.03; (c) 0.16; (d) 0.07; (e) 0.0125 g. (Sec. 2.6)
P2.25 (a) 24.9; (b) 63.5; (c) 50; (d) between 24.9 and 77.8 items. (Sec. 2.7)
P2.26 (a) 2500; (b) 3100 per liter; (c) 45.16 and 48.39. percentiles. (Sec. 2.7)
P2.27 (a) 0.009; (b) 0.025; (c) between 0.0060 and 0.0125 g; (d) P_{90}. (Sec. 2.7)
P2.28 (a) 70; (b) 0.087 percent; (c) between 0.031 percent and 0.108 percent; (d) 0.09. (Secs. 2.5 to 2.7)

Chapter 3

P3.1 15.11 h. (Sec. 3.1)
P3.2 (a) 5.98 V; (b) 5.98 V; (c) yes. (Sec. 3.2)
P3.3 (a) 6.0 V; (b) midpoint approximations used. (Sec. 3.2)
P3.4 1279 spectators. (Sec. 3.2)
P3.6 31 days. (Sec. 3.3)
P3.7 (a) 5.8 and 6.1 V; (b) 5.9–6.2 V; (c) yes. (Sec. 3.3)
P3.8 Best estimate is 1625 spectators. (Sec. 3.3)

P3.10 (a) True; (b) false; (c) false; (d) false. (Sec. 3.4)
P3.12 (a) and (b) 1.93°C; (c) 1.83°C, 5.2 percent underestimate. (Sec. 3.5)
P3.13 (a) 0.22 V; (b) 75 percent; (c) 0.05 (V)^2. (Sec. 3.5)
P3.16 s^2 = 212,181.56 (spectators)^2; s = 460.63 spectators. (Sec. 3.6)
P3.17 (a) s = 38.45 kg/cm^2; (b) yes with 26.7 percent out. (Sec. 3.6)
P3.20 1750 spectators. (Sec. 3.7)
P3.21 (a) $\bar{X}$ = 95.27 dBa, s = 7.78 dBa, R = 25 dBa; (b) no; (c) bi-modal or U-shaped. (Sec. 3.7)
P3.24 (a) $\bar{X}$ = 1279, M = 1375, m = 1625; (b) 357 to 2201; (c) 587 to 1971. (Sec. 3.8)

Chapter 4

P4.2 k = 2 and $\bar{X}$ = 0. (Sec. 4.1)
P4.3 (a) 310; (b) 133. (Sec. 4.1)
P4.4 (a) 5975; (b) 21. (Sec. 4.1)
P4.6 (a) Old: $a_3 = -0.28$, new: $a_3 = -0.09$; (b) old system shows negative skewness compared to new system. (Sec. 4.2)
P4.7 (a) -0.28, left. (Sec. 4.2)
P4.9 (a) 2.43; (b) less peaked than a normal. (Sec. 4.3)
P4.10 Old: a_4 = 2.63, new: a_4 = 2.21, new is flatter. (Sec. 4.3)
P4.11 (a) 7.01 m^2; (b) 8.76 m^2. (Sec. 4.4)
P4.13 Lathe 2. (Sec. 4.4)
P4.14 (a) Annual estimate = \$187.08, which is cheaper than \$250; (b) $\frac{1}{6}$ for each month for the cost data given. (Sec. 4.5)
P4.15 7.15 percent. (Sec. 4.5)
P4.18 (a) v_1 = 39.9 percent, v_2 = 45.7 percent. (Sec. 4.6)
P4.19 (a) $c = 10^{-6}$, X_0 = 0.008590, d = 6; (b) 500.195. (Sec. 4.7)
P4.20 (c) $\bar{X}$ = 136.13 g, s = 0.31 g. (Sec. 4.7)
P4.22 (a) $\bar{d}$ = 0.014, $\bar{X}$ = 0.514; (b) $\bar{d} = -0.086$, $\bar{X}$ = 0.514; (c) no. (Sec. 4.7)
P4.23 $\bar{X}$ = 129.762 m^2, s = 9.497 m^2. (Sec. 4.8)
P4.24 (a) and (b) $\bar{X}$ = 3.860 km/s^2. (Sec. 4.8)
P4.25 (a) and (b) $\bar{X}$ = 84.246 s, s = 1.93 s. (Sec. 4.8)

Chapter 5

P5.1 (a) A = {2, 3, 5, 7, 11, 13, 17, 19, 23}; (b) B = {4, 8, 12, 16, 24}; (c) C = {2, 12, 20, 21, 22, 23, 24, 25}. (Sec. 5.1)
P5.2 (a) A = {0, 1, ..., 10}; (b) B = {0, 1/10, ..., 1}; (c) C = {0, 1, 2, 3, 4, 5}. (Sec. 5.1)
P5.4 (a) $A \cup B \cup C$ = {2, 3, 4, 5, 7, 8, 11, 12, 13, 16, 17, 19 to 25}; (b) (1) $B \cap C$ = {12, 24}; (2) $(B \cup C)'$ = {1, 3, 5, 6, 7, 9, 10, 11, 13, 14, 15, 17, 18, 19}; (3) $(A \cup B \cup C)'$ = {1, 6, 9, 10, 14, 15, 18}; (4) $(A \cap B) = \phi$; (5) $(B \cup C) \cap C = C$ = {2, 12, 20, 21, 22, 23, 24, 25}; (6) $A \cap B \cap C = \phi$. (Sec. 5.2)
P5.7 (a) 17; (b) 11; (c) 22; (d) 2; (e) 39. (Sec. 5.2)
P5.9 (a) {0, 1, 2, 3, 5}; (b) {0, 1, ..., 8}. (Sec. 5.3)

FINAL ANSWERS TO SELECTED PROBLEMS **595**

P5.10 (a) No; (b) 0.23. (Sec. 5.5)
P5.11 (a) 1/2; (b) 1/10; (c) 1/2. (Sec. 5.5)
P5.13 (a) 0.88; (b) 0.75; (c) 0.10; (d) 0.88; (e) 0.78; (f) 0.13. (Sec. 5.5)
P5.15 (a) 0.8; (b) 0.25; (c) 0.95; (d) 0.75; (e) 0.05. (Sec. 5.5)
P5.16 (a) 0.57; (b) 0.30. (Sec. 5.6)
P5.17 (a) 0.40; (b) 0.95. (Sec. 5.6)
P5.19 0.025. (Sec. 5.6)
P5.21 (a) 0.79; (b) 0.65. (Sec. 5.6)
P5.22 (a) 0.38; (b) 0.45. (Sec. 5.7)
P5.24 0.29. (Sec. 5.7)
P5.25 Attendent B. (Sec. 5.7)
P5.27 (a) 0.14; (b) 0.46. (Sec. 5.8)
P5.28 (a) C and D; (b) A and E. (Sec. 5.8)

Chapter 6

P6.1 (a) 7/10; (b) 5/8. (Sec. 6.1)
P6.4 (Sec. 6.3)

x	$f(x)$ (a) $a = 3$	(b) $a = 5$
0	0	0
1	0.282	0.181
2	0.342	0.268
3	0.223	0.244
4	0.093	0.162
5	0.026	0.082
6	0.005	0.033

P6.6 (Sec. 6.3)

t	0	0.5	1	2	3	4	5
$f(t)$	1.0	0.606	0.368	0.135	0.050	0.018	0.001

P6.8 (b) For $\lambda = 2.7$, $P(x \leq 2) = 0.492$; For $\lambda = 2.0$, $P(x < 2) = 0.677$. (Sec. 6.4)
P6.9 1/10. (Sec. 6.4)
P6.11 (b) 6/10. (Sec. 6.4)
P6.12 (b) Show the sum of $f(x)$ is one using the fact that (Sec. 6.4)

$$\sum_{x=0}^{\infty} a^x/x! = e^a$$

P6.13 $\mu = 3$, $\sigma = 1$. (Sec. 6.5)

P6.15 (a) (Sec. 6.5)

x	0	1	2	3	4	5	6
$f(x)$	0.10	0.15	0.20	0.25	0.15	0.10	0.05

(b) 2.7

P6.16 (a) (Sec. 6.5)

x	$50	10	−5
$f(x)$	0.1	0.4	0.5

(b) $6.50.

P6.17 Project 1 with $\mu = \$16{,}250$. (Sec. 6.5)

P6.18 (a) (Sec. 6.5)

x	2	3	4	5	6	7	8	9	10	11	12
$f(x)$	$\frac{1}{36}$	$\frac{2}{36}$	$\frac{3}{36}$	$\frac{4}{36}$	$\frac{5}{36}$	$\frac{6}{36}$	$\frac{5}{36}$	$\frac{4}{36}$	$\frac{3}{36}$	$\frac{2}{36}$	$\frac{1}{36}$

(b) $\mu = 7.0$, $\sigma = 2.42$.

P6.19 $\frac{1}{2}$. (Sec. 6.6)

P6.21 0.632. (Sec. 6.6)

P6.22 (a) 6; (b) (Sec. 6.6)

p	0	0.2	0.4	0.5	0.6	0.8	1.0
$f(p)$	0	0.96	1.44	1.50	1.44	0.96	0

P6.23 0.09. (Sec. 6.6)
P6.24 (a) $\mu = 1/\lambda$, $\sigma = 1/\lambda$; (b) 0.865. (Sec. 6.7)
P6.25 $E[P] = \frac{1}{4}$, $V[P] = \frac{3}{80}$. (Sec. 6.7)
P6.27 (a) $\mu = 0.667$, $\sigma = 0.236$. (Sec. 6.7)
P6.28 $F(p, = 0$ if $p < 0$; $1 - (1-p)^3$ if $0 \le p \le 1$; and 1 if $p > 1$. (Sec. 6.8)
P6.30 (a) $F(x) = 0$ if $x < 1$; 1/10 if $1 \le x < 2$; 3/10 if $2 \le x < 3$; 6/10 if $3 \le x < 4$, and 1 if $x \ge 4$. (Sec. 6.8)
P6.32 (a) $F(t) = 1 - e^{-t}$ for $0 \le t < \infty$; (b) 0.174. (Sec. 6.8)
P6.34 (a) 0.55 (b) $f(t) = 0.008 e^{-0.008t}$ for $t > 0$. (Sec. 6.8)
*P6.37** (a) Use Eq. (6.23) to get $M(t) = (1 - t/\lambda)^{-1}$ for $t < \lambda$; (b) use Eq. (6.24) to get $E[Z] = 1/\lambda$. (Sec. 6.10)
*P6.38** $\frac{1}{5t}(e^{10t} - e^{5t})$. (Sec. 6.10)
*P6.39** $\frac{0.1}{t}(e^{20t} - e^{10t})$. (Sec. 6.10)

Chapter 7

P7.1 (a) 840; (b) 969; (c) 9,880; (d) 93,024. (Sec. 7.2)
P7.3 (a) −0.33 percent; (b) 0.33 percent; (c) 1.7 percent. (Sec. 7.2)
P7.4 120. (Sec. 7.3)
P7.6 2520. (Sec. 7.3)
P7.7 (a) 32,432,400; (b) 1.709×10^8. (Sec. 7.3)
P7.9 (a) 48; (b) 72. (Sec. 7.3)
P7.11 7200. (Sec. 7.3)
P7.13 6720. (Sec. 7.3)
P7.14 63,063,000. (Sec. 7.3)
P7.15 495. (Sec. 7.4)
P7.16 3528. (Sec. 7.4)
P7.18 350. (Sec. 7.4)
P7.19 (a) 80,730; (b) 142,155; (c) 61,776. (Sec. 7.4)
P7.21 36. (Sec. 7.4)
P7.23 (a) 15; (b) 90; (c) 9; (d) 6; (e) 35. (Sec. 7.4)
P7.24 (a) 16; (b) 14. (Sec. 7.5)
P7.25 $x = 0, 1, 2, 3, 4$. (Sec. 7.5)
P7.27 (a) (Sec. 7.6)

x	pdf, $h(x)$	cdf, $H(x)$
0	0.282	0.282
1	0.470	0.752
2	0.217	0.969
3	0.030	0.999
4	0.001	1.000

(b) 0.687.

P7.29 (Sec. 7.6)

x	1	2	3	4	5
$h(x)$	0.024	0.238	0.476	0.238	0.024

P7.30 (a) 3 (Sec. 7.7)
(b)

r	0	1	2	3
$h(r)$	0.292	0.525	0.175	0.008

(c) $\mu = 0.9$, $\sigma = 0.7$.
P7.32 $\mu = 3$, $\sigma = 0.816$. (Sec. 7.7)

598 FINAL ANSWERS TO SELECTED PROBLEMS

P7.34 (a) X = number incorrect adjustments in sample, $N = 20$, $n = 7$, (Sec. 7.7)
$p = 0.10$; (b) $\mu = 0.7$, $\sigma = 0.657$.
P7.35 0.202. (Sec. 7.8)
P7.37 (a) 0.008; (b) 0.882. (Sec. 7.8)
P7.38 (a) 0.889; (b) decrease by 0.119. (Sec. 7.8)
P7.40 $\mu = 1.67$, $\sigma = 1.016$. (Sec. 7.8)

Chapter 8

P8.2 $f(x) = 0.025$ for defective and 0.975 for acceptable. (Sec. 8.1)
P8.5 (Sec. 8.2)

y	0	1	2	3	4
$f(y)$	0.4096	0.4096	0.1536	0.0256	0.0016
$F(y)$	0.4096	0.8192	0.9728	0.9984	1.0000

P8.6 In general, binomial pdf is symmetric about the value $n/2$. (Sec. 8.2)
P8.7 (a) 0.7734; (b) 0.9297. (Sec. 8.2)
P8.8 (a) $\mu = 0.3$, $\sigma = 0.542$; (b) 0.261. (Sec. 8.3)
P8.9 (a) $n = 50$, $p = 0.30$; (b) $\mu = 15$, $\sigma^2 = 10.5$, $\sigma = 3.24$; (c) 4.40×10^{-6}. (Sec. 8.3)
P8.11 $\mu = 3.75$, $\sigma = 0.97$. (Sec. 8.3)
P8.13 0.096. (Sec. 8.3)
P8.14 (a) 0.9999; (b) 0.0246; (c) 0.9984; (d) 0.0245. (Sec. 8.4)
P8.16 (a) 0.3672; (b) 0.9844; (c) 0.7471; (d) 0.0146. (Sec. 8.4)
P8.17 (a) $p = 0.2$, $n = 10$; (b) 0.9936, 0.6242. (Sec. 8.4)
P8.18 (a) 0.1722; (b) 3.6; (c) 0.0163; (d) 40. (Sec. 8.5)
P8.20 (a) 0.3277; (b) 0.2560; (c) 1.0. (Sec. 8.5)
P8.21 0.2237, do not stop. (Sec. 8.6)
P8.24 (Sec. 8.6)

x	0	1	2	3	4	5
expected	2.3	11.7	23.5	23.5	11.7	2.3

Not very close.
P8.25 (a) 0.085; (b) $h(3; 10, 18, 150)$. (Sec. 8.7)
P8.27 (a) 0.323; (b) -4.4 percent error in the approximation. (Sec. 8.7)
P8.29 (b) 0.9914; (c) 0.10. (Sec. 8.8)
P8.31 (a) $f(y) = {}_{50}C_{50y}(0.30)^{50y}(0.70)^{50-50y}$, $y = 0, 1/50, 2/50, \ldots, 1$; (Sec. 8.8)
(b) $\mu = 0.30$, $\sigma^2 = 0.0042$, $\sigma = 0.065$; (c) 4.4×10^{-6}.

P8.33 (a) (Sec. 8.8)

y	0	1/10	2/10	3/10	4/10	5/10
f(y)	0.1074	0.2504	0.3020	0.2013	0.0881	0.0264

y	6/10	7/10	8/10	9/10	1
f(y)	0.0055	0.0008	0.0004	$<10^{-4}$	$<10^{-4}$

(b) $\mu = 0.20$, $\sigma = 0.126$, $\mu \pm \sigma = 0.074, 0.326$

*P8.36 $E[X] = M'(0) = n(p+q)^{n-1}\left(\dfrac{p}{n}\right) = p$, $M''(0) = \dfrac{p^2(n-1)}{n} + \dfrac{p}{n}$, (Sec. 8.9)
$V[X] = \dfrac{pq}{n}$

Chapter 9

P9.1 (a) $e^{-0.8}(0.8)^x/x!$ for $x = 0, 1, 2, \ldots$; (b) $\dfrac{e^{-2.1}(2.1)^d}{d!}$ for $d = 0, 1, 2, \ldots$. (Sec. 9.1)

P9.4 (a) (Sec. 9.2)

x	0	1	2	3	4	5	6	≥ 7
f(x)	0.449	0.360	0.144	0.038	0.008	0.001	0.0002	$<10^{-4}$

(b) 0.542.

P9.5 (Sec. 9.2)

d	0	1	2	3	4	5	6
f(d)	0.123	0.257	0.270	0.189	0.099	0.042	0.015
F(d)	0.123	0.380	0.650	0.839	0.938	0.980	0.995

f	7	8	≥ 9
f(d)	0.004	0.001	$<10^{-3}$
F(d)	0.999	1.000	1.000

P9.6 (a) $\hat{\lambda} = 7.30$; (b) $\mu = 7.30$, $\sigma = 2.70$. (Sec. 9.3)
P9.8 0.4768. (Sec. 9.3)
P9.9 2.2994. (Sec. 9.3)
P9.10 (a) 0.360; (b) 0.991; (c) 0.047; (d) 0.047. (Sec. 9.4)
P9.12 5. (Sec. 9.4)

P9.13 0.871. (Sec. 9.5)

P9.15 5. (Sec. 9.5)

P9.17 (Sec. 9.6)

x	0	1	2	3	4
observed	109	65	22	3	1
expected	108.7	66.3	20.2	4.1	0.6

P9.18 (a) No; (b) 5 is recommended. (Sec. 9.6)

P9.20 (a) Plant A: $h(x; 55, 0.12, 600)$, $\mu = 6.6$, $\sigma = 2.30$; Plant B: $h(x; 120, 0.075, 1500)$, $\mu = 9$, $\sigma = 2.77$; (Sec. 9.7)
  (b) Plant A: $b(x; 55, 0.12)$, $\mu = 6.6$, $\sigma = 2.41$; Plant B: $P(x; 9)$, $\mu = 9$, $\sigma = 3$.

P9.21 (a) $\mu = 15$, $\sigma^2 = 15$, $\sigma = 3.87$, $P(x \leq 2.5) = 3.93 \times 10^{-5}$; (b) no. (Sec. 9.7)

*P9.25 $P(y; 17)$, $\mu = 17$, $\sigma = 4.12$. (Sec. 9.8)

Chapter 10

P10.1 (a) 0.0048; (b) 200. (Secs. 10.1–10.3)

P10.3 (a) 25, less than; (b) 0.1847. (Secs. 10.1–10.3)

P10.4 (b) 0.294. (Secs. 10.1–10.3)

P10.5 (a) 5; (b) 0.5904. (Secs. 10.1–10.3)

P10.6 (a) 0.0604; (b) 0.1476. (Secs. 10.4–10.6)

P10.8 (a) 4; (b) 0.125, 0.0088. (Secs. 10.4–10.6)

P10.10 (a) $f(x) = 1/11$ for $x = 20, 21, \ldots, 30$; (b) $\mu = 25$, $\sigma = 14.14$, $\mu \pm 0.1\sigma = 23.59, 26.41$; (c) 3/11. (Secs. 10.7–10.9)

P10.12 (a) $\mu = \$0.50$, $\sigma = \$3.64$. (Secs. 10.7–10.9)

P10.14 (a) 10; (b) 3.94, 16.06; (c) 0.62. (Secs. 10.7–10.9)

P10.17 (a) uniform, $a = 0$, $b = 59$, (b) $f(x) = 1/60$ for $x = 0, 1, \ldots, 59$. (Sec. 10.10)

Chapter 11

P11.1 (Sec. 11.1)

$$f(x) = \frac{1}{\sqrt{(0.21)2\pi}} \exp\left[\frac{-(x - 9.01)^2}{2(0.21)}\right] \text{ for } -\infty \leq x \leq \infty$$

$$f(y) = \frac{1}{\sqrt{(0.0025)2\pi}} \exp\left[\frac{-(y - 0.76)^2}{2(0.0025)}\right] \text{ for } -\infty \leq y \leq \infty$$

FINAL ANSWERS TO SELECTED PROBLEMS **601**

P11.3 (a) 0.3521; (b) 0.0648; (c) 0.0022. (Sec. 11.1)
P11.4 (a) $\mu = 25$ cm, $\sigma = 0.3$ cm; (b) $\mu = 8.5$ percent, $\sigma = 1$ percent. (Sec. 11.2)
P11.5 $\hat{\mu} = 5.808$ h/month, $\hat{\sigma} = 2.466$ h/month. (Sec. 11.2)
P11.7 (Sec. 11.2)

x	μ	$\mu \pm 0.25\mu$	$\mu \pm 0.5\mu$	$\mu \pm 1\sigma$	$\mu \pm 2\sigma$	$\mu \pm 3\sigma$
points plotted	18,500	17,800 19,200	17,100 19,900	15,700 21,300	12,900 24,100	10,100 26,900
$f(x) \times 10^{-4}$	1.425	1.381	1.258	0.864	0.193	0.016

P11.9 (Sec. 11.3)

x	0	± 1	± 2	± 3
$f(x)$	0.3989	0.2420	0.0540	0.0044

P11.10 (a) 2.67; (b) -2.14. (Sec. 11.3)
P11.12 (a) 0.9773; (b) 0.6826; (c) 0.0454. (Sec. 11.4)
P11.14 (a) 2.27 percent; (b) 477.3 houses. (Sec. 11.4)
P11.16 (a) False, 614 tires; (b) true; (c) false, 0.3413; (d) true. (Sec. 11.4)
P11.17 (a) 0; (b) 99.914 percent; (c) 9.78 percent. (Sec. 11.5)
P11.18 (a) 13,880; (b) 23,106; (c) 14,552 and 22,448 mi. (Sec. 11.5)
P11.20 (a) 0.0062; (b) 0.15735; (c) 0.901; (d) 0.758. (Sec. 11.5)
P11.21 Yes. (Sec. 11.5)
P11.23 7.25 ± 0.0625. (Sec. 11.6)
P11.24 (a) 4.36; (b) 3.84 eggs. (Sec. 11.6)
P11.26 Upper: 2.84, lower: 2.32 oz. (Sec. 11.6)
P11.27 Upper: 25,013, lower: 11,987 mi. (Sec. 11.6)
P11.28 (Sec. 11.7)

cell limit	1	2	3	4	5	6	7
expected frequency	2.00	2.82	5.09	8.24	10.76	12.57	11.90

cell limit	8	9	10	11	12
expected frequency	10.05	6.88	4.20	2.08	0.93

P11.30 (Sec. 11.7)

cell limit	40	50	60	70	80	90	100
expected frequency	5.2	8.9	14.4	15.9	12.2	6.4	2.3

cell limit	110	120	130	140	150
expected frequency	0.6	0.1	0	0	0

P11.32 Normal approximation to binomial; (a) 0.5279; (b) 0.0183; (c) 0.021. (Sec. 11.8)
P11.33 (a) 0.3783; (b) 0.3632. (Sec. 11.8)
P11.35 (a) 0.0136; (b) 0.1428. (Sec. 11.8)
P11.36 (a) 0.1921; (b) 0; (c) 0.197 for Poisson. (Sec. 11.8)
P11.38 (a) 0.1444; (b) 0.1469, reasonable not to stop. (Sec. 11.8)
*P11.40 $\mu = 2.5, \sigma = 0.707$. (Sec. 11.9)
*P11.41 0.747. (Sec. 11.9)
*P11.42 0.0359. (Sec. 11.9)

Chapter 12

P12.1 $\mu = 10$ percent, $\sigma_{\bar{X}} = 0.4$ percent. (Sec. 12.1)
P12.3 (a) $\mu = 20$ percent; $\sigma = 9$ percent; (b) $\mu = 20$ percent, $\sigma = 11.63$ percent. (Sec. 12.1)
P12.4 Infinite: 0.158; finite: 0.157; factor: 0.9909. (Sec. 12.1)
P12.5 (a) $\mu = 40, \sigma = 14.577$; (b) $\bar{X} = 40, s_{\bar{X}} = 5.61$; (c) $\sigma_{\bar{X}} = 5.95$. (Sec. 12.1)
P12.8 156.64. (Sec. 12.4)
P12.9 9.85 percent. (Sec. 12.4)
P12.11 (a) 8; (b) 0.0446. (Sec. 12.4)
P12.12 25. (Sec. 12.4)
P12.14 0.0227. (Sec. 12.4)
P12.15 (a) 99.87 on the average; (b) 8.7 percent; (c) reject; (d) 1284.61 h. (Sec. 12.4)

Chapter 13

P13.1 (a) 16; (b) 19; (c) 11. (Sec. 13.2)
P13.2 (a) 0.0974; (b) 0.0031; (c) 0.1839. (Sec. 13.2)
P13.4 (a) $\mu = 3, \sigma = 2.45, \alpha_3 = 1.63$; (b) $\mu = 25, \sigma = 7.07, \alpha_3 = 0.57$. (Sec. 13.3)

P13.5 (a) Use $\Gamma(4.5) = 11.6317$ to compute $f(y)$. (Sec. 13.3)

y	0	2	4	6	7	8	10	15
f(y)	0	0.001	0.066	0.100	0.104	0.100	0.081	0.027

(b) $y \geq 5.06$.

P13.7 0.1131. (Sec. 13.3)
P13.9 (a) 0.150; (b) 0.898. (Sec. 13.4)
P13.11 $P(13.09 \leq \chi^2 \leq 35.17) = 0.90$. (Sec. 13.4)
P13.12 55.77, 95.94. (Sec. 13.4)
P13.13 2 percent, decreased. (Sec. 13.5)
P13.15 (a) 0.003, 0.029. (Sec. 13.5)
P13.17 (Sec. 13.5)

Sample	1	2	3
$P(s^2 > 40.0)$	0.31	0.35	0.02

*__P13.19__ $\mu = 14, \sigma^2 = 28$. (Sec. 13.6)
*__P13.20__ (a) $\mu = 9, \sigma = 4.24$; (b) 0.019. (Sec. 13.6)
*__P13.22__ (a) 0.950; (b) 0.9525. (Sec. 13.6)

Chapter 14

P14.1 (a) 1.57; (b) $-3.198 \leq t \leq 3.198$. (Sec. 14.1)
P14.3 137.99. (Sec. 14.1)
P14.4 (a) 0.1228 using $\Gamma(2.5) = 1.3293$; (b) 0.3873. (Sec. 14.1)
P14.5 (Sec. 14.2)

t or z	0	±1	±2	±3	±4
f(t)	0.3750	0.2147	0.0663	0.0197	0.0067
f(z)	0.3989	0.2419	0.0540	0.0044	0.0001

P14.6 (a) $\mu = 0, \sigma = 1.29$; (b) $\mu = 0, \sigma = 1.05$. (Sec. 14.2)
P14.8 $\mu = 0$ and σ approaches one like the SND. (Sec. 14.2)
P14.9 (a) 0.025; (b) 0.95; (c) 0.059. (Sec. 14.3)
P14.10 0.022. (Sec. 14.3)
P14.12 0.04, 0.08. (Sec. 14.3)
P14.13 No, probability is 0.007. (Sec. 14.4)
P14.15 (a) Yes, probability is 0.07. (Sec. 14.4)
P14.17 (a) 0.058, accept; (b) 0.0498, marginally reject. (Sec. 14.4)
*__P14.18__ (a) and (b) 2.384. (Sec. 14.5)

Chapter 15

P15.1 (a) 0.655; (b) 0.89. (Sec. 15.1)
P15.2 (a) 0.2304; (b) 0.0658. (Sec. 15.1)
P15.3 Plant 1: $\mu = 1.22$, $\sigma = 0.88$; Plant 2: $\mu = 1.22$, $\sigma = 0.79$. (Sec. 15.2)
P15.4 (a) 1.20; (b) 0.22, 2.18. (Sec. 15.2)
P15.6 (a) 4.06; (b) 0.10 (Sec. 15.3)
P15.8 0.935. (Sec. 15.3)
P15.10 (a) 8; (b) 0.198. (Sec. 15.3)
P15.12 Yes. (Sec. 15.4)
P15.13 Approximated at 0.033. (Sec. 15.4)
P15.14 $F = 5.91$, decreased. (Sec. 15.4)

Chapter 16

P16.1 (a) $f(t) = \frac{1}{4}$ for $1 \leq t \leq 5$ min; (b) $\mu = 3$; (c) 12.5 percent. (Sec. 16.1)
P16.3 (a) $F(x) = \dfrac{x - 150}{75}$ for $150 \leq x \leq 225$; (b) 0.67; (c) 0.58. (Sec. 16.1)
P16.4 (a) $M(t) = \dfrac{1}{t(\beta - \alpha)}[e^{\alpha t} - e^{\beta t}]$. (Sec. 16.1)
P16.6 Lower: 1600 h; upper: 3400 h. (Sec. 16.1)
P16.8 (Sec. 16.2)

t	0.01	0.05	0.50	0.75	1	2	5	10
$f(t)$	0.150	0.149	0.139	0.134	0.129	0.111	0.071	0.033

P16.10 $\hat{\lambda} = 0.25$, $\mu = \sigma = 4$ h per breakdown. (Sec. 16.3)
P16.12 (a) 10; (b) uniform 0.423, exponential 0.135. (Sec. 16.3)
P16.13 (a) 0.55; (b) 0.20; (c) 0.25. (Sec. 16.4)
P16.15 (Sec. 16.4)

$t(+1,000)$	1	5	10	20	25	50
$f(t)(\times 10,000)$	0.905	0.607	0.368	0.135	0.082	0.007
$R(t)$	0.905	0.607	0.368	0.135	0.082	0.007

P16.16 (a) 299.57 h; (b) 39 percent; (c) 55 percent. (Sec. 16.4)
P16.18 (a) 9.5 percent, return; (b) 527 h minimum life. (Sec. 16.4)
P16.20 Use $\hat{\lambda} = 0.25$ to fit the cdf. (Sec. 16.4)

t	1	2	3	4	5	6	7
$F(t)$	0.22	0.39	0.53	0.63	0.71	0.78	0.83

t	8	9	10	12	15
$F(t)$	0.87	0.89	0.92	0.95	0.98

*P16.22 (a) 22.5 buses; (b) 0.758. (Sec. 16.5)
*P16.23 (a) 4.8 per h; (b) 0.63, 0.09. (Sec. 16.5)
P16.25 (a) (Sec. 16.6)

t	0.25	0.50	1.0	1.5	2.0	4.0
$f(t)$	0.025	0.123	0.361	0.448	0.391	0.057

(b) $\mu = 2$ min, $m = 1.5$ min.
P16.26 (a) $n = 3$ (known), $\hat{\lambda} = 0.53$; (b) $\mu = 5.66$ h, $\sigma = 3.26$ h. (Sec. 16.6)
*P16.28 (a) $\mu = 200$ h, $\sigma = 200$ h; (b) $n = 1$ and $T \sim \exp(0.005)$. (Sec. 16.7)
*P16.31 (a) 8; (b) 1. (Sec. 16.7)

Chapter 17

P17.1 $\alpha = 0.0124$; $1 - \alpha = 0.9876$. (Sec. 17.1)
P17.3 (a) $1 - \alpha = 0.9245$ for $x = 1, 2, 3, 4, 5$; (b) 0.0016. (Sec. 17.1)
P17.5 861 cans. (Sec. 17.2)
P17.7 (a) 464; (b) 431. (Sec. 17.2)
P17.8 (b) 2078 for percentage, 96 for error. (Sec. 17.2)
P17.9 (a) $h = 3.5$ percent; (b) 99.44 percent. (Sec. 17.2)
P17.11 51.6 percent. (Sec. 17.2)
P17.13 (a) $P(156.39 \le \mu \le 182.09) = 0.95$; (b) $P(161.04 \le \mu) = 0.90$. (Sec. 17.4)
P17.15 $n = 20$: 534.5, $n = 200$: 163.2 Hz. (Sec. 17.4)
P17.16 $P(\mu \le 11.2) = 0.95$. (Sec. 17.4)
P17.17 (a) 2.25; (b) yes (Sec. 17.4)
P17.19 (b) $P(1236.5 \le \mu \le 1314.1) = 0.95$. (Sec. 17.4)
P17.20 (a) $P(23.25 \le \sigma \le 46.26) = 0.98$; (b) $P(540.78 \le \sigma^2 \le 2140.22) = 0.98$. (Sec. 17.5)
P17.21 (a) $P(3.68 \le \sigma \le 4.39) = 0.95$; (b) $P(0 \le \sigma \le 4.32) = 0.95$. (Sec. 17.5)
P17.23 (a) 6; (b) $P(9.23 \le \mu \le 19.11) = 0.95$, $P(2.94 \le \sigma \le 11.56) = 0.95$; (c) actual is 2.5 times wider. (Sec. 17.5)
P17.25 $P(0.90 \le \sigma \le 2.91) = 0.98$, yes. (Sec. 17.5)
P17.27 (a) $P(0.57 \le \pi \le 0.87) = 0.90$; (b) $P(0.6 \le \pi \le 1) = 0.90$. (Sec. 17.6)
P17.29 (a) $P(0 \le \pi \le 0.23) = 0.98$; (b) $P(0 \le \pi \le 0.22) = 0.95$. (Sec. 17.6)
P17.30 $P(0.057 \le \pi \le 0.103) = 0.90$, not increased. (Sec. 17.6)
P17.31 886. (Sec. 17.6)

Chapter 18

P18.2 $\frac{1}{12}$. (Sec. 18.1)
P18.3 (a) 1—yes, 2—no; (b) 162.5. (Sec. 18.1)
P18.5 $F(x', y') = (1 - e^{-x'})(1 - e^{-y'})$ x' and $y' > 0$. (Sec. 18.1)
P18.7 (Sec. 18.1)

x	y	$f(x, y)$
0	0	4/100
1	0	16/100
0	1	16/100
1	1	64/100

P18.9 $\frac{12}{29}$. (Sec. 18.1)

P18.10 (a) $\frac{1}{4}$; (b) $\frac{1}{2}$. (Sec. 18.1)

P18.12 (a) $f(p_1) = 2p_1$ for $0 \le p_1 \le 1$, $f(p_2) = 2(1 - p_2)$ for $0 \le p_2 \le 1$; (Sec. 18.2)

(b)

p_1 or p_2	0	0.5	1
$f(p_1)$	0	1	2
$f(p_2)$	2	1	0

(c) $E[P_1] = \frac{2}{3}$, $E[P_2] = \frac{1}{3}$.

P18.14 (a) $0.0015 \exp[-(0.05x + 0.03y)]$ for $x, y > 0$; (b) $0.03 \exp[-0.03y]$ for $y > 0$; (c) $0.003 \exp[-(0.03y + 0.10z)]$ for $y, z > 0$. (Sec. 18.2)

P18.16 $\frac{1}{4}$. (Sec. 18.2)

P18.17 No. (Sec. 18.3)

P18.19 Yes. (Sec. 18.3)

P18.20 (b) $\frac{1}{2}$; (c) $f(x) = 1 - \frac{x}{2}$ for $0 < x < 2$, $f(y) = \frac{y}{2}$ for $0 < y < 2$; (d) No; (e) $\frac{16}{25}$; (f) $\frac{7}{16}$. (Sec. 18.3)

P18.22 (a) $\frac{1}{4}$; (b) Yes; (c) 100 percent. (Sec. 18.3)

P18.24 (a) $f(x, y) = \frac{1}{15}$ for $1 \le x \le 4$, $5 \le y \le 10$; (b) $\frac{2}{3}$. (Sec. 18.3)

P18.26 (a) (Sec. 18.4)

		$f(y/x)$		
y x	0	1	2	3
0	0	1/6	2/6	3/6
1	1/10	2/10	3/10	4/10
2	2/14	3/14	4/14	5/14

(b) 2.

P18.27 (a) $E[X/Y] = \frac{4}{3}$ for $0 \le y \le 1$; (b) $E[Y/X] = \frac{3}{5}$ for $0 \le x \le 2$. (Sec. 18.4)

P18.29 (a) 2; (b) $\frac{8}{3}$ (Sec. 18.4)

P18.30 (a) $f(y/x) = \frac{1}{2 - x}$ for $0 < x < 2$, $x < y < 2$; (b) 0.8; (c) $\frac{3}{2}$. (Sec. 18.4)

P18.31 (b) $\frac{5}{8}$. (Sec. 18.4)

P18.32 (b) 0.026; (c) (Sec. 18.5)

X_i	1	2	3	4
$E[X_i]$	1	3	4	2

P18.35 (a) $E[Y/X] = 4 + 1.2x$, $V[Y/X] = 0.81$, $E[X/Y] = -0.333 + 0.533y$, $V[X/Y] = 0.36$; (b) 0.7096. (Sec. 18.5)

P18.37 (a) 10 min; (b) 7.5 min. (Sec. 18.6)

P18.39 (a) 5.07; (b) 24.33; (c) cannot do. (Sec. 18.6)

P18.41 33 min. (Sec. 18.6)

P18.42 $E[X]E[Y] = (\frac{2}{3})(\frac{4}{3}) = \frac{8}{9}$, $E[XY] = 1$. (Sec. 18.6)

P18.43 (a) 4.8 min; (b) 7.05 min^2. (Sec. 18.7)

P18.45 $E[Z] = 0.5$; $V[Z] = 1.08$. (Sec. 18.7)

P18.46 $E[3X + 2Y] = 31.5$ cm, $\sigma[3X + 2Y] = 2.28$ cm. (Sec. 18.7)

P18.48 (a) $E[Z] = 13$, $V[Z] = 20.67$; (b) 7.67. (Sec. 18.7)

*P18.50 (a) $f(y) = \frac{1}{5}$ $y = \frac{1}{6}, \frac{1}{7}, \frac{1}{8}, \frac{1}{9}, \frac{1}{10}$; (b) $f(y) = \frac{1}{5}$ $y = $ 72, 98, 128, 162, 200. (Sec. 18.8)

*P18.52 (a) $f(y) = \frac{e^y}{15}$ for $y = 0, \ln 2, \ln 3, \ln 4, \ln 5$; (b) $f(y) = \frac{y-2}{30}$ for $y = 4, 6, 8, 10, 12$. (Sec. 18.8)

*P18.54 (a) $f(y) = {}_nC_{y/3}p^{y/3}q^{n-y/3}$ for $y = 0, 3, 6, \ldots, 3n$; (b) and (c) $3np$ (Sec. 18.8)

*P18.55 (a) $f(y) = \frac{1}{6}$ for $y = 2, 4, 6, 8, 10, 12$; (b) $E[Y] = 7$; $V[Y] = 11.67$; (c) yes. (Sec. 18.8)

*P18.57 (a) $f(y) = 3(1 - y)^2$ for $0 \leq y \leq 1$; (b) $\frac{3}{4}$ month; (c) $\frac{1}{4}$ month. (Sec. 18.9)

*P18.59 (a) $f(y) = 0.0059 + 0.00079y$ for $7.5 \leq y \leq 45$; (b) $E[Y] = 29.72$ s, $\sigma[Y] = 10.25$ s. (Sec. 18.9)

*P18.60 (a) $f(y) = \frac{3}{8}e^{-3y}$ for $-\ln 2 \leq y \leq \infty$; (b) $f(y) = 20e^{-2y}(1 - e^{-y})^3$ for $0 \leq y \leq \infty$. (Sec. 18.9)

*P18.62 $f(z_1, z_2) = \frac{3}{20}z_1\left(1 + \frac{z_2^2}{2}\right)$ for $0 \leq z_1$ and $z_2 \leq 2$. (Sec. 18.10)

*P18.63 (a) $f(y_1, y_2) = \sqrt{y_1}/12\left(1 + \frac{y_2^2}{2}\right)$ for $y_1 = 0, 1, 4$; $y_2 = 0, 2$; (b) $f(y_1) = \frac{1}{3}\sqrt{y_1}$ for $y_1 = 0, 1, 4$; $f(y_2) = \frac{1}{4}\left(1 + \frac{y_2^2}{2}\right)$ for $y_2 = 0, 2$. (Sec. 18.10)

*P18.65 (a) $f(y_1, y_2) = \frac{1}{2}$; (b) $f(y_1, y_2) = \frac{1}{2}(y_1 + y_2)(y_1 - y_2)$, range for both pdf's is bounded by the limits $y_2 = y_1$, $y_2 = -y_1$, $y_2 = 2 - y_1$ and $y_2 = y_1 - 2$. (Sec. 18.10)

Chapter 19

P19.2 (a) $H_0: p_1 = p_2$, $H_1: p_1 < p_2$; (b) $H_0: |\bar{X}_1 - \bar{X}_2| = \$150$, $H_1: |\bar{X} - \bar{X}_2| < \$150$; (c) $H_0: v_1/v_2 = 1.5$, $H_1: v_1/v_2 \neq 1.5$. (Sec. 19.1)

P19.3 (a) $H_0: \mu = 2000$, $H_1: \mu > 2000$ units; (b) $-\infty \leq \bar{X} \leq 2100.3$. (Sec. 19.2)

P19.5 (b) 0.0071; (c) 0.0516. (Sec. 19.2)

P19.6 (a) 0.2061, 0.50. (Sec. 19.2)

P19.7 $\alpha = 0.353$, $\beta = 0.181$. (Sec. 19.2)

P19.8 (a) $4.42 \leq \bar{X} \leq 5.58$; (b) $\alpha = 0.02$, $1 - \alpha = 0.98$; (c) 0.035. (Sec. 19.3)

*P19.12 (a) 0.0054; (b) 0.0375. (Sec. 19.5)

*P19.13 8. (Sec. 19.5)

*P19.15 (a) 0.8173; (b) as the true μ value approaches the H_0 value, β increases toward $(1 - \alpha)$. (Sec. 19.5)

P19.16 (a) 0.15; (b) in P19.14a, $\beta = 0.1151$. (Sec. 19.6)
P19.17 (a) 24, 38; (b) 8, 11. (Sec. 19.6)
P19.20 (Sec. 19.6)

μ	4.0	4.5	5.0	5.5	6.0	6.5	7.0	7.5	8.0
β	0.009	0.03	0.24	0.84	0.95	0.84	0.24	0.03	0.009
$1 - \beta$	0.991	0.97	0.76	0.16	0.05	0.16	0.76	0.97	0.991

P19.21 (Sec. 19.6)

μ	20.0	20.5	21.0	21.5	22.0	22.5	23.0	23.5
$1 - \beta$	0.05	0.09	0.23	0.45	0.69	0.86	0.96	0.99

Chapter 20

P20.1 (a) $z = (\bar{X} - \mu_0)\frac{\sqrt{n}}{\sigma}$; (b) reject H_0: $\mu = 18.7$ per h; (c) using the t distribution $P(17.96 \leq \mu \leq 22.04) = 0.95$. (Sec. 20.2)
P20.2 (a) 0.25; (b) n is approximately 18 from Fig. 20.2. (Sec. 20.2)
P20.5 (a) Reject H_0: $\mu = 3500$ h with $z = -2.78$; (b) approximately 0.85; (c) 0.55, 0.05; (d) $P(-\infty \leq \mu \leq 3418.2 \text{ h}) = 0.95$. (Sec. 20.2)
P20.6 (a) Do not reject H_0: $\mu = 18.7$ per hour with $t = 1.47$; (b) yes; (c) Fig. 20.4 gives $\beta = 0.62$ for t and Fig. 20.1 gives $\beta = 0.57$ for SND. (Sec. 20.3)
P20.7 (a) M; (b) $P(3310.2 \leq \mu \leq 3719.8) = 0.90$. (Sec. 20.3)
P20.10 Do not reject H_0: $\mu_A = \mu_B$ with $z = -0.99$. (Sec. 20.4)
P20.11 (a) Reject H_0: $\mu_N = \mu_S$ with $z = 2.92$; (b) 0.28. (Sec. 20.4)
P20.13 (a) 18; (b) do not reject H_0: $\mu_1 - \mu_2 = 0.020$ with $z = 0.74$ not same conclusion as in P20.12b because H_0 is different. (Sec. 20.4)
P20.15 Do not reject H_0: $\mu_1 = \mu_2$ with $t = -1.09$. (Sec. 20.5)
P20.16 Do not reject H_0: $\mu_A = \mu_B$ with $t = -1.00$ and $\nu = 20$. (Sec. 20.5)
P20.17 (a) No ($t = -0.94$); (b) $\beta = 0.62$ approximately for $n' = 19$ and $d = 0.32$; (c) approximately 31. (Sec. 20.5)
P20.19 (a) Do not reject H_0: $\mu_A = \mu_B$ with $t = 0.195$; (b) approximately 0.88. (Sec. 20.5)
P20.20 Do not reject H_0: $\delta = 0$ with $t = 0.054$. (Sec. 20.6)
P20.22 Reject H_0: $\delta = 0$ with $z = -2.65$; trainer A is better. (Sec. 20.6)
P20.23 (a) Do not reject H_0: $\mu_C = 8$ percent with $t = -0.33$; (b) do not reject H_0: $\delta = 0$ with $t = 0.58$; (c) do not reject H_0: $\delta = 0$ for destructive versus chemical ($t = 0.68$) or destructive versus isolation ($t = 2.33$). (Sec. 20.6)

Chapter 21

P21.1 (a) One sample, normal population; (b) do not reject H_0: $\sigma^2 = 45$ h^2 with $\chi^2 = 20.0$; (c) 0.80. (Sec. 21.2)
P21.2 (a) Reject H_0: $\sigma^2 = 36$ db^2 with $\chi^2 = 19.64$; (b) approximately 0.75; (c) approximately 25; (d) $P(6.10 \leq s^2 \leq 16.18) = 0.95$. (Sec. 21.2)

P21.4 Do not reject H_0: $\sigma^2 = (0.05)^2$ with $\chi^2 = 21.6$ (Sec. 21.2)

P21.5 (a) Do not reject H_0: $\sigma^2 = 26$ (points)^2 with $\chi^2 = 11.32$; (b) approximately 0.80; (c) 45. (Sec. 21.2)

P21.7 Do not reject H_0: $\sigma^2 = 0.0025$ with $z = 2.30$. (Sec. 21.2)

P21.9 (a) Do not reject H_0: $\sigma_1^2 = \sigma_2^2$ with $F = 1.81$; (b) No (Sec. 21.3)

P21.11 Do not reject H_0: $\sigma_1^2 = \sigma_2^2$ with $F = 1.09$, wrong procedure in P20.18. (Sec. 21.3)

P21.12 Do not reject H_0: $\sigma_A^2 = \sigma_B^2$ with $F = 1.105$, assumption correct in P20.19. (Sec. 21.3)

P21.13 (a) Reject H_0: $\sigma_A^2 = \sigma_B^2$ with $F = 2.78$; (b) 0.22 approximately; (c) reject H_0: $\mu_A = \mu_B$ with $t = -2.57$. (Sec. 21.3)

*P21.15 (a) 0.50; (b) reject H_0: $\sigma_1^2 = \sigma_2^2$ with $F = 3$, $\beta = 0.55$ approximately. (Sec. 21.4)

Chapter 22

P22.1 (a) Do not reject H_0: $\pi = 0.41$ with $z = -1.57$; (b) reject H_0: $\pi = 0.41$ with $z = -1.57$. (Sec. 22.2)

P22.3 John: do not reject H_0: $\pi = 0.15$, incorrect conclusion; Carol: reject H_0: $\pi = 0.15$, correct conclusion. (Sec. 22.2)

P22.5 (a) Reject H_0: $\pi = 0.65$ with $z = 3.9$; (b) reject H_0: $n\pi = 1462.5$ with $z = 3.9$. (Sec. 22.2)

P22.6 (a) Do not reject H_0: $\pi = 0.4$; (b) 1.5, 2.0; (c) no, 8110 bottles needed. (Sec. 22.2)

P22.7 Reject H_0: $n\pi = 52(0.40) = 20.8$ with $z = 4.87$; the number increased. (Sec. 22.2)

P22.9 Reject H_0: with $z = -3.16$. (Sec. 22.3)

P22.11 Decision different, reject H_0: with $z = 11.58$. (Sec. 22.3)

P22.12 (a) Do not reject H_0: $\pi_P = \pi_C$ with $z = 1.48$; percent meat is equal; (b) marginally reject H_0 with $z = 1.67$. (Sec. 22.3)

P22.13 (a) Plant A: do not reject H_0: $\pi = 0.25$ with $z = 0.31$, plant B: reject H_0: $\pi = 0.25$ with $z = 2.18$; (b) do not reject H_0: $p_A = p_B$ with $z = -0.90$. (Sec. 22.3)

P22.14 (a) True with $z = 4.14$; (b) true with $z = -1.90$; (c) false with $z = -1.33$; (d) false with $z = -4.3$. (Sec. 22.3)

Chapter 23

Note: The χ^2 value you compute may be slightly different than the ones given here.

P23.1 (a) χ^2; (b) $K - S$; (c) χ^2. (Sec. 23.1)

P23.3 (a) Do not reject H_0: distribution is $b(x; 4, 0.5)$, ($\chi^2 = 0.933$); (b) $\alpha = 0.92$. (Sec. 23.2)

P23.5 Do not reject H_0: number of problems follows $f(x)$, ($\chi^2 = 5.25$). (Sec. 23.2)

P23.6 Reject H_0: sample from geometric pdf for $x = 1, 2, \ldots, 7$ ($\chi^2 = 23.84$). (Sec. 23.2)

P22.8 H_0: violations per day are from $P(x; 1.0)$, (a) reject H_0; (b) do not reject H_0 ($\chi^2 = 8.76$). (Sec. 23.3)

P23.9 (a) Yes; (b) H_0: accident sample from Poisson ($\chi^2 = 2.94$); for $\alpha = 0.10$, reject H_0; for $\alpha = 0.05$, do not reject H_0. (Sec. 23.3)

P23.11 Do not reject H_0: number of flaws are Poisson ($\chi^2 = 0.66$). (Sec. 23.3)

P23.12 Do not reject H_0: sample follows indicated frequencies ($\chi^2 = 3.73$); Judy's predictions are substantiated. (Sec. 23.3)

P23.14 (a) $a = 0$, $b = 8$; (b) do not reject H_0: sample from a uniform distribution ($\chi^2 = 3.28$). (Sec. 23.3)

P23.16 Using the values $x = 0, 1, 2, \ldots, 13$, do not reject H_0: sample from Poisson ($\chi^2 = 2.76$). (Sec. 23.3)

P23.17 Same conclusion: do not reject H_0: sample from binomial; (a) $\chi^2 = 5.67$; (b) $\chi^2 = 4.67$. (Sec. 23.3)

P23.18 Do not reject H_0: sample from normal ($\chi^2 = 3.781$). (Sec. 23.4)

P23.21 Do not reject H_0: sample from normal ($\chi^2 = 2.65$). (Sec. 23.4)

P23.22 Use the technique in Example 16.6 to fit the cumulative exponential distribution; reject H_0: sample from an exponential ($\chi^2 = 13.64$). (Sec. 23.4)

P23.23 Do not reject H_0: baking times from $U(15, 25)$ distribution ($\chi^2 = 2.50$). (Sec. 23.4)

P23.24 $\chi^2 = 6.52$; for $\alpha = 0.10$, reject H_0: factors are independent; for $\alpha = 0.05$, do not reject H_0. (Sec. 23.5)

P23.26 Do not reject H_0: factors are independent ($\chi^2 = 5.43$). (Sec. 23.5)

P23.30 Since $d = 0.15$, do not reject H_0: baking times from $U(15, 25)$. (Sec. 23.7)

P23.31 Since $d = 0.110$ do not reject H_0: processing times follow the specified $f(t)$. (Sec. 23.7)

P23.32 (a) Since $d = 0.491$, reject H_0: times from exp(0.3); (b) Since $d = 0.132$ do not reject H_0: times from exp(0.1). (Sec. 23.7)

Chapter 24

Note: Your parameter estimates may vary slightly from those given here when the plotting procedure is utilized.

P24.2 (Sec. 24.1)

Relation	Value for $i = 1$	Value for $i = 10$	Increment between $F(x_i)$ values
a	0.05	0.95	0.10
b	0.10	1.00	0.10
c	0.0909	0.909	0.0909

P24.4 (a) Normal is appropriate; (b) plot estimates: $\mu = 19.6$ min, $\sigma = 2.75$ min, formula estimates: $\mu = 19.7$ min, $\sigma = 2.68$ min. (Sec. 24.2)

P24.6 (a) Normal is appropriate; (b) $\mu = 56$, $\sigma = 28$. (Sec. 24.2)

P24.7 No, not too close to exponential. (Sec. 24.3)

P24.9 (a) Exponential is better; (b) $\lambda = 0.07$. (Sec. 24.3)

P24.11 No. (Sec. 24.3)

P24.12 $\lambda = 0.07$ from plot, so exp(0.1) gives a better explanation. (Sec. 24.3)

P24.13 No. (Sec. 24.4)

P24.16 Exponential is better. (Sec. 24.4)

P24.17 (a) Uniform seems appropriate; (b) no. (Sec. 24.4)
P24.18 Uniform is better. (Sec. 24.4)
P24.19 (a) (Sec. 24.5)

x_i	3	4	5	6	7	8	9	10
$F(x_i)$	0.038	0.124	0.242	0.392	0.522	0.683	0.823	0.909

x_i	11	12	13
$F(x_i)$	0.941	0.973	0.995

Poisson is appropriate; (b) Plot: $\lambda = 7.5$ orders per day; equation: $\hat\lambda = 7.3$ order per day; they are close.

P24.21 (Sec. 24.5)

x_i	0	1	2	3	4	5
$F(x_i)$	0.260	0.660	0.860	0.953	0.980	0.993

Chapter 25

P25.2 (a) Do not reject H_0: order is random ($z = 1.54$); (b) same conclusion as in (a). (Sec. 25.2)
P25.4 Do not reject H_0: random sample of flow rates ($r = 12$). (Sec. 25.2)
P25.5 (Sec. 25.2)

Part	H_0	Conclusion	r
a	Random about median	Do not reject	8
b	Random about mode	Do not reject	7
c	Random about 22	Do not reject	5

P25.6 Reject H_0: random sample of ratings ($r = 7$). (Sec. 25.2)
P25.9 (a) One sample sign test; (b) do not reject H_0: $\mu = 0.30$ with $s = 9$; (c) reject H_0: $\mu = 0.30$ with $t = 2.82$. (Sec. 25.3)
P25.12 (a) Do not reject H_0: $\mu = 60$ min with $z = 0.204$; (b) do not reject H_0: $\mu = 60$ min with $t = 1.05$. (Sec. 25.3)
P25.14 Do not reject H_0: $\mu = 15$ min with $z = -1.13$; reject H_0: $\mu = 20$ min with $z = -7.02$. (Sec. 25.3)
P25.15 Reject H_0: $\mu = 4.5$ years with $s = 6$ for $\alpha = 0.20$. (Sec. 25.3)
P25.16 Reject H_0: $\mu_1 = \mu_2$ with $s = 2$. (Sec. 25.4)
P25.18 Do not reject H_0: $\mu_N = \mu_C$ with $z = 0$; the hypothesis H_1 is H_1: $\mu_N > \mu_C$ which implies fewer minus signs, thus a high z value for large samples. (Sec. 25.4)

P25.20 (a) Do not reject H_0: $\mu_1 = 62$ days ($s = 8$), but reject H_0: $\mu_1 = 59$ days ($s = 10$); (b) do not reject H_0: $\mu_1 = \mu_2$ with $s = 6$. (Sec. 25.4)

P25.21 (a) Reject H_0: $\mu_{new} = 60$ min with $z = -3.47$; (b) reject H_0: $\mu_{new} = \mu_{old}$ with $z = 2.34$ and the signs determined by $X_{i(old)} - X_{i(new)}$. (Sec. 25.4)

P25.23 Do not reject H_0: $\mu_N = \mu_W$ ($u = 150.5$ and $z = 1.80$). (Sec. 25.5)

P25.24 Do not reject H_0: $\mu_N = \mu_C$ ($u = 265.5$ and $z = -0.91$). (Sec. 25.5)

P25.26 H_0: $\mu_H = \mu_L$ versus H_1: $\mu_H \neq \mu_L$; (a) do not reject H_0 ($s = 4$); (b) do not reject H_0 ($u = 55$, $z = -0.62$); (c) do not reject H_0 ($t = 0.67$). (Sec. 25.5)

P25.29 Do not reject H_0: $\mu_1 = \mu_2$ with $T = 23.5$. (Sec. 25.6)

P25.30 (a) Do not reject H_0: $\mu_b = \mu_a$ with $T = 14$; (b) same result with $s = 4$. (Sec. 25.6)

P25.32 (a) Do not reject H_0: $\mu_N = \mu_C$ with $T = 105.5$; (b) Yes. (Sec. 25.6)

P25.33 Do not reject H_0: $\mu_E = \mu_M$ with $T = 140.5$ and $z = -0.89$. (Sec. 25.6)

P25.34 (a) Yes, $r_1 = 8$ and $r_2 = 5$; (b) yes, $s = 5$; (c) yes, $z = 0.49$; (d) yes, $T = 29$. (Sec. 25.6)

Chapter 26

P26.2 If loss is profit amount less than \$3 per unit, the loss functions are: (Sec. 26.2)

(a) $L(d_1, \theta) = \begin{cases} 0 & \theta < 5000 \text{ units} \\ 1(\theta - 5000) & \theta \geq 5000 \text{ units} \end{cases}$

(b) $L(d_2, \theta) = \begin{cases} 1.5\theta & \theta < 5000 \text{ units} \\ 2.5(\theta - 5000) & \theta \geq 5000 \text{ units} \end{cases}$

P26.3 (b) (Sec. 26.2)

θ	Contracted weeks, d				
	0	1	2	3	4
	$L(d, \theta)$ and $P(d/\theta)$ values				
0	0	1	2	3	4
	$\frac{1}{5}$	$\frac{1}{5}$	$\frac{1}{5}$	$\frac{1}{5}$	$\frac{1}{5}$
1	2	0	1	2	3
	0	$\frac{1}{4}$	$\frac{1}{4}$	$\frac{1}{4}$	$\frac{1}{4}$
2	4	2	0	1	2
	0	0	$\frac{1}{3}$	$\frac{1}{3}$	$\frac{1}{3}$
3	6	4	2	0	1
	0	0	0	$\frac{1}{2}$	$\frac{1}{2}$
4	8	6	4	2	0
	0	0	0	0	1

(c)

θ	0	1	2	3	4
$R(d/\theta)$ weeks	2.0	1.5	1.0	0.5	0

(d) $r(d, \theta) = 1.4$ weeks, \$7000.

FINAL ANSWERS SELECTED PROBLEMS **613**

P26.5 (a)   (Sec. 26.2)

θ	10,000	20,000	5,000
$R(d, \theta)$	$4,250	$3,000	$21,750

(b) $9,250
P26.6 (b) $5906.25.   (Sec. 26.2)
P26.8 $\mu/\overline{X} \sim N(27.4, 0.35)$.   (Sec. 26.3)
P26.10 (a) $\Pi/r \sim \beta(13, 117)$; (b) $E[\Pi/r] = 0.10$, $V[\Pi/r] = 0.0007$ (Sec. 26.3)
P26.13 (a) 0.65; (b) 0.10.   (Sec. 26.4)
P26.14 (a) 0.577; (b) 0.482.   (Sec. 26.4)
*__P26.16__ (a) 4.94; (b) 0.20 h per arrival.   (Sec. 26.4)

Chapter 27

Note: Your answers may vary slightly from these due to round-off, calculator differences, etc.

P27.3 (a) and (b) Y-16.77$-$0.0365X; (c) 13.67 km/1; (d) 62.2 km/h.

(Sec. 27.3)
P27.4 (a) $Y = 56.186 - 0.258X$; (b) 415.9; (c) 43.3 percent.   (Sec. 27.3)
P27.7 (a) 0.047; (b) no ($t = 10.28$); (c) no ($t = 4.13$).   (Sec. 27.4)
P27.9 (a) 0.049; (b) reject H_0: $a = 12$ with $t = 1.97$.   (Sec. 27.4)
P27.10 (a) Reject H_0: $b = 4.0$ with $t = 2.30$; (b) do not reject H_0: $a = 1255$ with $t = 0.33$. (Sec. 27.4)
P27.11 (a)   (Sec. 27.5)

X	60	70	80	90	100
lower	14.40	14.07	13.73	13.35	12.96
upper	14.76	14.35	13.97	13.61	13.28

(b) 13.07 to 13.89, wider.
P27.12 (a) 16.23 to 17.31; (b) -0.043 to -0.030.   (Sec. 27.5)
P27.13 (a) -0.380 to -0.136; (b) 39.56 to 72.81, yes; (c) no; (d) yes.   (Sec. 27.5)
P27.15 Using the regression line: sales/10,000 $= 18.035 - 0.53$(month)   (Sec. 27.5)
(a) 71.64; (b) 13.86 to 16.92; (c) -0.936 to -0.124.
P27.16 (a) $\hat{Y} = 19.22 + 22.17X_1 - 0.1675X_2$; (b) 22.97 percent.   (Sec. 27.6)
P27.18 (a) $\hat{Y} = 15.87 + 0.121X_1 + 0.039X_2$; (b) 0.24.   (Sec. 27.6)
*__P27.20__ (a) $\hat{Y} = 8.26 + 158.0/X$; (b) hyperbolic $\Sigma\epsilon_i^2 = 12.35$, linear   (Sec. 27.8)
$\Sigma\epsilon_i^2 = 12.15$; linear just slightly better.
*__P27.23__ (a) $\hat{Y} = 30,760X^{-1.53}$; (b) power $\Sigma\epsilon_i^2 = 130.54$, linear $\Sigma\epsilon_i^2 = $ (Sec. 27.8)
415.9; power is better.
*__P27.26__ Linear: $\hat{Y} = 18.035 - 0.53X$, $\Sigma\epsilon_i^2 = 71.64$; exponential: $\hat{Y} = $ (Sec. 27.8)
$18.42(0.96)^X$, $\Sigma\epsilon_i^2 = 75.47$; linear is better.

P27.27 Linear: $\hat{Y} = 36.08 + 0.16X$; exponential: $\hat{Y} = 43.75(1.002)^X$; exponential is better. (Sec. 27.8)

P27.30 (a) 5.44; (b) 4.97; (c) $MS_T = 0.49$, $MS_R = 4.97$. (Sec. 27.9)

P27.31 (a) 0.75; (b) 0.91. (Sec. 27.9)

Chapter 28

P28.3 -0.956. (Sec. 28.2)

P28.4 -0.865. (Sec. 28.2)

P28.6 Do not reject H_0: $\rho = 0.8$ with $z = 1.22$. (Sec. 28.3)

P28.9 (a) Plant A: reject H_0: $\rho = 0$ with $z = 2.30$; Plant B: do not reject H_0: $\rho = 0$ with $t = 1.85$; (b) yes with $z = -0.35$. (Sec. 28.3)

P28.10 (a) 7; (b) 19. (Sec. 28.3)

P28.11 (a) There may be a small round-off error. The minus sign is due to the negative slope value; (b) 91 percent of variation explained by linear fit. (Sec. 28.4)

P28.13 (a) Units per hour = 5.012 (units produced)$^{0.3906}$; (b) power: $r = 0.943$, linear: $r = 0.928$; (c) power; (d) power: $r^2 = 0.89$, linear: $r^2 = 0.86$. (Sec. 28.4)

P28.15 Note: numbers are sensitive, so yours may vary slightly. Let T = time and X = age (Sec. 28.4)

Model	Equation	r value
(a) Linear	$T = -0.043 + 0.0305X$	0.808
(b) Exponential	$T = 0.083(1.0877)^X$	0.627
(c) Hyperbolic	$T = 0.4267 - 0.372/X$	0.430

Linear offers best fit.

P28.19 0.5. (Sec. 28.5)

P28.20 (a) 0.92; (b) 0.846; (c) 0.827. (Sec. 28.6)

P28.22 (b) Multiple linear removes 72.25 percent, X_1 alone removes 4 percent, and X_2 alone removes 50 percent. (Sec. 28.6)

Chapter 29

P29.3 (a) 21.46 and 21.54 cm; (b) 0, 0.148 cm. (Sec. 29.2)

P29.5 (b) $LCL_{\bar{X}} = 598.14$; $UCL_{\bar{X}} = 603.36$; no. (Sec. 29.2)

P29.6 (a) Warning limits: 599.40 and 602.44; (b) probability limits: 598.52 and 603.32; (c) same $\bar{X}$ values are out on both. (Sec. 29.2)

P29.7 (a) $\bar{X}$ limits: $5.68 and $6.02, R limits: $0.06 and $0.84; (b) 76.7 percent; (c) increase specification limit band by $0.55. (Sec. 29.3)

P29.9 (a) From P29.5, $\hat{\mu} = 600.75$, $\hat{\sigma} = 1.51$, probability is 0.2033; (b) 596.22 and 605.28 V. (Sec. 29.3)

P29.10 (a) No for the 2.4 percent limit and yes for the 2.6 percent limit; (b) 2.33 percent, 2.63 percent, wider than specification limit band by 0.10 percent. (Sec. 29.3)

FINAL ANSWERS SELECTED PROBLEMS **615**

P29.13 (a) $CL_p = 0$ and 0.139, in control; (b) using $\bar{p} = 0.065$, Thursday ($p = 0.200$) is above the $UCL_p = 0.170$. (Sec. 29.4)

P29.15 (a) After three revisions, $\bar{p} = 0.0457$ and $CL_p = 0.0014$ and 0.0900 for $n = 200$; (b) all in control for $n = 100$. (Sec. 29.4)

P29.17 For $\overline{np} = 315.20$, $CL_{np} = 267.15$ and 363.25, six semesters are outside these limits; exact limits needed for some semesters. (Sec. 29.4)

P29.18 0.004. (Sec. 29.4)

P29.20 $CL_c = 42.24$ and 91.26 per 10 bushel baskets; out of control for days 3 (92), 8 (102) and 12 (31). (Sec. 29.5)

P29.21 Different values outside limits now: days 4 (32.7 per 10 baskets) and 9 (118.3 per 10 baskets). (Sec. 29.5)

P29.24 (a) 0.809; (b) 0.176; (c) 0.942. (Sec. 29.6)

P29.25 (a) (Sec. 29.6)

p	0.001	0.003	0.01	0.02	0.04	0.06
$P_a, n = 75$	0.997	0.978	0.827	0.558	0.199	0.061
$P_a, n = 50$	0.999	0.990	0.894	0.736	0.406	0.199

(b) $n = 75$, $\beta = 0.017$; $n = 50$, $\beta = 0.092$.

P29.26 (a) 0.966; (b) 0.0276; (c) 4.6 percent; (d) 8.35 percent. (Sec. 29.6)

P29.28 For $n = 50$, $c = 1$, $P_a = 0.910$ at AQL and $P_a = 0.1365$ at $LTPD$; does not meet either criteria. (Sec. 29.6)

P29.30 (a) $n = 200$, $c = 14$; (b) $n = 200$, $c = 10$. (Sec. 29.7)

P29.32 (a) Depot A: $n = 50$, $c = 2$; depot B: $n = 125$, $c = 5$; (b) depot B plan with $\alpha = 0.0125$. (Sec. 29.7)

P29.33 Liberal. (Sec. 29.7)

Chapter 30

Note: Only the mean square and final F values are given for ANOVA problems.

P30.2 (Sec. 30.2)

Source	Mean square	F
Regions	0.65	0.104
Error	6.28	

Do not reject H_0: $T_I = T_{II} = \ldots = T_V$.

P30.5 (a) (Sec. 30.2)

Source	Mean square	F
Lots	1.6	0.038
Error	42.1	

Do not reject H_0: $T_A = T_B$.

P30.7 (Sec. 30.3)

Source	Mean square	F
Regression	4.97	105.74
Error	0.047	

Reject $H_0: b = 0$.

P30.9 (a) (Sec. 30.3)

Source	Mean square	F
Regression	39.12	41.62
Error	0.94	

Reject $H_0: b = 0$; (b) $t = 6.45$.

P30.10 (Sec. 30.3)

Source	Mean square	F
Regression	701.34	19.56
Error	35.85	

Reject $H_0: b_1 = b_2 = 0$.

P30.12 (Sec. 30.4)

Source	Mean square	F
Catalyst	19.57	1.23
Location	157.87	9.95
Error	15.86	

(a) Do not reject $H_0: T_j = 0$ for all catalysts (no treatment effect); (b) reject $H_0: B_i = 0$ for all locations (block effect present).

P30.13 (a) (Sec. 30.4)

Source	Mean square	F
Region	0.65	0.085
Division	0.94	0.123
Error	7.62	

Do not reject H_0: $T_j = 0$ for all regions (treatments) or H_0: $B_i = 0$ for all divisions (block); (b) yes.

P30.15 (a) (Sec. 30.4)

Variation	Mean square	F
Technique	198.06	1.13
Number of weeks	448.90	2.56
Error	175.56	

Do not reject H_0: $T_j = 0$ for all techniques; (b) no; (c) one possible design follows.

Number of weeks (blocks)	Technique (treatments)			
	1	2	3	4
2	MT	FT	FA	MA
3	FA	MT	MA	FT
4	MT	FA	FT	MA
5	FT	FA	MT	MA

P30.17 (a) Plot of cell averages are close to parallel, so there is no interaction affect. (Sec. 30.5)
    (b)

Source	Mean square	F
A	4.31	1.40
B	1.71	0.56
AB	0.27	0.09
Error	3.07	

Do not reject any of the null hypotheses.

P30.20 (b) (Sec. 30.5)

Source	Mean square	F
Catalyst (A)	24.28	1.18
Location (B)	216.30	10.54
AB	22.88	1.11
Error	20.53	

Do not reject the null hypotheses of no factor A or interaction effect, but there is a factor B (location) effect.

APPENDIX
B

STATISTICAL TABLES

B-1 Cumulative Distribution Function for the Binomial Distribution
B-2 Cumulative Distribution Function for the Poisson Distribution
B-3 Cumulative Distribution Function for the Standard Normal Distribution (SND)
B-4 The χ^2 Distribution
B-5 The t Distribution
B-6 The F Distribution
B-7 D Distribution for the Kolmogorov-Smirnov Goodness-of-Fit Test
B-8 r Distribution for the Runs Test of Randomness
B-9 T Distribution of the Sum of Signed Ranks for the Wilcoxon Test of Means

Table B-1 Cumulative distribution function for the binomial distribution

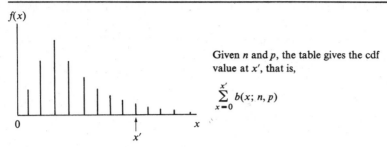

Given n and p, the table gives the cdf value at x', that is,

$$\sum_{x=0}^{x'} b(x; n, p)$$

n	x'	0.025	0.05	0.10	0.20	0.25	0.30	0.40	0.50
1	0	.9750	.9500	.9000	.8000	.7500	.7000	.6000	.5000
2	0	.9506	.9025	.8100	.6400	.5625	.4900	.3600	.2500
	1	.9994	.9975	.9900	.9600	.9375	.9100	.8400	.7500
3	0	.9269	.8574	.7290	.5120	.4219	.3430	.2160	.1250
	1	.9982	.9928	.9720	.8960	.8438	.7840	.6480	.5000
	2	1.0000	.9999	.9990	.9920	.9844	.9730	.9360	.8750
4	0	.9073	.8145	.6561	.4096	.3164	.2401	.1296	.0625
	1	.9964	.9860	.9477	.8192	.7383	.6517	.4752	.3125
	2	.9999	.9995	.9963	.9728	.9492	.9163	.8208	.6875
	3	1.0000	1.0000	.9999	.9984	.9961	.9919	.9744	.9375
5	0	.8811	.7738	.5905	.3277	.2373	.1681	.0778	.0312
	1	.9941	.9774	.9185	.7373	.6328	.5282	.3370	.1875
	2	.9998	.9988	.9914	.9421	.8965	.8369	.6826	.5000
	3	1.0000	1.0000	.9995	.9933	.9344	.9692	.9130	.8125
	4			1.0000	.9997	.9990	.9976	.9898	.9688
6	0	.8591	.7351	.5314	.2621	.1780	.1176	.0467	.0156
	1	.9912	.9672	.8857	.6554	.5339	.4202	.2333	.1094
	2	.9997	.9978	.9842	.9011	.8306	.7443	.5443	.3438
	3	1.0000	.9999	.9987	.9830	.9624	.9295	.8208	.6562
	4		1.0000	.9999	.9984	.9954	.9891	.9590	.8906
	5			1.0000	.9999	.9998	.9993	.9959	.9844
7	0	.8376	.6983	.4783	.2097	.1335	.0824	.0280	.0078
	1	.9879	.9556	.8503	.5767	.4449	.3294	.1586	.0625
	2	.9995	.9962	.9743	.8520	.7564	.6471	.4199	.2266
	3	1.0000	.9998	.9973	.9667	.9294	.8740	.7102	.5000
	4		1.0000	.9998	.9953	.9871	.9712	.9037	.7734
	5			1.0000	.9996	.9987	.9962	.9812	.9375
	6				1.0000	.9999	.9998	.9984	.9922
8	0	.8167	.6634	.4305	.1678	.1001	.0576	.0168	.0039
	1	.9842	.9428	.8131	.5033	.3671	.2553	.1064	.0352
	2	.9992	.9942	.9619	.7969	.6785	.5518	.3154	.1445
	3	1.0000	.9996	.9950	.9437	.8862	.8059	.5941	.3633
	4		1.0000	.9996	.9896	.9727	.9420	.8263	.6367

All blank spaces below 1.0000 for a particular p and n may be interpreted to be 1.0000.

Table B-1 Cumulative distribution function for the binomial distribution

n	x'	0.025	0.05	0.10	0.20	0.25	0.30	0.40	0.50
	5			1.0000	.9988	.9958	.9887	.9502	.8555
	6				.9999	.9996	.9987	.9915	.9648
	7				1.0000	1.0000	.9999	.9993	.9961
9	0	.7962	.6302	.3874	.1342	.0751	.0404	.0101	.0020
	1	.9800	.9288	.7748	.4362	.3003	.1960	.0705	.0195
	2	.9988	.9916	.9470	.7382	.6007	.4628	.2318	.0898
	3	1.0000	.9994	.9917	.9144	.8343	.7297	.4826	.2539
	4		1.0000	.9991	.9804	.9511	.9012	.7334	.5000
	5			.9999	.9969	.9900	.9747	.9006	.7461
	6			1.0000	.9997	.9987	.9957	.9750	.9102
	7				1.0000	.9999	.9996	.9962	.9805
	8					1.0000	1.0000	.9997	.9980
10	0	.7763	.5987	.3487	.1074	.0563	.0282	.0060	.0010
	1	.9754	.9139	.7361	.3758	.2440	.1493	.0464	.0107
	2	.9984	.9885	.9298	.6778	.5256	.3828	.1673	.0547
	3	.9999	.9990	.9872	.8791	.7759	.6496	.3823	.1719
	4	1.0000	.9999	.9984	.9672	.9219	.8497	.6331	.3770
	5		1.0000	.9999	.9936	.9803	.9527	.8338	.6230
	6			1.0000	.9991	.9965	.9894	.9452	.8281
	7				.9999	.9996	.9984	.9877	.9453
	8				1.0000	1.0000	.9999	.9983	.9893
	9						1.0000	.9999	.9990
12	0	.7380	.5404	.2824	.0687	.0317	.0138	.0022	.0002
	1	.9651	.8816	.6590	.2749	.1584	.0850	.0196	.0032
	2	.9971	.9804	.8891	.5583	.3907	.2528	.0834	.0193
	3	.9998	.9978	.9744	.7946	.6488	.4925	.2253	.0730
	4	1.0000	.9998	.9957	.9274	.8424	.7237	.4382	.1938
	5		1.0000	.9995	.9806	.9456	.8822	.6652	.3872
	6			.9999	.9961	.9857	.9614	.8418	.6128
	7			1.0000	.9994	.9972	.9905	.9427	.8062
	8				.9999	.9996	.9983	.9847	.9270
	9				1.0000	1.0000	.9998	.9972	.9807
	10						1.0000	.9997	.9968
	11							1.0000	.9998
14	0	.7016	.4877	.2288	.0440	.0178	.0068	.0008	.0001
	1	.9534	.8470	.5846	.1979	.1010	.0475	.0081	.0009
	2	.9954	.9699	.8416	.4481	.2811	.1608	.0398	.0065
	3	.9997	.9958	.9559	.6982	.5213	.3552	.1243	.0287
	4	1.0000	.9996	.9908	.8702	.7415	.5842	.2793	.0898
	5		1.0000	.9985	.9561	.8883	.7805	.4859	.2120
	6			.9998	.9884	.9617	.9067	.6925	.3953
	7			1.0000	.9976	.9897	.9685	.8499	.6047
	8				.9996	.9978	.9917	.9417	.7880
	9				1.0000	.9997	.9983	.9825	.9102

Table B-1 Cumulative distribution function for the binomial distribution

n	x'	0.025	0.05	0.10	0.20	0.25	0.30	0.40	0.50
	10					1.0000	.9998	.9961	.9713
	11						1.0000	.9994	.9935
	12							.9999	.9991
	13							1.0000	.9999
15	0	.6840	.4633	.2059	.0352	.0134	.0047	.0005	.0000
	1	.9471	.8290	.5490	.1671	.0802	.0353	.0052	.0005
	2	.9943	.9638	.8159	.3980	.2361	.1268	.0271	.0037
	3	.9996	.9945	.9444	.6482	.4613	.2969	.0905	.0176
	4	1.0000	.9994	.9873	.8358	.6365	.5155	.2173	.0592
	5		.9999	.9978	.9389	.8516	.7216	.4032	.1509
	6		1.0000	.9997	.9819	.9434	.8689	.6098	.3036
	7			1.0000	.9958	.9827	.9500	.7869	.5000
	8				.9992	.9958	.9848	.9050	.6964
	9				.9999	.9992	.9963	.9662	.8491
	10				1.0000	.9999	.9993	.9907	.9408
	11					1.0000	.9999	.9981	.9824
	12						1.0000	.9997	.9963
	13							1.0000	.9995
	14								1.0000
16	0	.6669	.4401	.1853	.0281	.0100	.0033	.0003	.0000
	1	.9405	.8108	.5147	.1407	.0635	.0261	.0033	.0003
	2	.9931	.9571	.7892	.3518	.1971	.0994	.0183	.0021
	3	.9994	.9930	.9316	.5981	.4050	.2459	.0651	.0106
	4	1.0000	.9991	.9830	.7982	.6302	.4499	.1666	.0384
	5		.9999	.9967	.9183	.8103	.6598	.3288	.1051
	6		1.0000	.9995	.9733	.9204	.8247	.5272	.2272
	7			.9999	.9930	.9729	.9256	.7161	.4018
	8			1.0000	.9985	.9925	.9743	.8577	.5982
	9				.9998	.9984	.9929	.9417	.7728
	10				1.0000	.9997	.9984	.9809	.8949
	11					1.0000	.9997	.9951	.9616
	12						1.0000	.9991	.9894
	13							.9999	.9979
	14							1.0000	.9997
	15								1.0000
18	0	.6340	.3972	.1501	.0180	.0056	.0016	.0001	.0000
	1	.9266	.7735	.4503	.0991	.0395	.0142	.0013	.0001
	2	.9904	.9419	.7338	.2713	.1353	.0600	.0082	.0007
	3	.9991	.9891	.9018	.5010	.3057	.1646	.0328	.0038
	4	.9999	.9985	.9718	.7164	.5187	.3327	.0942	.0154
	5	1.0000	.9998	.9936	.8671	.7175	.5344	.2088	.0481
	6		1.0000	.9988	.9487	.8610	.7217	.3743	.1189
	7			.9998	.9837	.9431	.8593	.5634	.2403
	8			1.0000	.9957	.9807	.9404	.7368	.4073
	9				.9991	.9946	.9790	.8653	.5927

Table B-1 Cumulative distribution function for the binomial distribution

n	x'	0.025	0.05	0.10	0.20	0.25	0.30	0.40	0.50
	10				.9998	.9988	.9939	.9424	.7597
	11				1.0000	.9998	.9986	.9797	.8811
	12					1.0000	.9997	.9942	.9519
	13						1.0000	.9987	.9846
	14							.9998	.9962
	15							1.0000	.9993
	16								.9999
	17								1.0000
20	0	.6027	.3585	.1216	.0115	.0032	.0008	.0000	.0000
	1	.9118	.7358	.3917	.0692	.0243	.0076	.0005	.0000
	2	.9870	.9245	.6769	.2061	.0913	.0355	.0036	.0002
	3	.9986	.9841	.8670	.4114	.2252	.1071	.0160	.0013
	4	.9999	.9974	.9568	.6296	.4148	.2375	.0510	.0059
	5	1.0000	.9997	.9887	.8042	.6172	.4164	.1256	.0207
	6		1.0000	.9976	.9133	.7858	.6080	.2500	.0577
	7			.9996	.9679	.8982	.7723	.4159	.1316
	8			.9999	.9900	.9591	.8867	.5956	.2517
	9			1.0000	.9974	.9861	.9520	.7553	.4119
	10				.9994	.9961	.9829	.8725	.5881
	11				.9999	.9991	.9949	.9435	.7483
	12				1.0000	.9998	.9987	.9790	.8684
	13					1.0000	.9997	.9935	.9423
	14						1.0000	.9984	.9793
	15							.9997	.9941
	16							1.0000	.9987
	17								.9998
	18								1.0000

Table B-2 Cumulative distribution function for the Poisson distribution

$P(x; \lambda)$

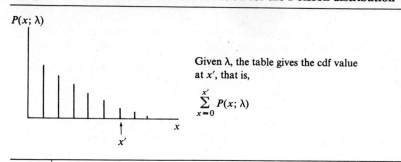

Given λ, the table gives the cdf value at x', that is,

$$\sum_{x=0}^{x'} P(x; \lambda)$$

					λ					
x'	0.02	0.04	0.05	0.06	0.08	0.10	0.15	0.20	0.25	0.30
0	.980	.961	.951	.942	.923	.905	.861	.819	.779	.741
1	1.000	.999	.999	.998	.997	.995	.990	.982	.974	.963
2		1.000	1.000	1.000	1.000	1.000	.999	.999	.998	.996
3							1.000	1.000	1.000	1.000

x'	0.35	0.40	0.45	0.50	0.55	0.60	0.65	0.70	0.75	0.80
0	.705	.670	.638	.607	.577	.549	.522	.497	.472	.449
1	.951	.938	.925	.910	.894	.878	.861	.844	.827	.809
2	.994	.992	.989	.986	.982	.977	.972	.966	.959	.953
3	1.000	.999	.999	.998	.998	.997	.996	.994	.993	.991
4		1.000	1.000	1.000	1.000	1.000	.999	.999	.999	.999
5							1.000	1.000	1.000	1.000

x'	0.85	0.90	0.95	1.00	1.05	1.10	1.15	1.20	1.25	1.30
0	.427	.407	.387	.368	.350	.333	.317	.301	.287	.273
1	.791	.772	.754	.736	.717	.699	.681	.663	.645	.627
2	.945	.937	.929	.920	.910	.900	.890	.879	.868	.857
3	.989	.987	.984	.981	.978	.974	.970	.966	.962	.957
4	.998	.998	.997	.996	.996	.995	.993	.992	.991	.989
5	1.000	1.000	1.000	.999	.999	.999	.999	.998	.998	.998
6				1.000	1.000	1.000	1.000	1.000	1.000	1.000

x'	1.35	1.40	1.45	1.50	1.55	1.60	1.65	1.70	1.75	1.80
0	.259	.247	.235	.223	.212	.202	.192	.183	.174	.165
1	.609	.592	.575	.558	.541	.525	.509	.493	.478	.463
2	.845	.833	.821	.809	.796	.783	.770	.757	.744	.731
3	.952	.946	.940	.934	.928	.921	.914	.907	.899	.891
4	.988	.986	.984	.981	.979	.976	.973	.970	.967	.964
5	.997	.997	.996	.996	.995	.994	.993	.992	.991	.990
6	.999	.999	.999	.999	.999	.999	.998	.998	.998	.997
7	1.000	1.000	1.000	1.000	1.000	1.000	1.000	1.000	1.000	.999
8										1.000

All blank spaces below 1.000 for a particular λ may be interpreted to be 1.000.

Table B-2 Cumulative distribution function for the Poisson distribution

x'	λ									
	1.85	1.90	1.95	2.00	2.20	2.40	2.60	2.80	3.00	3.20
0	.157	.150	.142	.135	.111	.091	.074	.061	.050	.041
1	.448	.434	.420	.406	.355	.308	.267	.231	.199	.171
2	.717	.704	.690	.677	.623	.570	.518	.469	.423	.380
3	.883	.875	.866	.857	.819	.779	.736	.692	.647	.603
4	.960	.956	.952	.947	.928	.904	.877	.848	.815	.781
5	.988	.987	.985	.983	.975	.964	.951	.935	.916	.895
6	.997	.997	.996	.995	.993	.988	.983	.976	.966	.955
7	.999	.999	.999	.999	.998	.997	.995	.992	.988	.983
8	1.000	1.000	1.000	1.000	1.000	.999	.999	.998	.996	.994
9						1.000	1.000	.999	.999	.998
10								1.000	1.000	1.000

x'	λ									
	3.40	3.60	3.80	4.00	4.20	4.40	4.60	4.80	5.00	5.20
0	.033	.027	.022	.018	.015	.012	.010	.008	.007	.006
1	.147	.126	.107	.092	.078	.066	.056	.048	.040	.034
2	.340	.303	.269	.238	.210	.185	.163	.143	.125	.109
3	.558	.515	.473	.433	.395	.359	.326	.294	.265	.238
4	.744	.706	.668	.629	.590	.551	.513	.476	.440	.406
5	.871	.844	.816	.785	.753	.720	.686	.651	.616	.581
6	.942	.927	.909	.889	.867	.844	.818	.791	.762	.732
7	.977	.969	.960	.949	.936	.921	.905	.887	.867	.845
8	.992	.988	.984	.979	.972	.964	.955	.944	.932	.918
9	.997	.996	.994	.992	.989	.985	.980	.975	.968	.960
10	.999	.999	.998	.997	.996	.994	.992	.990	.986	.982
11	1.000	1.000	.999	.999	.999	.998	.997	.996	.995	.993
12			1.000	1.000	1.000	.999	.999	.999	.998	.997
13						1.000	1.000	1.000	.999	.999
14									1.000	1.000

x'	5.40	5.60	5.80	6.00	6.20	6.40	6.60	6.80	7.00	7.50
0	.005	.004	.003	.002	.002	.002	.001	.001	.001	.001
1	.029	.024	.021	.017	.015	.012	.010	.009	.007	.005
2	.095	.082	.072	.062	.054	.046	.040	.034	.030	.020
3	.213	.191	.170	.151	.134	.119	.105	.093	.082	.059
4	.373	.342	.313	.285	.259	.235	.213	.192	.173	.132
5	.546	.512	.478	.446	.414	.384	.355	.327	.301	.241
6	.702	.670	.638	.606	.574	.542	.511	.480	.450	.378
7	.822	.797	.771	.744	.716	.687	.658	.628	.599	.525
8	.903	.886	.867	.847	.826	.803	.780	.755	.729	.662
9	.951	.941	.929	.916	.902	.886	.869	.850	.830	.776
10	.977	.972	.965	.957	.949	.939	.927	.915	.901	.862
11	.990	.988	.984	.980	.975	.969	.963	.955	.947	.921
12	.996	.995	.993	.991	.989	.986	.982	.978	.973	.957
13	.999	.998	.997	.996	.995	.994	.992	.990	.987	.978
14	1.000	.999	.999	.998	.998	.997	.997	.996	.994	.990

Table B-2 Cumulative distribution function for the Poisson distribution

x'	λ 5.40	5.60	5.80	6.00	6.20	6.40	6.60	6.80	7.00	7.50
15		1.000	1.000	.999	.999	.999	.999	.998	.998	.995
16				1.000	1.000	1.000	1.000	.999	.999	.998
17								1.000	1.000	.999
18										1.000

x'	λ 8.00	8.50	9.00	9.50	10.00	10.50	11.00	11.50	12.00	12.50
0	.000	.000	.000	.000	.000	.000	.000	.000	.000	.000
1	.003	.002	.001	.001	.000	.000	.000	.000	.000	.000
2	.014	.009	.006	.004	.003	.002	.001	.001	.001	.000
3	.042	.030	.021	.015	.010	.007	.005	.003	.002	.002
4	.100	.074	.055	.040	.029	.021	.015	.011	.008	.005
5	.191	.150	.116	.089	.067	.050	.038	.028	.020	.015
6	.313	.256	.207	.165	.130	.102	.079	.060	.046	.035
7	.453	.386	.324	.269	.220	.179	.143	.114	.090	.070
8	.593	.523	.456	.392	.333	.279	.232	.191	.155	.125
9	.717	.653	.587	.522	.458	.397	.341	.289	.242	.201
10	.816	.763	.706	.645	.583	.521	.460	.402	.347	.297
11	.888	.849	.803	.752	.697	.639	.579	.520	.462	.406
12	.936	.909	.876	.836	.792	.742	.689	.633	.576	.519
13	.966	.949	.926	.898	.864	.825	.781	.733	.682	.628
14	.983	.973	.959	.940	.917	.888	.854	.815	.772	.725
15	.992	.986	.978	.967	.951	.932	.907	.878	.844	.806
16	.996	.993	.989	.982	.973	.960	.944	.924	.899	.869
17	.998	.997	.995	.991	.986	.978	.968	.954	.937	.916
18	.999	.999	.998	.996	.993	.988	.982	.974	.963	.948
19	1.000	.999	.999	.998	.997	.994	.991	.986	.979	.969
20		1.000	1.000	.999	.998	.997	.995	.992	.988	.983
21				1.000	.999	.999	.998	.996	.994	.991
22					1.000	.999	.999	.998	.997	.995
23						1.000	1.000	.999	.999	.998
24								1.000	.999	.999
25									1.000	.999
26										1.000

x'	λ 13.00	13.50	14.00	14.50	15.00	16.00	17.00	18.00	19.00	20.00
1	.000	.000	.000	.000	.000	.000	.000	.000	.000	.000
2	.000	.000	.000	.000	.000	.000	.000	.000	.000	.000
3	.001	.001	.000	.000	.000	.000	.000	.000	.000	.000
4	.004	.003	.002	.001	.001	.000	.000	.000	.000	.000

Table B-2 Cumulative distribution function for the Poisson distribution

x'	\multicolumn{10}{c}{λ}									
	13.00	13.50	14.00	14.50	15.00	16.00	17.00	18.00	19.00	20.00
5	.011	.008	.006	.004	.003	.001	.001	.000	.000	.000
6	.026	.019	.014	.010	.008	.004	.002	.001	.001	.000
7	.054	.041	.032	.024	.018	.010	.005	.003	.002	.001
8	.100	.079	.062	.048	.037	.022	.013	.007	.004	.002
9	.166	.135	.109	.088	.070	.043	.026	.015	.009	.005
10	.252	.211	.176	.145	.118	.077	.049	.030	.018	.011
11	.353	.304	.260	.220	.185	.127	.085	.055	.035	.021
12	.463	.409	.358	.311	.268	.193	.135	.092	.061	.039
13	.573	.518	.464	.413	.363	.275	.201	.143	.098	.066
14	.675	.623	.570	.518	.466	.368	.281	.208	.150	.105
15	.764	.718	.669	.619	.568	.467	.371	.287	.215	.157
16	.835	.798	.756	.711	.664	.566	.468	.375	.292	.221
17	.890	.861	.827	.790	.749	.659	.564	.469	.378	.297
18	.930	.908	.883	.853	.819	.742	.655	.562	.469	.381
19	.957	.942	.923	.901	.875	.812	.736	.651	.561	.470
20	.975	.965	.952	.936	.917	.868	.805	.731	.647	.559
21	.986	.980	.971	.960	.947	.911	.861	.799	.725	.644
22	.992	.989	.983	.976	.967	.942	.905	.855	.793	.721
23	.996	.994	.991	.986	.981	.963	.937	.899	.849	.787
24	.998	.997	.995	.992	.989	.978	.959	.932	.893	.843
25	.999	.998	.997	.996	.994	.987	.975	.955	.927	.888
26	1.000	.999	.999	.998	.997	.993	.985	.972	.951	.922
27		1.000	.999	.999	.998	.996	.991	.983	.969	.948
28			1.000	.999	.999	.998	.995	.990	.980	.966
29				1.000	1.000	.999	.997	.994	.988	.978
30						.999	.999	.997	.993	.987
31						1.000	.999	.998	.996	.992
32							1.000	.999	.998	.995
33								1.000	.999	.997
34									.999	.999
35									1.000	.999
36										1.000

Table B-3 Cumulative distribution function for the standard normal distribution (SND)

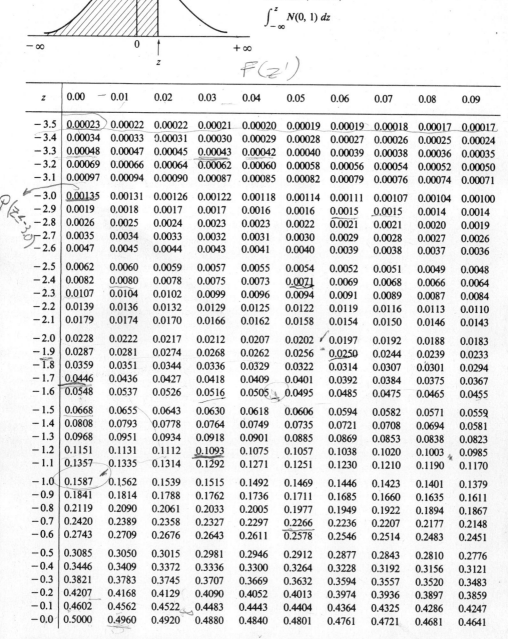

For the SND, the table gives the cdf value at z, that is,

$$\int_{-\infty}^{z} N(0, 1)\, dz$$

z	0.00	0.01	0.02	0.03	0.04	0.05	0.06	0.07	0.08	0.09
−3.5	0.00023	0.00022	0.00022	0.00021	0.00020	0.00019	0.00019	0.00018	0.00017	0.00017
−3.4	0.00034	0.00033	0.00031	0.00030	0.00029	0.00028	0.00027	0.00026	0.00025	0.00024
−3.3	0.00048	0.00047	0.00045	0.00043	0.00042	0.00040	0.00039	0.00038	0.00036	0.00035
−3.2	0.00069	0.00066	0.00064	0.00062	0.00060	0.00058	0.00056	0.00054	0.00052	0.00050
−3.1	0.00097	0.00094	0.00090	0.00087	0.00085	0.00082	0.00079	0.00076	0.00074	0.00071
−3.0	0.00135	0.00131	0.00126	0.00122	0.00118	0.00114	0.00111	0.00107	0.00104	0.00100
−2.9	0.0019	0.0018	0.0017	0.0017	0.0016	0.0016	0.0015	0.0015	0.0014	0.0014
−2.8	0.0026	0.0025	0.0024	0.0023	0.0023	0.0022	0.0021	0.0021	0.0020	0.0019
−2.7	0.0035	0.0034	0.0033	0.0032	0.0031	0.0030	0.0029	0.0028	0.0027	0.0026
−2.6	0.0047	0.0045	0.0044	0.0043	0.0041	0.0040	0.0039	0.0038	0.0037	0.0036
−2.5	0.0062	0.0060	0.0059	0.0057	0.0055	0.0054	0.0052	0.0051	0.0049	0.0048
−2.4	0.0082	0.0080	0.0078	0.0075	0.0073	0.0071	0.0069	0.0068	0.0066	0.0064
−2.3	0.0107	0.0104	0.0102	0.0099	0.0096	0.0094	0.0091	0.0089	0.0087	0.0084
−2.2	0.0139	0.0136	0.0132	0.0129	0.0125	0.0122	0.0119	0.0116	0.0113	0.0110
−2.1	0.0179	0.0174	0.0170	0.0166	0.0162	0.0158	0.0154	0.0150	0.0146	0.0143
−2.0	0.0228	0.0222	0.0217	0.0212	0.0207	0.0202	0.0197	0.0192	0.0188	0.0183
−1.9	0.0287	0.0281	0.0274	0.0268	0.0262	0.0256	0.0250	0.0244	0.0239	0.0233
−1.8	0.0359	0.0351	0.0344	0.0336	0.0329	0.0322	0.0314	0.0307	0.0301	0.0294
−1.7	0.0446	0.0436	0.0427	0.0418	0.0409	0.0401	0.0392	0.0384	0.0375	0.0367
−1.6	0.0548	0.0537	0.0526	0.0516	0.0505	0.0495	0.0485	0.0475	0.0465	0.0455
−1.5	0.0668	0.0655	0.0643	0.0630	0.0618	0.0606	0.0594	0.0582	0.0571	0.0559
−1.4	0.0808	0.0793	0.0778	0.0764	0.0749	0.0735	0.0721	0.0708	0.0694	0.0581
−1.3	0.0968	0.0951	0.0934	0.0918	0.0901	0.0885	0.0869	0.0853	0.0838	0.0823
−1.2	0.1151	0.1131	0.1112	0.1093	0.1075	0.1057	0.1038	0.1020	0.1003	0.0985
−1.1	0.1357	0.1335	0.1314	0.1292	0.1271	0.1251	0.1230	0.1210	0.1190	0.1170
−1.0	0.1587	0.1562	0.1539	0.1515	0.1492	0.1469	0.1446	0.1423	0.1401	0.1379
−0.9	0.1841	0.1814	0.1788	0.1762	0.1736	0.1711	0.1685	0.1660	0.1635	0.1611
−0.8	0.2119	0.2090	0.2061	0.2033	0.2005	0.1977	0.1949	0.1922	0.1894	0.1867
−0.7	0.2420	0.2389	0.2358	0.2327	0.2297	0.2266	0.2236	0.2207	0.2177	0.2148
−0.6	0.2743	0.2709	0.2676	0.2643	0.2611	0.2578	0.2546	0.2514	0.2483	0.2451
−0.5	0.3085	0.3050	0.3015	0.2981	0.2946	0.2912	0.2877	0.2843	0.2810	0.2776
−0.4	0.3446	0.3409	0.3372	0.3336	0.3300	0.3264	0.3228	0.3192	0.3156	0.3121
−0.3	0.3821	0.3783	0.3745	0.3707	0.3669	0.3632	0.3594	0.3557	0.3520	0.3483
−0.2	0.4207	0.4168	0.4129	0.4090	0.4052	0.4013	0.3974	0.3936	0.3897	0.3859
−0.1	0.4602	0.4562	0.4522	0.4483	0.4443	0.4404	0.4364	0.4325	0.4286	0.4247
−0.0	0.5000	0.4960	0.4920	0.4880	0.4840	0.4801	0.4761	0.4721	0.4681	0.4641

Table B-3 is reprinted, with permission, from E. L. Grant and R. S. Leavenworth, *Statistical Quality Control*, McGraw-Hill Book Company, New York, 1972.

Table B-3 Cumulative distribution function for the standard normal distribution (SND)

z	0.00	0.01	0.02	0.03	0.04	0.05	0.06	0.07	0.08	0.09
+0.0	0.5000	0.5040	0.5080	0.5120	0.5160	0.5199	0.5239	0.5279	0.5319	0.5359
+0.1	0.5398	0.5438	0.5478	0.5517	0.5557	0.5596	0.5636	0.5675	0.5714	0.5753
+0.2	0.5793	0.5832	0.5871	0.5910	0.5948	0.5987	0.6026	0.6064	0.6103	0.6141
+0.3	0.6179	0.6217	0.6255	0.6293	0.6331	0.6368	0.6406	0.6443	0.6480	0.6517
+0.4	0.6554	0.6591	0.6628	0.6664	0.6700	0.6736	0.6772	0.6808	0.6844	0.6870
+0.5	0.6915	0.6950	0.6985	0.7019	0.7054	0.7088	0.7123	0.7157	0.7190	0.7224
+0.6	0.7257	0.7291	0.7324	0.7357	0.7389	0.7422	0.7454	0.7486	0.7517	0.7549
+0.7	0.7580	0.7611	0.7642	0.7673	0.7704	0.7734	0.7764	0.7794	0.7823	0.7852
+0.8	0.7881	0.7910	0.7939	0.7967	0.7995	0.8023	0.8051	0.8079	0.8106	0.8133
+0.9	0.8159	0.8186	0.8212	0.8238	0.8264	0.8289	0.8315	0.8340	0.8365	0.8389
+1.0	0.8413	0.8438	0.8461	0.8485	0.8508	0.8531	0.8554	0.8577	0.8599	0.8621
+1.1	0.8643	0.8665	0.8686	0.8708	0.8729	0.8749	0.8770	0.8790	0.8810	0.8830
+1.2	0.8849	0.8869	0.8888	0.8907	0.8925	0.8944	0.8962	0.8980	0.8997	0.9015
+1.3	0.9032	0.9049	0.9066	0.9082	0.9099	0.9115	0.9131	0.9147	0.9162	0.9177
+1.4	0.9192	0.9207	0.9222	0.9236	0.9251	0.9265	0.9279	0.9292	0.9306	0.9319
+1.5	0.9332	0.9345	0.9357	0.9370	0.9382	0.9394	0.9406	0.9418	0.9429	0.9441
+1.6	0.9452	0.9463	0.9474	0.9484	0.9495	0.9505	0.9515	0.9525	0.9535	0.9545
+1.7	0.9554	0.9564	0.9573	0.9582	0.9591	0.9599	0.9608	0.9616	0.9625	0.9633
+1.8	0.9641	0.9649	0.9656	0.9664	0.9671	0.9678	0.9686	0.9693	0.9699	0.9706
+1.9	0.9713	0.9719	0.9726	0.9732	0.9738	0.9744	0.9750	0.9756	0.9761	0.9767
+2.0	0.9773	0.9778	0.9783	0.9788	0.9793	0.9798	0.9803	0.9808	0.9812	0.9817
+2.1	0.9821	0.9826	0.9830	0.9834	0.9838	0.9842	0.9846	0.9850	0.9854	0.9857
+2.2	0.9861	0.9864	0.9868	0.9871	0.9875	0.9878	0.9881	0.9884	0.9887	0.9890
+2.3	0.9893	0.9896	0.9898	0.9901	0.9904	0.9906	0.9909	0.9911	0.9913	0.9916
+2.4	0.9918	0.9920	0.9922	0.9925	0.9927	0.9929	0.9931	0.9932	0.9934	0.9936
+2.5	0.9938	0.9940	0.9941	0.9943	0.9945	0.9946	0.9948	0.9949	0.9951	0.9952
+2.6	0.9953	0.9955	0.9956	0.9957	0.9959	0.9960	0.9961	0.9962	0.9963	0.9964
+2.7	0.9965	0.9966	0.9967	0.9968	0.9969	0.9970	0.9971	0.9972	0.9973	0.9974
+2.8	0.9974	0.9975	0.9976	0.9977	0.9977	0.9978	0.9979	0.9979	0.9980	0.9981
+2.9	0.9981	0.9982	0.9983	0.9983	0.9984	0.9984	0.9985	0.9985	0.9986	0.9986
+3.0	0.99865	0.99869	0.99874	0.99878	0.99882	0.99886	0.99889	0.99893	0.99896	0.99900
+3.1	0.99903	0.99906	0.99910	0.99913	0.99915	0.99918	0.99921	0.99924	0.99926	0.99929
+3.2	0.99931	0.99934	0.99936	0.99938	0.99940	0.99942	0.99944	0.99946	0.99948	0.99950
+3.3	0.99952	0.99953	0.99955	0.99957	0.99958	0.99960	0.99961	0.99962	0.99964	0.99965
+3.4	0.99966	0.99967	0.99969	0.99970	0.99971	0.99972	0.99973	0.99974	0.99975	0.99976
+3.5	0.99977	0.99978	0.99978	0.99979	0.99980	0.99981	0.99981	0.99982	0.99983	0.99983

Table B-4 The χ^2 Distribution

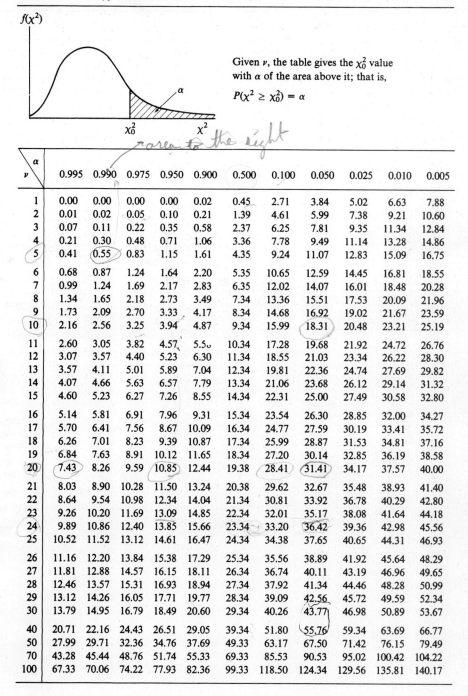

Given ν, the table gives the χ_0^2 value with α of the area above it; that is,

$$P(\chi^2 \geq \chi_0^2) = \alpha$$

α ν	0.995	0.990	0.975	0.950	0.900	0.500	0.100	0.050	0.025	0.010	0.005
1	0.00	0.00	0.00	0.00	0.02	0.45	2.71	3.84	5.02	6.63	7.88
2	0.01	0.02	0.05	0.10	0.21	1.39	4.61	5.99	7.38	9.21	10.60
3	0.07	0.11	0.22	0.35	0.58	2.37	6.25	7.81	9.35	11.34	12.84
4	0.21	0.30	0.48	0.71	1.06	3.36	7.78	9.49	11.14	13.28	14.86
5	0.41	0.55	0.83	1.15	1.61	4.35	9.24	11.07	12.83	15.09	16.75
6	0.68	0.87	1.24	1.64	2.20	5.35	10.65	12.59	14.45	16.81	18.55
7	0.99	1.24	1.69	2.17	2.83	6.35	12.02	14.07	16.01	18.48	20.28
8	1.34	1.65	2.18	2.73	3.49	7.34	13.36	15.51	17.53	20.09	21.96
9	1.73	2.09	2.70	3.33	4.17	8.34	14.68	16.92	19.02	21.67	23.59
10	2.16	2.56	3.25	3.94	4.87	9.34	15.99	18.31	20.48	23.21	25.19
11	2.60	3.05	3.82	4.57	5.58	10.34	17.28	19.68	21.92	24.72	26.76
12	3.07	3.57	4.40	5.23	6.30	11.34	18.55	21.03	23.34	26.22	28.30
13	3.57	4.11	5.01	5.89	7.04	12.34	19.81	22.36	24.74	27.69	29.82
14	4.07	4.66	5.63	6.57	7.79	13.34	21.06	23.68	26.12	29.14	31.32
15	4.60	5.23	6.27	7.26	8.55	14.34	22.31	25.00	27.49	30.58	32.80
16	5.14	5.81	6.91	7.96	9.31	15.34	23.54	26.30	28.85	32.00	34.27
17	5.70	6.41	7.56	8.67	10.09	16.34	24.77	27.59	30.19	33.41	35.72
18	6.26	7.01	8.23	9.39	10.87	17.34	25.99	28.87	31.53	34.81	37.16
19	6.84	7.63	8.91	10.12	11.65	18.34	27.20	30.14	32.85	36.19	38.58
20	7.43	8.26	9.59	10.85	12.44	19.38	28.41	31.41	34.17	37.57	40.00
21	8.03	8.90	10.28	11.50	13.24	20.38	29.62	32.67	35.48	38.93	41.40
22	8.64	9.54	10.98	12.34	14.04	21.34	30.81	33.92	36.78	40.29	42.80
23	9.26	10.20	11.69	13.09	14.85	22.34	32.01	35.17	38.08	41.64	44.18
24	9.89	10.86	12.40	13.85	15.66	23.34	33.20	36.42	39.36	42.98	45.56
25	10.52	11.52	13.12	14.61	16.47	24.34	34.38	37.65	40.65	44.31	46.93
26	11.16	12.20	13.84	15.38	17.29	25.34	35.56	38.89	41.92	45.64	48.29
27	11.81	12.88	14.57	16.15	18.11	26.34	36.74	40.11	43.19	46.96	49.65
28	12.46	13.57	15.31	16.93	18.94	27.34	37.92	41.34	44.46	48.28	50.99
29	13.12	14.26	16.05	17.71	19.77	28.34	39.09	42.56	45.72	49.59	52.34
30	13.79	14.95	16.79	18.49	20.60	29.34	40.26	43.77	46.98	50.89	53.67
40	20.71	22.16	24.43	26.51	29.05	39.34	51.80	55.76	59.34	63.69	66.77
50	27.99	29.71	32.36	34.76	37.69	49.33	63.17	67.50	71.42	76.15	79.49
70	43.28	45.44	48.76	51.74	55.33	69.33	85.53	90.53	95.02	100.42	104.22
100	67.33	70.06	74.22	77.93	82.36	99.33	118.50	124.34	129.56	135.81	140.17

Table B.5 The t distribution

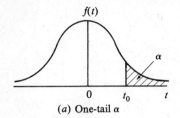
(a) One-tail α

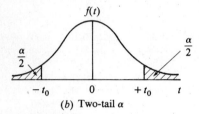
(b) Two-tail α

Given ν, the table gives (a) the one-tail t_0 value with α of the area above it, that is, $P(t \geq t_0) = \alpha$, or (b) the two-tail $+t_0$ and $-t_0$ values with $\alpha/2$ in each tail, that is, $P(t \leq -t_0) + P(t \geq +t_0) = \alpha$

	\multicolumn{6}{c}{One-tail α}
	0.10	0.05	0.025	0.01	0.005	0.001
	\multicolumn{6}{c}{Two-tail α}
ν	0.20	0.10	0.05	0.02	0.01	0.002
1	3.078	6.314	12.706	31.821	63.657	318.300
2	1.886	2.920	4.303	6.965	9.925	22.327
3	1.638	2.353	3.182	4.541	5.841	10.214
4	1.533	2.132	2.776	3.747	4.604	7.173
5	1.476	2.015	2.571	3.305	4.032	5.893
6	1.440	1.943	2.447	3.143	3.707	5.208
7	1.415	1.895	2.365	2.998	3.499	4.785
8	1.397	1.860	2.306	2.896	3.355	4.501
9	1.383	1.833	2.262	2.821	3.250	4.297
10	1.372	1.812	2.228	2.764	3.169	4.144
11	1.363	1.796	2.201	2.718	3.106	4.025
12	1.356	1.782	2.179	2.681	3.055	3.930
13	1.350	1.771	2.160	2.650	3.012	3.852
14	1.345	1.761	2.145	2.624	2.977	3.787
15	1.341	1.753	2.131	2.602	2.947	3.733
16	1.337	1.746	2.120	2.583	2.921	3.686
17	1.333	1.740	2.110	2.567	2.898	3.646
18	1.330	1.734	2.101	2.552	2.878	3.611
19	1.328	1.729	2.093	2.539	2.861	3.579
20	1.325	1.725	2.086	2.528	2.845	3.552
21	1.323	1.721	2.080	2.518	2.831	3.527
22	1.321	1.717	2.074	2.508	2.819	3.505
23	1.319	1.714	2.069	2.500	2.807	3.485
24	1.318	1.711	2.064	2.492	2.797	3.467
25	1.316	1.708	2.060	2.485	2.787	3.450

Table B-5 The *t* distribution

ν	One-tail α					
	0.10	0.05	0.025	0.01	0.005	0.001
	Two-tail α					
	0.20	0.10	0.05	0.02	0.01	0.002
26	1.315	1.706	2.056	2.479	2.779	3.435
27	1.314	1.703	2.052	2.473	2.771	3.421
28	1.313	1.701	2.048	2.467	2.763	3.408
29	1.311	1.699	2.045	2.462	2.756	3.396
30	1.310	1.697	2.042	2.457	2.750	3.385
40	1.303	1.684	2.021	2.423	2.704	3.307
60	1.296	1.671	2.000	2.390	2.660	3.232
80	1.292	1.664	1.990	2.374	2.639	3.195
100	1.290	1.660	1.984	2.365	2.626	3.174
∞	1.282	1.645	1.960	2.326	2.576	3.090

Table B-6 The F distribution ($\alpha = 0.10$, 0.05, and 0.01)

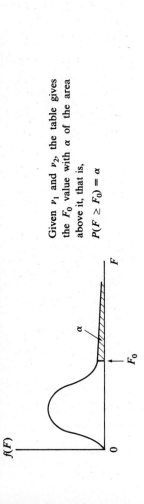

Given v_1 and v_2, the table gives the F_0 value with α of the area above it, that is, $P(F \geq F_0) = \alpha$

v_1 (numerator)

v_2	α	1	2	3	4	5	6	7	8	9	10	11	12	14	15	19	20	24	30	50	100	500	∞
1	.10	39.9	49.5	53.6	55.8	57.2	58.2	58.9	59.4	59.9	60.2	60.5	60.7	61.1	61.2	61.6	61.7	62.0	62.3	62.7	63.0	63.3	63.3
	.05	161	200	216	225	230	234	237	239	241	242	243	244	245	246	248	248	249	250	252	253	254	254
2	.10	8.53	9.00	9.16	9.24	9.29	9.33	9.35	9.37	9.38	9.39	9.40	9.41	9.42	9.42	9.44	9.44	9.45	9.46	9.47	9.48	9.49	9.49
	.05	18.5	19.0	19.2	19.2	19.3	19.3	19.4	19.4	19.4	19.4	19.4	19.4	19.4	19.4	19.4	19.4	19.5	19.5	19.5	19.5	19.5	19.5
	.01	98.5	99.0	99.2	99.2	99.3	99.3	99.4	99.4	99.4	99.4	99.4	99.4	99.4	99.4	99.4	99.4	99.5	99.5	99.5	99.5	99.5	99.5
3	.10	5.54	5.46	5.39	5.34	5.31	5.28	5.27	5.25	5.24	5.23	5.22	5.22	5.20	5.20	5.18	5.18	5.18	5.17	5.15	5.14	5.14	5.13
	.05	10.1	9.55	9.28	9.12	9.10	8.94	8.89	8.85	8.81	8.79	8.76	8.74	8.71	8.70	8.67	8.66	8.64	8.62	8.58	8.55	8.53	8.53
	.01	34.1	30.8	29.5	28.7	28.2	27.9	27.7	27.5	27.3	27.2	27.1	27.1	26.9	26.9	26.7	26.7	26.6	26.5	26.4	26.2	26.1	26.1
4	.10	4.54	4.32	4.19	4.11	4.05	4.01	3.98	3.95	3.94	3.92	3.91	3.90	3.88	3.87	3.84	3.84	3.83	3.82	3.80	3.78	3.76	3.76
	.05	7.71	6.94	6.59	6.39	6.26	6.16	6.09	6.04	6.00	5.96	5.94	5.91	5.87	5.86	5.81	5.80	5.77	5.75	5.70	5.66	5.64	5.63
	.01	21.2	18.0	16.7	16.0	15.5	15.2	15.0	14.8	14.7	14.5	14.4	14.4	14.2	14.2	14.0	14.0	13.9	13.8	13.7	13.6	13.5	13.5
5	.10	4.06	3.78	3.62	3.52	3.45	3.40	3.37	3.34	3.32	3.30	3.28	3.27	3.25	3.24	3.21	3.21	3.19	3.17	3.15	3.13	3.11	3.10
	.05	6.61	5.79	5.41	5.19	5.05	4.95	4.88	4.82	4.77	4.74	4.71	4.68	4.64	4.62	4.57	4.56	4.53	4.50	4.44	4.41	4.37	4.36
	.01	16.26	13.27	12.06	11.39	10.97	10.67	10.46	10.29	10.16	10.05	9.96	9.89	9.77	9.72	9.58	9.55	9.47	9.38	9.24	9.13	9.04	9.02

The degrees of freedom are v_1 for the numerator and v_2 for the denominator.

STATISTICAL TABLES 633

Table B-6 The F distribution (α = 0.10, 0.05, and 0.01)

v_1 (numerator)

v_2	α	1	2	3	4	5	6	7	8	9	10	11	12	14	15	19	20	24	30	50	100	500	∞
6	.10	3.78	3.46	3.29	3.18	3.11	3.05	3.01	2.98	2.96	2.94	2.92	2.90	2.88	2.87	2.84	2.84	2.82	2.80	2.77	2.75	2.73	2.72
	.05	5.99	5.14	4.76	4.53	4.39	4.28	4.21	4.15	4.10	4.06	4.03	4.00	3.96	3.94	3.88	3.87	3.84	3.81	3.75	3.71	3.68	3.67
	.01	13.74	10.92	9.78	9.15	8.75	8.47	8.26	8.10	7.98	7.87	7.79	7.72	7.60	7.56	7.42	7.40	7.31	7.23	7.09	6.99	6.90	6.88
7	.10	3.59	3.26	3.07	2.96	2.88	2.83	2.78	2.75	2.72	2.70	2.68	2.67	2.64	2.63	2.60	2.59	2.58	2.56	2.52	2.50	2.48	2.47
	.05	5.59	4.74	4.35	4.12	3.97	3.87	3.79	3.73	3.68	3.64	3.60	3.57	3.53	3.51	3.46	3.44	3.41	3.38	3.32	3.27	3.24	3.23
	.01	12.25	9.55	8.45	7.85	7.46	7.19	6.99	6.84	6.72	6.62	6.54	6.47	6.36	6.31	6.18	6.16	6.07	5.99	5.86	5.75	5.67	5.65
8	.10	3.46	3.11	2.92	2.81	2.73	2.67	2.62	2.59	2.56	2.54	2.52	2.50	2.47	2.46	2.43	2.42	2.40	2.38	2.35	2.32	2.30	2.29
	.05	5.32	4.46	4.07	3.84	3.69	3.58	3.50	3.44	3.39	3.35	3.31	3.28	3.24	3.22	3.16	3.15	3.12	3.08	3.02	2.97	2.94	2.93
	.01	11.26	8.65	7.59	7.01	6.63	6.37	6.18	6.03	5.91	5.81	5.73	5.67	5.56	5.52	5.38	5.36	5.28	5.20	5.07	4.96	4.88	4.86
9	.10	3.36	3.01	2.81	2.69	2.61	2.55	2.51	2.47	2.44	2.42	2.40	2.38	2.35	2.34	2.31	2.30	2.28	2.25	2.22	2.19	2.17	2.16
	.05	5.12	4.26	3.86	3.63	3.48	3.37	3.29	3.23	3.18	3.14	3.10	3.07	3.03	3.01	2.95	2.94	2.90	2.86	2.80	2.76	2.72	2.71
	.01	10.56	8.02	6.99	6.42	6.06	5.80	5.61	5.47	5.35	5.26	5.18	5.11	5.00	4.96	4.83	4.81	4.73	4.65	4.52	4.42	4.33	4.31
10	.10	3.28	2.92	2.73	2.61	2.52	2.46	2.41	2.38	2.35	2.32	2.30	2.28	2.25	2.24	2.21	2.20	2.18	2.16	2.12	2.09	2.06	2.06
	.05	4.96	4.10	3.71	3.48	3.33	3.22	3.14	3.07	3.02	2.98	2.94	2.91	2.86	2.85	2.78	2.77	2.74	2.70	2.64	2.59	2.55	2.54
	.01	10.04	7.56	6.55	5.99	5.64	5.39	5.20	5.06	4.94	4.85	4.77	4.71	4.60	4.56	4.43	4.41	4.33	4.25	4.12	4.01	3.93	3.91
11	.10	3.23	2.86	2.66	2.54	2.45	2.39	2.34	2.30	2.27	2.25	2.23	2.21	2.18	2.17	2.13	2.12	2.10	2.08	2.04	2.00	1.98	1.97
	.05	4.84	3.98	3.59	3.36	3.20	3.09	3.01	2.95	2.90	2.85	2.82	2.79	2.74	2.72	2.66	2.65	2.61	2.57	2.51	2.46	2.42	2.40
	.01	9.65	7.21	6.22	5.67	5.32	5.07	4.89	4.74	4.63	4.54	4.46	4.40	4.29	4.25	4.12	4.10	4.02	3.94	3.81	3.71	3.62	3.60
12	.10	3.18	2.81	2.61	2.48	2.39	2.33	2.28	2.24	2.21	2.19	2.17	2.15	2.11	2.10	2.07	2.06	2.04	2.01	1.97	1.94	1.91	1.90
	.05	4.75	3.89	3.49	3.26	3.11	3.00	2.91	2.85	2.80	2.75	2.72	2.69	2.64	2.62	2.56	2.54	2.51	2.47	2.40	2.35	2.31	2.30
	.01	9.33	6.93	5.95	5.41	5.06	4.82	4.64	4.50	4.39	4.30	4.22	4.16	4.05	4.01	3.88	3.86	3.78	3.70	3.57	3.47	3.38	3.36
14	.10	3.10	2.73	2.52	2.39	2.31	2.24	2.19	2.15	2.12	2.10	2.08	2.05	2.02	2.01	1.97	1.96	1.94	1.91	1.87	1.83	1.80	1.80
	.05	4.60	3.74	3.34	3.11	2.96	2.85	2.76	2.70	2.65	2.60	2.57	2.53	2.48	2.46	2.40	2.39	2.35	2.31	2.24	2.19	2.14	2.13
	.01	8.86	6.51	5.56	5.04	4.69	4.46	4.28	4.14	4.03	3.94	3.86	3.80	3.70	3.66	3.53	3.51	3.43	3.35	3.22	3.11	3.03	3.00
15	.10	3.07	2.70	2.49	2.36	2.27	2.21	2.16	2.12	2.09	2.06	2.04	2.02	1.98	1.97	1.93	1.92	1.90	1.87	1.83	1.79	1.76	1.76
	0.5	4.54	3.68	3.29	3.06	2.90	2.79	2.71	2.64	2.59	2.54	2.51	2.48	2.42	2.40	2.34	2.33	2.29	2.25	2.18	2.12	2.08	2.07
	.01	8.68	6.36	5.42	4.89	4.56	4.32	4.14	4.00	3.89	3.80	3.73	3.67	3.56	3.52	3.40	3.37	3.29	3.21	3.08	2.98	2.89	2.87
16	.10	3.05	2.67	2.46	2.33	2.24	2.18	2.13	2.09	2.06	2.03	2.01	1.99	1.95	1.94	1.90	1.89	1.87	1.84	1.79	1.76	1.73	1.72
	.05	4.49	3.63	3.24	3.01	2.85	2.74	2.66	2.59	2.54	2.49	2.46	2.42	2.37	2.35	2.29	2.28	2.24	2.19	2.12	2.07	2.02	2.01
	.01	8.53	6.23	5.29	4.77	4.44	4.20	4.03	3.89	3.78	3.69	3.62	3.55	3.45	3.41	3.28	3.26	3.18	3.10	2.97	2.86	2.78	2.75

Table B-6 The F distribution ($\alpha = 0.10$, 0.05, and 0.01)

v_1 (numerator)

v_2	α	1	2	3	4	5	6	7	8	9	10	11	12	14	15	19	20	24	30	50	100	500	∞
18	.10	3.01	2.62	2.42	2.29	2.20	2.13	2.08	2.04	2.00	1.98	1.96	1.93	1.90	1.89	1.85	1.84	1.81	1.78	1.74	1.70	1.67	1.66
	.05	4.41	3.55	3.16	2.93	2.77	2.66	2.58	2.51	2.46	2.41	2.37	2.34	2.29	2.27	2.20	2.19	2.15	2.11	2.04	1.98	1.93	1.92
	.01	8.29	6.01	5.09	4.58	4.25	4.01	3.84	3.71	3.60	3.51	3.43	3.37	3.27	3.23	3.10	3.08	3.00	2.92	2.78	2.68	2.59	2.57
19	.10	2.99	2.61	2.40	2.27	2.18	2.11	2.06	2.02	1.98	1.96	1.94	1.91	1.87	1.86	1.82	1.81	1.79	1.76	1.71	1.67	1.64	1.63
	.05	4.38	3.52	3.13	2.90	2.74	2.63	2.54	2.48	2.42	2.38	2.34	2.31	2.26	2.23	2.17	2.16	2.11	2.07	2.00	1.94	1.89	1.88
	.01	8.18	5.93	5.01	4.50	4.17	3.94	3.77	3.63	3.52	3.43	3.36	3.30	3.19	3.15	3.03	3.00	2.92	2.84	2.71	2.60	2.51	2.49
20	.10	2.97	2.59	2.38	2.25	2.16	2.09	2.04	2.00	1.96	1.94	1.92	1.89	1.85	1.84	1.80	1.79	1.77	1.74	1.69	1.65	1.62	1.61
	.05	4.35	3.49	3.10	2.87	2.71	2.60	2.51	2.45	2.39	2.35	2.31	2.28	2.22	2.20	2.14	2.12	2.08	2.04	1.97	1.91	1.86	1.84
	.01	8.10	5.85	4.94	4.43	4.10	3.87	3.70	3.56	3.46	3.37	3.29	3.23	3.13	3.09	2.96	2.94	2.86	2.78	2.64	2.54	2.44	2.42
24	.10	2.93	2.54	2.33	2.19	2.10	2.04	1.98	1.94	1.91	1.88	1.85	1.83	1.79	1.78	1.74	1.73	1.70	1.67	1.62	1.58	1.54	1.53
	.05	4.26	3.40	3.01	2.78	2.62	2.51	2.42	2.36	2.30	2.25	2.21	2.18	2.13	2.11	2.04	2.03	1.98	1.94	1.86	1.80	1.75	1.73
	.01	7.82	5.61	4.72	4.22	3.90	3.67	3.50	3.36	3.26	3.17	3.09	3.03	2.93	2.89	2.76	2.74	2.66	2.58	2.44	2.33	2.24	2.21
30	.10	2.88	2.49	2.28	2.14	2.05	1.98	1.93	1.88	1.85	1.82	1.79	1.77	1.73	1.72	1.68	1.67	1.64	1.61	1.55	1.51	1.47	1.46
	.05	4.17	3.32	2.92	2.69	2.53	2.42	2.33	2.27	2.21	2.16	2.13	2.09	2.04	2.01	1.95	1.93	1.89	1.84	1.76	1.70	1.64	1.62
	.01	7.56	5.39	4.51	4.02	3.70	3.47	3.30	3.17	3.07	2.98	2.91	2.84	2.74	2.70	2.57	2.55	2.47	2.39	2.25	2.13	2.03	2.01
50	.10	2.81	2.41	2.20	2.06	1.97	1.90	1.84	1.80	1.76	1.73	1.70	1.68	1.64	1.63	1.58	1.57	1.54	1.50	1.44	1.39	1.34	1.33
	.05	4.03	3.18	2.79	2.56	2.40	2.29	2.20	2.13	2.07	2.03	1.99	1.95	1.89	1.87	1.80	1.78	1.74	1.69	1.60	1.52	1.46	1.44
	.01	7.17	5.06	4.20	3.72	3.41	3.19	3.02	2.89	2.79	2.70	2.63	2.56	2.46	2.42	2.29	2.27	2.18	2.10	1.95	1.82	1.71	1.68
100	.10	2.76	2.36	2.14	2.00	1.91	1.83	1.78	1.73	1.70	1.66	1.63	1.61	1.57	1.56	1.50	1.49	1.46	1.42	1.35	1.29	1.23	1.21
	.05	3.94	3.09	2.70	2.46	2.31	2.19	2.10	2.03	1.97	1.93	1.89	1.85	1.79	1.77	1.69	1.68	1.63	1.57	1.48	1.39	1.31	1.28
	.01	6.90	4.82	3.98	3.51	3.21	2.99	2.82	2.69	2.59	2.50	2.43	2.37	2.26	2.22	2.09	2.07	1.98	1.89	1.73	1.60	1.47	1.43
500	.10	2.72	2.31	2.10	1.96	1.86	1.79	1.73	1.68	1.64	1.61	1.58	1.56	1.52	1.50	1.45	1.44	1.41	1.36	1.28	1.21	1.12	1.09
	.05	3.86	3.01	2.62	2.39	2.23	2.12	2.03	1.96	1.90	1.85	1.81	1.77	1.71	1.69	1.61	1.59	1.54	1.48	1.38	1.28	1.16	1.11
	.01	6.69	4.65	3.82	3.36	3.05	2.84	2.68	2.55	2.44	2.36	2.28	2.22	2.12	2.07	1.94	1.92	1.83	1.74	1.56	1.41	1.23	1.16
∞	.10	2.71	2.30	2.08	1.94	1.85	1.77	1.72	1.67	1.63	1.60	1.57	1.55	1.51	1.49	1.43	1.42	1.38	1.34	1.26	1.18	1.08	1.00
	.05	3.84	3.00	2.60	2.37	2.21	2.10	2.01	1.94	1.88	1.83	1.79	1.75	1.69	1.67	1.59	1.57	1.52	1.46	1.35	1.24	1.11	1.00
	.01	6.63	4.61	3.78	3.32	3.02	2.80	2.64	2.51	2.41	2.32	2.25	2.18	2.08	2.04	1.90	1.88	1.79	1.70	1.52	1.36	1.15	1.00

Table B-7 D distribution for the Kolmogorov-Smirnov goodness-of-fit test

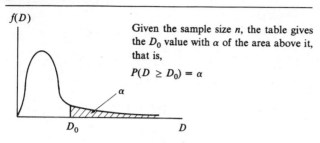

Given the sample size n, the table gives the D_0 value with α of the area above it, that is,

$$P(D \geq D_0) = \alpha$$

n	α level		
	0.10	0.05	0.01
1	0.95	0.98	0.995
2	0.78	0.84	0.93
3	0.64	0.71	0.83
4	0.56	0.62	0.73
5	0.51	0.56	0.67
6	0.47	0.52	0.62
7	0.44	0.49	0.58
8	0.41	0.46	0.54
9	0.39	0.43	0.51
10	0.37	0.41	0.49
11	0.35	0.39	0.47
12	0.34	0.38	0.45
13	0.33	0.36	0.43
14	0.31	0.35	0.42
15	0.30	0.34	0.40
16	0.30	0.33	0.39
17	0.29	0.32	0.38
18	0.28	0.31	0.37
19	0.27	0.30	0.36
20	0.26	0.29	0.36
25	0.24	0.27	0.32
30	0.22	0.24	0.29
35	0.21	0.23	0.27
40	0.19	0.21	0.25
50	0.17	0.19	0.23
> 50	$\dfrac{1.22}{\sqrt{n}}$	$\dfrac{1.36}{\sqrt{n}}$	$\dfrac{1.63}{\sqrt{n}}$

Table B-7 is reprinted from F. J. Massey, Jr., "The Kolmogorov-Smirnov Test for Goodness of Fit," *Journal of the American Statistical Association*, vol. 46, no. 253, 1951, with permission of the American Statistical Association.

Table B-8 r distribution for the runs test of randomness for $\alpha = 0.05$

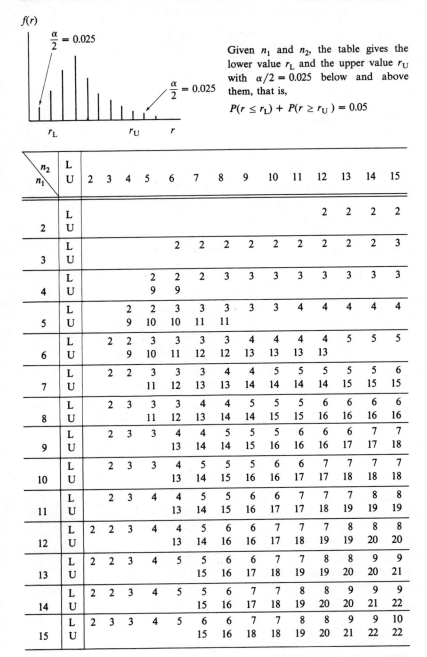

Given n_1 and n_2, the table gives the lower value r_L and the upper value r_U with $\alpha/2 = 0.025$ below and above them, that is,

$$P(r \leq r_L) + P(r \geq r_U) = 0.05$$

n_1 \ n_2	L/U	2	3	4	5	6	7	8	9	10	11	12	13	14	15
2	L									2	2	2	2		
	U														
3	L					2	2	2	2	2	2	2	2	2	3
	U														
4	L			2	2	2	3	3	3	3	3	3	3	3	3
	U			9	9										
5	L			2	2	3	3	3	3	3	4	4	4	4	4
	U			9	10	10	11	11							
6	L		2	2	3	3	3	3	4	4	4	4	5	5	5
	U			9	10	11	12	12	13	13	13	13			
7	L		2	2	3	3	3	4	4	5	5	5	5	5	6
	U				11	12	13	13	14	14	14	14	15	15	15
8	L		2	3	3	3	4	4	5	5	5	6	6	6	6
	U				11	12	13	14	14	15	15	16	16	16	16
9	L		2	3	3	4	4	5	5	5	6	6	6	7	7
	U					13	14	14	15	16	16	16	17	17	18
10	L		2	3	3	4	5	5	5	6	6	7	7	7	7
	U					13	14	15	16	16	17	17	18	18	18
11	L		2	3	4	4	5	5	6	6	7	7	7	8	8
	U					13	14	15	16	17	17	18	19	19	19
12	L	2	2	3	4	4	5	6	6	7	7	7	8	8	8
	U					13	14	16	16	17	18	19	19	20	20
13	L	2	2	3	4	5	5	6	6	7	7	8	8	9	9
	U					15	16	17	18	19	19	20	20	21	
14	L	2	2	3	4	5	5	6	7	7	8	8	9	9	9
	U					15	16	17	18	19	20	20	21	22	
15	L	2	3	3	4	5	6	6	7	7	8	8	9	9	10
	U					15	16	18	18	19	20	21	22	22	

Table B-8 is adapted from F. S. Swed and C. Eisenhart, "Tables for Testing Randomness of Grouping in a Sequence of Alternatives," *The Annals of Mathematical Statistics*, vol. 14, n. 1, 1943, with permission of the Institute of Mathematical Statistics.

Table B-9 T Distribution of the sum of signed ranks for the Wilcoxon test of means

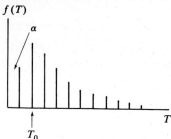

For a specified value of sample size n, the table gives the value of T with α of the area in the lower tail, that is,

$$P(T \leq T_0) = \alpha$$

	Significance level		
	One-tail α		
	0.025	0.010	0.005
	Two-tail α		
n	0.050	0.020	0.010
6	0	—	—
7	2	0	—
8	4	2	0
9	6	3	2
10	8	5	3
11	11	7	5
12	14	10	7
13	17	13	10
14	21	16	13
15	25	20	16
16	30	24	20
17	35	28	23
18	40	33	28
19	46	38	32
20	52	43	38
21	59	49	43
22	66	56	49
23	73	62	55
24	81	69	61
25	89	77	68

Table B-9 is reprinted from F. Wilcoxon and R. A. Wilcox, *Some Rapid Approximate Statistical Procedures*, Lederle Laboratories, 1964, with permission of the American Cyanamid Co.

BIBLIOGRAPHY

1. Beck, J. V., and K. J. Arnold: *Parameter Estimation in Engineering and Science*, John Wiley & Sons, Inc., New York, 1977.
2. Belz, M. H.: *Statistical Methods for the Process Industries*, John Wiley & Sons, Inc., New York, 1973.
3. Benjamin, J. R., and C. A. Cornell: *Probability, Statistics and Decision for Civil Engineers*, McGraw-Hill Book Company, New York, 1970.
4. Bowker, A. H., and G. J. Lieberman: *Engineering Statistics*, 2d ed., Prentice-Hall, Inc., Englewood Cliffs, N. J., 1972.
5. Brownlee, K. A.: *Statistical Theory and Methodology in Science and Engineering*, 2d ed., John Wiley & Sons, Inc., New York, 1965.
6. Draper, N. R., and H. Smith: *Applied Regression Analysis*, John Wiley & Sons, Inc., New York, 1966.
7. Duncan, A. J.: *Quality Control and Industrial Statistics*, 4th ed., Richard D. Irwin, Inc., Homewood, Ill., 1974.
8. Gibra, I. N.: *Probability and Statistical Inference for Scientists and Engineers*, Prentice-Hall, Inc., Englewood Cliffs, N. J., 1973.
9. Grant, E. L., and R. Leavenworth: *Statistical Quality Control*, 4th ed., McGraw-Hill Book Company, New York, 1972.
10. Hahn, G. J., and S. S. Shapiro: *Statistical Models in Engineering*, John Wiley & Sons, Inc., New York, 1967.
11. Hicks, C. R.: *Fundamental Concepts in the Design of Experiments*, Holt, Rinehart and Winston, Inc., New York, 1964.
12. Hines, W. M., and D. C. Montgomery: *Probability and Statistics in Engineering and Management Science*, The Ronald Press Company, New York, 1972.
13. Hoel, P. G.: *Introduction to Mathematical Statistics*, 4th ed., John Wiley & Sons, Inc., New York, 1971.
14. Hogg, R. V., and A. T. Craig: *Introduction to Mathematical Statistics*, 2d ed., The Macmillan Company, New York, 1965.
15. Kirkpatrick, E. G.: *Introductory Statistics and Probability for Engineering, Science and Technology*, Prentice-Hall, Inc., Englewood Cliffs, N.J., 1974.
16. LaValle, I. H.: *An Introduction to Probability, Decision and Inference*, Holt, Rinehart and Winston, Inc., New York, 1970.

17. Lipson, C., and N. J. Sheth: *Statistical Design and Analysis of Engineering Experiments*, McGraw-Hill Book Company, New York, 1973.
18. Miller, I., and J. E. Freund: *Probability and Statistics for Engineers*, 2d ed., Prentice-Hall, Inc., Englewood Cliffs, N.J., 1977.
19. Neville, A. M., and J. B. Kennedy: *Basic Statistical Methods for Engineers and Scientists*, Intext Publishing Co., New York, 1964.
20. Newton, B. L.: *Statistics for Business*, Science Research Associates, Inc., Chicago, 1973.
21. Siegel, S.: *Nonparametric Statistics for the Behaviorial Sciences*, McGraw-Hill Book Company, 1956.
22. Snedecor, G. W., and W. G. Cochran: *Statistical Methods*, 6th ed., The Iowa State University Press, Ames, Iowa, 1967.
23. Steel, R. G. D., and J. H. Torrie: *Principles and Procedures of Statistics with Special Reference to the Biological Sciences*, McGraw-Hill Book Company, New York, 1960.
24. Volk, W.: *Applied Statistics for Engineers*, 2d ed., McGraw-Hill Book Company, New York, 1969.
25. Walpole, R. E., and R. H. Myers: *Probability and Statistics for Engineers and Scientists*, 2d ed., Macmillan Publishing Co., Inc., New York, 1978.
26. Wilks, S. S.: *Mathematical Statistics*, John Wiley & Sons, Inc., New York, 1962.
27. Winkler, R. L.: *An Introduction to Bayesian Inference and Decision*, Holt, Rinehart and Winston, Inc., New York, 1972.

INDEX

Absolute error loss function, 472
Acceptable quality level (AQL), 554
Acceptance region, 358, 362, 364
Acceptance sampling, 551–559
Acceptance sampling plan:
  definition, single, 552
  Dodge-Romig, 556
  double, 556
  MIL-STD-105D, 555–559
  MIL-STD-414, 556
  normal inspection, 555
  operating characteristic curve for, 552–555
  probability of acceptance, 552
Additivity property:
  of χ^2 distribution, 262
  of Poisson distribution, 204
  of probability, 90, 111
Alternative hypothesis, 358, 361
Analysis of variance (ANOVA):
  factorial, 577–581
  interaction effect, 577
  for linear regression, 571–573
  rationale, 566
  single factor: completely randomized, 568–571
    randomized blocks, 573–576
  table format, 568
Approximations:
  binomial to hypergeometric, 175–177
  normal to binomial, 231–233
  normal to χ^2, 257

Approximations:
  normal to Poisson, 233
  Poisson to binomial, 188–189, 198–199
  standard normal to t, 269
Average:
  of coded data, 77–80
  computation of, 47–49
  for frequency data, 47–49
  plotted on frequency polygon, 60–61
  weighted, 73–75, 81–82
  (*See also* Expected value)
Average ($\bar{X}$) chart:
  limits, 540
  obtaining control on, 542–543
  procedure to construct, 541–543
Axioms of probability, 90

Bayes' theorem:
  computation using probability trees, 99–100
  computations for, 96–98
  formula, 96
  prior-posterior relations and, 476
Bayesian approach, 471
Bayesian estimate, 479–481
Bernoulli distribution, 162
Bernoulli trial, 162
Beta distribution, 296, 299
  prior-posterior distributions, 477–478
Bimodal frequency data, 50–51
Binomial distribution:
  approximation to hypergeometric, 175–177

641

Binomial distribution:
  assumptions, 162
  coefficient, 164
  comparing observed and predicted
    outcomes for, 173–175, 182
  cumulative distribution function, 165
  derivation, 162–164, 179
  expected value, 167, 168, 179
  in goodness-of-fit tests, 414, 418
  graph, 165–166
  graphical probability plot (small p), 439–441
  moment generating function, 183
  normal approximation to, 231–233
  parameter estimation, 168
  Poisson approximation to, 196–198
  problem solving procedure for, 171
  relation to multinomial distribution, 335–336
  standard deviation, 167, 168, 180
  table of values, 619
  test for parameter $p=\frac{1}{2}$, 449
  use: in number defective chart, 548
    of table, 169–171
Binomial theorem, 179
Bivariate distribution:
  expected value formula, 338
  properties of any, 327–328
  sample space, 326
  transformation (see Transformation of
    variable procedure)
Bivariate normal distribution, 336–337
Block effect, 573

Cell:
  boundaries, 18–20
  midpoint, 19, 20
  procedure to determine, 19
  width, 18
Central limit theorem:
  procedure to demonstrate, 245–246
  statement of, 243
  uses of, 247–250
Central moment, 66
Central tendency:
  definition, 47
  measures of, 47–53
Change of variable (see Transformation of
  variable procedure)
χ^2 distribution:
  additivity property, 262
  derivation, 262
  expected value, 257
  formulation procedure, 255–256
  graph, 257
  moment generating function, 262
  normal distribution approximation to, 257
  problem solving procedure for s^2, 260–261
  relation to gamma distribution, 296
  standard deviation, 257
  table of values, 629
  use in goodness-of-fit tests, 411–418
  use of table, 258–260

χ^2 goodness-of-fit test:
  for contingency tables, 415–418
  derivation of test statistic, 418
  minimum frequency correction, 413
  statistic for, 411
χ^2 statistic, 255
χ^2 test for variances, 391–395
Class (see Cell)
Classical definition of probability, 6
CLT (see Central limit theorem)
Clustering, 47, 53
Coded data:
  computation of: for grouped data, 78
    for ungrouped data, 76
  procedure for computing $\bar{X}$ and s using,
    77–80
Coefficient:
  of determination, 523–526
  of multiple determination, 529–531
  of peakedness, 70–71, 81, 222
  of skewness, 68–70, 222
  of variation, 75
Combinations:
  addition and multiplication of, 145
  computation, 144–145
  relation to permutations, 144
Conditional distribution, 331–335
  definition, 331–332
  expected value, 334
Conditional expectation, 334
Conditional probability, 92–94
Confidence interval (see Interval estimate)
Confidence level, 306, 361
Consistent estimator, 129
Consumer's risk, 555
Contingency table, 415–416
Continuity correction:
  in normal approximation to binomial,
    231–232
  in one-sample sign test, 450
Continuous variable:
  cumulative distribution function (cdf), 127
  definition, 8–9
  expected value, 125
  moment generating function (mgf) for, 130
  probability density function, 114–117,
    122–123
  standard deviation, 125
  transformation (see Transformation of
    variable procedure)
  variance, 125
Control chart:
  c, 549
  np, 548
  p, 547–549
  probability limits, 540
  R, 541–543
  type I error, 540
  u, 550
  warning limits, 540
  $\bar{X}$, 538–543
  (See also Average ($\bar{X}$) chart)

INDEX **643**

Correlation analysis, 516–531
Correlation coefficient:
　in bivariate normal distribution, 336
　cause and effect implication, 517
　computation, linear, 518–521
　definition using covariance, 528
　generalized, 523–526
　hypothesis test, 521–523
　multiple, 528
　partial, 529–530
　relation to coefficient of determination, 523
　in variance of dependent variables, 352
Correlation matrix, 529
Counting rule, basic, 141
Covariance, 528
Critical region, 358, 364
Cumulative distribution function (cdf):
　for binomial distribution, 165
　definition, 127
　for exponential distribution, 289
　for hypergeometric distribution, 148, 149
　for Poisson distribution, 190
　relation to probability density function (pdf), 127
　for standard normal distribution, 225
　for uniform distribution, 287
Cumulative frequency distribution:
　for continuous variable data, 25–28
　definition, 22
　for discrete variable data, 24–25, 37–40
　procedure to construct, 24
Cumulative relative frequency:
　definition, 29
　plot of, 31–33
　relation to cumulative frequency, 29
Curve fitting (*see* Curvilinear regression; Distribution fitting; Linear regression)
Curvilinear regression, 503–507
　correlation coefficient for, 524
　summary of equations, 505

Data coding (*see* Coded data)
Decile, 33
Degrees of freedom:
　in analysis of variance, 567–568
　for χ^2 distribution, 256
　of χ^2 goodness-of-fit test, 411, 416
　definition, 254
　of F distribution, 279
　of t distribution, 269
Density function (*see* Distribution)
Dependent events, 92
Dependent variable:
　distribution (*see* Conditional distribution)
　expected value of, 352
　variance of, 352
Deterministic data, 4, 55
Discrete uniform distribution:
　application, 212–213
　probability density function, 211
　properties, 212

Discrete variable:
　cumulative distribution function (cdf), 127, 134–135
　definition, 89
　expected value, 120
　moment generating function (mgf) for, 130
　probability density function, 114–119
　standard deviation, 121
　transformation (*see* Transformation of variable procedure)
　variance, 121
Dispersion, measures of, 53–60
Distribution:
　Bernoulli, 162
　beta, 296, 298
　bimodal, 50–51
　binomial, 162–180
　bivariate, 326–337
　bivariate normal (BVN), 336
　χ^2, 253–264
　χ^2 goodness-of-fit test of, 411–415
　conditional, 331–335
　cumulative, 22, 24–28, 127
　definition, 112–114
　　(*See also* Probability density function)
　discrete uniform, 211–215
　F, 277–283
　of function of variable(s), 340–346
　gamma, 293–296
　Gaussian (*see* Normal distribution)
　geometric, 206–208
　hypergeometric, 146–155
　joint, 326–337
　K-S goodness-of-fit test of, 419–421
　log normal, 296, 299
　marginal, 328–330
　multinomial, 335–336
　negative binomial, 208–211
　normal, 218–237
　Pascal, 208–211
　Poisson, 187–202
　population correlation coefficient, 526
　posterior, 476
　prior, 471, 476
　problem solving procedure for, 153, 171
　of proportions, 177–179
　Rayleigh, 296, 299
　standard normal (SND), 221–227
　t, 267–274
　uniform, 287–288
　Weibull, 296, 299
Distribution fitting:
　exponential distribution, 297
　graphical (*see* Probability plotting)
　procedure for normal distribution, 229–231
Distribution-free methods (*see* Nonparametric statistical inference)
Distribution function (*see* Cumulative distribution function)

Efficient estimator, 129

Equally likely events, 4–5
Erlang distribution, 296
Error sum of squares, 488, 501, 567
Estimate:
　Bayes', 479–481
　best, 129
　interval (see Interval estimate)
　least-square regression, 491–494
　point, 310
　of population mean, 120, 220
　of population proportion, 168
　of population standard deviation, 128, 220
　properties of, 129
Events:
　conditional, 92
　definition, 4
　equally likely, 4–5
　independent, 94
　mutually exclusive, 88, 91, 95
　in sample spaces, 89–90
Expected value:
　of conditional distributions, 334
　formulas: for continuous distributions, 124–125
　　for discrete distributions, 120–121
　of functions: of one variable, 337
　　of two variables, 338
　from moment generating function, 130
　of risk function, 472–473
Exponential distribution:
　cumulative distribution function, 289
　derivation, 289
　distribution fitting, 297
　expected value, 290
　graphical probability plot, 435–437
　parameter estimation, 290
　relation to gamma distribution, 296
　relation to Poisson distribution, 293
　standard deviation, 290–291
　use in reliability analysis, 291–293
Exponential failure law, 290
Exponential model regression, 503–506

F distribution:
　in analysis of variance, 567–568
　area in left tail of, 281
　derivation, 283
　expected value, 279
　formulation procedure, 278–279
　graph, 280
　problem solving for variance ratio, 282
　relation to t distribution, 283
　standard deviation, 279
　table of values, 632
　use of table, 280–282
F statistic, 278
F test for variances, 392, 395–397
Factorial:
　definition, 141
　distributions and, 256, 269, 279
　Stirling's approximation to, 142, 155

Factorial analysis of variance, 577–581
Factorial experiment, 577–581
Failure rate, 291
Fractiles, 33–36
Fraction defective (p) chart, 547–549
Frequency:
　assumption for grouped, cell data, 48, 49
　cumulative, 24–28
　cumulative relative, 29
　definition, 15
　on histograms, 17
　relative, 28
Frequency definition of probability, 6
Frequency distribution (histogram):
　grouped data, 21–22
　ungrouped data, 17–18
Frequency polygon, 21–22
Frequency tally sheet, 15–17
Functions of variables:
　distributions of (see Transformation of variable procedure)
　expected value rules for, 337–338
　variance rules for, 339–340

Gamma distribution:
　graph, 295
　parameter estimation, 294
　prior-posterior distributions, 478
　properties, 294
　relation to other distributions, 296
　use of gamma function in, 294
Gamma function, 256, 294
Geometric distribution:
　application, 207–208
　probability density function, 206
　properties, 207
　relation to Pascal distribution, 208
Goodness-of-fit tests:
　classification, 410
　for contingency tables, 415–418
　continuous distributions, 414–415, 419–421
　discrete distributions, 411–414
　graphical, 428–441
　Kolmogorov-Smirnov (K-S), 419–421
Grouped data:
　definition, 15
　determination of cells, 19
　fractiles for, 35–36
　frequency of, 21
　normal distribution fit to, 229–231
　summary of graphical presentation formats, 32

Histogram (see Frequency distribution)
Hyperbolic model regression, 505
Hypergeometric distribution:
　assumptions, 146
　binomial approximation to, 175–177
　cumulative distribution function, 148, 149

Hypergeometric distribution:
  derivation, 146–147
  expected value, 151, 154–155
  graph, 148–150
  parameter estimation, 151–152
  probability density function, 147
  problem-solving procedure for, 153–154
  standard deviation, 151, 154
Hypothesis:
  types of, 358
  types of errors in testing, 359
Hypothesis testing:
  acceptance region in, 358
  critical region in, 358
  one- and two-sided tests, 361
  procedure, 362
  rationale, 358
  for sample means, 372–386
  for sample populations, 401–405
  for sample variances, 391–398
  (*See also* Statistical inference)

Independence of factors, test for, 415–418
Independent events, 92, 94
Independent variables:
  definition, 330
  expected value: of product of, 338
    of sum of, 338
  methods of testing for, 330–331
  variance: of linear combination of, 339
    of sum or difference of, 339
Interaction of factors, 577
Interval estimate:
  definition and interpretation, 310
  for population mean, 311–313
  for population proportion, 317–318
  for population standard deviation, 313–317
  for population variance, 314
  for regression coefficients, 497
  for regression line, 496

Jacobian of a transformation, 343, 345
Joint distribution (*see* Bivariate distribution)

Kolmogorov-Smirnov (K-S) goodness-of-fit test:
  procedure, 419–420
  statistic, 419
  table for, 635
Kurtosis (*see* Peakedness)

Least squares regression rationale, 488–489
Linear combinations of random variables:
  expected value, 338
  variance, 339
Linear regression:
  analysis of variance, 571–573
  coefficient computation, 491–494

Linear regression:
  confidence interval: on coefficients, 497
    on line, 496
    on single value, 497
  linearizing data for, 503, 505
  normal equations, 490
  relation to correlation analysis, 527
  sum of squares partitioning, 507–509
  test for coefficients, 494–496
  variance around line, 494
Log normal distribution, 296, 299
Loss function, 472, 474, 479
Lot tolerance percent defective (LTPD), 555

MAD (*see* Mean absolute deviation)
Mann-Whitney U test, 453–456
Marginal distribution:
  for bivariate normal distribution, 336
  computation of, 328–330
  conditional distribution and, 332
  independence of variables and, 330
Maximum likelihood estimators, 129
Mean:
  interval estimate of, 311–313
  sample, 47–49
    (*See also* Average)
  sum of squares about, 507, 567
Mean absolute deviation, 71–73
Mean risk, 472–473
Mean square:
  in analysis of variance, 568, 571
  in regression, 508
Mean squared deviation, 54, 55
Mean time between failures, 291
Means, statistical inference for, 372–386
Measured unit accuracy, 20
Median:
  definition, 51
  for grouped data, 51, 52
  plotted on frequency polygon, 60–61
  for ungrouped data, 51
Median cell, 51, 52
MIL-STD-105D sampling plan, 555–559
Modal cell, 49–51
Mode:
  definition, 49
  plotted on frequency polygon, 60–61
  for ungrouped data, 50
Moment, 66
Moment generating function (mgf):
  for binomial distribution, 183
  for χ^2 distribution, 262
  computing expected values from, 130–131
  computing variances from, 130–131
  definition, 129–130
  for distribution of proportions, 186
  for normal distribution, 234
  for Poisson distribution, 201–202
  for uniform distribution, 302
MTBF (mean time between failures), 291

Multifactor experiment, 577–581
Multinomial distribution:
 in goodness-of-fit tests, 418
 probability density function, 335, 336
 properties, 335
Multiple correlation coefficient, 528
Multiple linear regression:
 analysis of variance, 571–573
 coefficient determination, 500–501
 multiple and partial correlation, 528–531
 stepwise, 502–503
 sum of squares partitioning, 507–509
 variance around surface, 501
Multivariate distribution, 326, 328
Mutually exclusive events, 88, 91, 95

Natural conjugate distributions, 476–479
Natural process limits, 544, 547
Negative binomial distribution, 208–211
Nonlinear regression (see Curvilinear regression)
Nonparametric statistical inference:
 χ^2 goodness-of-fit test, 411–415
 compared to parametric inference, 446
 K-S goodness-of-fit test, 419–421
 Mann-Whitney U test, 453–456
 one-sample runs test, 447–449
 one-sample sign test, 449–452
 two-sample sign test, 452–453
 Wilcoxon test, 456–459
Normal distribution:
 approximation: to binomial distribution, 231–233
  to χ^2 distribution, 257
  to Poisson distribution, 233
 central limit theorem, 242–246
 computations of area under, 226–227
 continuity correction for binomial approximation, 231–232
 correlation and, 527–528
 definition, 219
 derivation of standard normal distribution from, 221–222
 expected value, 220, 234
 graph, 223
 graphical probability plot, 431–435
 linear regression and, 488–489, 527
 moment generating function, 234
 parameter estimation, 220, 221, 230
 prior-posterior distributions, 476–477
 probability paper, 431, 433
 procedure for distribution fitting, 229–231, 235–236
 solving problems for, 227–229
 standard deviation, 220
 table of values, 627
 for two variables, 336–337
 use: in nonparametric tests, 448–450, 454, 458
  in $\bar{X}$ chart, 538
 (See also Standard normal distribution)

Normal equations in regression, 490, 501, 504
Normal probability paper, 431–435
Normal (z) test, 373–376, 378–381
Null hypothesis, 358
Number of defects, (c, u) charts, 549–551
Number of observations, 308–309

OC curve (see Operating characteristic curve)
Ogive curve (see Cumulative frequency distribution)
One-sample runs test, 447–449
 table of critical values, 634
One-sample sign test, 449–452
One-sided test, 361–362
One-way (single factor) analysis of variance, 568–571
Operating characteristic curve:
 acceptance sampling: consumer's risk, 555
  ideal, 552
  procedure for construction, 552–553
  producer's risk, 554
 statistical inference: definition, 366–367
  finding probability of type II error from, 367–369
  ideal, 367
  one-sample variance, χ^2 distribution, 392–394
  relation to power, 367, 369
  sample means: normal distribution, 374
   t distribution, 237–238
  two-sample variances, F distribution, 396

Paired data test of means, 384–386
Parameter:
 best estimate, 128–129
 meaning, 113
 unbiased estimate of, 129
Parameter estimation, graphical: exponential distribution, 436
 normal distribution, 431
 Poisson distribution, 440
 procedure, 429–430
 uniform distribution, 438
Partial correlation coefficient, 529
Pascal distribution, 208–211
Peakedness:
 coefficient of, 70–71, 81, 222
 definition, 70
Percentile, 33
Permutations:
 computation, 142–143
 for k classes, 144, 155–156
 relation to combinations, 144
 with replacement, 143
Phases:
 of probability analysis, 7
 of statistics, 9–10
Point estimate, 310

Poisson distribution:
  additivity property of, 204
  approximation to binomial, 196–198
  assumptions, 189
  χ^2 goodness-of-fit, 413–414
  comparing observed and predicted
    outcomes for, 195–196
  cumulative distribution function, 190
  derivation from binomial, 188–189, 198–199
  expected value, 191, 199
  graph, 190–191
  graphical goodness-of-fit, 439–441
  moment generating function, 201–202
  normal approximation to, 233
  operating characteristic curve and, 552–554
  parameter estimation, 192
  probability paper, 439
  probability plot, 439–441
  problem-solving procedure for, 194–195
  relation to exponential distribution, 293
  standard deviation, 191, 199
  table of values, 623
  use: in number of defects chart, 549
    of table, 193
Poisson probability paper, 439
Polynomial model regression, 504–507
Population, 7
Posterior distribution, 476
Power model regression, 505
Power of a test, 367, 369
Precision, use in sample size determination, 307
Prior distribution, 471, 476
Prior-posterior distributions:
  beta-beta, 477–478
  definition, 476
  gamma-gamma, 478
  normal-normal, 476–477
Probabilistic data, 4
Probability:
  addition law of, 90, 111
  axioms of, 90
  conditional, 92–94
  definition, 5–6
  for independent events, 94
  limits on $\bar{X}$ chart, 540
  prior, 473
  quite low value, 172, 174–175
  rare event, 188
  reasonably high value, 172, 174–175, 182
  rules for computing, 90–91
  summary table for computing, 95
  of type I and II errors, 359–361
Probability density function:
  bivariate (see Bivariate distribution)
  continuous: expected value, 125, 126
    properties, 122–123
    standard deviation, 125, 126
    variance, 125
  discrete: expected value, 120, 122
    properties, 117

Probability density function:
  discrete:
    standard deviation, 121
    variance, 121, 122
  graphs for discrete and continuous variable, 115–116
  parameters in, 113–114
  probability computations for continuous, 125, 126
  terminology, 112–114
Probability distribution (see Probability density function)
Probability plotting:
  for exponential distribution, 435–437
  general procedure, 429–430
  for normal distribution, 431–435
  for Poisson distribution, 439–441
  for uniform distribution, 437–438
Probability tree, 98–100
Process capability analysis, 544–547
Process control charts (see Control chart)
Producer's risk, 554
Proportions:
  distribution of: moment generating function, 186
    properties of, 177
    relation to binomial distribution, 177
    use in fraction defective chart, 547
  statistical inference for, 401–405
Pseudorandom numbers (see Random numbers)

Quadratic loss function:
  Bayesian estimate with, 479–481
  definition, 472
Quality control, 536–551
Quartile, 33

Random numbers, 213–215
Random sample, 7
Random variables:
  definition, 110
  independence of, 330
  notation, 110–111
  (See also Variables)
Randomness test, 447–449
  table of critical values, 636
Range, 59–60
Range (R) chart, 541–543
  use in process capability analysis, 544
Rank sum tests, 453–459
Rayleigh distribution, 296, 299
Rectangular distribution (see Uniform distribution)
Regression coefficients:
  curvilinear, 505
  linear, 490–492
  multiple linear, 500–501
Regression sum of squares, 507–509

Relative frequency:
    definition, 28
    plot for, 31–33
Reliability:
    computation using exponential distribution, 291–293
    definition, 289
Residual sum of squares, 488, 501
Risk function, 472

Sample:
    average of, 47–49
    median of, 51–53
    mode of, 49–51
    random, 7
    range of, 59–60
    standard deviation of, 53–59
    statistic, 8
    variance of, 55–59
Sample mean, statistical inference for, 372–386, 449–459
    (*See also* Nonparametric statistical inference; Statistical inference)
Sample size determination, 308–309
Sample space, 89–90
Sampling plan (*see* Acceptance sampling plan)
Scattergram, 517
Set operations, 87–89
Sets, 87–88
Significance level, 306, 361
Significance tests (*see* Statistical inference)
Skewed data, 67–70
Skewness:
    coefficient of, 68–70, 222
    definition, 67
Standard deviation:
    approximation using mean absolute deviation, 71–72
    biased form of, 56, 58
    of coded data, 77, 79
    definition, 53–54
    estimate from sample range, 540, 544
    expected value formula for, 121, 125
    formula for computing, 54, 56, 58
    for frequency data, 57–59
    interval estimate for, 313–317
    of mean, 243
    plotted on frequency polygon, 60–61
    relation to moment of inertia, 55
    relation to variance, 55
    unbiased form of, 56, 58
Standard deviation (s) chart, 544
Standard normal distribution:
    areas under probability curve for, 222–225
    derivation from normal distribution, 221–222
    expected value, 222
    graph, 223
    procedure for using table of, 225–226
    quick reference table, 228
    standard deviation, 222

Standard normal distribution:
    sum of squares of independent, 262
    table for, 627–628
Statistic, sample, 8
Statistical inference:
    analysis of variance, 565–581
    classification: for means tests, 373
        for proportions tests, 401
        for variance tests, 391–392
    correlation coefficient, 521–523
    for distributions, 409–421
        (*See also* Goodness-of-fit tests)
    nonparametric (*see* Nonparametric statistical inference)
    one sample mean: σ known, 373–376
        σ unknown, 237–238
    one sample proportion, 402–403
    one sample variance, 392–394
    procedure for performing tests, 362
    in quality control, 537, 538
    regression coefficients, 494–496
    relation between F and t tests, 397–398
    two sample means: paired data, 384–386
        σ known, 387–381
        σ unknown, 381–383
    two-sample proportions, 403–405
    two-sample variances, 395–397
Statistics, definition, 9
Step function (*see* Cumulative frequency distribution)
Stepwise multiple linear regression, 502–503
Stirling's approximation, 142, 155
Stochastic, 4
Straight line regression, 488–489
Student t distribution (see t distribution)
Subjective probability, 6
Sum of squares:
    block, 574–576
    in correlation analysis, 523, 528
    due to regression, 507, 572–573
    error, 488, 494, 507, 567–581
    interaction, 577, 578
    partitioning of, 507–509
    total, 507, 567
    treatment, 567–581

t distribution:
    derivation, 274
    expected value, 269
    formulation procedure, 268–269
    graph, 270
    problem solving procedure for μ, 273
    relation to F distribution, 283
    relation to standard normal distribution, 269
    standard deviation, 269
    table of values, 630
    use: of gamma function in, 269
        in regression analysis, 494, 496–497
        of table, 271–272
t statistic, 269

t test for means, 373, 377–378, 381–386
   relation to nonparametric tests, 454, 457
Tables:
   contingency, 415
   for control chart limits, 541
   for distribution values, 618–637
   random digits, 214
   for standard normal, 228, 625–626
Transformation of functions of variables (*see* Transformation of variable procedure)
Transformation of variable procedure:
   Jacobian for, 343, 345
   one continuous variable, 343–344
   one discrete variable, 340–342
   one-to-one transformations, 341, 344
   two continuous variables, 344–346
   two discrete variables, 344–346
Treatment effect, 566
Two-sample rank sum test, 453–456
   table of critical values, 635
Two-sample sign test, 452–453
Two-sample signed rank test, 456–459
Two-sided tests, 361–362
Two-way analysis of variance, 573–576
Type I and II errors:
   definition, 359
   on operating characteristic (OC) curve, 366–369
   probability of, 359–361
   relations to sample size, 364–366

U test, 453–456
Unbiased estimator, 129
Uncorrelated variables, 518, 527
Ungrouped data:
   fractiles for, 34–36
   frequency of, 15
   summary of graphical presentation formats, 32
Uniform distribution:
   discrete (*see* Discrete uniform distribution)
   graph, 288
   graphical probability plot, 437–438
   moment generating function, 302
   properties, 287

Uniform distribution:
   random numbers from, 297
   relation to beta distribution, 298
Unimodal frequency data, 50
Union of sets, 87, 89
Universe, 7, 87–88

Variables:
   conditional, 331–335
   continuous, 8–9
   correlation and independence, 526–528
   definition, 110
   discrete, 8–9
   independence of, 330
   notation, 110–111
   range of, 113, 326
   summing probabilities, 111
   transformation of (*see* Transformation of variable procedure)
Variance:
   analysis of, 565–581
   around regression line, 494
   around regression surface, 501
   biased form of, 56
   expected value formula for, 121, 125
   formula for computing, 55, 56, 58
   for frequency data, 58
   of functions: of one variable, 339
      of two variables, 339, 352
   from moment generating function, 130
   relation to standard deviation, 55
   statistical inference for, 391–397
   unbiased form of, 56, 58
Variation, measures of, 53–60
Venn diagrams, 87–89, 91

Weibull distribution, 296, 299
Weighted average, 73–75, 81–82
Wilcoxon test, 456–459
   table of critical values, 637

z (normal) test for means, 373–376, 378–381
Z variable, 221–222
   (*See also* Standard normal distribution)

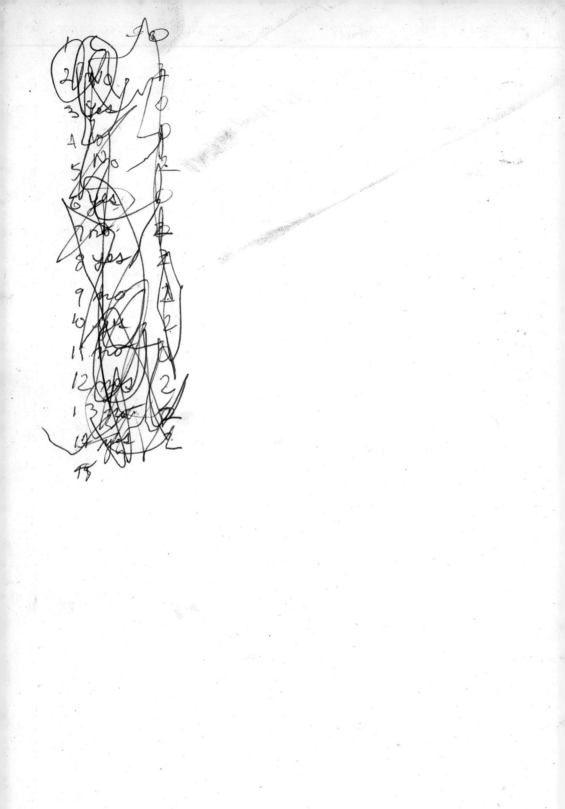